EDEXCEL A LEVEL BOOK 1

GEOGRAPHY 1

Includes AS level

Third Edition

CAMERON DUNN, KIM ADAMS, DAVID HOLMES, SIMON OAKES, SUE WARN, MICHAEL WITHERICK

HODDER EDUCATION
LEARN MORE

In order to ensure that this resource offers high-quality support for the associated Pearson qualification, it has been through a review process by the awarding body. This process confirms that this resource fully covers the teaching and learning content of the specification or part of a specification at which it is aimed. It also confirms that it demonstrates an appropriate balance between the development of subject skills, knowledge and understanding, in addition to preparation for assessment.

Endorsement does not cover any guidance on assessment activities or processes (e.g. practice questions or advice on how to answer assessment questions), included in the resource nor does it prescribe any particular approach to the teaching or delivery of a related course.

While the publishers have made every attempt to ensure that advice on the qualification and its assessment is accurate, the official specification and associated assessment guidance materials are the only authoritative source of information and should always be referred to for definitive guidance.

Pearson examiners have not contributed to any sections in this resource relevant to examination papers for which they have responsibility.

Examiners will not use endorsed resources as a source of material for any assessment set by Pearson.

Endorsement of a resource does not mean that the resource is required to achieve this Pearson qualification, nor does it mean that it is the only suitable material available to support the qualification, and any resource lists produced by the awarding body shall include this and other appropriate resources.

Orders: please contact Bookpoint Ltd, 130 Milton Park, Abingdon, Oxon OX14 4SB. Telephone:
+44 (0)1235 827720. Fax: +44 (0)1235 400454. Email education@bookpoint.co.uk Lines are open from 9 a.m. to 5 p.m., Monday to Saturday, with a 24-hour message answering service. You can also order through our website: www.hoddereducation.co.uk

ISBN: 978 1 4718 5654 9

First published in 2016 by

Hodder Education,

An Hachette UK Company

Carmelite House

50 Victoria Embankment

London EC4Y 0DZ

www.hoddereducation.co.uk

Impression number 10 9 8 7 6 5 4

Year 2020 2019 2018 2017

Cover photo © Francesco R Iacomino – Fotolia

Illustrations by Aptara, Inc. and Barking Dog Art

Typeset in Bembo Std Regular 8/9pt by Aptara, Inc.

Printed in Dubai

A catalogue record for this title is available from the British Library.

CONTENTS

INTRODUCTION

This book has been written specifically for the new Edexcel specification introduced for first teaching in September 2016. The writers are all experienced authors, teachers and subject specialists with examining experience, who provide comprehensive and up-to-date information that is both accessible and informative.

This title (Book 1) covers the AS or your first year of A level, including all the options. There is also a Place Investigations section, which will support you with your two in-depth place studies: of the place in which you live or study and of a contrasting place. Year 2 of A level, including all the options, is covered in Book 2.

Edexcel AS Geography – Paper 1

Content	Covered in
AS Compulsory content	
Topic 1: Tectonic Processes and Hazards	*Edexcel A level Geography Book 1*
AS Optional content	
Topic 2: Landscape Systems, Processes and Change One of these topics must be studied: • Topic 2A: Glaciated Landscapes and Change • Topic 2B: Coastal Landscapes and Change	*Edexcel A level Geography Book 1*
AS Fieldwork	*Edexcel A level Geography Book 1* Guidance on fieldwork is given in Chapter 23

Summary of the specification and its coverage in *Edexcel A level Geography Book 1*

Edexcel AS Geography – Paper 2

Content	Covered in
AS Compulsory content	
Topic 3: Globalisation	*Edexcel A level Geography Book 1*
AS Optional content	
Topic 4: Shaping Places One of these topics must be studied: • Topic 4A: Regenerating Places • Topic 4B: Diverse Places	*Edexcel A level Geography Book 1*
AS Fieldwork	*Edexcel A level Geography Book 1* Guidance on fieldwork is given in Chapter 23

Summary of the specification and its coverage in *Edexcel A level Geography Book 1*

Edexcel A level Geography – Paper 1

Content	Covered in
A level Compulsory content	
Topic 1: Tectonic Processes and Hazards	*Edexcel A level Geography Book 1*
Topic 5: The Water Cycle and Water Insecurity	*Edexcel A level Geography Book 2*
Topic 6: The Carbon Cycle and Water Insecurity	*Edexcel A level Geography Book 2*
A level Optional content	
Topic 2: Landscape Systems, Processes and Change One of these topics must be studied: • Topic 2A: Glaciated Landscapes and Change • Topic 2B: Coastal Landscapes and Change	*Edexcel A level Geography Book 1*

Summary of the specification and its coverage in *Edexcel A level Geography Books 1 and 2*

Edexcel A level Geography – Paper 2

Content	Covered in
A level Compulsory content	
Topic 3: Globalisation	*Edexcel A level Geography Book 1*
Topic 7: Superpowers	*Edexcel A level Geography Book 2*
A level Optional content	
Topic 4: Shaping Places One of these topics must be studied: • Topic 4A: Regenerating Places • Topic 4B: Diverse Places	*Edexcel A level Geography Book 1*
Topic 8: Global Development and Connections One of these topics must be studied: • Topic 8A: Health, Human Rights and Intervention • Topic 8B: Migration, Identity and Sovereignty	*Edexcel A level Geography Book 2*

Summary of the specification and its coverage in *Edexcel A level Geography Books 1 and 2*

Edexcel A level Geography – Paper 3

Content	Covered in
Synoptic investigation	*Edexcel A level Geography Book 2* Chapter 18 provides detail on the synoptic investigation
Coursework: Independent investigation	*Edexcel A level Geography Book 1* Guidance on fieldwork is given in Chapter 23

Summary of the specification and its coverage in *Edexcel A level Geography Books 1 and 2*

Each chapter in this book covers one enquiry question in a topic. Within a chapter, each key idea from the specification is addressed and there is full coverage of what the specification indicates that you need to learn.

There is a range of features designed to give you confidence and to present the content of your course in a clear and accessible way, as well as supporting you in your revision and exam preparation.

- An **introduction** to each chapter gives you an overview of what is covered in that chapter.
- **Key terms** are defined throughout to increase your geographical vocabulary.
- **Key concepts** explain important ideas and how to apply them.
- **Skills focus** provides opportunities for you to learn and practise the skills required as indicated within the detailed content for each topic in the specification.
- **Synoptic themes** in the compulsory topics indicate that the content relates to the three over-arching synoptic themes in the specification: players, attitudes and actions, and futures and uncertainties.
- **Place contexts** are indicated with a globe symbol and provide detailed content in context, as suggested by the place contexts within the specification.
- **Fieldwork opportunities** suggest ideas for how you could carry out fieldwork for that topic.
- A range of photographs, maps and graphs to develop your data-response skills.
- **Further research** provide information on useful websites relevant to the chapter.
- **Review questions** at the end of each chapter enhance understanding of key ideas and provide extension activities.
- **Exam-style questions** have been designed to offer study practice and to develop your exam technique skills.

The Publishers would like to thank the following for permission to reproduce copyright material.

Photo credits

p.1 © dawn brealey / Alamy Stock Photo; p.10 U.S. Geological Survey photograph by Brian Collins; p.13 © dawn brealey / Alamy Stock Photo; p.26 (left) © PHAS / UIG via Getty Images; p.26 (right) © Majid / Getty Images; p.34 © Alessio Moiola / 123RF.com; p.38 © David Holmes; p.43 © ARCTIC IMAGES / Alamy Stock Photo; p.50 and p.55 © Sue Warn; p.63 (top) CPOM / Leeds / ESA; p.63 (bottom) and p.70 © Sue Warn; p.81 © Ashley Cooper / Alamy Stock Photo; p.83 © Sue Warn; p.92 © Getty Images / iStockphoto / Thinkstock; p.96 © ARCTIC IMAGES / Alamy Stock Photo; p.97 (top) © Hemis / Alamy Stock Photo; p.97 (bottom) and p.98 © Sue Warn; p.101 © Ashley Cooper pics / Alamy Stock Photo; p.102 © ARCTIC IMAGES / Alamy Stock Photo; p.109 © David Woods – Fotolia; p.111 © Getty Images / iStockphoto / Thinkstock; p.115 © Cameron Dunn; p.117 (top) contains British Geological Survey materials © NERC 2016; p.117 (bottom) Google and the Google logo are registered trademarks of Google Inc., used with permission. Imagery ©2016 Getmapping plc, Data SIO NOAA, US Navy, NGA, GEBCO, Map data ©2016 Google; p.119 © Peter Boardman / Alamy Stock Photo; p.124 © David Woods – Fotolia; p.127 Google and the Google logo are registered trademarks of Google Inc., used with permission. Imagery ©2016 Google, Map data ©2016 Google; p.129 © David Bird / Alamy Stock Photo; p.132 (left and right) © Cameron Dunn; p.138 Google and the Google logo are registered trademarks of Google Inc., used with permission. Imagery ©2016 Data SIO NOAA, US Navy, NGA, GEBCO, Landsat, DigitalGlobe, CNES/Astrium, Map data ©2016 Google; p.144 © robertharding / Alamy Stock Photo; p.148 © Mick House / Alamy Stock Photo; p.150 (top) © IUCN / MFF; p.150 (bottom) © NRM / Pictorial Collection / Science & Society Picture Library; p.154 Google and the Google logo are registered trademarks of Google Inc., used with permission. Imagery ©2016 Getmapping plc, DigitalGlobe, Infoterra Ltd & Bluesky, Map data ©2016 Google; p.155 © Anirut Rassameesritrakool / Alamy Stock Photo; p.157 © Cameron Dunn; p.159 © Dan Kitwood / Getty Images; p.164 © Imaginechina / Corbis; p.165 (top) © Eckel/ullstein bild via Getty Images; p.165 (bottom) © NASA / Glenn Research Center Collection; p.166 © ChinaFotoPress via Getty Images; p.168 © Fender Musical Instruments Corporation; p.169 © Justin Kase z12z / Alamy; p.176 © Motoring Picture Library / Alamy; p.182 © david pearson / Alamy Stock Photo; p.187© Getty Images; p.188 © Charles O. Cecil / Alamy Stock Photo; p.194 (top) © imageBROKER / Alamy Stock Photo; p.194 (bottom) © Bob Daemmrich / Alamy Stock Photo; p.203 © Richard Naude / Alamy Stock Photo; p.210 © Mamunur Rashid / Alamy Stock Photo; p.215 © Geoffrey Robinson / Alamy Stock Photo; p.216 (left) © Kim Adams; p.216 (right) © IRStone – Fotolia; p.230 and p.234 © Kim Adams; p.238 (top left) © Alex Segre / Alamy Stock Photo; p.238 (top right and bottom), p.242 and p.250 © Kim Adams; p.251 © Geoffrey Robinson / Alamy Stock Photo; p.253 (top) © Andrew Michael / Alamy Stock Photo; p.253 (bottom) © G&D Images / Alamy Stock Photo; p.256 Courtesy of Peel Land And Property Group; p.261 © Kim Adams; p.266 (left) © Skyscan.co.uk; p.266 (right) © Kim Adams; p.275 © Lance Bellers – Fotolia; p.276 © Rawpixel – Fotolia; p.289 (top) © Jonathan Goldberg / Alamy Stock Photo; p.289 (bottom) © London Street / Alamy Stock Photo; p.294 © Andrew Holt / Alamy Stock Photo; p.295 (top left) Wellcome Library, London / http://creativecommons.org/licenses/by/4.0; p.295 (bottom left) © Pictorial Press Ltd / Alamy Stock Photo; p.295 (top right) © Heritage Image Partnership Ltd / Alamy Stock Photo; p.295 (bottom right) © Lordprice Collection / Alamy Stock Photo; p.298 © Getty Images / Photodisc / Thinkstock; p.300 © Trevor Smithers ARPS / Alamy Stock Photo; p.301 © Greg Balfour Evans / Alamy Stock Photo; p.302 © Peter Lane / Alamy Stock Photo; p.306 (top) Courtesy National Gallery of Art, Washington; p.306 (bottom) Wellcome Library, London / http://creativecommons.org/licenses/by/4.0; p.307 © ROPI / Alamy Stock Photo; p.312 © Tessa Bunney / Corbis; p.314 © Tal Cohen / Photoshot; p.315 (top left) © Jeffrey Blackler / Alamy Stock Photo; p.315 (top right) © Mike Kemp / In Pictures / Corbis; p.315 (bottom left) © roger parkes / Alamy Stock Photo; p.315 (bottom right) © Greg Balfour Evans / Alamy Stock Photo; p.319 © Lana Rastro / Alamy Stock Photo; p.320 (bottom) © SSPL / Getty Images; p.320 (top) © Jimmy James / Associated Newspapers / REX / Shutterstock; p.321 © Art Directors & TRIP / Alamy Stock Photo; p.329 (left) © Getty Images / iStockphoto / Thinkstock; p.329 (right) © A.P.S. (UK) / Alamy Stock Photo; p.331 © Skyscan Photolibrary / Alamy Stock Photo; p.334 © geogphotos / Alamy Stock Photo; p.336 (left) © Kim Adams; 336 (right) © Mark Waugh / Alamy Stock Photo; p.354 (left) © David Holmes; p.354 (right) © Arena Photo UK / 123RF.com.

Acknowledgements

Every effort has been made to trace all copyright holders, but if any have been inadvertently overlooked, the Publishers will be pleased to make the necessary arrangements at the first opportunity.

p.7 www.visionlearning.com/en/library/Earth-Science/6/Plates-Plate-Boundaries-and-Driving-Forces/66; p.9 *Environmental Hazards: Assessing Risk and Reducing Disaster*, Keith Smith and David N. Petley, Routledge; p.12 *Environmental Hazards: Assessing Risk and Reducing Disaster*, Keith Smith and David N. Petley, Routledge; p.21 *Environmental Hazards: Assessing Risk and Reducing Disaster*, Keith Smith and David N. Petley, Routledge; p.31 www.economist.com/news/briefing/21647958-two-hundred-years-ago-most-powerful-eruption-modern-history-made-itself-felt-around; p.39 *Environmental Hazards: Assessing Risk and Reducing Disaster*, Keith Smith and David N. Petley, Routledge; p.40 http://documents.wfp.org/stellent/groups/public/documents/newsroom/wfp246342.jpg; p.41 www.economist.com/blogs/graphicdetail/2012/01/daily-chart-4?fsrc=gn_ep; p.46 *An Introduction to Physical Geography and the Environment*, 3rd Ed., Joseph Holden ©2012. Reprinted by permission of Pearson Education, Inc., New York, New York; p.65 *Glaciers and Glaciation*, D. Benn, Routledge; p.74 *Geography Factsheets 99*, David Holmes, Curriculum Press; p.87 *Geography Factsheets 99*, David Holmes, Curriculum Press; p.88 *Process and Landform*, Peter Comfort & Alan Clowes, Oliver and Boyd; p.91 www.bas.ac.uk; p.134 www.ipcc.ch/publications_and_data/ar4/wg1/en/faq-5-1-figure-1.html; p.137 (left) This work is based on data provided through www.VisionofBritain.org.uk and uses historical material which is copyright of the Great Britain Historical GIS Project and the University of Portsmouth; p.137 (right) © Crown copyright 2016 Ordnance Survey. Licence number 100036470; p.162 © Chris Gray 2007; p.206 adapted from data from the office for National Statistics Licensed under the Open Government Licence v.3.0; p.211 www.newscenter.philips.com/pwc_nc/main/shared/assets/nl/Newscenter/2015/SR-AR/Infographic-Philips-Circular-Economy.jpg; p.217 adapted from data from the Office for National Statistics licensed under the Open Government Licence v.3.0; p.285 adapted from data from the Office for National Statistics licensed under the Open Government Licence v.3.0; p.308 adapted from data from the Office for National Statistics licensed under the Open Government Licence v.3.0; p.309 adapted from data from the Office for National Statistics licensed under the Open Government Licence v.3.0; p.310 adapted from data from the Office for National Statistics licensed under the Open Government Licence v.3.0; p.311 adapted from data from the Office for National Statistics licensed under the Open Government Licence v.3.0; p.315 www.visitsouthall.co.uk; p.326 http://media.gallup.com/WorldPoll/PDF/GMSLonMuslims-BeyondMultiCvsAssim41307.pdf; p.327 www.quantum.international/house-prices-most-and-least-affordable-cities-in-the-uk.

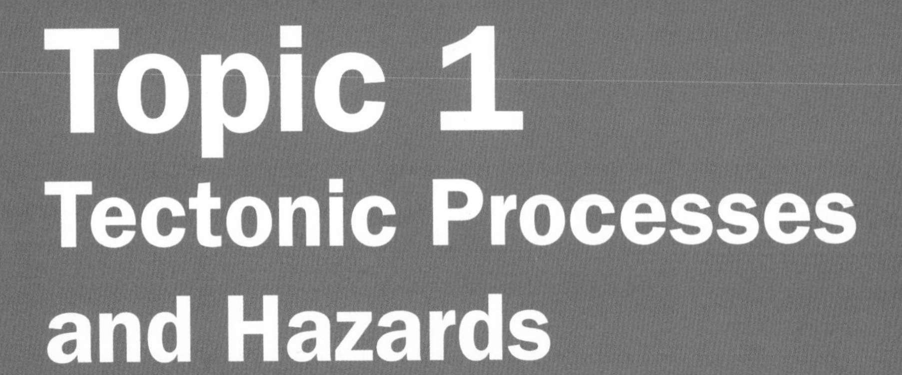

Topic 1
Tectonic Processes and Hazards

1 Locations at risk from tectonic hazards

Why are some locations more at risk from tectonic hazards?

By the end of this chapter you should:

- understand how the global distribution of tectonic hazards can be explained by plate boundaries and other tectonic processes
- understand the theoretical frameworks that attempt to explain plate motion and movement
- be able to understand and explain the physical causes of tectonic hazards.

1.1 The global distribution of tectonic hazards

Tectonic hazards include earthquakes and volcanic eruptions, as well as secondary hazards such as tsunami. These represent a significant risk in some parts of the world in terms of loss of life, livelihoods and economic impact. This is especially the case where active tectonic plate boundaries interact with areas of high population density, and medium and high levels of development. Tectonic hazards can be classified as either seismic or volcanic.

The global distribution of tectonics hazards: earthquakes

The global distribution of tectonic hazards is far from random. Figure 1.1 clearly shows that the main earthquake zones are found (often in clusters) along plate boundaries. About 70 per cent of all earthquakes are found in the 'Ring of Fire' in the Pacific Ocean. The most powerful earthquakes are associated with convergent or conservative boundaries, although rare intra-plate earthquakes can occur. This distribution of earthquakes reveals the following pattern of tectonic activity:

- The oceanic fracture zone (OFZ) – a belt of activity through the oceans along the mid-ocean ridges, coming ashore in Africa, the Red Sea, the Dead Sea rift and California.
- The continental fracture zone (CFZ) – a belt of activity following the mountain ranges from Spain, via the Alps, to the Middle East, the Himalayas to the East Indies and then circumscribing the Pacific.
- Scattered earthquakes in continental interiors. A small minority of earthquakes can also occur along old fault lines and the hazard is associated with the reactivation of this weakness, for example the Church Stretton Fault in Shropshire.

Earthquakes are a common hazard and can develop into a major disaster, especially when they are both high magnitude and occur in a densely populated area. Earthquakes are primary hazards (ground movement and ground shaking) but also cause secondary hazards such as landslides and tsunamis. The distribution of tsunamis is discussed later in the chapter, on page 11.

The global distribution of tectonics hazards: volcanoes

The violence of a volcanic eruption is determined by the amount of dissolved gases in the magma and how easily the gases can escape. There are about 500 active volcanoes throughout the world (Figure 1.2) and, on average, around 50 of them erupt each year.

Key terms

Seismic hazards: Generated when rocks within 700 km of the Earth's surface come under such stress that they break and become displaced.

Volcanic hazards: Associated with eruption events.

Intra-plate earthquakes: These occur in the middle or interior of tectonic plates and are much rarer than boundary earthquakes.

Volcano: A landform that develops around a weakness in the Earth's crust from which molten magma, volcanic rock, and gases are ejected or extruded.

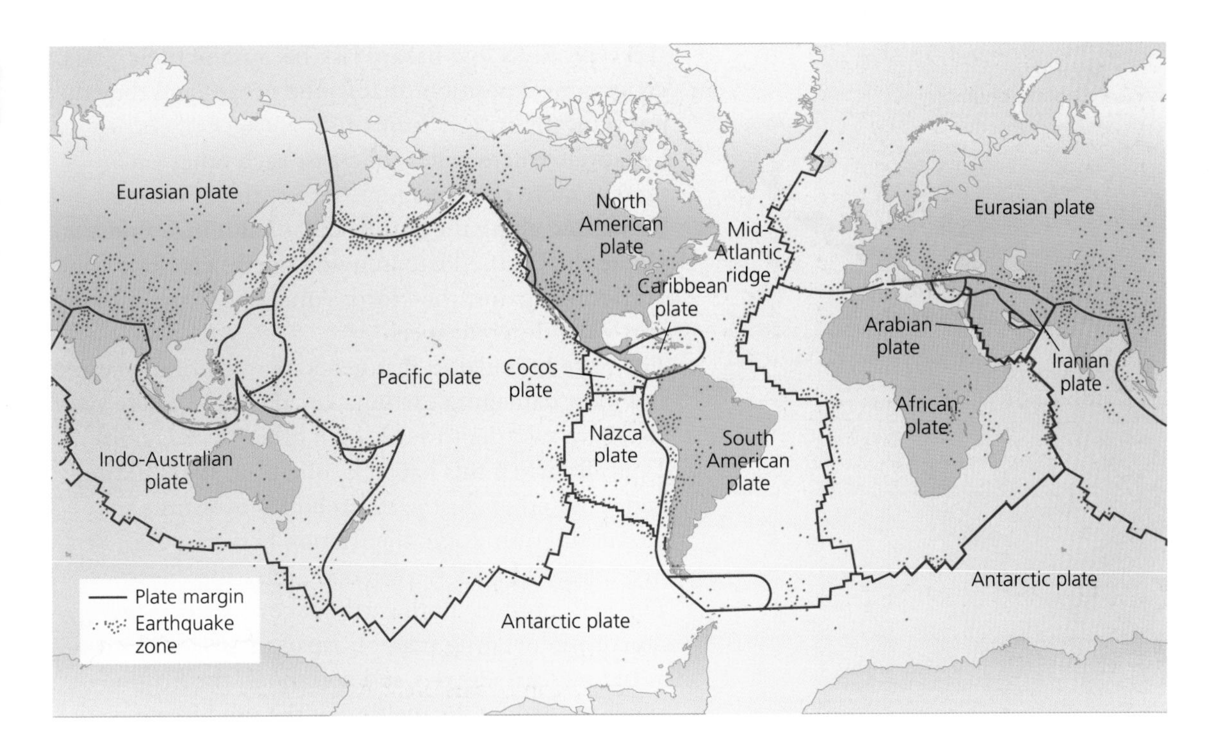

Figure 1.1 The global distribution of earthquakes and their associated plate margins

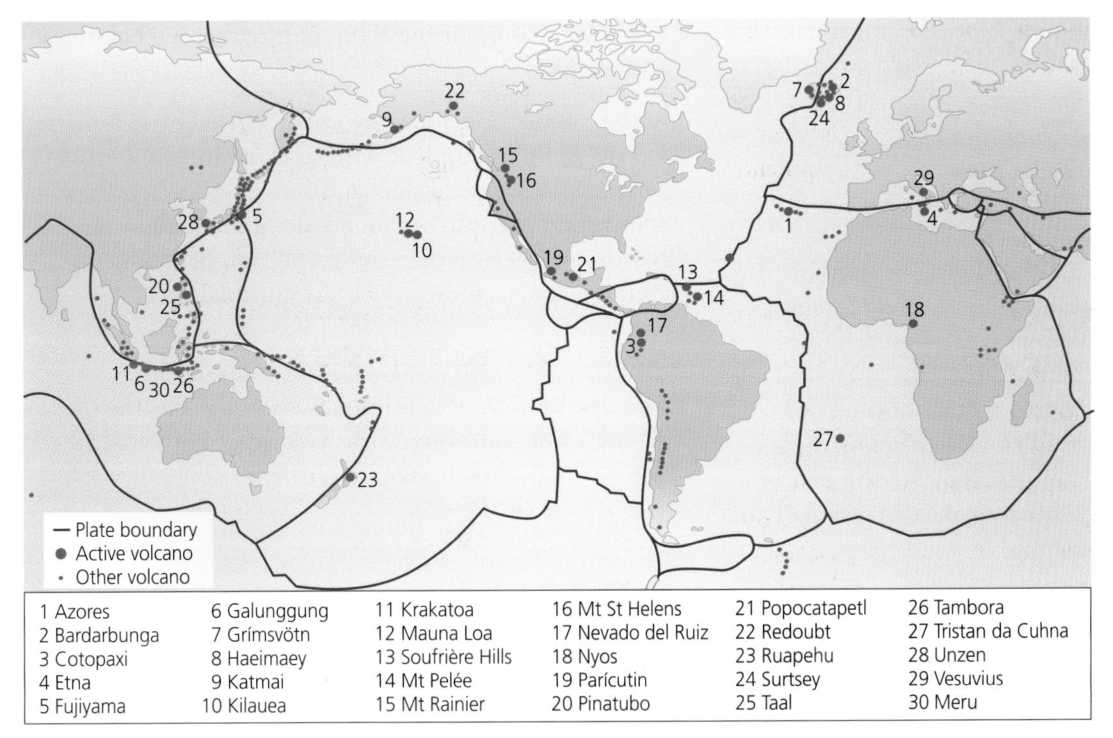

Figure 1.2 The global distribution of active volcanoes

Plate boundary types and their distribution

There are three types of plate boundary (Figure 1.3):

- **Divergent** (constructive) margins, most clearly displayed at mid-ocean ridges. At these locations there are large numbers of shallow focus and generally low magnitude earthquake events. Most are submarine (under the sea).
- **Convergent** (where plates move together): these are actively deforming collision locations with plate material melting in the mantle, causing frequent earthquakes and volcanoes.

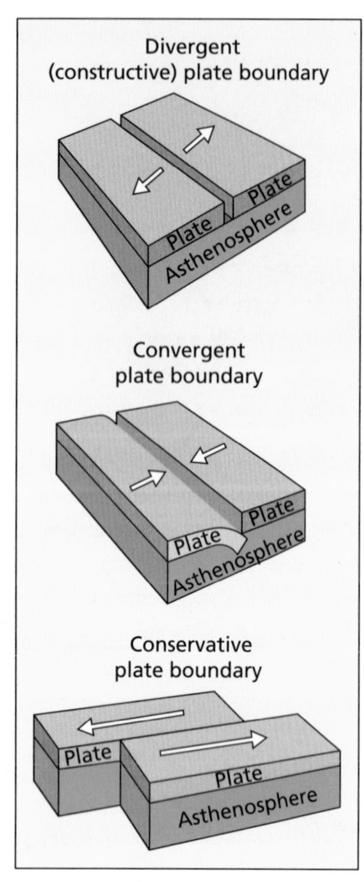

Figure 1.3 Different plate boundaries

- **Conservative** (oblique-slip, sliding or transform) margins, where one plate slides against another. Here the relative movement is horizontal and classified as either sinistral (to the left) or dextral (to the right). Lithosphere is neither created nor subducted, and while conservative plate margins do not result in volcanic activity, they are the sites of extensive shallow focus earthquakes, occasionally of considerable magnitude.

Plate movement and earthquake type

There are three ways plates can move with respect to one another: they can pull away from each other, slide past each other or crunch into each other. Each possibility offers different focal depths and typical magnitudes.

- The places where they move away from each other are the divergent 'spreading ridges' in the oceans. New oceanic crust, which is thinner and denser than the continental crust, is created. The earthquakes seen at these boundaries tend to be frequent, small and typically a low hazard risk because of their geographical position (that is, the ocean) and they do not typically trigger tsunamis.
- Locations where plates slide past each other can present more risk. In simple terms, this is what is happening along the San Andreas Fault in California, where the Pacific Plate (moving north) creates a zone of friction against the North American Plate (moving north at a different speed).
- The plate boundaries that generate some of the largest and most damaging earthquakes are those where two plates are moving towards each other (convergent). Typically when this happens, one plate starts sliding under the other. As the strain builds over time in the subduction zone, the friction between the two masses of rock is overcome, releasing energy. This will produce both earthquakes – such as the tsunami-generating ones off Japan in 2011 and Aceh in Indonesia in 2004 – and volcanoes, the magma of which are fed by the melting of the subducting plate. The subduction zones at the edge of the Pacific Plate are the reason for the Ring of Fire that is a feature of this ocean.

Active subduction zones are characterised by magmatic activity, a mountain belt with thick continental crust, a narrow continental shelf and active seismicity. Passive continental margins are found along the remaining coastlines. Because there is no collision or subduction taking place, tectonic activity is minimal here.

Plate movement and volcanic activity

The distribution of volcanoes is controlled by the global geometry of plate tectonics. Volcanoes are found in a number of different tectonic settings:

1. **Destructive** plate boundaries (Figure 1.4). These occur at locations where two plates are moving together. Here they form either a subduction zone

Key term

Plate tectonics: A theory developed more than 60 years ago to explain the large-scale movements of the lithosphere (the outermost layer of the Earth). It was based around the evidence from sea floor spreading and ocean topography, marine magnetic anomalies, paleomagnetism and geomagnetic field reversals. A knowledge of Earth's interior and outer structure is essential for understanding plate tectonics (see Figure 1.6).

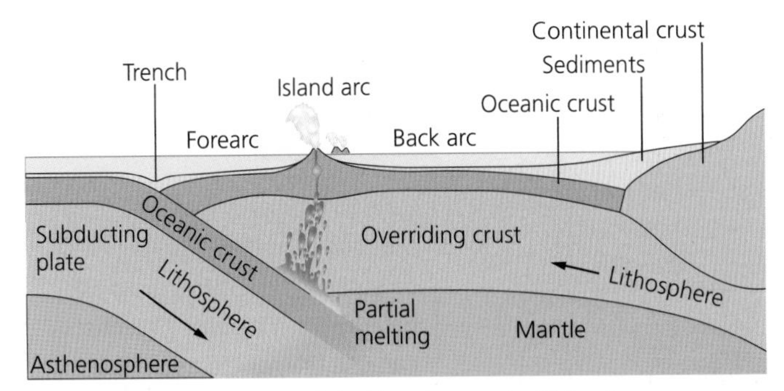

Figure 1.4 Subduction and the development of island arc volcanoes at a destructive boundary

or a continental collision, depending on the type of plates. When a dense oceanic plate collides with a less-dense continental plate, the oceanic plate is typically thrust underneath because of the greater buoyancy of the continental lithosphere, forming a subduction zone. Surface volcanism (volcanoes at the ocean floor or the Earth's surface) typically appears above the magma that forms directly above down-thrust plates. During collisions between two continental plates, however, large mountain ranges are formed, such as the Himalayas. Destructive boundaries comprise a large proportion of the world's active volcanoes and create the most explosive type, characterised by a composite cone associated with a number of hazards. These volcanic eruptions tend to be more infrequent but more destructive.

2. **Divergent** boundaries create rift volcanoes where plates diverge from one another at the site of a thermally buoyant mid-ocean ridge. These are generally less explosive and more effusive, especially when they occur under water deep in the ocean floor, for example the Mid-Atlantic Ridge. Here there is basaltic magma, which has low viscosity.
3. **Hotspot volcanoes** are found in the middle of tectonic plates and are thought to be fed by underlying mantle plumes that are unusually hot compared with the surrounding mantle.

Key term

Lithosphere: The surface layer of the Earth is a rigid outer shell composed of the crust and upper mantle. It is on average 100 km deep. The lithosphere is always moving, but very slowly, fuelled by rising heat from the mantle which creates convection currents. The distinction between lithosphere and asthenosphere is one of physical strength rather than a difference in physical composition. The lithosphere is broken into huge sections, which are the tectonic plates.

Volcanoes at different types of boundary will of course present a risk for people and property, both in the shorter and longer term.

Hotspot volcanoes and mantle plumes

The vast majority of volcanic eruptions occur near plate boundaries, but there are some exceptions: hotspot volcanoes. The presence of a hotspot is inferred by anomalous volcanism (that is, not at a plate boundary), such as the Hawaiian volcanoes within the Pacific Plate.

A volcanic hotspot is an area in the mantle from which heat rises as a hot thermal plume from deep in the Earth. High heat and lower pressure at the base of the lithosphere enable melting of the rock. This molten material, magma, rises through cracks and erupts to form active volcanoes on the Earth's surface. As the tectonic plate moves over the stationary hotspot, the volcanoes are rafted away and new ones form in their place. As oceanic volcanoes move away from the hotspot, they cool and subside, producing older islands, atolls and seamounts. Over long periods of time this can also create chains of volcanoes, such as the Hawaiian Islands (Figure 1.5).

1.2 Theoretical frameworks and plate movements

There are two different types of *crust*, which are are made up of different types rock:

- thin oceanic crust, which underlies the ocean basins, is composed primarily of basalt
- thicker continental crust, which underlies the continents, is composed primarily of granite.

The low density of the thick continental crust allows it to 'float' high on the much higher density mantle below.

The Earth's *mantle* has a temperature gradient (geothermal gradient). The highest temperatures occur where the mantle material is in contact with the heat-producing core so there is a steady increase of temperature with depth. Rocks in the upper mantle are cool and brittle, while rocks in the lower mantle are hot and plastic (but not molten). Rocks in the upper mantle are brittle enough to break under stress and produce earthquakes. However, rocks in the lower mantle are plastic and flow when subjected to forces instead of breaking. The lower limit of brittle behaviour is the boundary between the upper and lower mantle.

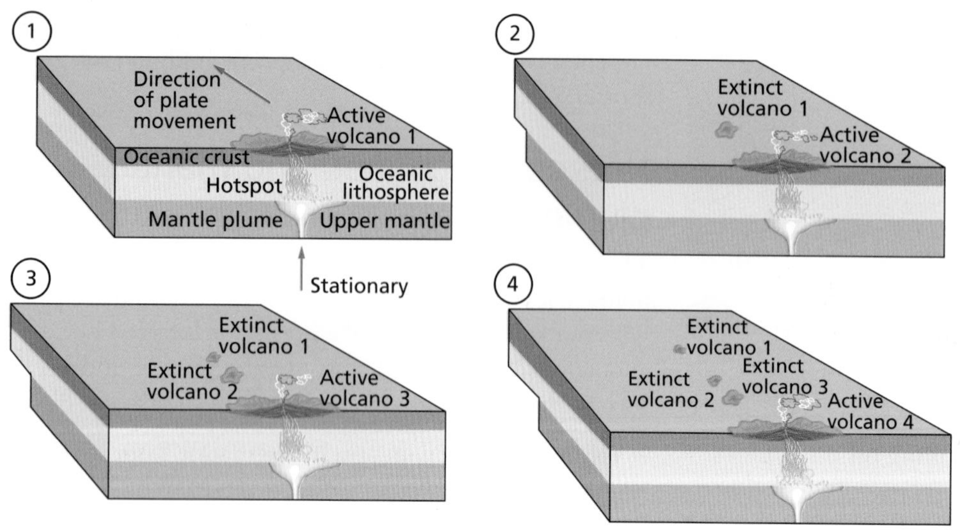

Figure 1.5 The formation of volcanic hotspots over time

Key term

Paleomagnetism results from the zone of magma 'locking in' or 'striking' the Earth's magnetic polarity when it cools. Scientists can use this tool to determine historic periods of large-scale tectonic activity through the reconstruction of relative plate motions. They create a geo-timeline.

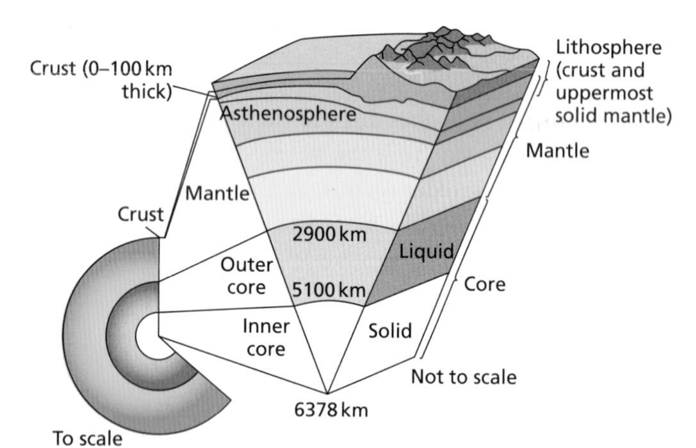

Figure 1.6 Section through the upper layers of the Earth as a schematic diagram

Heat which is derived from the Earth's core (radioactive decay) rises within the mantle to drive *convection* currents, which in turn move the tectonic plates. These convection currents operate as cells (Figure 1.7). We already know that plates can move in a number of directions when in contact with each other, and that the type of movement can be translated into a particular hazard risk.

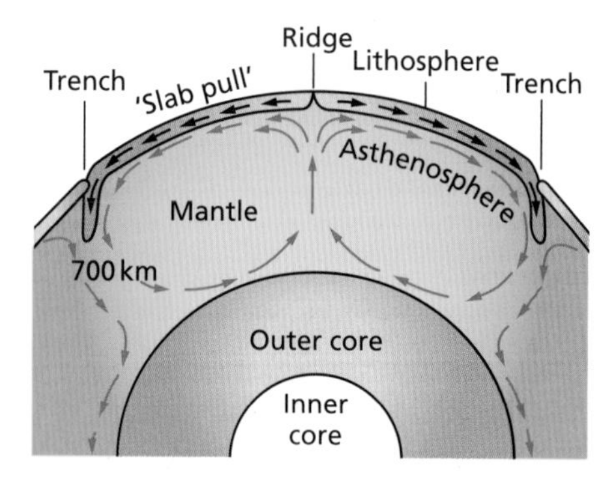

Figure 1.7 The role of convection currents in 'slab pull'

Sea floor spreading occurs at divergent boundaries under the oceans. This is a continuous input of magma forming a mid-ocean ridge, for example the Mid-Atlantic Ridge. On land a rift valley forms. A technique involving the reconstruction of paleomagnetic reversals (called paleomagnetism) can be used to date the age of new tectonic crust (Figure 1.8).

The importance of convection cells is disputed, however. There is likely to be a combined force of convection and gravity driving tectonic plate movement. Gravity in particular causes the denser oceanic crust to be pulled down at the site of subduction. At constructive margins (i.e. spreading ridges), magma is simply 'gap filling', rather than the main driver pushing the plates in opposite directions away from each other.

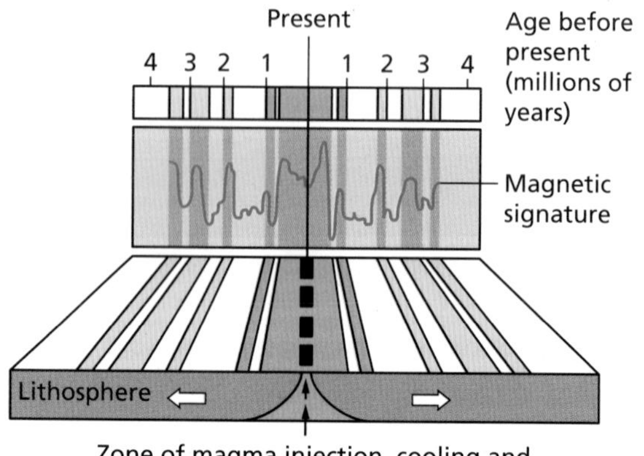

Figure 1.8 Changes in the direction of magnetic 'signatures' allow crust to be dated

The Benioff Zone and subduction processes

The Benioff Zone of is an area of seismicity corresponding with the slab being thrust downwards in a subduction zone. The different speeds and movements of rock at this point produce numerous earthquakes. It is the site of intermediate/deep-focused earthquakes. This theoretical framework is therefore an important factor in determining earthquake magnitude, since it determines the position and depth of the hypocentre (page 8).

Key terms

Subduction zones are broad areas where two plates are moving together, often with the thinner, more dense oceanic plate descending beneath a continental plate. The contact between the plates is sometimes called a thrust or megathrust fault. Where the plates are locked together, frictional stress builds. When that stress exceeds a given threshold, a sudden failure occurs along the fault plane that can result in a 'mega-thrust' earthquake, releasing strain energy and radiating seismic waves. It is common for the leading edge to lock under high friction. The locked fault can hold for hundreds of years, building up enormous stress before releasing. The process of strain, stress and failure is referred to as the 'elastic-rebound theory'.

Locked fault: A fault that is not slipping because the frictional resistance on the fault is greater than the shear stress across the fault, that is, it is stuck. Such faults may store strain for extended periods that is eventually released in a large magnitude earthquake when the frictional resistance is eventually overcome. The 2004 Indian Ocean tsunami was the result of a mega-thrust locked fault (subducting Indian Plate) with strain building up at around 20 mm per year. It generated huge seismic waves and the devastating tsunami.

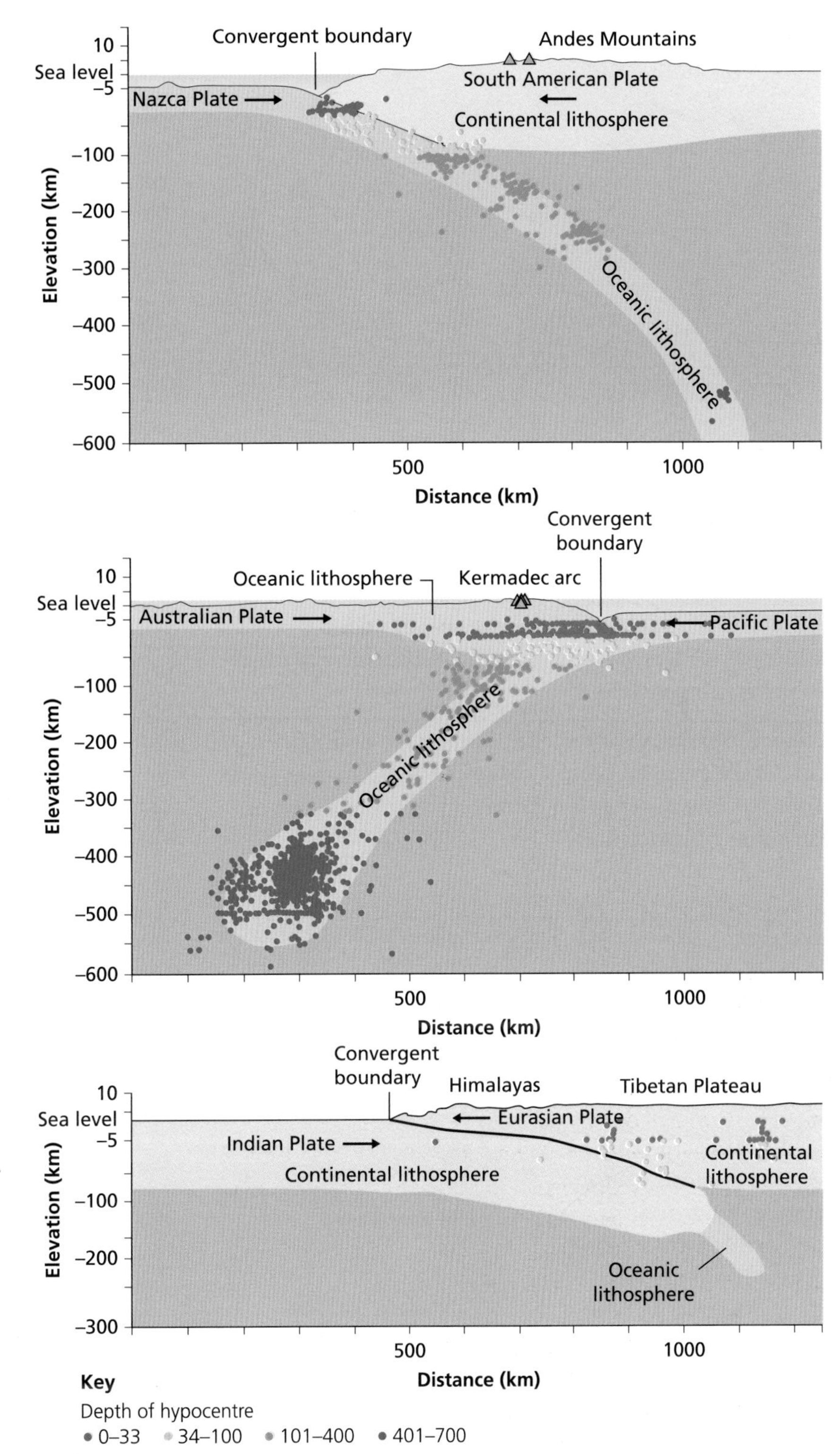

Figure 1.9 Variations in earthquake hypocentres at convergent boundaries

Zones where there are locked faults can present a significant tectonic hazard. The Andes owe their existence to a subduction zone on the western edge of the South American Plate; in fact, this type of boundary is often called an Andean boundary since it is the primary example.

Figure 1.9 shows an idealised distribution of earthquakes at a convergent boundary, along a subduction zone and a continental–continental

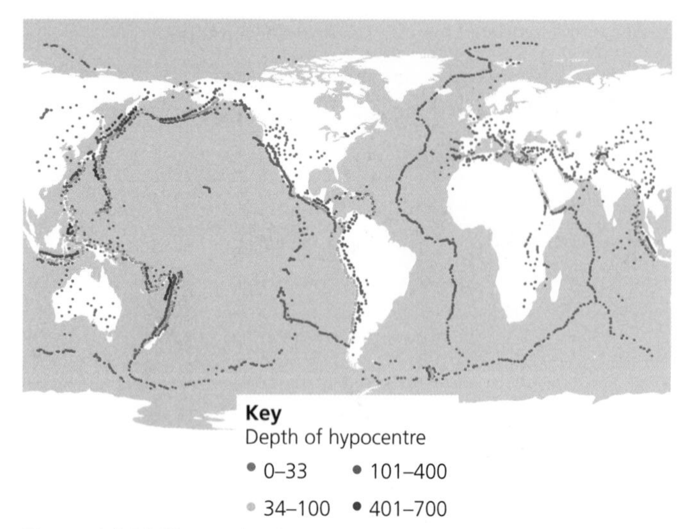

Figure 1.10 Different depth tectonic earthquake boundaries mapped on to a world map

boundary. Earthquake centres are colour coded according to depth. Green triangles represent volcanoes on the Earth's surface.

These different depth tectonic earthquake boundaries can also be mapped on to a complete world map to give an interesting distribution pattern in terms of depth of earthquake hypocentres.

1.3 Physical processes and tectonic hazards

Earthquakes, crustal fracturing and ground shaking

Earthquakes are caused by sudden movements comparatively near to the Earth's surface along a fault. Faults are zones of pre-existing weakness in the Earth's crust. A sequence of events occurs in the generation of an earthquake:

1 The movements are preceded by a gradual build-up of tectonic strain, which stores elastic energy in crustal rocks.

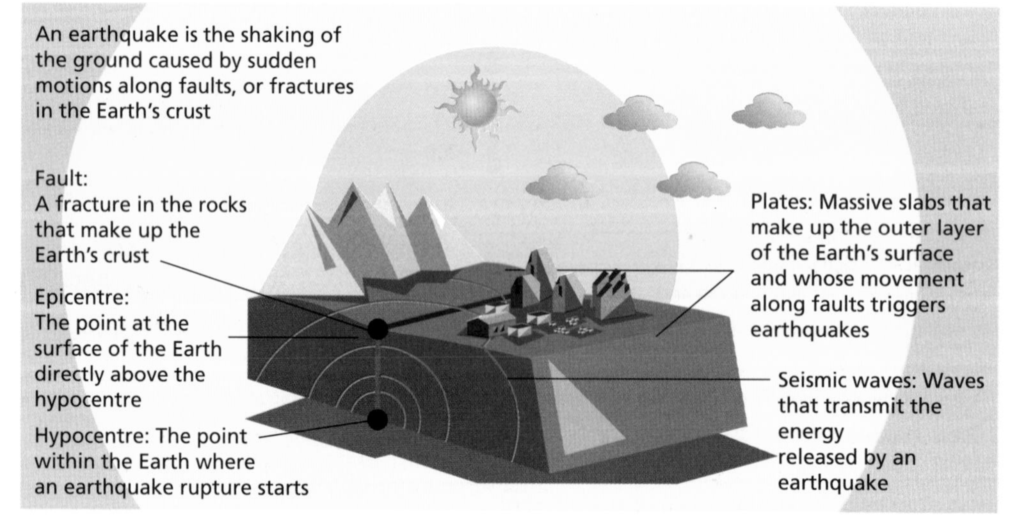

Figure 1.11 Anatomy of an earthquake

> **Key term**
>
> Hypocentre is the 'focus' point within the ground where the strain energy of the earthquake stored in the rock is first released. The distance between this and the epicentre on the surface is called focal length.

2 When the pressure exceeds the strength of the fault, the rock fractures.

3 This produces the sudden release of energy, creating seismic waves that radiate away from the point of fracture.

4 The brittle crust then rebounds either side of the fracture, which is the ground shaking, that is, the earthquake felt on the surface.

The point of rupture, the hypocentre, can occur at any depth between the Earth's surface and about 700 km. Usually, the rupture of the fault propagates along the fault with the earthquake waves coming from both the hypocentre and the fault plain itself (Figure 1.11). The most damaging events are usually shallow focus, with a hypocentre of less than 40 km.

Table 1.1 shows the ten largest recorded magnitude earthquakes since 1900.

You can find out more at: http://earthquake.usgs.gov/earthquakes/world/10_largest_world.php

Table 1.1 Largest recorded magnitude earthquakes since 1900

Location	Date	Magnitude
Chile	22 May 1960	9.5
Great Alaska Earthquake, USA	28 March 1964	9.2
Off the west coast of northern Sumatra, Indonesia	26 December 2004	9.1
Near the east coast of Honshu, Japan	11 March 2011	9.0
Kamchatka, Russia	4 November 1952	9.0
Offshore Maule, Chile	27 February 2010	8.8
Off the coast of Ecuador	31 January 1906	8.8
Rat Islands, Alaska, USA	4 February 1965	8.7
Northern Sumatra, Indonesia	28 March 2005	8.6
Assam, Tibet	15 August 1950	8.6

Skills focus: Earthquake analysis

Research and download data on recent earthquake events from the US Geological Survey (USGS) (www.usgs.gov). Using a spreadsheet, calculate the median and inter-quartile ranges. Bear in mind that magnitude is a non-linear unit, so there is a real difference of ten times between each point on the scale.

Seismic waves

A device called a seismometer measures the amount of ground shaking during an earthquake, recording both the vertical and horizontal movements of the ground. Analysis of the data shows that an earthquake produces different seismic waves.

Seismic waves

Primary or P waves are vibrations caused by compression, like a shunt through a line of connected train carriages. They spread quickly from the fault at a rate of about 8 km/sec.

Secondary or S waves move more slowly, however, at around 4 km/sec. They vibrate at right angles to the direction of travel and cannot travel through liquids (unlike P waves). .

Love waves or L waves (also known as Q waves) are surface waves with the vibration occurring in the horizontal plain. They have a high amplitude.

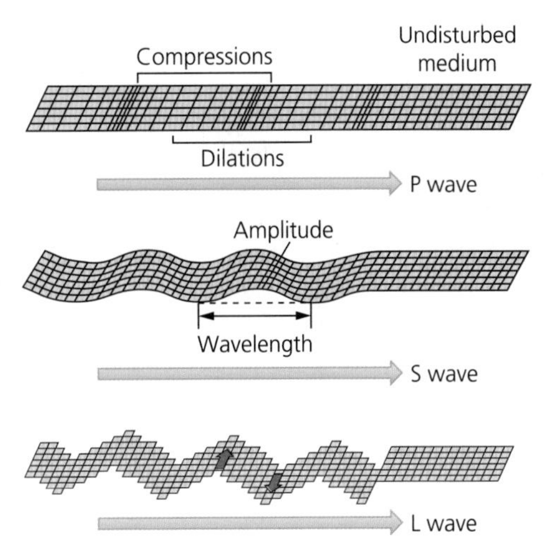

Figure 1.12 Differences between P, S and L waves

The overall severity of an earthquake is linked to the amplitude and frequency of these wave types. The ground surface may be displaced horizontally, vertically or obliquely during an earthquake depending on the strength of individual waves. The S and L waves are more destructive than the P waves as they have a larger amplitude and energy force.

Secondary hazards of earthquakes: liquefaction and landslides

Secondary hazards are side effects of an earthquake but should be considered no less significant that the primary hazards. A serious secondary hazard from earthquakes, especially where there is loose rock and sediment, is soil liquefaction.

Liquefaction can cause buildings to settle, tilt and eventually collapse in the most serious of events (Figure 1.13). In some earthquakes tilts of up to 60 degrees have been recorded, for example in Japan. Land adjacent to rivers and sloping ground can present a hazard by sliding under low-friction conditions

Key term

Soil liquefaction: The process by which water-saturated material can temporarily lose normal strength and behave like a liquid under the pressure of strong shaking. Liquefaction occurs in saturated soils (ones in which the pore space between individual particles is completely filled with water). An earthquake can cause the water pressure to increase to the point where the soil particles can move easily, especially in poorly compacted sand and silt.

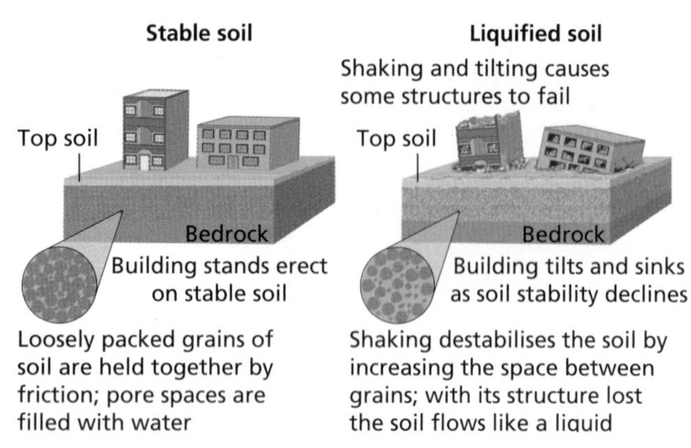

Figure 1.13 The process of soil liquefaction

across a liquefied soil layer. This is called lateral spreading, sometimes creating large fissures and cracks in the ground surface. The consequence of such hazards can be considerable: damage to roads and bridges as well as telecommunication and other services (gas, electricity, sewerage) which run through the upper sections of the ground. The short-term impact on the delivery of aid and the longer-term rebuild costs can be substantial as a direct result of this secondary impact.

Landslides are another important secondary hazard from earthquakes, where slopes weaken and fail. As many destructive earthquakes occur in mountainous areas, landslides (as well as rock falls and avalanches) can be major secondary impacts. Studies linking earthquake intensity to landslides show that they rarely occur when magnitudes are less than 4, but are significant problems when they are larger. For example, more than half the earthquake deaths in Japan are linked to events with

Key terms

Intensity: A measure of the ground shaking. It is the ground shaking that causes building damage and collapse, and the loss of life from the hazard.

Magnitude: The magnitude of an earthquake is related to the amount of movement, or displacement, in the fault, which is in turn a measure of energy release. The 2004 earthquake in Indonesia was very large (M = 9.1) because a large vertical displacement (15 m) occurred along a very long fault distance, approximately 1500 km. (Earthquake magnitude is measured at the epicentre, the point on the Earth's surface directly above the hypocentre.)

Epicentre: the location on the Earth's surface that is directly above the earthquake focus, i.e. the point where an earthquake originates.

Figure 1.14 The rockslide near the Kali Gandaki River in Nepal in May 2015 that buried the village of Baisari and blocked the flow of the river after the main earthquake shock

a magnitude greater than 6.9. This can be especially hazardous to people and property as landslides can travel several miles from their source, growing in size as they pick up trees, boulders, cars and other materials.

A report on the 2015 Nepal earthquake by the USGS suggests that the landslides created by this event could have been made worse by summer monsoon rainfall (Figure 1.14). The annual wet season in Nepal triggers landslides on the highly susceptible slopes in many parts of the country in normal conditions. Landscape disturbance caused by the 2015 earthquake could significantly worsen landslide susceptibility in future monsoons, for a period of at least a few years.

Research by the USGS suggests that, over the last 40 years, around 70 per cent of all deaths caused by earthquakes globally (excluding those from shaking, building collapse and tsunami) are attributable to the secondary impacts of landslides. In the 2005 Kashmir and 2008 Sichuan earthquakes, for example, landslides account for around a third of all deaths.

Key term

Tsunami: The word comes from two Japanese words: *tsu* (port or harbour) and *nami* (wave or sea). Tsunamis are initiated by undersea earthquakes, landslides, slumps and, sometimes, volcanic eruptions. They are characterised by:

- long wavelengths, typically 150–1000 km
- low amplitude (wave height), 0.5–5 m
- fast velocities, up to 600 kph in deep water.

Tsunamis

Tsunamis are one of the most distinctive earthquake-related hazards.

Tsunami waves do not resemble normal sea waves as their wavelength is much longer. Out to sea they do not represent a hazard since they are generally low in height (often below 300 mm) and generally go unnoticed. It is only as they approach a coastline that they grow in height as the water becomes shallower. A tsunami is not a single wave but a series of waves, also known as a wave train, caused by seabed displacement. The first wave in a tsunami is not necessarily the most destructive, so often there is an escalation effect in terms of damage and loss of life. The amount of time between successive waves (the wave period) is often only a few minutes but, in rare instances, waves can be over an hour apart. This represents a greater risk: people have lost their lives after returning home in between the waves of a tsunami, thinking that the waves had stopped coming.

The global distribution of tsunamis is fairly predictable in terms of source areas, with around 90 per cent of all events occurring within the Pacific Basin, associated with activity at the plate margins. Most are generated at subduction zones (convergent boundaries), particularly off the Japan–Taiwan island arc, South America and Aleutian Islands (25 per cent of all historical events have been recorded in this geographic region).

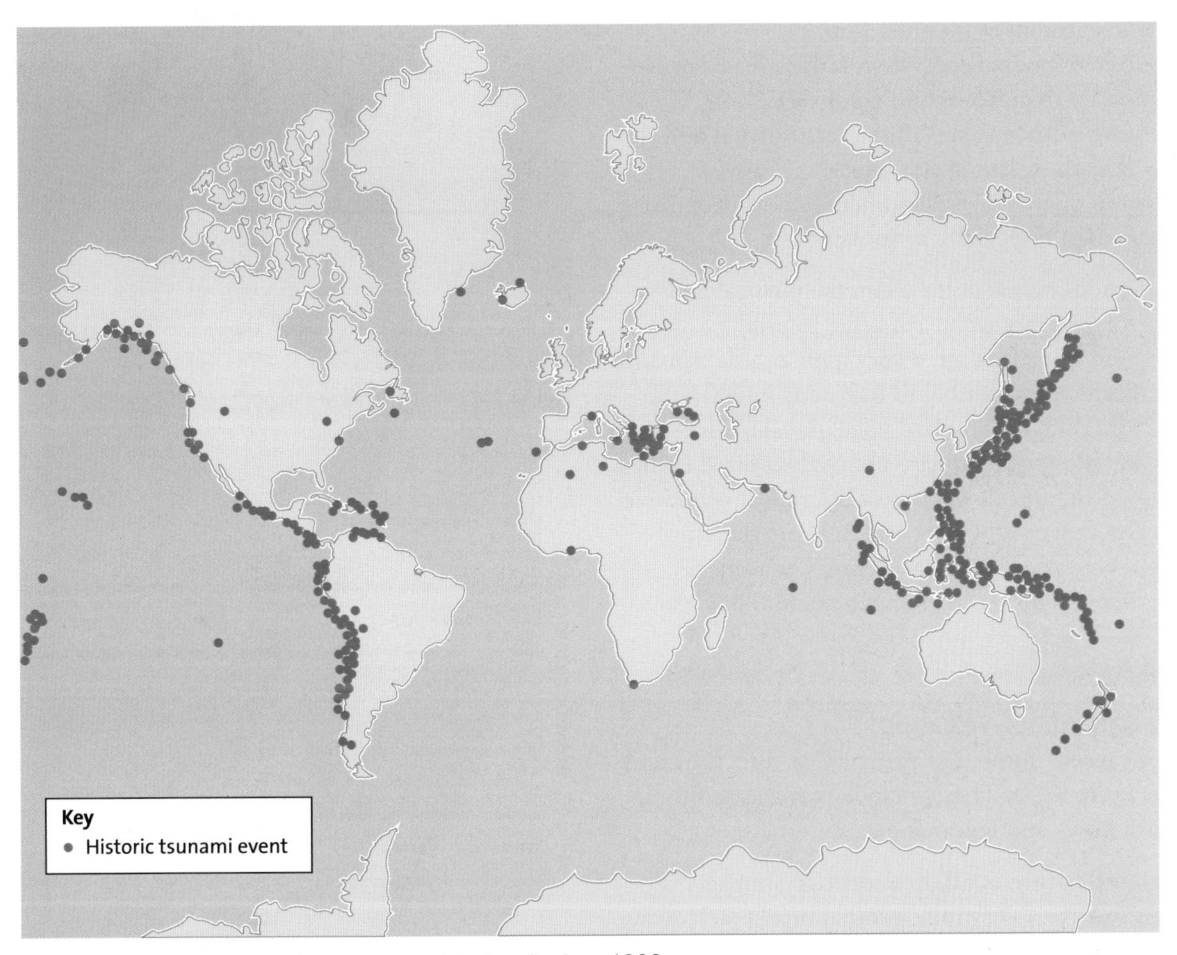

Figure 1.15 Notable tsunami events on a global scale since 1900

Skills focus: Analysis of tsunami travel time maps

Tsunami travel time maps give a better understanding of how an event may occur, and the potential risks for people living within a tsunami's reach. Figure 1.16, for example, shows the predicted time it takes for a tsunami to travel to Hawaii from various points of origin. The blue dots and white triangles are tsunami monitoring stations. How can people prepare for this type of hazard? Where are the gaps in the monitoring stations and why? Compare this model to the 2004 Indian Ocean tsunami.

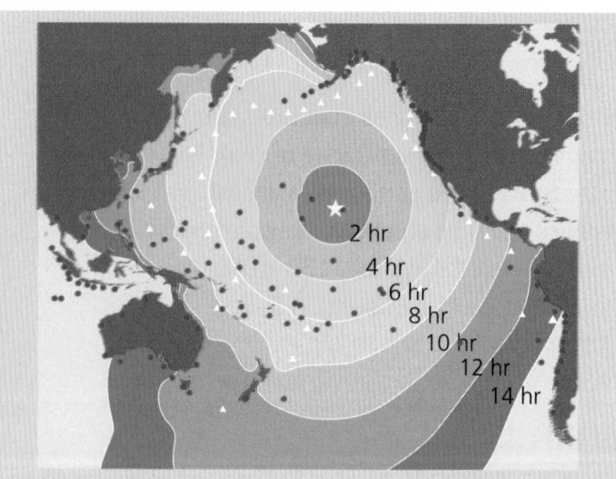

Figure 1.16 A tsunami travel time map centred on an event in Hawaii

The impact of a tsunami depends on a number of physical and human factors:

1. The duration of the event.
2. The wave amplitude, water column displacement and the distance travelled.
3. The physical geography of the coast, especially water depth and gradient at the shoreline.
4. The degree of coastal ecosystem buffer, for example protection by mangroves and coral reefs.
5. The timing of the event – night versus day – and the quality of early warning systems.
6. The degree of coastal development and its proximity from the coast, especially in tourist areas.

The most serious events occur when the physical and human factors interact with each other to produce a disaster. There have been some very high-profile tsunami events in recent years, notably in Indonesia in 2004 and Japan in 2011. Both of these had wide global media coverage, but often tsunamis are not well reported as they typically involve much lower loss of life and/or economic damage. The tsunamis of 2004 and 2011 are therefore what might be classed as 'mega-events'. On average, however, there might be one notable tsunami per year.

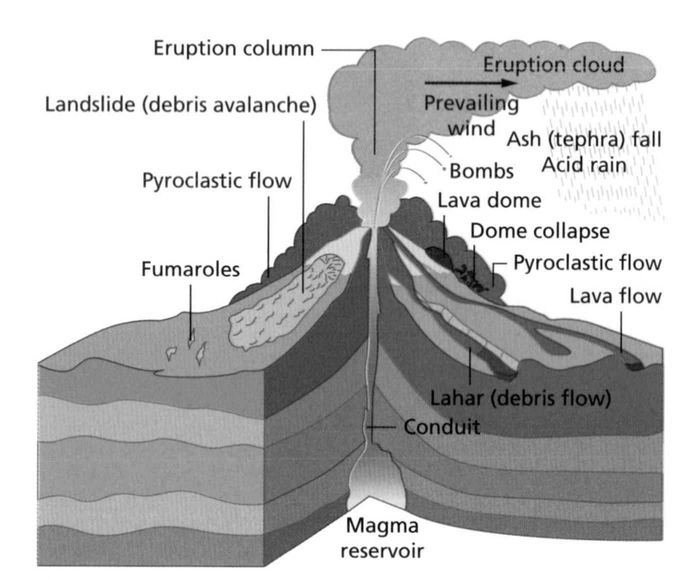

Figure 1.17 Primary hazard impacts of volcanoes

Volcanoes

Primary hazards of volcanoes

Volcanoes cause a number of primary, or direct, hazard impacts (Figure 1.17). These include pyroclastic flows, tephra, lava flows and volcanic gases.

An important feature of all these primary impacts is that they can have a very long geographical reach away from the source (Figure 1.18). Table 1.2 considers the different primary volcanic hazards.

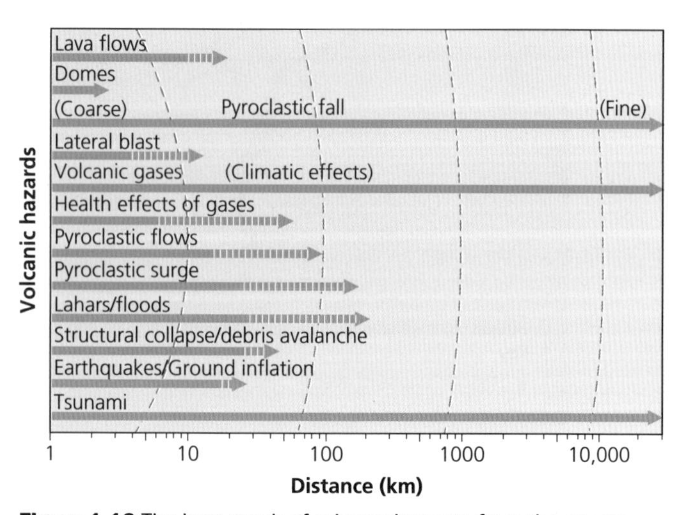

Figure 1.18 The long reach of primary impacts from the source

Table 1.2 The range of primary volcanic hazards

Pyroclastic flows	Responsible for most volcanic related deaths. 'Nuées ardentes' as they are sometimes called, result from the frothing of molten magma in the vent of the volcano. The bubbles burst explosively to eject hot gases and pyroclastic material, which contains glass shards, pumice, crystals and ash. These clouds can be up to 1000 °C. They are most hazardous when they come out sideways from the volcano, close to the ground.
Tephra	When a volcano erupts it will sometimes eject material such as rock fragments into the atmosphere – this is tephra. It can vary in size from 'bombs' (>32 mm in diameter) to fine dust (<4 mm). This ash and larger materials can cause building roofs to collapse as well as start fires on the ground. Dust can reduce visibility and affect air travel.
Lava flows	These pose a big threat to human life if they are fast moving. The viscosity of the lava is determined by the amount of silicon dioxide it contains. On steep slopes some lava flows can reach 15 m/sec. The greatest lava-related disaster of all time was in 1873 when molten material issued from the Lakagígar fissure, Iceland, for five months. An estimated 22 per cent of the country's population died in the resulting famine.
Volcanic gases	Gases are associated with explosive eruptions and lava flows. The mix normally includes water vapour, sulphur dioxide, hydrogen and carbon monoxide. Most deaths have been associated with carbon dioxide; it is dangerous because it is colourless and odourless and can accumulate in valleys undetected by people. In 1986, emissions of carbon dioxide from Lake Nyos in Cameroon killed 1700 people.

Secondary impacts of volcanoes

Volcanoes have a number of secondary impacts. The most significant of these are volcanic mudflows (lahars) and catastrophic floods (jökulhlaups).

- Lahars are volcanic mudflows generally composed of relatively fine sand and silt material. The degree of hazard varies depending on the steepness of slopes, the volume of material and particle size. As a secondary hazard they are associated with heavy rainfall as a trigger as old tephra deposits on steep slopes can be re-mobilised into mudflows.
- Jökulhlaups are type of catastrophic glacial outburst flood. They are a hazard to people and infrastructure, and can cause widespread landform modification through erosion and deposition. These floods occur very suddenly with rapid discharge of large volumes of water, ice and debris from a glacial source (Figure 1.19). They can occur anywhere where water accumulates in a subglacial lake beneath a glacier. The flood is initiated following the failure of an ice or moraine dam.

Figure 1.19 Jökulhlaups can be a very destructive force with huge, but often shortly lived, discharges

In comparison with other hazards, such as droughts, earthquakes and floods, volcanoes have historically killed far fewer people. Nevertheless, they claim a significant number of lives. More than 250,000 people have died in volcanic eruptions in the last 300 years. In any single decade, up to 1 million people may be affected by volcanic activity. This figure is likely to rise as vulnerability increases in populations living close to volcanoes. Catastrophic eruptions occur irregularly in both space and time, which makes the hazard all the more dangerous.

Review questions

1 Explain why earthquakes and volcanoes have both a predictable, similar and yet different distribution.

2 Which type of plate boundary does *not* lead to volcanic activity? Give reasons for this.

3 Describe two areas of active volcanoes that are associated with plumes from hotspots rather than inter-plate boundaries.

4 Explain the significance of the Benioff Zone in relation to the hypocentre and, therefore, earthquake risk.

5 Examine the different processes operating at different plate margins.

6 Examine the different roles of P, S and L waves in crustal fracturing and ground shaking. How does this lead to stress on buildings?

7 Describe the factors that influence the degree of impact of a tsunami.

8 Describe the range of impacts, within different spatial zones, that occur as a result of volcanic activity.

Further research

Look at this map of active volcanoes and write down their distribution linked to either inter-plate or intra-plate locations: http://earthquakes.volcanodiscovery.com

Look at Figure 1 on this website and account for the different depths of earthquakes shown: www.visionlearning.com/en/library/Earth-Science/6/Plates-Plate-Boundaries-and-Driving-Forces/66

Download the spreadsheet of volcanoes from this website (click on 'Holocene spreadsheet') then use a GIS package to plot the volcanoes according to location, that is, from the latitude and longitude column: http://volcano.si.edu

Find out more about the role of USGS's Volcano Hazards Program: http://volcanoes.usgs.gov

Have a look at this video from YouTube, which gives a wider context to paleomagnetism: www.youtube.com/watch?v=NfVNnk8FHcU

2 Tectonic hazards and disasters

Why do some tectonic hazards develop into disasters?

By the end of this chapter you should:

- understand how disasters come about because of the interaction between hazards, vulnerability and resilience
- recognise the significance of tectonic hazard profiles as a tool in understanding different hazard impacts
- know how development and governance are important in understanding disaster impact and vulnerability.

2.1 Hazards, vulnerability, risk, resilience and disaster

The terms hazard and disaster are often used synonymously but they actually mean very different things.

The UN's International Strategy for Disaster Reduction (ISDR) states that a disaster is:

> *'A serious disruption of the functioning of a community or a society involving widespread human, material, economic or environmental losses and impacts, which exceeds the ability of the affected community or society to cope using its own resources.'*

Alternative interpretations of disaster are provided by some large insurers, which define it as economic losses of over $1.5 million.

Degg's Model (Figure 2.1) shows the interaction between hazards, disaster and human vulnerability. Importantly, disaster may only occur when a vulnerable population is exposed to a hazard.

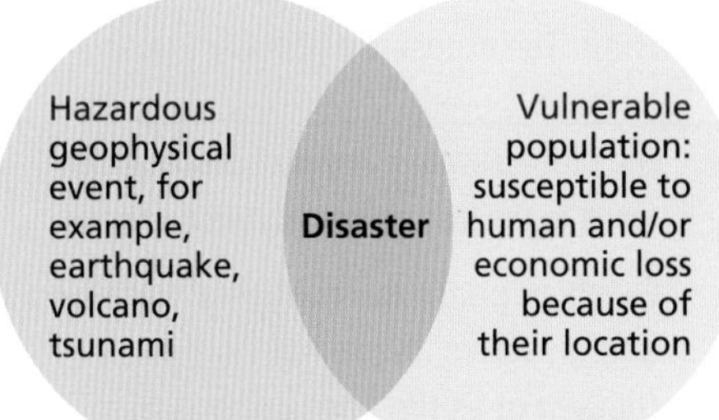

Figure 2.1 Degg's Model

Understanding risk

There is a complex relationship between risk, hazards and people. This is due to several factors:

1. **Unpredictability** – many hazards are not predictable; people may be caught out by either the timing or magnitude of an event.
2. **Lack of alternatives** – people may stay in a hazardous area due to a lack of options. This may be for economic reasons (work), because of a lack of space to move, or a lack of skills or knowledge.
3. **Dynamic hazards** – the threat from hazards is not a constant one, and it may increase or decrease over time. Human influence may also change the location or increase the frequency or magnitude of hazardous events.

Key terms

Hazard: 'A perceived natural/geophysical event that has the potential to threaten both life and property' (Whittow). Yet a geophysical hazard event would not be such without, for example, people at or near its location. That is to say, earthquakes would not be hazards if people did not live in buildings that collapse as a result of ground shaking. Many hazards occur at the interface between natural and human systems.

Disaster: The realisation of a hazard, when it 'causes a significant impact on a vulnerable population' (Degg). The Centre for Research on the Epidemiology of Disasters (CRED) states that a hazard becomes as disaster when:

- 10 or more people are killed, and/or
- 100 or more people are affected.

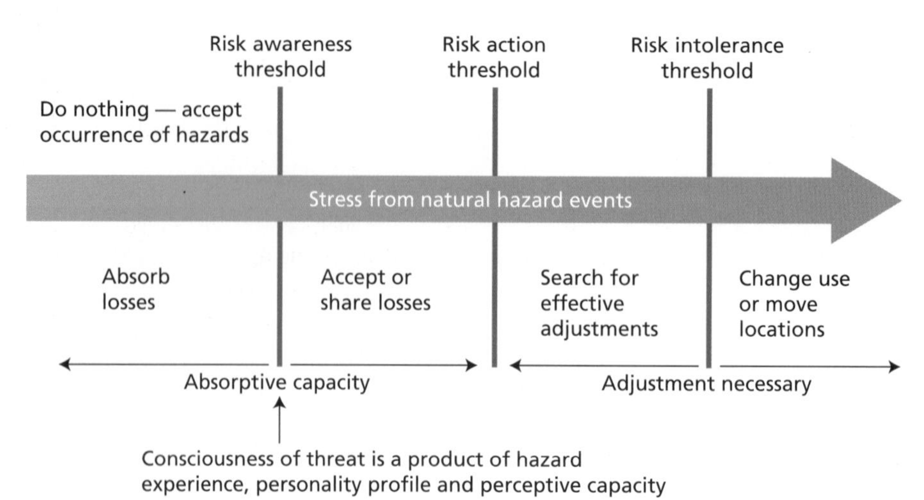

Figure 2.2 The risk-perception process

4. **Cost-benefit** – the benefits of a hazardous location may well outweigh the risks involved in staying there. Perception of risk may also play a role here.
5. **'Russian roulette reaction'** – the acceptance of the risks as something that will happen whatever you do, that is, one of fatalism.

The hazard-risk formula attempts to capture the various components that influence the amount of risk that a hazard may produce for a community or population.

$$\text{Risk} = \text{hazard} \times \text{exposure} \times \frac{\text{vulnerability}}{\text{manageability}}$$

Figure 2.3 The hazard-risk formula

Risk can also be understood through perception. In Figure 2.2, for example, when there is increasing stress from natural hazards, there may come a point when the population or community has to 'adjust'. What is interesting is that the balance between absorption and adjustment will vary according to the type of hazard, as well as the attitudes of decision makers.

People and populations also vary in terms of their resilience. Resilience is an important concept. Resilience is also about the ability to 'spring back' from a hazard event or disaster shock. According to the United Nations Office for Disaster Risk Reduction (UNISDR) the resilience of a community in respect to potential hazard events is determined by the degree to which the community has the necessary resources and is capable of organising itself both prior to and during times of need.

Disaster Risk and Age Index

The Disaster Risk and Age Index highlights two important trends:

1. ageing populations
2. the acceleration of risk in a world that is increasingly exposed to a range of hazard types.

Age is a significant factor in people's resilience, with children and the elderly likely to suffer much more from a range of hazards, including those of a tectonic origin. Around 66 per cent of the world's population aged over 60 live in less-developed regions. By 2050, this is expected to rise to 79 per cent.

A comparison of Myanmar and Japan in terms of a disaster risk and age index produces clear differences (Figure 2.4).

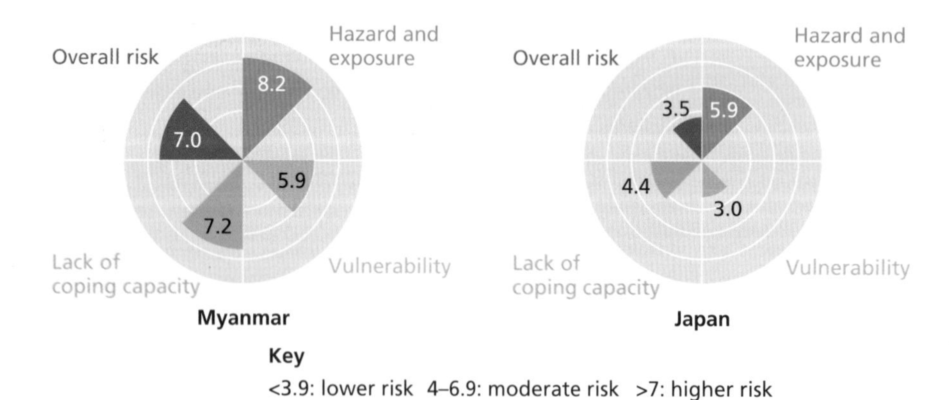

Figure 2.4 Comparing disaster risk for the elderly in Myanmar and Japan

Key terms

Risk: The exposure of people to a hazardous event. More specifically, it is the probability of a hazard occurring that leads to the loss of lives and/or livelihood.

Resilience: In the context of hazards and disasters, resilience can be thought of as the ability of a system, community or society exposed to hazards to resist, absorb and recover from the effects of a hazard.

Table 2.1 Comparing Myanmar and Japan

	Myanmar	Japan
Hazard and exposure score	Myanmar has a significantly high natural hazard component due to the potential for tsunami and earthquakes (as well as floods and storms).	Japan is subject to a range of natural hazards and is highly exposed.
Vulnerability	Moderate risk though a relatively low score – there have been few natural shocks in recent years.	Vulnerability is high compared to other wealthy nations due to the ageing population, but it is still low risk.
Coping capacity	Poor coping capacity; low level of internet/mobile phone access for older people; education is poor.	Coping capacity is good; the elderly tend to be educated, have high internet connectivity, effective government and low gender inequality.
Overall risk	Myanmar is ranked 7th out of 190 nations, which means that the disaster risk to elderly citizens is very high.	Although Japan is highly exposed to natural hazards, it is ranked 133rd out of 190 nations thanks to its strong coping capacity and lower levels of vulnerability.

However older people in Japan are still relatively vulnerable to hazards, at least in the context of their own county. The tsunami of 2011 killed 15,000 people and 9500 were either injured or missing; 56 per cent of those who died in the tsunami were aged 65 and over, even though this age group comprised just 23 per cent of the population in the area affected.

This index, developed by the UNISDR, is a way of signalling how age should be an important factor in understanding both vulnerability and the coping capacity of the older generation.

The Pressure and Release Model

We know that the risk faced by people must be seen as a combination of vulnerability and hazard. There cannot be a disaster if there are hazards but vulnerability is (theoretically) nil, or if there is a vulnerable population but no actual hazard event.

The basis for the Pressure and Release (PAR) Model (also known as the Disaster Crunch Model) is that a disaster is the intersection of two processes:

1 processes generating vulnerability on one side, and
2 the natural hazard event on the other.

The authors of the PAR model suggest that it resembles a 'nutcracker', with increasing pressure on people arising from either side – from their vulnerability and from the impact (and severity) of the hazard for those people. The 'release' idea is incorporated to conceptualise the reduction of disaster: to relieve the pressure, vulnerability has to be reduced (Figure 2.5).

Understanding the PAR Model

Root causes, such as limited access to power and resources, create vulnerability through different pressures such as inadequacies in training, local institutional systems, or capacity and standards in

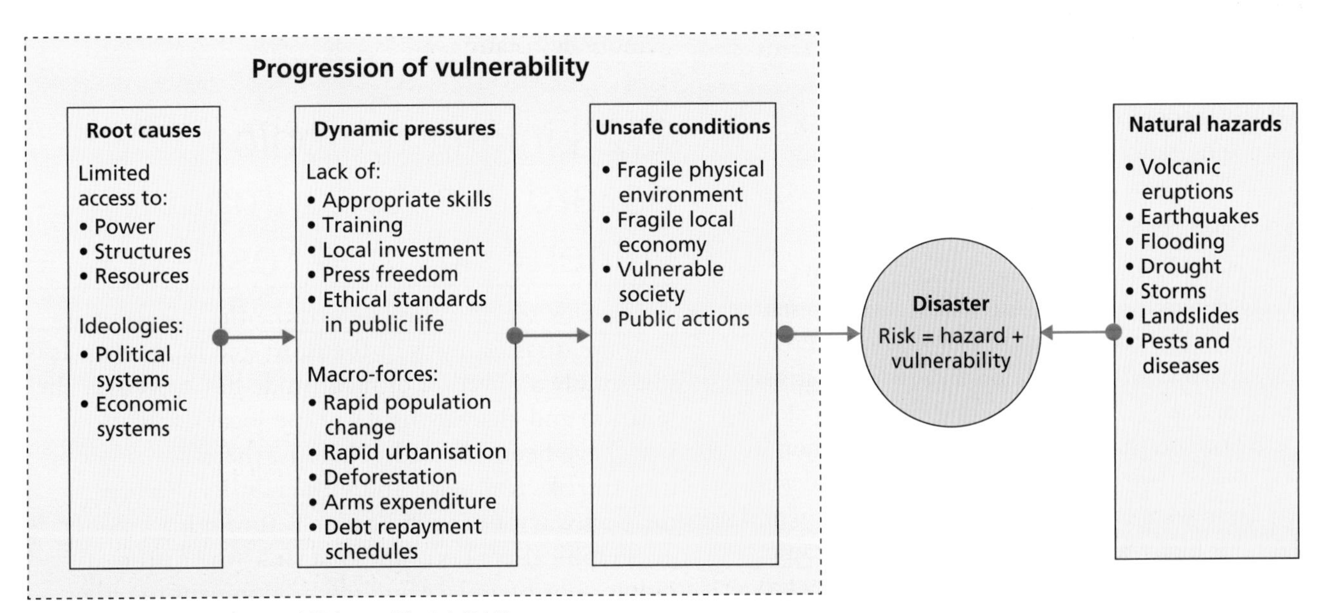

Figure 2.5 The Pressure and Release Model (PAR)

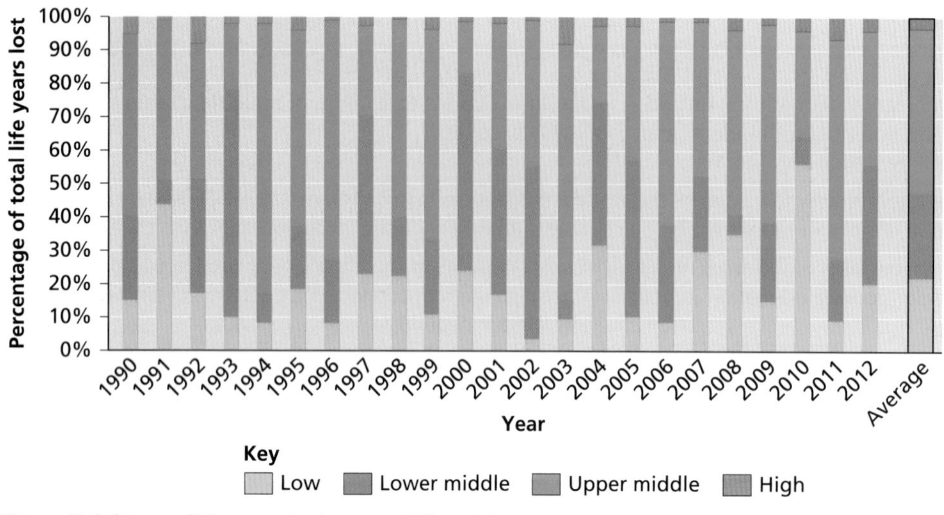

Figure 2.6 Share of life years lost across different income groups

> **Key term**
>
> Development: Development is linked to an improving society, enabling people to achieve their aspirations. It includes the provision of social services, acquisition of economic assets, improved productivity and reducing vulnerability to natural disasters. Low levels of development are closely associated with high levels of risk and vulnerability to natural disasters.

government. These dynamic pressures produce unsafe conditions in the physical and social environments of the people and groups most susceptible to vulnerability and risk. Physically unsafe conditions include dangerous locations and buildings with low resilience to the hazard (that is, unprotected). Socially unsafe conditions include risks to local economies as well as inadequacies in disaster-preparedness measures.

As an example, Figure 2.6 shows how income is an important factor in vulnerability. Upper-middle and lower-middle income groups account for the greatest degree of lives lost.

The social and economic impacts of tectonic hazards

The social and economic impacts of tectonic hazards vary considerably in terms of time and, more importantly, geographical region. They may vary from minor nuisances to major disasters involving a considerable impact on people in terms of loss of livelihood or even death.

The impacts of earthquakes (and linked secondary effects) are generally much greater than those presented by volcanoes. The concentration of volcanoes in relatively narrow belts means not only that a relatively small proportion of the land area of the world is close to a volcano but also that a relatively small proportion of the human population has direct exposure to volcanic activity. Somewhat less than one per cent of the world's population is likely to experience risk from volcanic activity, whereas the figure for earthquakes (directly) is estimated to be five per cent. That figure rises considerably when secondary impacts are considered (landslides and tsunami, for instance).

The economic impacts are roughly proportional to the land area exposed to the relevant hazard. Again, the earthquake hazard wins out. But economic impacts need to be considered more carefully set against the context, for example:

1. level of development (region or country)
2. insured impacts versus non-insured losses
3. total numbers of people affected and the speed of economic recovery following the event (a measure of resilience)
4. degree of urbanisation and, linked to this, land values, and the county or region's degree of interdependence
5. absolute versus relative impacts on a country's gross domestic product (GDP); higher relative impacts are more devastating.

2.2 Linking tectonic hazard events to impact, vulnerability and resilience

A complex set of factors determine the effects of a tectonic disaster, both in the short and long term. It is not simply a question of magnitude but moreover the location and characteristics of the local population who have been affected. There is, therefore, a strong link to risk, resilience and vulnerability. Perhaps the best way of describing this is through a broader geographical context, one that links the historical dimensions of a place and its development together with its physical, cultural and societal characteristics.

The magnitude and intensity of tectonic hazards

A number of tools and techniques can be used to measure the magnitude and intensity of tectonic hazards. Magnitude and intensity are objective; numerical descriptors of the size and intensity of tectonic events are usually based on measurements recorded from instrumentation. Different scales are typically used for different types of hazard (see Table 2.2). Each has its own advantages and disadvantages.

Unfortunately many of these scales are imperfect in that they typically measure just one or two physical processes that might cause damage. In the case of earthquakes, for example, the Richter Scale was developed in the 1930s as a mathematical tool to compare the size of earthquakes based on the amplitude of waves recorded by seismographs. The nature of the impact depends on both the event itself (size, duration and so on) but also the nature of the environment in which it is happening. We know that the impact depends on the degree of physical

Table 2.2 Scales used for different types of tectonic hazard

	Hazard	Scale	Overview
Richter Scale	Earthquake	0–9	A measurement of the height (amplitude) of the waves produced by an earthquake. The Richter Scale is an absolute scale; wherever an earthquake is recorded, it will measure the same on the Richter Scale.
Mercalli Scale (modified)	Earthquake	I–XII	Measures the experienced impacts of an earthquake. It is a relative scale, because people experience different amounts of shaking in different places. It is based on a series of key responses, such as people awakening, the movement of furniture and damage to structures (see Figure 2.7).
Moment Magnitude Scale (MMS)	Earthquake	0–9	A modern measure used by seismologists to describe earthquakes in terms of energy released. The magnitude is based on the 'seismic moment' of the earthquake, which is calculated from: the amount of slip on the fault, the area affected and an Earth-rigidity factor. The USGS uses MMS to estimate magnitudes for all large earthquakes.
Volcanic Explosivity Index (VEI)	Volcanic eruption	0–8	A relative measure of the explosiveness of a volcanic eruption, which is calculated from the volume of products (ejecta), height of the eruption cloud and qualitative observations (see Figure 2.8). Like the Richter Scale and MMS, the VEI is logarithmic: an increase of one index indicates an eruption that is ten times as powerful.

I	Not felt	Not felt except by a very few under especially favourable conditions.
II	Weak	Felt by only a few persons, especially on upper floors of buildings.
III	Weak	Felt quite noticeably by persons indoors, especially on upper floors of buildings. Many people do not recognise it as an earthquake. Vibrations similar to the passing of a truck. Duration estimated.
IV	Light	Felt indoors by many, outdoors by few during the day. At night, some awakened. Dishes, windows, doors disturbed; walls make cracking sound. Sensation like a heavy truck striking the building. Standing motor cars rocked noticeably.
V	Moderate	Felt by nearly everyone; many awakened. Some dishes and windows broken. Unstable objects overturned. Pendulum clocks may stop.
VI	Strong	Felt by all, many frightened. Some heavy furniture moved; a few instances of fallen plaster. Damage slight.
VII	Very Strong	Damage negligible in buildings of good design and construction; slight to moderate damage in well-built ordinary structures; considerable damage in poorly built or badly designed structures; some chimneys broken.
VIII	Severe	Damage slight in specially designed structures; considerable damage in ordinary substantial buildings with partial collapse. Damage great in poorly built structures. Fall of chimneys, factory stacks, columns, monuments, walls. Heavy furniture overturned.
IX	Violent	Damage considerable in specially designed structures; well-designed frame structures thrown out of plumb. Damage great in substantial buildings, with partial collapse. Buildings shifted off foundations.
X	Extreme	Some well-built wooden structures destroyed; most masonry and frame structures destroyed with foundations. Rails bent.
XI	Extreme	Few, if any, (masonry) structures remain standing. Bridges destroyed. Broad fissures in ground. Underground pipe lines completely out of service. Earth slumps and landslips in soft ground. Rails bent greatly.
XII	Extreme	Damage total. Waves seen on ground surfaces. Lines of sight and level distorted. Objects thrown upwards into the air.

Figure 2.7 Modified Mercalli intensity scale, with the scale I–XII

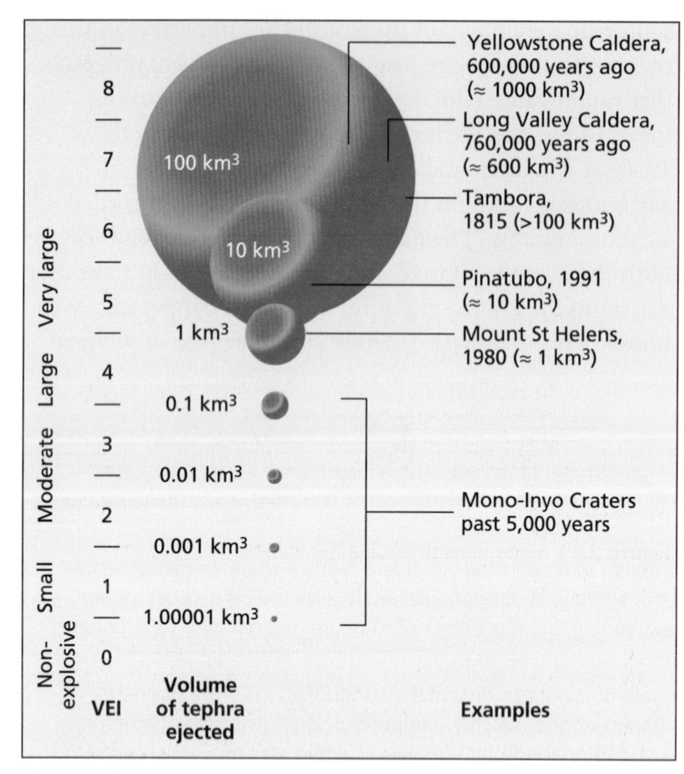

Figure 2.8 Volcanic Explosivity Index and ejecta volume correlation

exposure and human vulnerability of the communities that might be threatened by the event.

Increasingly hazards managers are also considering magnitude frequency relationships as a tool to help understand risk. These probability-based estimates help engineers to plan and design key infrastructure in hazard-prone areas. Such modelling is based on the general assumption that magnitude is often inversely proportional to the frequency of a particular event, that is, large earthquakes are much rarer than small ones.

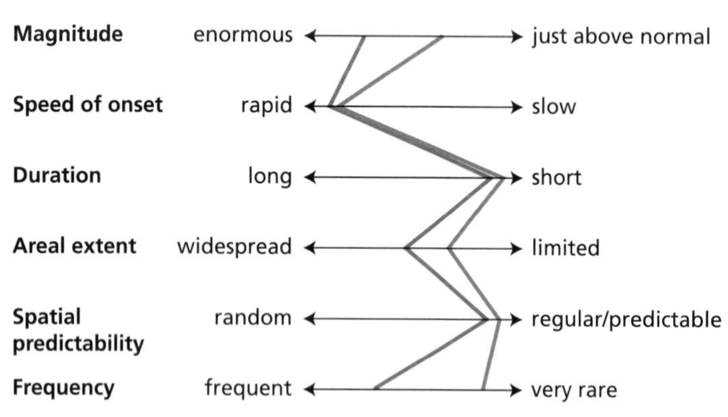

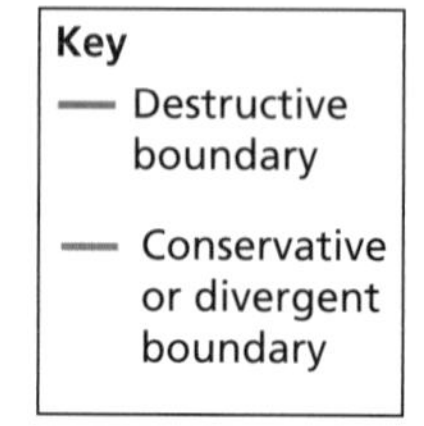

Figure 2.9 Examples of earthquake hazard profiles

Key term

Tectonic hazard profile: A technique used to try to understand the physical characteristics of different types of hazards, for example earthquakes, tsunamis and volcanoes. Hazard profiles can also be used to analyse and assess the same hazards which take place in contrasting locations or at different times. Hazard profiles are developed for each natural hazard and are based on criteria such as frequency, duration and speed of onset. Figure 2.9, for example, compares the features of earthquakes at two different boundaries.

Understanding tectonic hazard profiles

Figure 2.9 shows one style of tectonic hazard profile.

Figure 2.10 is a different style of hazard profile for California, showing the differences between volcanoes, earthquakes, landslides and tsunamis.

A hazard profile compares the *physical processes* that all hazards share and helps decision makers to identify and rank the hazards that should be given the most attention and resources.

One of the difficulties with hazard profiling, however, is the degree of reliability when comparing different event types. It is relatively easy to compare, say, an earthquake in Nepal to an earthquake in California because they are measured using similar scales or metrics and cause similar types of damage. However, it is much more difficult to compare across hazards, for example an earthquake to a tsunami or a volcanic eruption, as they all have different impacts on society and have varying spatial and temporal distributions. To accurately rank multiple hazards on one scale certain elements of the hazard become inaccurately displayed or must be omitted from the profile itself.

The traditional strategy for hazard planning has been on an individual hazard-by-hazard basis. Each hazard was treated as unique and, therefore, mitigation strategies should also be unique. This type of hazard planning can be problematic, however, due to conflicts between cost and government willingness to pay, and the resources available.

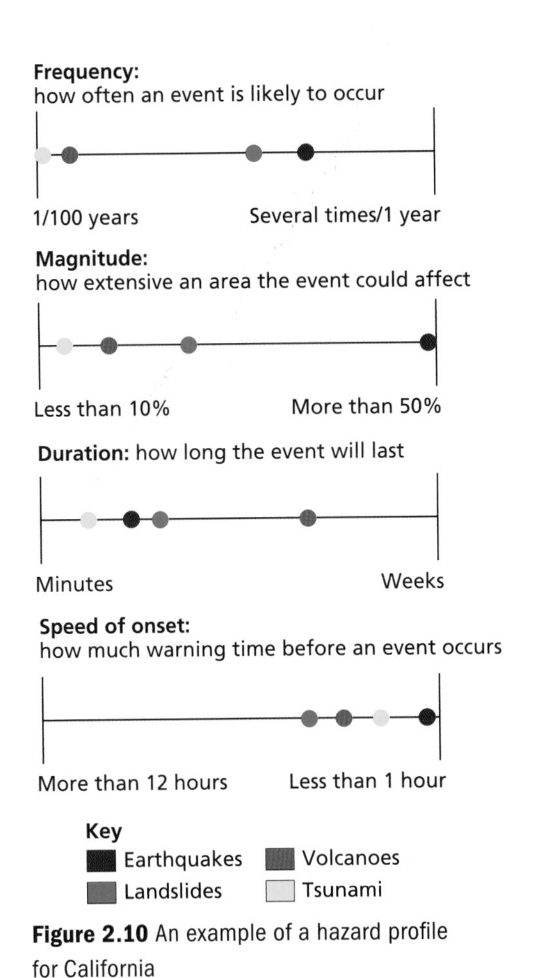

Figure 2.10 An example of a hazard profile for California

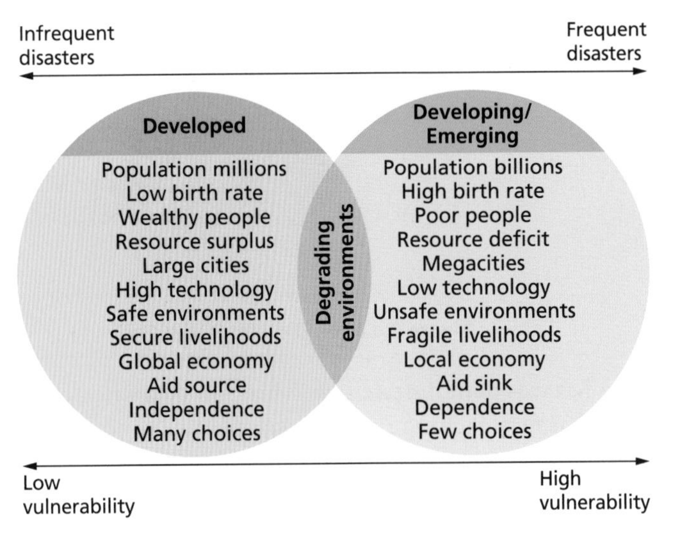

Figure 2.11 A comparison of disaster vulnerability between developed and developing/emerging countries

Skills focus: Manipulating hazards data

Hazards data relating to numbers of deaths, numbers affected, levels of development and type of event contains a number complex patterns and trends. Using a spreadsheet it should be possible to analyse some of this data to find out means, modes and medians, as well ranges and inter-quartile distributions. There isn't a pattern based on level of development and deaths or numbers of people affected. Why should this be the case?

A comparison of hazard events in developing, developed and emerging economies

Figure 2.11 shows that, at the international scale, there are wide differences in disaster vulnerability.

These variations can also been seen in data from countries at different levels of development. Table 2.3 shows data for the number of people reported killed and affected by tectonic hazard event and level of human development (HD) between 2004 and 2013.

Disaster Risk Index (DRI)

This index seeks to combine physical exposure to hazards with vulnerability. Figure 2.12 shows how risk varies globally, with the darker red colours indicating a greater degree of risk. The top countries at risk in terms of numbers of people killed per year (absolute) are the most populated countries (China, India, Indonesia, Bangladesh), whereas small island states (Vanuatu, Dominica, Mauritius, Antigua and Barbuda, St Kitts and Nevis, Solomon Islands, Grenada, and so on) come first in terms of numbers killed per million inhabitants per year (relative).

Table 2.3 Number of people killed and affected 2004–2013

Tectonic risk	Very high HD	High HD	Medium HD	Low HD	Total
Earthquakes and tsunamis	21,036	38,019	29,3941	29,7328	650,321
Volcanic eruptions	0	21	330	12	363
Earthquakes and tsunamis	4,010,000	2,476,000	67,972,000	9,495,000	83,953,000
Volcanic eruptions	12,000	348,000	568,000	323,000	1,215,000

Red = killed; orange = affected.

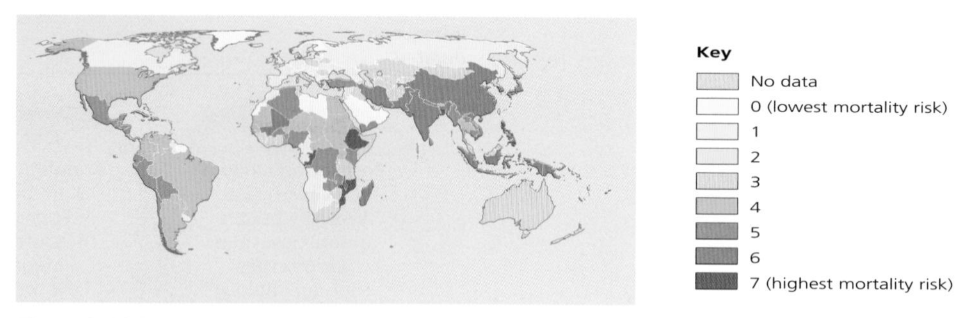

Figure 2.12 The Disaster Risk Index

Some hazards demonstrate that there can be a disaster which has far-reaching impacts, irrespective of its geographical source area, however. This is discussed further at the beginning of Chapter 3.

2.3 Development and governance: disaster impact and vulnerability

In the 'root cause' phase of the PAR Model, the most important causes are those which have an economic, demographic or a political foundation. In developing and recently emerging countries people tend to have less power over their socio-political and physical environments than the more wealthy. As a result of this difference, risk vulnerability is greater for them. This can be explained as follows:

- People and communities in developing and recently emerging countries only have access to livelihoods and resources that are insecure and difficult.
- They are likely to be a low priority for government interventions intended to deal with hazard mitigation.
- People who are economically and politically 'on the edge' are more likely to stop trusting their own methods for self-protection and to lose confidence in their own local knowledge. This means they rely more on government help, which may actually not work very well for them or their families.

Development, disaster impact and vulnerability

People's basic health and nutritional status correlates strongly with their ability to survive disruptions to their livelihood and normal well-being. It is an important measure of their resilience when dealing with the external shock from hazard events.

There is also a clear relationship between nutrition and disease, which is often evident after a hazard impact (especially when people are forced to find shelter and come into close contact with one another). People who are undernourished and sick are at greater risk of disease as they have weaker immune systems.

There are several elements of development that relate to vulnerability and disaster risk, which broadly fit into a sustainability framework:

1. An *economic* component dealing with the creation of wealth and the improvement of quality of life which is equitably distributed.
2. A *social* dimension in terms of health, education, housing and employment opportunities.
3. An *environmental* strand which has a duty of care for resource usage and distribution, now and in the future.
4. A *political* component including values such as human rights, political freedom and democracy.

Level of development and other human activities related to development may contribute towards disasters by increasing vulnerability as well as creating new hazard risk. But development can also *reduce* disaster risk. Table 2.4 (taken from a United Nations Development Programme report) summarises the complex development-disaster relationship.

In the aftermath of the devastating Haiti earthquake in 2010, for example, an estimated 9000 people died from cholera, and around 700,000 were thought to be affected (Figure 2.13). The source of the 2010 outbreak is disputed but it centres around the Artibonite River, from which most of the affected people had drunk water. There was suspicion among Haitians that a UN military base, located on a tributary of the river and home to peacekeepers from Nepal (who had come to help with the recovery) was actually the source of the disease. They thought that the base could have caused the epidemic, and this was confirmed in 2011 by the UN who stated that there was 'substantial evidence that the Nepalese troops

Table 2.4 The development–disaster relationship

Disasters limit or destroy development	• Destruction of physical assets and loss of production capacity, market access and input materials. • Damage to infrastructure and erosion of livelihoods and savings. • Destruction of health or education infrastructure and key workers. • Deaths, disablement or migration of productive labour force.
Development causes disaster risk	• Unsustainable development practices that create unsafe working conditions and reduce environmental quality. • Development paths generating inequality, promoting social isolation or political exclusion.
Development reduces disaster risk	• Access to safe drinking water, food and secure dwelling places increase community resilience. • Fair trade and technology can reduce poverty; social security can reduce vulnerability. • Development can build communities and broaden the provision of opportunities for participation and involvement in decision making, recognising excluded groups such as women, and enhancing education, health and well-being.
Disasters create development opportunities	• Favourable environment for advocacy for disaster-risk reduction measures. • Decision makers are more willing to allocate resources in the wake of a disaster. • Rehabilitation and reconstruction activities create opportunities for integrating disaster-risk measures.

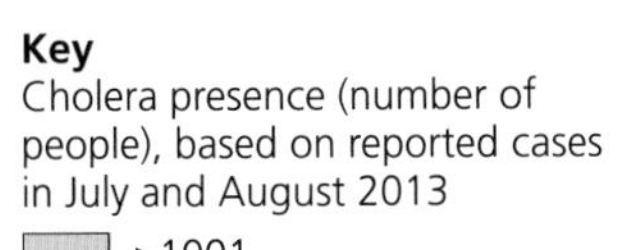

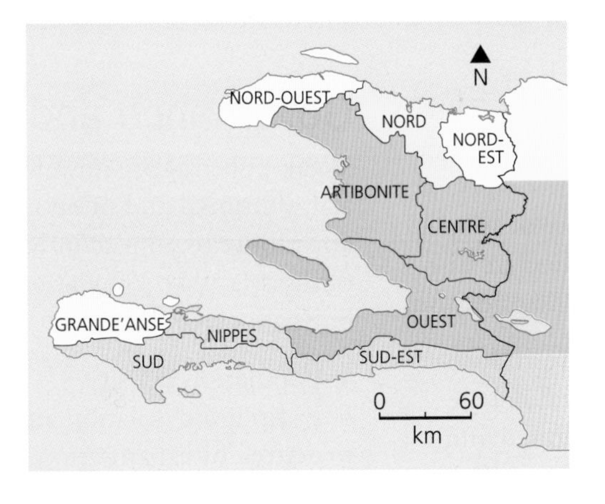

Figure 2.13 Haiti cholera map

Key term

Inequality: Usually refers to an unfair situation or distribution of assets and resources. It may also be used when people, nations and non-state players (ranging from transnational corporations to international agencies) have different levels of authority, competence and outcomes.

had brought the disease to Haiti'. Other scientists believe that the outbreak may in fact have been triggered by a more complex set of factors, including above-average temperatures and precipitation in 2010, coupled with destroyed water and sanitation infrastructure as a result of the earthquake. Ultimately, low levels of development are often at the roots of such disasters.

Development and cross-cutting factors

Drought, violence and armed conflict may turn natural hazards into disasters. In addition, the incidence and risks of diseases such as malaria and HIV/AIDS may interact with human vulnerability, worsening disaster risks brought about by urbanisation, climate change, violence and armed conflict, and marginalisation.

Cross-cutting factors may therefore be internal or external to the region or country in the context of disaster risk. Internal factors are often politically derived, whereas external factors may be longer term and much harder to manage or control, for example climate change and the risk of drought.

The 'risk-poverty nexus'

We recognise that low-income households and communities suffer a disproportionate share of disaster losses and impacts. The social processes and power dynamics that drive the disaster risk-poverty nexus are strongly linked with inequality.

In the context of tectonics, inequality has a number of dimensions and many of these aspects have a more significant impact on disaster risk levels than simply income inequality alone.

1. **Asset inequality** relates to housing and security of tenure, as well as agricultural productivity (in farming communities) or goods and savings in trading communities.
2. **Inequality of entitlements** refers to unequal access to public services and welfare systems, as well as inequalities in the application of the rule of law (policing, judging and sentencing).

3 **Political inequality** exists worldwide in the unequal capacities for political agency possessed by different groups and individuals in any society.

4 **Social status inequality** is often directly linked to space (for example, informal settlements in urban settings) and has a bearing on other dimensions of inequality, including the ability of individuals and groups to secure regular income and access services.

Table 2.5 shows how there were huge differences in the vulnerability of the population in two comparable earthquake events, Haiti and Chile (2010). Inequality was a key factor in the different rates of survival.

Disaster risk inequality is therefore characteristic of broader social, economic and political inequalities, rather than just one dimension of disparity (Figure 2.14).

The Global Assessment Report on Disaster Risk Reduction (2015) also suggests that urban segregation can generate new patterns of disaster risk. Low-income households are often forced to occupy hazard-exposed areas where there are low land values. Such places have poor infrastructure and social protection; they are also likely to have high levels of environmental degradation.

People living in such areas often have low resilience as they have little 'voice' in terms of political debate and influence, as well as being socially excluded and marginalised. Their lack of secure tenure discourages planners from investing in better housing, and they are less likely to benefit from the services or measures, such as earthquake protection measures, provided for other neighbourhoods. When these communities suffer losses as a result of disasters, their exposure to risk may be used to justify their relocation to even less suitable sites, far from sources of employment, rendering them more vulnerable in the longer term.

Table 2.5 Death tolls for two similarly sized (magnitude) earthquake events, 2010

	Haiti 2010	Chile 2010
Quake magnitude	7.0	8.8
Death toll	160,000* people	550* people
GDP per capita (2010 data)	$608	$12,640

*Best estimates

Governance and hazard vulnerability

Weak political organisation and political corruption are additional and often compounding factors that contribute to a more vulnerable population in terms of disaster risk. They are also linked to other factors at both a local and national scale, including:

- population density
- geographic isolation and accessibility
- degree of urbanisation.

All of this contributes to a community's resilience. At all scales, inequality also quickens the transfer of disaster

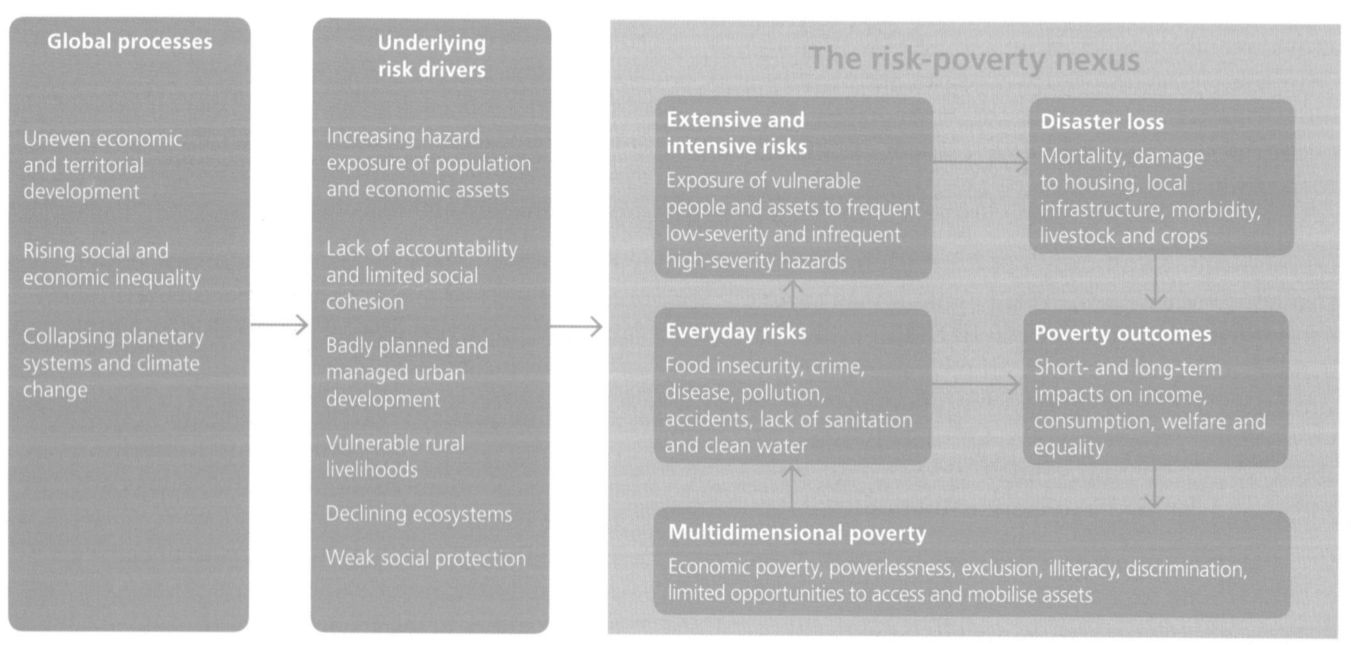

Figure 2.14 The risk–poverty nexus model

risks, through ineffective accountability and increased corruption, from those who benefit from taking the risks to other sectors, people and communities, who bear the costs.

Governance encompasses a number of formal and informal arrangements and procedures, which can change over time. Governance, and its impacts, also varies in scale from local to regional and national. Figure 2.15 shows the interaction of the three main components of governance.

However, modern thinking about governance suggests that there is no longer a single higher or sovereign authority. There are instead a range of stakeholders and blurred boundaries between the public, private and voluntary sectors. This means that risk governance is the result of the interaction of several socio-political forms of governing. In particular, governance should be considered against the following wider characteristics and processes:

- an increase in economic activity on a global, transnational scale
- increased activity of institutions such as the European Union (EU) operating across national boundaries in their scope and operation
- the rise of neo-liberal ideology values (market-led, smaller state, and so on) and its tackling of what were perceived to be the inefficiencies of centralised state control and the overly bureaucratic public sector, replacing these with a market-based logic of service provision through the private sector
- the spread of information technology, which made it easier to link different organisations and introduce changes.

Comparing natural disasters in countries with different levels of development

The characteristics of a natural disaster are the result of the changing pattern of social and physical factors that influence the event, and their degree of interaction. Many people involved in trying to better understand disasters have recognised that disasters need to be viewed through a lens of complexity, one that recognises the systemic linkage of physical and human factors.

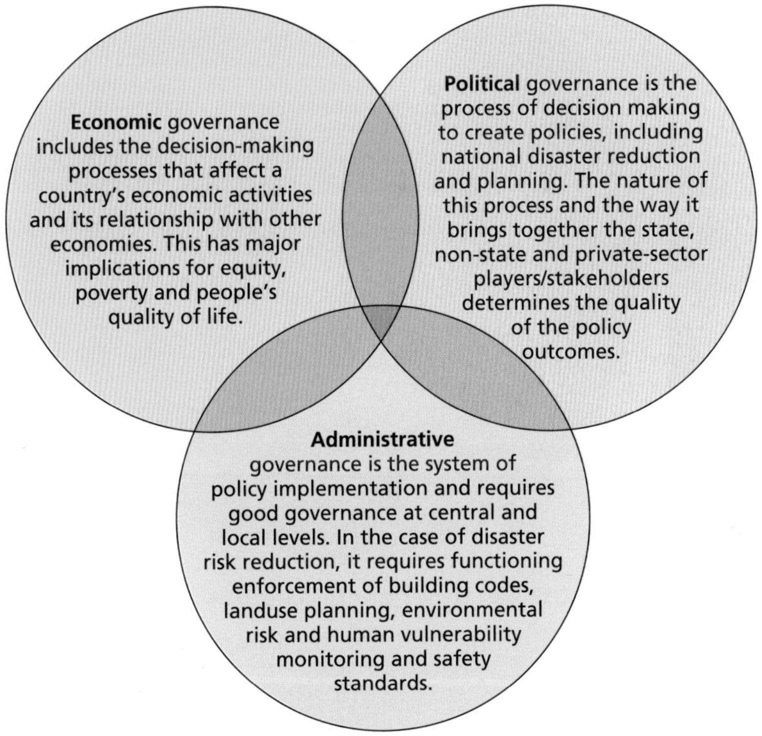

Economic governance includes the decision-making processes that affect a country's economic activities and its relationship with other economies. This has major implications for equity, poverty and people's quality of life.

Political governance is the process of decision making to create policies, including national disaster reduction and planning. The nature of this process and the way it brings together the state, non-state and private-sector players/stakeholders determines the quality of the policy outcomes.

Administrative governance is the system of policy implementation and requires good governance at central and local levels. In the case of disaster risk reduction, it requires functioning enforcement of building codes, landuse planning, environmental risk and human vulnerability monitoring and safety standards.

Figure 2.15 Interactions of governance

Synoptic themes:

Players

Local and national government

Key term

Governance: The sum of the many ways individuals and institutions, public and private, manage their common affairs. It is a continuing process through which conflicting or diverse interests may be accommodated and co-operative action may be taken. It includes formal institutions and regimes empowered to enforce compliance, as well as informal arrangements that people and institutions have either agreed to or perceive to be in their interest (The Commission on Global Governance, 1995).

Bam 2003: a tool to understanding the 'complexity perspective'

The relevance of these ideas can be shown by reference to an example: the January 2003 earthquake in Bam, southern Iran. Iran is classified as a country with an upper middle income (World Bank classification) and is ranked 75 of 187 countries (2015).

The earthquake had a magnitude of 6.6 (on average about one earthquake of this magnitude occurs every week worldwide) yet 26,000 people were killed in the ancient city of Bam. What were the reasons for this unexpectedly high impact?

1. The earthquake was shallow, with a hypocentre depth of 7 km.
2. It occurred at 5.26 a.m. local time, when most people would have been in their homes, asleep.
3. Research suggests that the release of energy in the form of seismic waves was directly under the city and the intensity of shaking was very high. It was also shaking in a vertical direction – causing maximum damage to the buildings.
4. The buildings were exceptionally vulnerable to shaking due to their age (Bam is an ancient citadel); some were 2400 years old. These 'adobe' buildings (made from earth and other organic materials) have very heavy roofs.
5. More recent construction in the city and recent reconstructions had been of poor quality. This was compounded further by the fact that the Iranian seismic building code had not been effectively enforced.
6. Many wooden structures in the city had previously experienced extensive damage from termite activity that had pre-weakened them.
7. The three main hospitals were destroyed in the earthquake, twenty per cent of health professionals were killed, and the remainder were incapable of giving care often due to their own injuries. Medical provision was also hindered by a lack of specialised medical training to deal with large-scale trauma care.
8. In the initial search and rescue phase, emergency services struggled with the destruction of their own facilities and infrastructure.
9. The cold winter temperatures in January meant that a large number of trapped victims died from hypothermia rather than direct crush injuries.

(Adapted from *Environmental Hazards: Assessing Risk and Reducing Disaster,* Smith and Petley, Routledge)

In summary, the Bam earthquake became a disaster because of a complex set of interactions, which are summarised in Figure 2.17.

Figure 2.16 Bam – before and after the earthquake

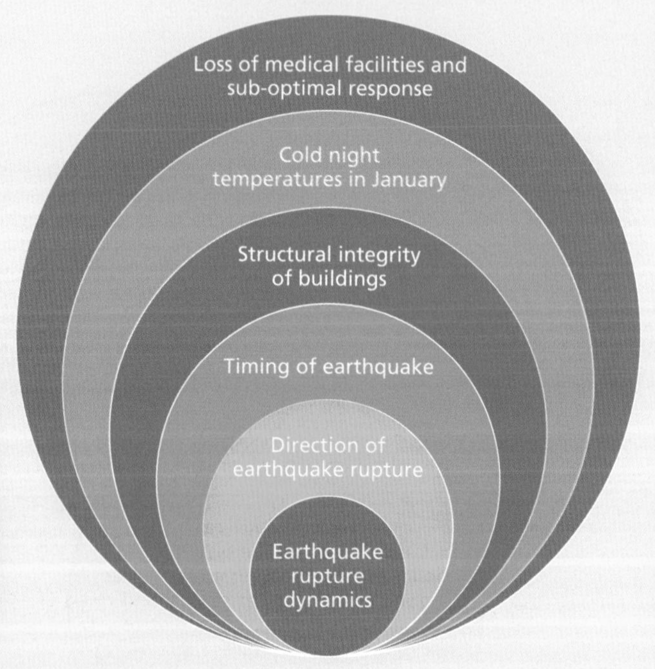

Figure 2.17 The complex interactions at work in the 2003 Bam earthquake

The Nepal earthquake of 2015: a true disaster

Nepal is a developing country with a population of about 26.5 million people. On 25 April 2015 a magnitude 7.8 earthquake struck Nepal and caused massive destruction. Approximately 9000 people lost their lives and more than 22,000 people were injured. Estimates suggest that more than half a million houses collapsed or were seriously damaged. There are several drivers which gave this particular disaster a context different to that of Bam:

- Nepal is a multiple hazard zone with a steep mountain landscape; it is exposed to landslides, debris and floods, as well as earthquakes.
- The low level of development means that much of the local earthquake science is out of date – the current seismic hazard map is around twenty years old.
- Kathmandu Valley has a population of 2.5 million people and a very high population density (about 13,000 per km^2). It is also growing at four per cent a year, making it one of the fastest-urbanising areas in South Asia. Around 85 per cent of the country's population is rural and much of the country's economy is primary industry.
- Nepal's population is vulnerable. Poor and socially excluded groups are less able to absorb shocks than well-positioned and better-off households. Because of poverty, many people build their own houses, which are often built without following the correct building code (Figure 2.18).

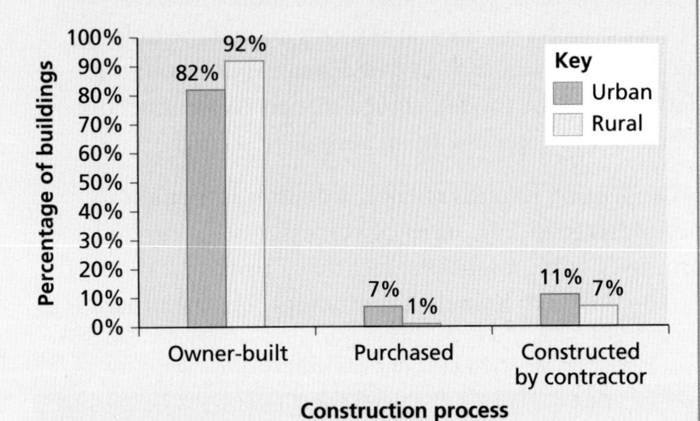

Figure 2.18 Building construction process for residential buildings in the Kathmandu Valley (Source: UNISDR 25 April 2015 Gorkha Earthquake Disaster Risk Reduction Situation Report)

New Zealand earthquakes 2010 and 2011: a resilient developed economy

In September 2010 and February 2011, the Canterbury region of New Zealand's South Island endured a series of major earthquakes. These earthquakes caused deaths and considerable destruction in Christchurch and the surrounding area. The Canterbury economy was resilient in the aftermath of the earthquakes. In spite of considerable damage to properties (residential and commercial), public infrastructure, and a large amount of relocation, business activity rebounded rapidly after the initial disruption. There are a number of reasons why the Canterbury as well as the wider New Zealand economy remained resilient to this hazard:

1. Disruption to industrial production, goods exports and activity was relatively short lived as the region's manufacturing hub escaped significant damage.
2. The agricultural sector was largely unaffected.
3. Rebuild costs of around NZ$20 billion (US$15 billion) were largely insured losses.
4. Financial markets largely ignored the earthquake impacts.
5. Indicators suggest that business activity has been quite resilient. Although business confidence dropped nationwide in the immediate aftermath of the February 2011 quake, they recovered quickly.

However, the tourism industry did suffer badly. The city of Christchurch had been the hub of tourist activity and many of its attractions were demolished. International visitors were down 40 per cent in the 2011–2012 period.

Summary: Reason's Swiss Cheese Model of disaster causation

The Swiss Cheese Model of disaster causation is also known as the *cumulative act effect model* (Figure 2.19) and is widely used in risk management and analysis, especially by the aviation industry. This industry in particular is very conscious of safety so there are many barriers put in place to minimise accidents – the idea of layered security or duplicate back-up systems. In the model the layers of cheese represent these safety systems and the holes the weaknesses in each line of defence. J. Reason, the developer of the model, argued that an accident occurs when all the holes line up in a single trajectory.

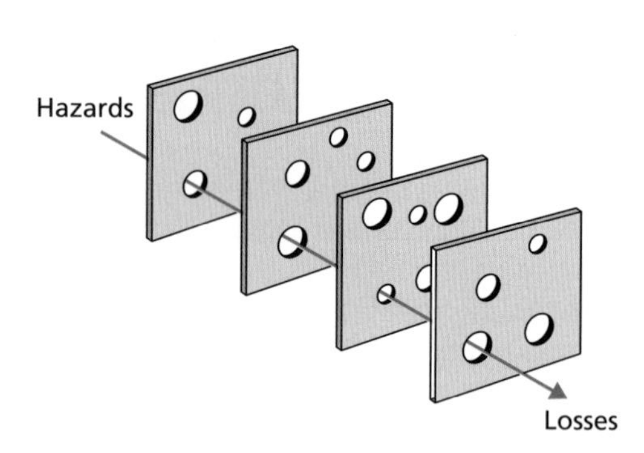

Figure 2.19 Reason's Swiss Cheese Model applied to hazards and disasters

So, in natural hazards science, a disaster is thought to occur as a result of a series of coincidental events and processes. It highlights the fact that a particular disaster can be linked to a single hazard event, but then there is a cascade of other events (possibly through the 'holes') that provide a context for the hazard. Generally the hazard becomes a disaster when several holes line up (a trajectory of accident opportunity), which creates the conditions for loss of life, property and livelihood.

Review questions

1 Explain how a natural hazard can become a disaster.

2 What is the difference between high- and low-resilience communities? Support your answer with examples.

3 How does the PAR Model help us to understand more about vulnerability hazard impact?

4 Explain how the social and economic impacts of tectonic hazards might affect people, the economy and the environment in different parts of the world.

5 Explain how unsustainable development rebalances the risk equation. Give examples of places and regions where you think this is happening.

6 Examine the most important root causes in the PAR Model.

7 Summarise the places globally where there are the highest degrees of vulnerability and state why, grouping into social, economic and political.

Further research

Research data from the International Disaster Database to investigate patterns and trends, and to look for links between magnitude of events, deaths and economic damage: www.emdat.be

Use International Red Cross World Disasters Reports to compare hazard impacts (loss of life, numbers affected and so on) between hazard types and regions: www.ifrc.org/en/publications-and-reports/world-disasters-report/world-disasters-report

Find out more about the role of UNISDR and what it does: www.unisdr.org

Research the most significant earthquakes in the last 30 days using the USGS website. If possible use GIS to show their distribution and then add a layer to show vulnerability in terms of wealth: http://earthquake.usgs.gov

Munich Re is a Swiss re-insurer. Research its connection to natural hazards online.

Research different tectonic hazard events in areas of varying development and explain the ways in which the context of each disaster is different.

3 Management of tectonic hazards and disasters

How successful is the management of tectonic hazards and disasters?

By the end of this chapter you should:

- understand the complex trends in disasters over time and how some disasters can become mega-events and have impacts over a very wide geographical area
- recognise the hazard models and frameworks that can be used to understand the prediction, impact and management of hazards
- know how tectonic hazard impacts can be managed through a range of mitigation and adaptation strategies which have varying successes.

It is worth remembering that seismic tectonic and volcanic processes cannot be prevented, and it's unlikely that they ever will be. Yet we have found out that the risks seem to be increasing for many people, especially those in the middle income and poorest groups. This increase in hazard vulnerability is mostly due to human factors rather than physical factors, as the trends in tectonic hazards reveal a pattern that does not indicate a significant increase in the last 50 years. This idea is true but complex and needs additional explanation.

3.1 Understanding tectonic and other disaster trends since 1960

In comparison with other natural hazards, few tectonic hazards manifest themselves into disasters. Figure 3.1, for example, shows that in the period 2004–2014 tectonic hazards had a low occurrence compared to hydrological and meteorological hazards, and also *much lower* numbers of victims compared to the other three hazards (climatological, hydrological and meteorological).

A look at the overall patterns

The overall longer-term natural hazard trends, since about 1960, show a number of key points:

- The total (aggregate) number of recorded hazards has increased over the last 50 years.
- The number of reported disasters seems to be falling, having peaked in the early 2000s (but that appears to be an anomaly to the longer-term trend).
- Number of deaths is also lower than in the recent past, but there are spikes with mega-events.

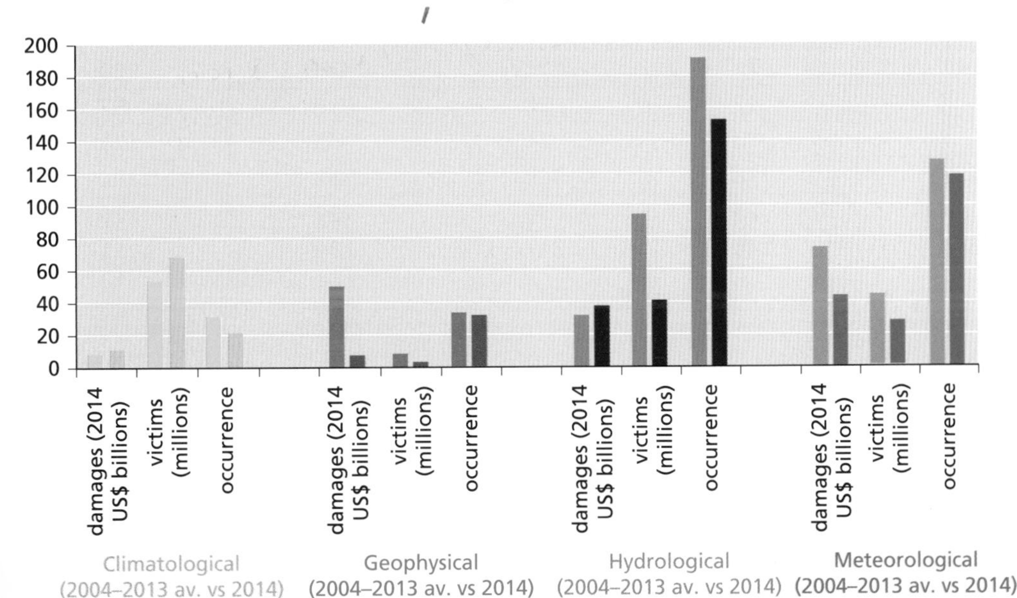

Figure 3.1 Natural hazards, 2004–14 (Source: CRED Annual Disaster Statistical Review 2014: The numbers and trends. http://cred.be/sites/default/files/ADSR_2014.pdf)

- The total number of people affected is increasing for some hazard and disaster types, especially meteorological and hydrological (Figure 3.2).
- The economic costs associated with both hazards and disasters of all types have increased significantly since 1960.

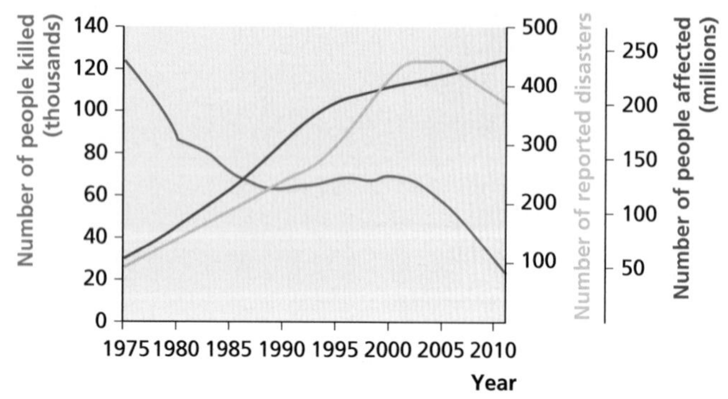

Figure 3.2 Natural disaster trends (all types), 1975–2011

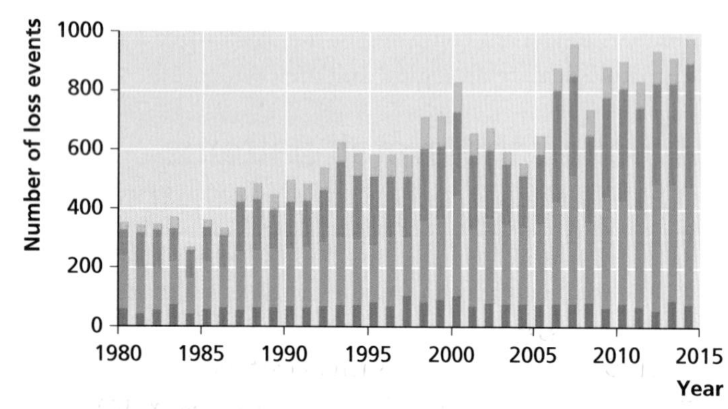

Figure 3.3 Number of hazard loss events (all types), 1980–2014

Note that tectonics (purple) remains stable compared to the other types, especially meteorological and hydrological, which appear to be increasing

But trends relating to tectonic (geophysical) hazards only show a *different* overall trend, one which is much more stable in terms of the number of events (Figure 3.3). However, somewhat hidden within that overall pattern is one that shows that the number of people affected and number of deaths does vary considerably year on year.

Spatial variation of tectonic disasters

Another important aspect of disaster geography is the spatial variation of tectonic impacts. It is wrong to assume that the locations of hazard impacts always translate into simple distributions. Data from the Centre for Research on the Epidemiology of Disasters (CRED) and the International Red Cross shows that the number of disasters reveals a complex pattern when either viewed by world region or by level of development.

Table 3.1 Total number of reported disasters grouped by type and level of economic development, 2004–13

	Very high human development	High human development	Medium human development	Low human development	Total
Earthquakes/tsunami	41	71	121	36	**269**
Volcanoes	5	12	30	10	**57**
Mass movements	7	33	84	49	**173**
Floods	237	378	585	552	**1751**
Droughts	14	25	57	129	**225**
Windstorms	384	146	347	134	**1011**

Table 3.2 Total number of reported disasters grouped by type and global region, 2004–13

	Africa	Americas	Asia	Europe	Oceania
Earthquakes/tsunami	18	39	174	27	11
Volcanoes	5	20	22	1	9
Mass movements	12	34	115	8	4
Floods	443	343	702	215	49
Droughts	124	51	37	9	4
Windstorms	82	336	400	139	54
Total	**684**	**823**	**1450**	**399**	**131**

How good are disaster statistics?

There is neither a universally agreed definition of a disaster nor a universally agreed numerical threshold for disaster designation. Reporting disaster impacts, especially deaths, is therefore controversial for a number of reasons:

1. It depends on whether direct (primary) deaths or indirect (secondary) deaths from subsequent hazards or associated diseases are counted.
2. Location is significant because local or regional events in remote places are often under-recorded.
3. Declaration of disaster deaths and casualties may be subject to political bias. The 2004 Asian tsunami was almost completely ignored in Myanmar but perhaps initially overstated in parts of Thailand, where foreign tourists were killed, and then played down to protect the Thai tourist industry.
4. Statistics on major disasters are difficult to collect, particularly in remote rural areas of low human development countries (LHDs), for example the earthquake in Kashmir in 2005, or in densely populated squatter settlements, for example the Caracas landslides in 2003–2004.
5. Time-trend analysis (interpreting historical data to produce trends) is difficult. Much depends on the intervals selected and whether the means of data collection have remained constant. Trends (deaths, numbers affected, economic impacts) can be upset by a cluster of mega-disasters, as happened in the 2004 Asian tsunami or the 2011 Haiti earthquake, or even in the 2015 Kathmandu earthquake.

Tectonic mega-disasters

Tectonic mega-disasters have several key characteristics:

- They are usually large-scale disasters on either an aerial/spatial scale or in terms of their economic and/or human impact.
- Because of their scale, they pose serious problems for effective management to minimise the impact of the disaster (both in the short and longer term).
- The scale of their impact may mean that communities, but usually government as well, often require international support in the immediate aftermath as well as during longer-term recovery. This may be at a regional level (for example the Asian tsunami of 2004) or globally (for example Japan 2011). These events can affect more than one country either directly or indirectly.

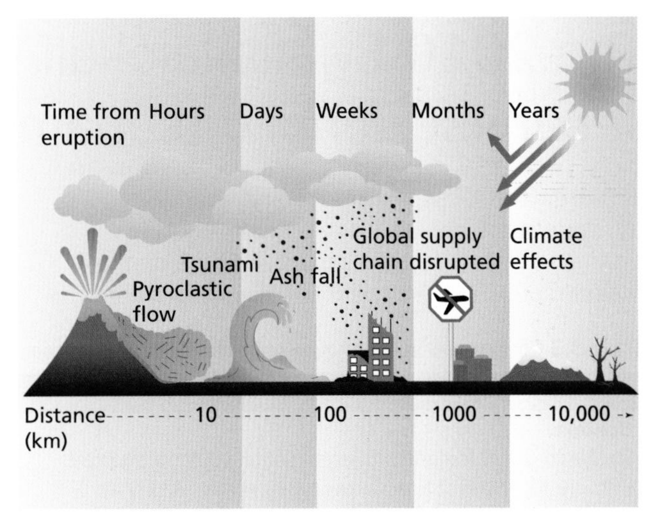

Figure 3.4 Diagrammatic representation of the likely range of impacts following a large VEI 6 (or above) eruption

Figure 3.4 illustrates how a large volcano, for example, can have significant impacts in both time and space.

Tectonic mega-events and disasters are often classified as high-impact, low-probability (HILP) events. So, one-off high-profile crises such as the 2010 Haiti earthquake and the 2011 Japanese earthquake and tsunami were all mega-disasters requiring rapid responses at a global level. But known hazards such as earthquakes and volcanoes (as well as floods and hurricanes), which, owing to the low likelihood of occurrence or the high cost of mitigating action, often remain ill- or under-prepared for in many parts of the world. These events are impossible to predict but very likely to occur over long time scales.

The globalisation of production and supply chains has increased manufacturing efficiencies, but it has also reduced resilience in the case of some events. High-value manufacturing is often most at risk because of its just-in-time (JIT) business model. The consequences of HILP events spread rapidly across both economic and geographic boundaries, creating other impacts (economists might call these negative externalities) that are difficult to plan for. The Japanese earthquake in 2011, for example, led to a five per cent reduction in the country's GDP. There were much wider knock-on impacts for global transnational corporations (TNCs) however, such as Toyota and Sony, which were forced to halt production.

The 2010 Eyjafjallajökull volcano and 2011 Japanese Tohoku tsunami

Two examples of significant tectonic events in recent years are the Iceland Eyjafjallajökull eruption in 2010 and the Japanese tsunami mega-disaster of 2011. They both had significant, but different, impacts at a global scale. Table 3.3 considers the effects on the global supply chain (Eyjafjallajökull) as well as the wider concerns about nuclear power (Tohoku).

Table 3.3 Context and impacts of two recent high-profile tectonic events

Eyjafjallajökull, Iceland	Tohoku tsunami, Japan
Context In March 2010 Iceland's Eyjafjallajökull volcano erupted into life for the first time in over 190 years. By 15 April 2010 the ash plume generated from the eruption had begun to affect much of Europe, spreading as far as northern Italy. The ash cloud grounded flights in most of Europe for several days. More than 100,000 air-journeys were cancelled, leading to the worst disruptions in air travel since the 9/11 terrorist attack in 2001. However, this was a relatively small eruption 'in the wrong place', with no direct deaths. It was high profile due to the impact on the air movements (passenger and freight).	**Context** A magnitude 9.0 earthquake in March 2011 produced a great tsunami that wreaked destruction along the Tohoku (eastern) coast of Japan, including to the Fukushima nuclear power station. It was the largest earthquake recorded in Japan and the combined impacts of the earthquake and tsunami left 15,749 dead and 3962 missing; 63 per cent of the dead were aged 60 and over. The event eroded public trust in the Japanese government and its nuclear energy policies. This was a very large magnitude event causing widespread deaths and large-scale destruction along the coast to properties, infrastructure and communities. It was particularly high profile because of the nuclear impact.
Evaluating the 2010 volcano's effect on the global supply chain Imports and exports in and out of Europe were greatly affected by the air travel shutdown in 2010. Although airfreight accounts for a tiny amount of world trade by weight, it accounts for a much higher proportion of trade by value. For example, airfreight accounts for approximately 0.5 per cent of UK trade by weight but a much bigger 25 per cent of trade by value. **Example 1: Car manufacturing disruption** The disruption to airfreight by the eruption highlighted how important airfreight is in supplying high-value key components to many manufacturers. The Nissan plant in Japan, for example, had to stop production of the Cube, Murano SUV and Rogue crossover models because they ran out of a critical sensor produced in Ireland. Airfreight is only used for a small quantity of high-value but vital electronic components where there are few alternative suppliers. **Example 2: Impacts on the transport of perishable goods** There were impacts on the producers of flowers, fruit and vegetables in African countries such as Kenya, Zambia and Ghana, with delays in transportation meaning large quantities of fast-perishing produce rotted, leading to losses for producers. The World Bank estimated that, in total, African countries may have lost US$65 million due to the effect of the airspace shutdown on perishable exports.	**Evaluating the earthquake and tsunami's impact on costs and attitudes to nuclear energy** The tsunami hit the Fukushima Daiichi nuclear power plant on the east coast of the island of Honshu, about 200 km northeast of Tokyo, and disabled the power supply. This affected the cooling of three reactors, causing high radioactive releases. Contaminated water leaked from the plant into the Pacific Ocean and into fishing grounds. The effects of the accident on energy security were not restricted to Japan. **Example 1: LNG price rises** The worldwide availability and affordability of liquefied natural gas (LNG) were affected by Japan's increased demand. This had the biggest impact in the Asian market, where they had the quickest rate of increasing energy consumption. **Example 2: Public acceptability of nuclear power and rising costs** The accident itself resulted in the loss of public acceptability of nuclear power and led countries, such as Germany and Italy to immediately shutdown some of their nuclear reactors or abandon plans to build new ones. The accident has also contributed to the escalating capital costs associated with the construction of new nuclear reactors because of the additional safety measures required.

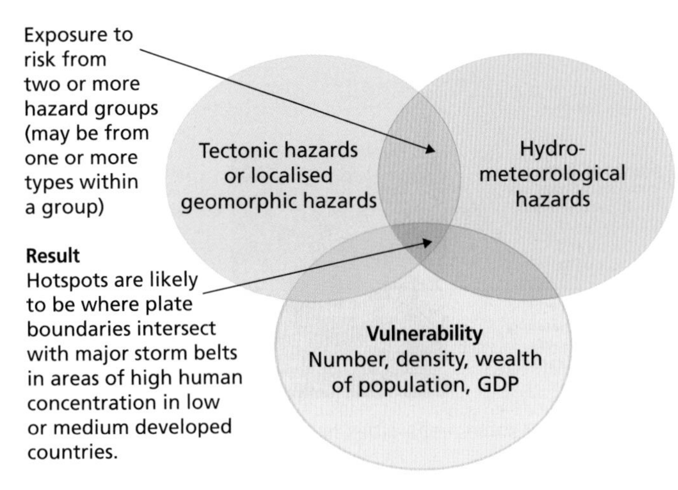

Figure 3.5 The characteristics of a hazard hotspot

Table 3.4 The countries most exposed to multiple hazards (Source: International Red Cross World Disasters Report www.ifrc.org/Global/Documents/Secretariat/201410/WDR%202014.pdf)

Country	Total area exposed (%)	Population exposed (%)	Number of different hazards the country is exposed to
Taiwan	73.1	73.1	4
Costa Rica	36.8	41.1	4
Vanuatu	28.8	20.5	3
Philippines	22.3	36.4	5
Guatemala	21.3	40.8	5
Ecuador	13.9	23.9	5
Chile	12.9	54.0	4
Japan	10.5	15.3	4

Multiple-hazard zones

Multiple-hazard zones are places where a number of physical hazards combine to create an increased level of risk for the country and its population. This is often made worse if the country's population is vulnerable (wealth/GDP, population density, and so on) or suffers repeated events, often on an annual basis, so that there is never any time for an extended period of recovery. Such places are often seen as disaster hotspots.

Hazards in multiple-hazard zones are, in fact, part of a wider picture of more complex geography linked to vulnerability over both space and time. This often makes their impact greater and more challenging to manage. The magnitude of the hazard event together with the human geography of the area in which it occurs are important factors, but hazards generally form part of a much more complex web of socio-economic-environmental issues that makes the impact greater and harder to manage. Table 3.4 lists the countries that are most exposed to multiple hazards globally. Figure 3.6 is a global summary of the multiple-hazard pattern.

Key term

Disaster hotspot: A country or area that is extremely disaster prone for a number of reasons, as shown in Figure 3.5.

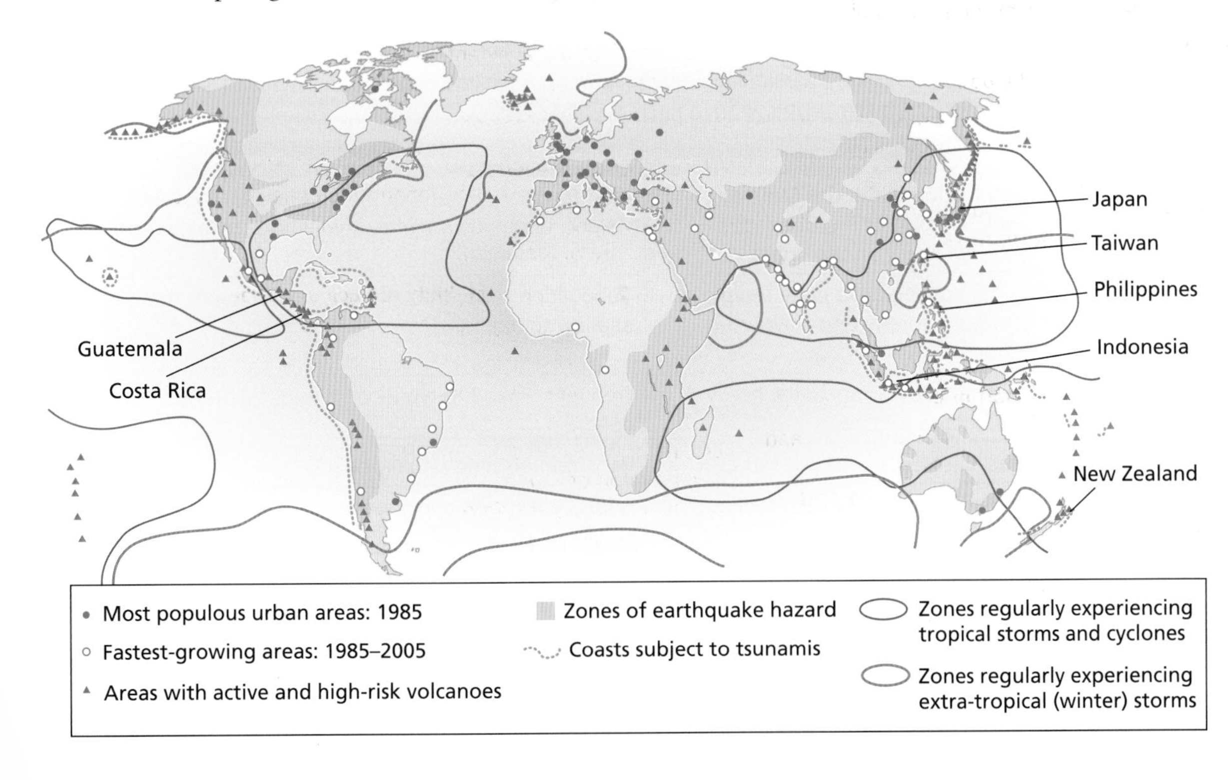

Figure 3.6 Global summary of the multiple-hazard pattern

There may also be variation in disaster risk within smaller geographical areas. Large urban areas are often zones of multiple-hazard risk (Figure 3.7). Cities are centres of economic development (economic cores) as they represent a natural focus for investment and development. They are also frequently centres of growing population as a result of the rapid urbanisation occurring in most developing countries. Many cities have huge areas of unplanned, poor-quality housing where growing numbers of the urban poor live, often located on marginal, potentially dangerous sites such as river banks and steep slopes.

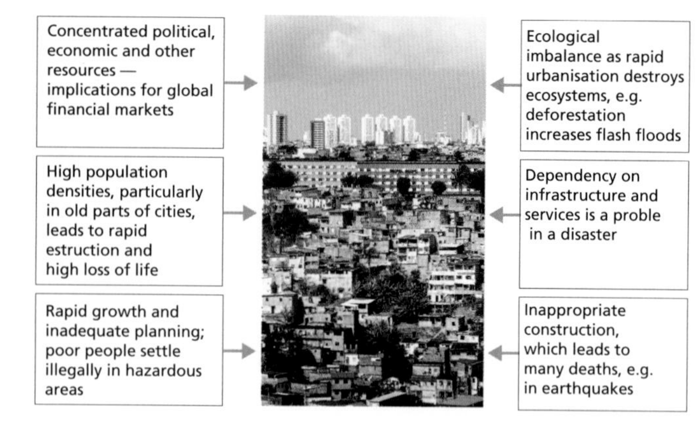

Figure 3.7 Why some mega-cities have low hazard resilience

Analysis of the global distribution of these rapidly growing mega-cities shows that many of them are located in hazard-prone areas. With such high densities of people, up to 25,000 per km^2, hazard management in large urban areas is both expensive and complex, making disasters inevitable, both socially (high concentration of vulnerable people) and economically (for example, loss of infrastructure).

Comparing the Philippines and California – classic multiple-hazard geography

There is a tendency to assume that all hazards occur in both places, and that the hazards may have the same root causes. In the Philippines, the main risk is typhoons with typically five or six storms a year as it lies on a major storm track. Annual deaths far exceed the long-term average for California (the last time more than 100 people died in a Californian natural disaster was the 1933 Long Beach earthquake). The Philippines has to spend around two per cent of its annual GDP cleaning up after typhoons. Table 3.5 summarises the hazard similarities and differences.

Table 3.5 Hazard similarities and differences (Source: International Red Cross World Disasters Report www.ifrc.org/Global/Documents/Secretariat/201410/WDR%202014.pdf)

	Californian coast	Philippines
Volcanoes	Rarely in northern California (Mount Shasta, Lassen Peak), which is part of the Cascades subduction zone – not really on the coast.	Very common; Pinatubo, Mount Mayon. Frequent and violent; andesitic magma, ash, lahars, pyroclastic flows.
Earthquakes	Frequent, within the conservative plate margin that includes the San Andreas and Hayward faults; usually shallow.	Subduction zone; frequent but vary in depth from shallow to deep.
Landslides	Frequent; associated with earthquakes, heavy rain, coastal erosion and wildfires.	Frequent; linked to typhoons and deforestation; often deadly.
Cyclones	Never occur here.	Very frequent and usually deadly.
Flood	Rarely; can be associated with El Niño cycles.	A frequent result of typhoons.
Drought	Very common, e.g. 2008–11 and 2012–15.	Rare, but El Niño does cause these, e.g. 1999 and 2010.

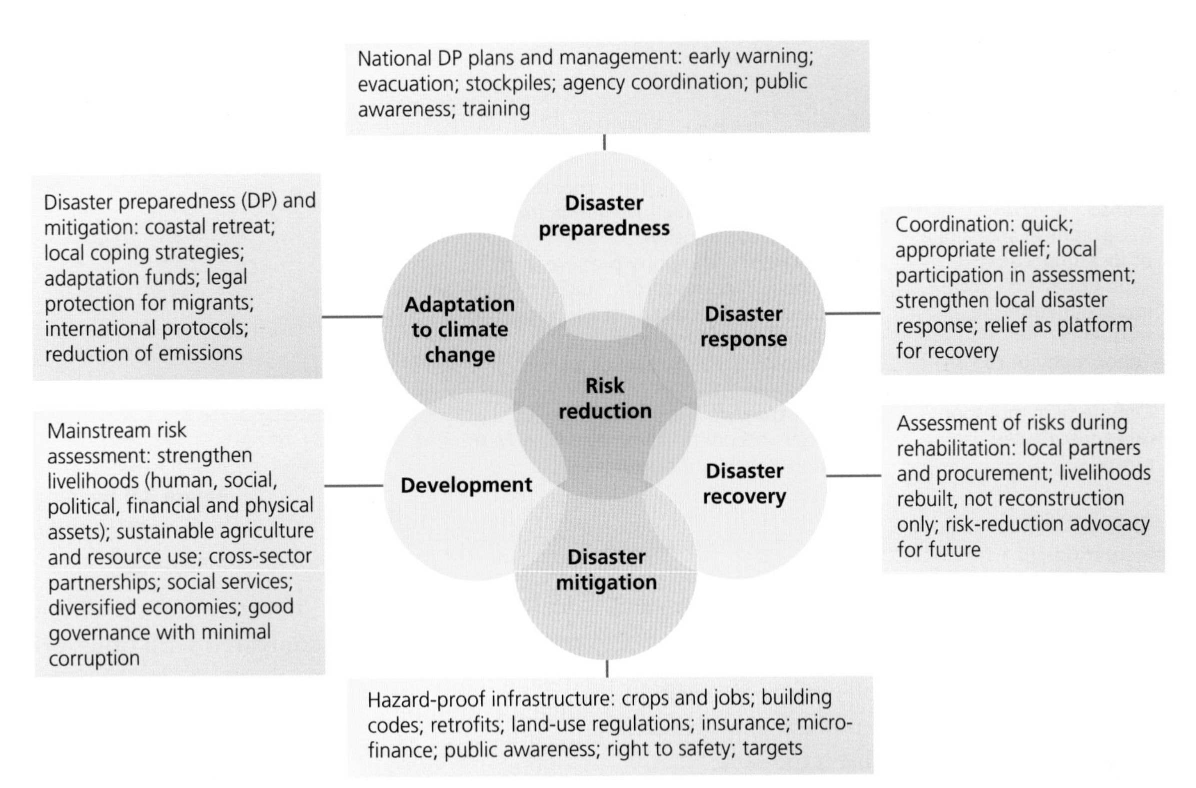

Figure 3.8 The risk disk – a model to help better understand disaster management

3.2 Managing tectonic hazards

Prediction and forecasting frameworks

The 'risk disc' (Figure 3.8) is one model that attempts to explain the reasons for the decline in deaths in terms of disaster preparedness, disaster mitigation (hazard proofing), disaster response and disaster recovery. The next section of this book will examine these different areas, together with the associated models that help to explain their purpose.

Getting closer to earthquake forecasting and prediction?

Earthquake forecasting and prediction is an active topic of geological research.

Earthquake risk can be forecast since it is based on a statistical likelihood of an event happening at a particular location. These forecasts are based on data and evidence gathered through global seismic monitoring networks, as well as from historical records. Long-term forecasts (years to decades) are currently much more reliable than short- to medium-term forecasts (days to months). Forecasting is very important as it can encourage governments to enforce better building regulations in areas of high stress, or create improved evacuation procedures in areas of highest risk.

Currently it is not possible to make accurate predictions of when and where earthquakes will happen. For this to be possible, it would be necessary to identify a 'diagnostic precursor' – a characteristic pattern of seismic activity or some other physical, chemical or biological change – which would indicate a high probability of an earthquake happening in a small window of space and time. So far, the search for diagnostic precursors has been unsuccessful.

Some geophysicists are trying to improve prediction based on calculating the underground movement of magma. Their models try to predict where the plates are running together with the most stresses, often a tell-tale sign of where an earthquake might hit. They map underground patterns of activity in the Earth's mantle across underground grid points. The models then predict where stress points will occur by simulating different rocky mantle flows. The calculations can ascertain where, as a result of these flows, the plates are likely to run together, and automatically detect these stressed zones.

Such models are still in their infancy and need considerable refinement as the link between earthquakes and underground mantle flows is complex and hard to model. Other scientists are working on predicting earthquakes

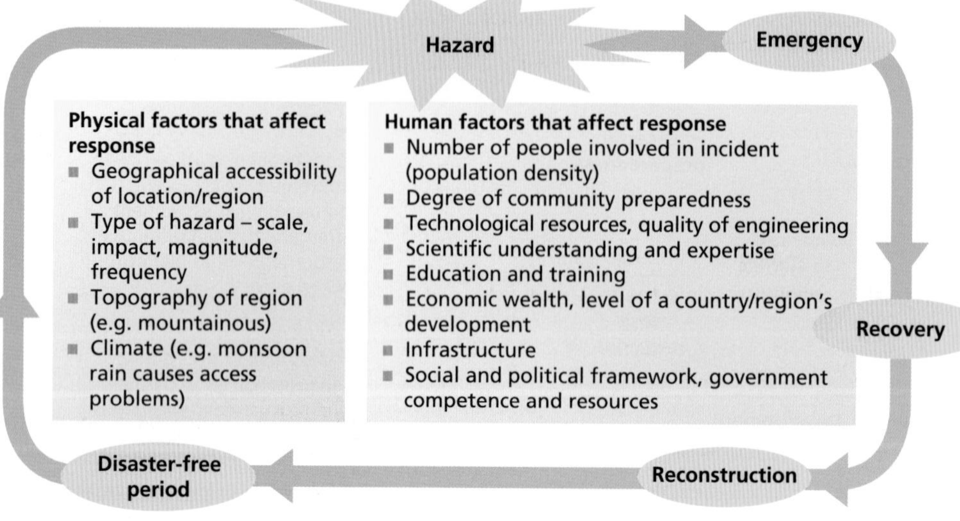

Figure 3.9 The range of factors affecting the response to hazards

based on animal behaviours, changes in radon emissions and electromagnetic variation, but with very limited success.

However, for predictions to be useful – that is, to enable evacuation of affected areas – they must be highly accurate, both spatially and temporally. And that is the issue. At present the science makes this impossible and most geoscientists do not believe that there is a realistic prospect of accurate prediction in the foreseeable future. They suggest that the main focus of research instead should be based on improving the forecasting of earthquakes.

Understanding the hazard management cycle

Figure 3.9 shows how the choice of response depends on a complex and interlinked range of physical and human factors. As people, communities and organisations have limited resources and time to make decisions, the relative importance of the physical risk from natural hazards, compared with other priorities (such as providing jobs, education, health services and defence) will be a major factor in influencing how resources are devoted to reducing hazard impacts.

Park's Model and levels of development

Park's Disaster Response Curve (1991) (Figure 3.10) can be used as a framework to help better understand the time dimensions of resilience: from a hazard striking to when a place, community or country returns to normal operation.

Each stage on the *x*-axis shows the different stages of time during which either relief, rehabilitation or reconstruction is started. The words on the *y*-axis describe quality of life, stability and infrastructure.

The model can be used to help plan and understand risk and resilience, as well as to better prepare for future events, for example through modification of the responses to the event.

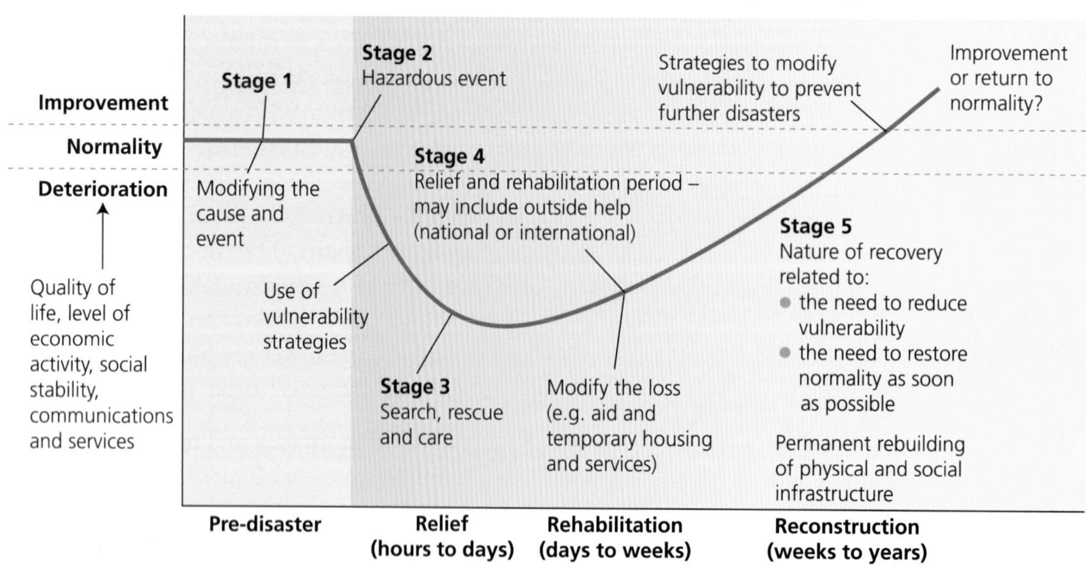

Figure 3.10 Park's Model – the Disaster Response Curve

Table 3.6 Understanding the stages in the Disaster Response Curve

Pre-disaster	Quality of life is normal before a disaster strikes; people do their best to prevent and prepare for such events happening, for example by educating the public on how to act when disaster strikes, preparing supplies, putting medical teams on standby, and so on.
Relief	The hazard event has occurred – immediate relief is the priority with medical attention, rescue services and emergency care provided. This period of time can last from hours to a number of days. The quality of life has seemingly stopped decreasing and is beginning to move up slowly.
Rehabilitation	Groups (for example, the government) try to return the state of things back to normal, by providing food, water and shelter to those who are without these basic needs.
Reconstruction	In the longer term rehabilitation moves into the reconstruction period during which infrastructure, crops and property are invested in. During this time organisations may use preparation and prevention to improve from the mistakes of this disaster to respond better to the next one.

3.3 Hazard management: a variety of approaches

In theory, the best response to tectonic hazards is to avoid all danger. In reality, however, this is impossible because of the conflicting development pressures on land. Hazard management is, therefore, always a series of imperfect solutions. The Swiss Cheese Model (page 28) suggests that hazard and disaster risk can be reduced by:

1. reducing the number of holes in each layer (that is, the number of systemic weaknesses), or,
2. reducing the size of the holes in each layer (that is, the 'gaps' in the system, or the scale of the system weaknesses).

A better understanding of the complexity of tectonic risk through systematic analysis is important. The Swiss Cheese Model, for example, provides a framework for tectonic hazard management which links a number of areas. These include where possible:

- modifying the hazard event
- modifying vulnerability and resilience (at an individual, community and country scale)
- modifying the loss (a component of resilience).

It is important to develop a number of frameworks when taking an overview of hazard events and their ability to develop into disasters so that descriptive accounts of suffering and damage are avoided. Figure 3.11, for example, provides a framework for response analysis.

Modifying the hazard event

During the 1970s and 1980s there was a general feeling that the technological capability and engineering skills to control earthquakes would soon be developed, for example by lubricating the fault plates. But it is now realised that seismic activity cannot be controlled, so efforts instead focus on science and civil engineering solutions to reduce the hazard by either *micro* or *macro* protection techniques:

- Micro: strengthening individual buildings and structures against hazardous stress.
- Macro: large-scale protective measures designed to protect whole communities.

Earthquakes

Micro approaches are generally used in the case of tectonic hazards. For earthquakes, most energy has been focused on public buildings and facilities, especially those expected to remain operational during a disaster: hospitals, police stations and pipelines. Schools and factories were also strengthened so that people could shelter in them. More recently there has been a shift towards improving the planning

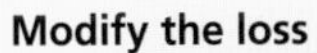

Modify the loss
- Aid vital for poor people
- Insurance more useful for people in richer communities and countries

Modify vulnerability
- Prediction and warning
- Community preparedness
- Education to change behaviour and prevent hazards realising into disasters

Modify the event
- Further environmental control
- Hazard avoidance by land use zoning
- Hazard-resistant design (e.g. building design to resist earthquakes)
- Engineering defences useful for coastal tsunami risk
- Retro fitting of homes is possible for protection

Modify the cause
- Environmental control
- Hazard prevention
- Only really possible for small-scale hazards, e.g. landslides

Increasingly technological

Figure 3.11 A framework for response analysis

Figure 3.12 Mangroves may provide an important coastal buffer both against rising sea levels and tsunamis

frameworks for private houses. Some authorities insist on the strengthening (or demolition) of existing hazardous buildings through a retrofit programme.

Tsunamis

No technologies can prevent tectonic disturbance but some regions and communities have put engineering solutions in place, for example tsunami walls that work for a given amplitude or threshold of wave. Research shows that replanting of coasts (Figure 3.12) may be a way of affording better protection and therefore modification of the event. In the great Asian tsunami of 2004, for example, science has indicated that fewer people might have died if coasts had been protected by mangroves or other types of dense coastal forest.

Yet there is still considerable debate as to the effectiveness of these so-called buffer zones. Mangroves are known to be effective at dissipating energy from waves whipped up by the wind. Modelling studies also suggest that shore vegetation can reduce the flow speed and height of an oncoming tsunami, but there is limited field evidence to back this claim up.

Volcanoes

In some instances it may be possible to modify a volcanic event once the eruption and lava flows have started, either by diverting or chilling the flows. The 1973 volcanic eruption on the island of Heimaey, off the southwest coast of Iceland, threatened to destroy a whole community (Figure 3.13). Seventy homes and farms were buried under tephra and 300 buildings were burned by fires or buried under lava flows.

The lava flow was heading towards the fishing port and harbour – the economic lifeline of the island. Loss of the harbour would have resulted in financial ruin for the community. The Icelanders sprayed seawater on to the lava to slow its movement by chilling. More than 30 km of pipe pumped 6 million cubic metres of water on to the flows. The effort saved the port and the residents returned to rebuild their town.

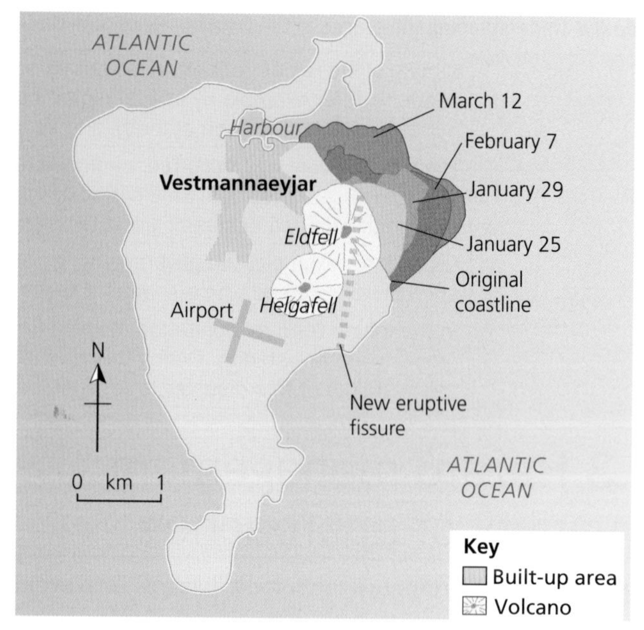

Figure 3.13 The classic 1973 volcanic eruption 'timeline' on the Icelandic island of Heimaey

Modifying vulnerability and resilience

Modification of the hazard event, as shown in Figure 3.11, can involve a number of approaches and adaptations including:

- prediction, forecasting and warnings
- improvements in community preparedness
- working with groups and individuals to change behaviours (to reduce the disaster risk), for example, better land-use planning.

With the advent of better technology, prediction, forecasting and warning are becoming increasingly important parts of disaster preparation and management. For example, a tsunami warning system (TWS) is used to detect tsunamis in advance and issue warnings to prevent loss of life and damage to movable possessions. It is made up of two components: a network of sensors to detect tsunamis and a communications infrastructure to quickly issue alerts to allow evacuation of the coastal areas.

Adaptation and preparedness is essential to ensure an effective response to disaster. It usually involves the planning and testing of hazard reduction systems on timescales that may operate from seconds (for example, tsunami warnings) to years (for example, improvements in land-use planning and zoning).

One of the complexities with adaptation and increasing resilience is the fact that there are a range of interest groups that have some role to play in modification of vulnerability (Figure 3.14).

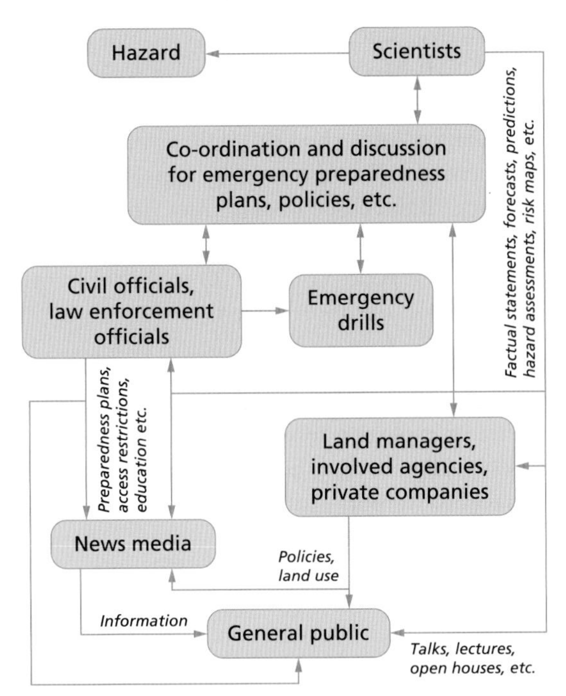

Figure 3.14 Interest groups that have roles to play in modification of vulnerability

Governance, for example, is important but it has limitations in terms of the affordability of prediction and prevention measures, especially in the management of mega-disasters immediately after the event. This means that other factors such as poverty may hamper any top-down efforts to reduce impact and adapt.

Volcanoes: a hazard vulnerability success story?

Volcanologists have an advantage over seismologists in that volcanoes usually do not erupt without warning. The warning signs typically take the form of numerous small earthquakes and a swelling of the ground surface, which reflect the passage of magma to the surface. But it is difficult to predict exactly when activity will take place, especially the timing of a major eruption. Technology in the form of a network of sensors is now being used to help predict eruptions and allow more sophisticated modelling to be undertaken. Monitoring may give time for the area under threat to be evacuated.

Looking at data on deaths, the volcanic hazard threat seems to have been successfully mitigated: only two eruptions since 1980 have caused more than 1000 deaths. Eruptions still affect large numbers of people but prediction and evacuation have reduced the death toll enormously.

An exception to this is was the Mount Ontake eruption in Japan on 27 September 2014. It is a popular area with hikers and walkers, who became the victims. There was no warning and the VEI 3 eruption killed 56 people, the first deaths in Japan from eruptions since 1991 (Mount Unzen). It was the highest death toll from an eruption in Japan since 1902.

Science to reduce earthquake vulnerability

Following the 1995 Kobe earthquake in Japan, the National Research Institute for Earth Science and Disaster Prevention (NIED) deployed 1000 strong-motion accelerometers throughout the country. This is the Kyoshin Network, or K-NET. The average distance between stations is 25 km. During an earthquake, primary and secondary wave velocities are measured at each site and logged. Data is then sent to the local municipality (via the internet). The municipality can use the information for local emergency management and response.

Figure 3.15 shows how the UN World Food Programme uses a range of strategies and players to modify loss and vulnerability. Importantly, new technology and communication, for example social media such as Facebook and Twitter, are used to help people adapt and improve their resilience.

So, adaptation is really about modifying resilience, which is a measure of how well an individual, community or country might absorb and recover from a hazard. This approach to disaster reduction is seen as very important nowadays, especially in the world's poorer communities where the disaster focus had traditionally been on seeing people as vulnerable victims and therefore recipients of external support (a 'top-down' model).

Now attention is focusing on supporting affected communities to prepare and manage themselves, and strengthening this local capability before, during and after a hazard event.

Modifying the loss

Mitigation is about modification of the loss burden. Insurance to cover the cost of earthquake damage, for example, is an important part of wider earthquake protection. Seismologists work with computer risk analysts to help the insurance industry calculate premiums and risks. Computer simulations are used to estimate the probability of damage from different earthquake events, based on:

- seismicity: the raw information about how frequently earthquakes affect a particular location
- seismic hazard: the probability that a certain strength of shaking will occur
- seismic risk: the probability that a certain amount of risk will occur.

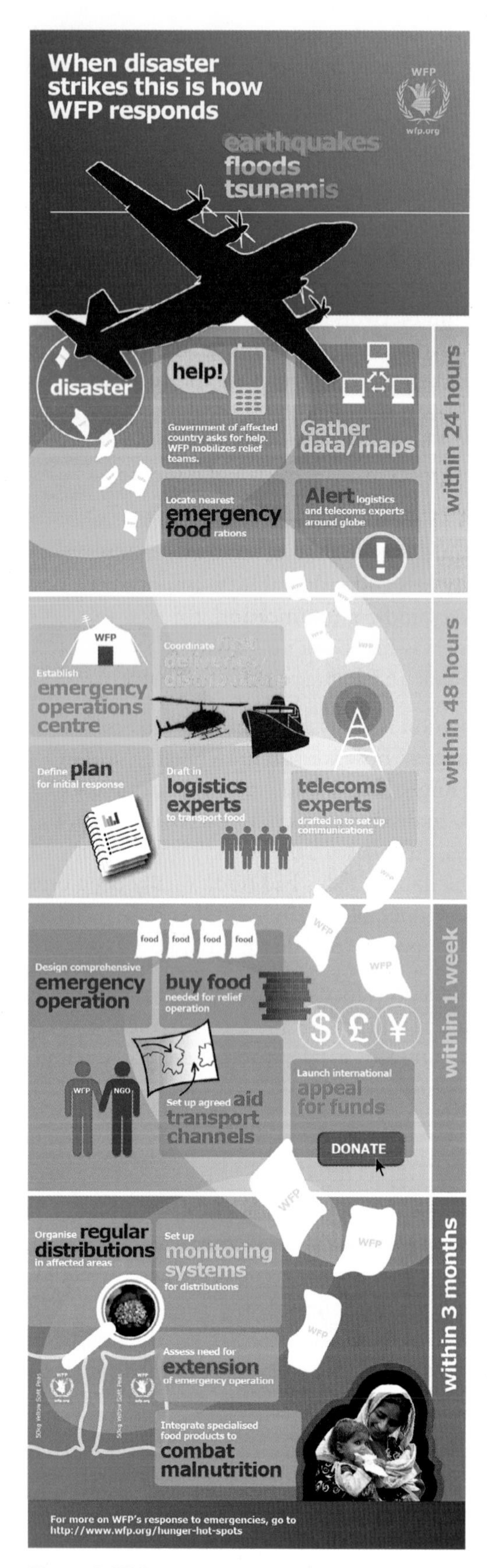

Figure 3.15 Response to hazards – the stages over time

Figure 3.16 shows that, since 1980, Japan's earthquakes (1995, 2004 and 2011) have been costly, with a total cost of over $280 billion. Other tectonic events are also shown to be expensive. Since 2000, the UNISDR estimates the total economic cost of all disasters to be approximately $1.3 trillion. Even in developed economies such as the USA and Japan, insured losses for tectonic events tend to be relatively low, at approximately 25 to 30 per cent, meaning many people are unprotected.

Most insurance policies provide cover for property loss caused by a volcanic blast, airborne shock waves, ash, dust or lava flow. Fire or explosion resulting from volcanic eruption is also usually covered, however some may not cover shock waves or tremors caused by volcanic eruption.

Disaster aid and internal governmental aid

Disaster aid is the result of humanitarian concern following severe loss. This aid is all about protecting life, health, subsistence and a person's physical security. Table 3.7 compares disaster aid and internal governmental aid.

Disaster aid is often criticised, however: there may be poor or corrupt distributions systems, and it doesn't encourage self-help or a more bottom-up management of the disaster at a local level. In the 2010 Haiti earthquake the Nepalese disaster relief workers were implicated in the introduction of cholera – see Chapter 2, page 22.

A complex risk environment

Despite considerable efforts to improve scientific understanding and better risk-management approaches, in general governments and businesses remain insufficiently prepared to confront many tectonic hazards and effectively manage their economic, social, political and humanitarian consequences. This is often true irrespective of a country's level of development. In certain high-threshold events, governments are the responders of last resort, but they may not have the resources or technical expertise to deal with such events.

Table 3.7 Disaster aid compared to internal governmental aid

Disaster aid	Internal governmental aid
Aid flows to countries and victims via governments, NGOs and private donors. In the longer term aid is used for relief, rehabilitation and reconstruction. This type of aid is often appropriate to middle- and lower-income countries.	This is typically used in emerging and developing countries where the disaster mitigation is achieved by spreading the financial load throughout the tax payers of the country. This may include a national disaster fund and release of funds may require a political declaration.

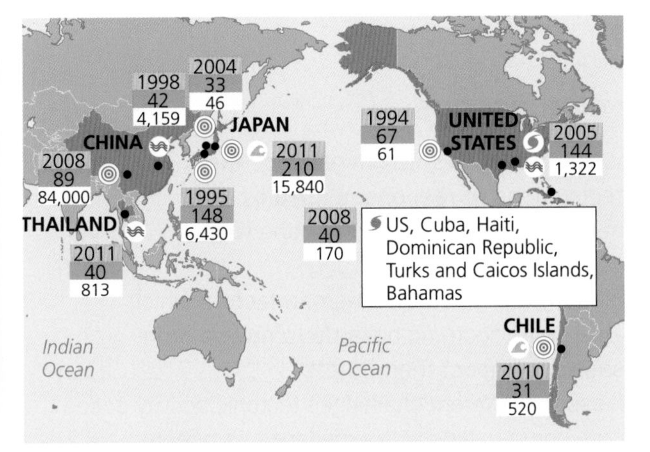

Key
Leading natural disasters, by overall economic losses, since 1980

- Earthquake
- Floods
- Hurricane
- Tsunami
- 000 Year
- 000 Cost, $bn (2011 dollars)
- 000 Number of deaths

Figure 3.16 Major natural disasters – costs ($bn) and loss of life

Synoptic themes:

Players

NGOs and insurers have a role to play in helping to modify the loss.

To get the right balance between planning for specific 'known' events and creating generic responses for events that are rare or unexpected, governments and agencies must strengthen planning processes to anticipate and manage shock events. This includes clarity of process in the chains of command (especially where multi-jurisdictions are involved), activating and connecting independent local experts with policymakers, as well as building common approaches in the management of complex risks.

Hyogo and Sendai approach to disaster management

The World Conference on Disaster Reduction was held in 2005 in Hyogo (Kobe), Japan, and established a 'Framework for Action'. Its aims were to promote a strategic and systematic approach to reducing vulnerability and risks to hazards through building the resilience of nations and communities to disasters.

This was replaced by the Sendai Framework in March 2015. It set out four priorities:

1. Understanding disaster risk.
2. Strengthening governance to manage disaster risk.
3. Investing in disaster-risk reduction for resilience and enhancing disaster preparedness for effective response.
4. 'Build back better' in recovery, rehabilitation and reconstruction.

Importantly the framework emphasised the need to tackle disaster risk reduction and climate change adaptation when setting Sustainable Development Goals, particularly in light of an insufficient focus on risk reduction and resilience in the original Millennium Development Goals (MDGs). There is also a focus on emergency preparedness. In the case of international disaster relief, the framework recognises that distribution is complex, fragmented and disorganised. This is because there are various separate institutions, mechanisms and approaches defining where the funding is directed and how it is spent.

Review questions

1. Summarise the main trends in tectonic natural hazards in recent decades, including deaths, numbers affected and economic costs.
2. What are the characteristics of a mega-disaster?
3. Outline the spatial variation in disasters by both region and level of economic development.
4. Explain the differences in the modification of hazard events for different tectonic hazards.
5. Compare the hazard management cycle to the Disaster Response Curve.
6. Why is there a complex risk environment?
7. Summarise the reasons why volcanoes pose a much lower disaster risk than they have historically.

Further research

Research and select a *Credcrunch* relevant to tectonic disasters and develop a case study from its findings: http://cred.be/publications?field_publication_type_tid=66&field_cred_staff_authors_nid=All

Use this link to research and summarise the key points relating to the 2010 Haiti earthquake: www.gfdrr.org/sites/default/files/publication/Haiti_August_2014_Summary.pdf

Research the most expensive tectonic hazards in the last 30 years online. How do tectonic hazards compare to hydrological, meteorological and climatological hazards?

Research the differences between the Hyogo and Sendai approaches to disaster management.

Exam-style questions

AS questions

1 Define what is meant by the term **disaster**. [1]

2 Study Figure 1. State the mode and range of earthquake risk magnitude. [2]
   a Mode:
   b Range:

3 Explain **two** reasons how a government might influence a community's resilience. [4]

4 Explain why some earthquakes generate secondary hazards. [6]

5 Assess the factors that contribute to increased impacts from some tectonic hazard events. [12]

A level questions

1 Assess the factors which influence the effectiveness of responses used by different groups of people to cope with tectonic hazards. [12]

2 Assess the physical and human factors which cause some tectonic hazards to have a more disastrous impact than others. [12]

3 Assess the different challenges tectonic activity poses for the communities who experience its effects. [12]

4 Study Table 1. Calculate the mean, median and interquartile range of **property damage** for the earthquake data. [4]

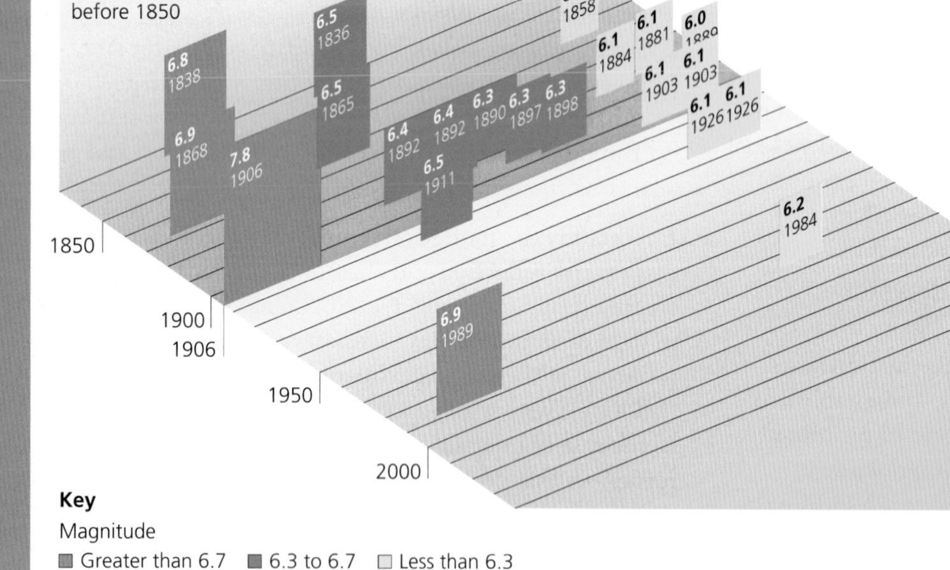

Figure 1 Earthquakes in the San Francisco area since 1850 (Source: Munich Re, based on USGS Earthquake Hazard Program, 2004)

Table 1 Earthquake data

Year	Name	Magnitude	Property damage: $billion (inflation-adjusted)
1923	Great Kanto earthquake	7.9	8
1976	Tangshan earthquake	7.8	42
1989	Loma Preita	7.1	21
1994	Northridge earthquake	6.7	67
1995	Great Hanshin earthquake	6.9	312
1998	Sichuan earthquake	8.0	95
1999	921 earthquake Taiwan	7.6	10
1999	Izmit earthquake	7.6	29
2010	Chile earthquake	8.8	30
2011	Japan earthquake and tsunami	9.0	249
2015	Nepal earthquake	7.8	11

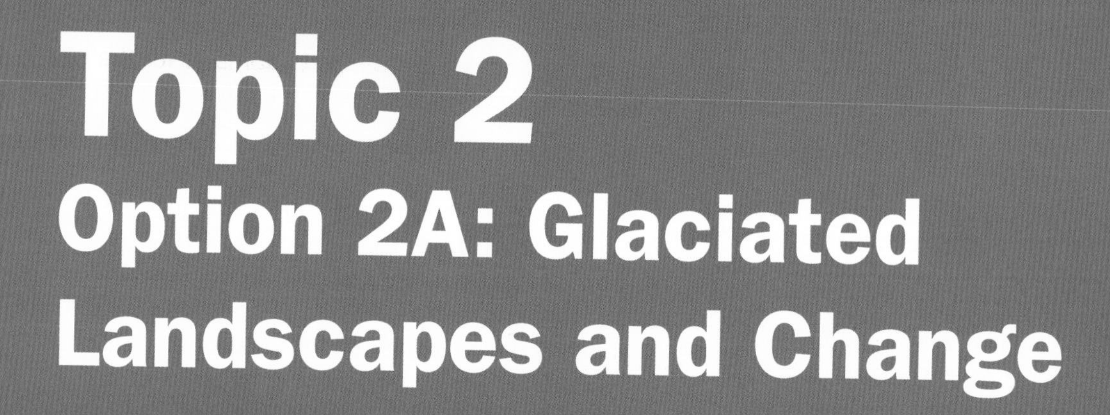

Topic 2
Option 2A: Glaciated Landscapes and Change

Past and present distribution of glacial and periglacial environments

How has climate change influenced the formation of glaciated landscapes over time?

By the end of this chapter you should:

- know the chronology of the Pleistocene ice Age
- understand the causes of the climate changes that have led to:
  - longer-term glacial and interglacial cycles
  - shorter-term stadial and interstadial fluctuations
- know and understand the present and past distribution of ice cover and periglacial areas
- understand the nature and importance of the cryosphere and its role in global systems
- understand what periglacial environments are and how their often-unique landforms contribute to distinctive landscapes.

4.1 The causes of longer- and short-term climate change

The Swiss Scientist Louis Agassiz can be credited with the 'discovery of the Ice Age'. He made scientific observations on existing Alpine glaciers in the 1830s and then identified evidence of glaciation all over Europe in areas such as the North European plain (where glaciers no longer existed) thus introducing the concept of 'continental glaciation'.

Subsequent scientific research using ice cores in Greenland and Antarctica has provided a record of the climate over the last 800,000 years. Air bubbles trapped in the ice contain atmospheric carbon dioxide and the ice itself preserves a record of oxygen isotopes. Low concentrations of carbon dioxide occur during glacial periods and much higher ones during interglacials.

We are currently living in the Quaternary, which began around 2 million years ago as the Tertiary period ended with the onset of global cooling and ice-house conditions. Recent theories suggest that plate tectonics created suitable conditions to 'kick start' the Pleistocene by positioning Antarctica as an isolated continent at the South Pole. As Figure 4.1 shows, the Quaternary Ice Age is just the latest of several ice ages over geological time.

Multiple glacials and interglacials

As can be seen from Figure 4.1 the Quaternary period is divided into two epochs of geological time:

- The Pleistocene covers the time span from the beginning of the Quaternary to about 11,500 years ago when the most recent continental glacial (UK Devensian) ended.
- The Holocene interglacial (the period we now live in) is similar climatically to previous interglacials, but is distinctive as it is noted for the growth of human civilisation, the development of agriculture and industrialisation.

Figure 4.2 summarises the characteristics of the Pleistocene: on a geological timescale it can be regarded as a single ice age, but in reality there are multiple periods of glacials (colder ice-house conditions) and interglacials (greenhouse or warmer conditions). Variations in the Earth's climate, with repeated shifts between colder and warmer conditions, have led to the oscillations shown, which can be correlated globally.

The multiple glacial phases have left evidence of erosional and depositional features created by glaciers and their meltwaters. However, the landforms produced

Key terms

Glacials: Cold, ice-house periods within the Pleistocene.

Interglacials: Warmer periods similar to the present, i.e. greenhouse periods.

Ice-house conditions: Very cold glacial conditions.

Greenhouse conditions: Much warmer interglacial conditions.

Pleistocene: A geological period from about 2 million years ago to 11,700 years ago, the early part of the quaternary which included the most recent ice age.

Age (billion years)	Eon	Glaciations
0.0–0.6	Phanerozoic	Late Proterozoic glaciations ↑
0.6–2.5	Proterozoic	Warm ↓ Huttonian glaciations ↑
2.5–4.0	Archaean	Warm ↓
4.0–4.6	Hadean	

Era	Period	Epoch	Glaciations	Age (million years)
Cenozoic	Quaternary	Holocene Pleistocene	Pleistocene glaciations ↑	2.6
	Tertiary	Pliocene Miocene Oligocene Eocene Palaeocene	Warm ↓	
Mesozoic	Cretaceous			66
	Jurassic		Permo-Carboniferous glaciations ↑	144
	Triassic			206
Upper Palaeozoic	Permian		Warm ↓	251
	Carboniferous			286
	Devonian		Late Ordovician glaciations ↑	360
Lower Palaeozoic	Silurian			410
	Ordovician		Warm ↓	440
	Cambrian			644

Figure 4.1 Ice ages in geological time

in earlier glaciations have usually been reworked and reshaped, and even destroyed, by later glaciations, so adding to complexities in reconstructing past patterns of glaciation. Even today the features from the later phases of the most recent glacial period (UK Devensian) are being modified by post-glacial geomorphic processes such as weathering, mass movement and fluvial action.

Figure 4.2 further shows numerous fluctuations within the major glacial/interglacial cycles, operating over a number of timescales. These shorter periods of intense cold are called stadial periods, with shorter periods of relative warmth known as interstadials. Recent ice core data suggests that the onset of some of the more severe fluctuations in temperature was quite abrupt.

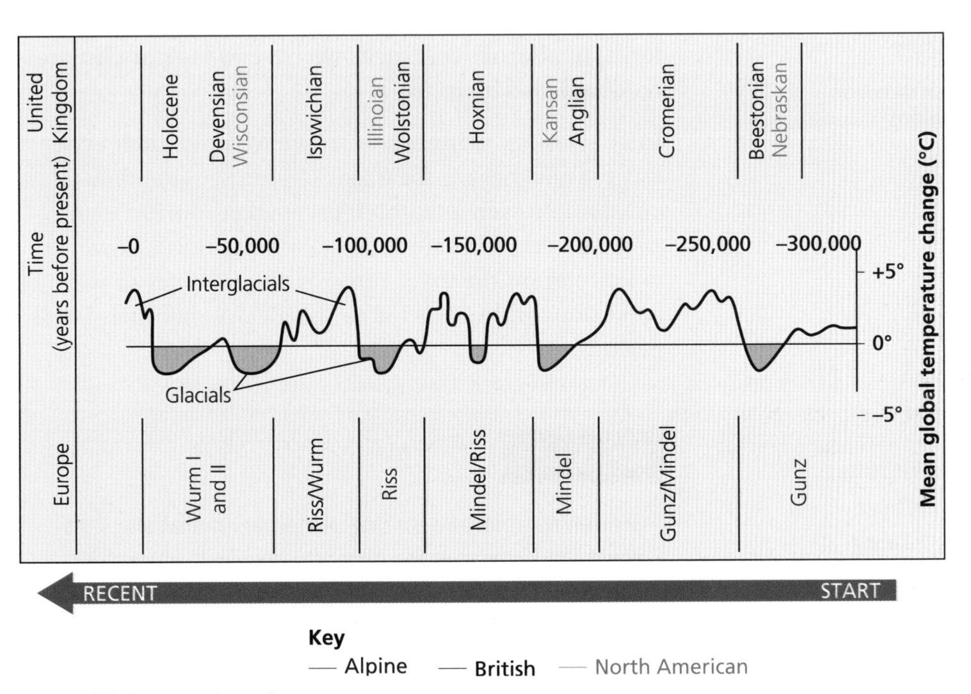

Figure 4.2 Ice age chronology

Key term

Stadials and interstadials: Short-term fluctuations within ice-house-greenhouse conditions; stadials are colder periods that lead to ice re-advances.

Causes of longer-term glacial/interglacial cycles

Long-term changes in the Earth's orbit around the Sun are currently seen as the primary cause of the oscillations between glacial and interglacial conditions. The Milankovitch theory based on orbital/astronomic forcing of glacial periods takes into account three main characteristics of the Earth's orbit (Figure 4.3):

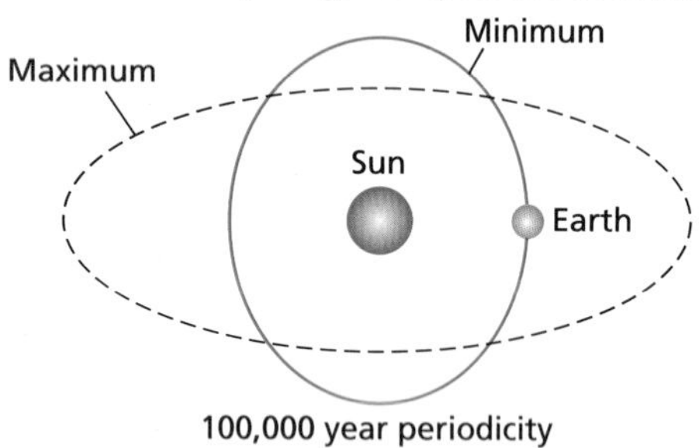

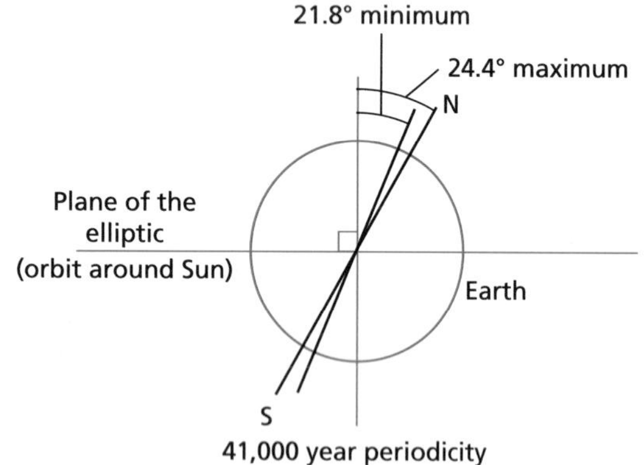

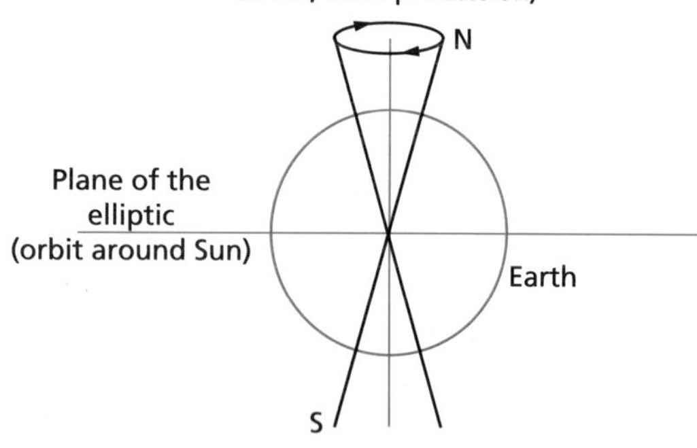

Figure 4.3 The combined effects of the three Milankovitch cycles (Source: Holden, Joseph, *An Introduction to Physical Geography and the Environment*, 3rd ed., ©2012. Reprinted by permission of Pearson Education, Inc., New York, New York.)

1. **Eccentricity** of the orbit – it changes from being more elliptical to more circular and back again over a period of around 100,000 years, so changing the amount of radiation received from the Sun (this is considered the dominant factor).
2. **Axial tilt** varies from 21.8° to 24.4° (currently the tilt is 23.5°) over a timescale of around 41,000 years. This changes the intensity of sunlight received at the poles and, therefore, the seasonality of the Earth's climate. The greater the tilt, the greater the difference between summer and winter.
3. **Wobble** – the Earth wobbles on its axis (just like a spinning top) changing the point in the year at which the Earth is closest to the Sun (axial precession) over a 21,000 year time cycle. This causes long-term changes to when different seasons occur along the Earth's orbital path.

The three orbital cycles can combine together to minimise the amount of solar energy reaching the Northern hemisphere during summer (leading to cooler summers overall).

In support of Milankovitch's theory is the fact that glacials seem to have occurred at regular intervals of approximately 100,000 years. However the actual impact of the combined orbital changes on solar radiation amount and distribution is small, probably only enough to change global temperatures by between 0.5 °C and 1 °C.

To explain the larger temperature changes of up to 5 °C that were required for the vast expanses of ice to form, or alternatively melt, we have to look at climate feedback mechanisms.

In conclusion, many scientists see Milankovitch cycles as a possible trigger for major ice-house–greenhouse changes, or even as a good 'pacemaker' during each cycle. It is the climate feedback mechanisms, however, which sustain the drive towards either colder or warmer conditions and which led to the glacial and interglacial periods.

Key term

Orbital/astronomic forcing: A mechanism that alters the global energy balance and forces the climate to change in response.

Key concept: Climate feedback

Feedback effects are those that can either amplify a small change and make it larger (positive feedback) or diminish the change and make it smaller (negative feedback). A number of interacting Earth systems are involved, as shown in Table 4.1 below.

Positive feedback: increasing the warming or cooling rates	Negative feedback: decreasing the warming or cooling rates
• Snow and ice cover. Small increases in snow/ice raise surface albedo (reflectivity) so more solar energy is reflected back into space, leading to further cooling, which could lead to further snowfall and ice cover. • The melting of snow/ice cover by carbon dioxide emissions decreases albedo; methane is emitted as permafrost melts, and warming seas lead to calving of ice sheets, which all lead to loss of snow/ice cover and of surface albedo, decreasing reflectivity and accelerating further warming.	• Increasing global warming leads to more evaporation and, over time, pollution from industrialisation adds to global cloud cover. Increasingly cloudy skies could reflect more solar energy back to space and diminish the effect of warming – so called 'global warming' may be less intense because of this global dimming. • Ice sheet dynamics can disrupt the thermohaline circulation (THC). Warming water in the Arctic disrupts ocean currents; less warm water from the Gulf Stream is drawn north, which could lead to global cooling in northern Europe.

Table 4.1 Feedback effects

Possible explanations for shorter-term fluctuations

As can be seen from Figure 4.2 both glacial and interglacial periods show fluctuations, with frequent warming and cooling periods (stadials). A number of factors, other than combinations of effects in the Milankovitch cycles, have been cited for these shorter-term fluctuations.

Key terms

Albedo: The reflective coefficient of a surface, i.e. the proportion of incident radiation reflected by a surface (very high in the case of snow or ice).

Calving: The breaking up of chunks of ice at the glacier snout or ice sheet front to form icebergs as the glacier reaches a lake or the ocean.

Thermohaline circulation: A global system of surface and deep-water ocean currents driven by differences in temperature (thermo) and salinity (haline) between areas of the oceans. Sometimes known as the ocean conveyor.

Solar forcing

The amount of energy emitted by the Sun varies as a result of the number and density of sunspots (dark spots on the Sun's surface caused by intense magnetic storms). There are a number of cycles of varying length including the 'eleven-year sunspot cycle'. There are reliable records of sunspot activity for the last 400 years, with some information for the last 2000 years. A longer period with no sunspot activity, known as the Maunder Minimum, occurred between 1645 and 1715, at the height of the Little Ice Age, to which it has often been linked, whereas the preceding medieval warm period has been linked to more intense sunspot activity. The big problem is that total variation in solar radiation caused by sunspot activity is only 0.1 per cent and is not by itself enough to explain the climate fluctuations. Even so, some scientists suggest that around twenty per cent of twentieth-century warming may be attributed to solar output variation.

Volcanic causes

Violent volcanic activity can alter global climate. Eruptions with a high volcanic explosivity index (VEI) eject huge volumes of ash, sulphur dioxide, water vapour and carbon dioxide into the atmosphere (volcanic aerosols), which high-level winds distribute around the globe. In 1815, Tambora in Indonesia ejected 200 million tonnes of sulphur dioxide into the atmosphere and, in the following two to three years, temperatures were recorded 0.4–0.7 °C lower, so short-lived global cooling occurred.

Another example is the Laki Fissure eruption (VEI 4) in Iceland from 1783–84. The size of the volcanic aerosol and length of the eruption were linked to the exceptionally cold winter in the northern hemisphere in 1785. However, a combined negative phase of North Atlantic Oscillation (NAO) and El Niño Southern Oscillation (ENSO) was also cited, especially as similar conditions led to exceptionally cold winters in 2009–10 in Europe and eastern North America.

Attributing causal relationships in science can be very problematic, especially as both of the possible causes discussed could not by themselves explain the size of the temperature changes associated with the short-term stadial and interstadial fluctuations, so yet again feedback mechanisms would be needed to amplify the changes.

Although these positive and negative feedback mechanisms are not yet fully understood, there is no doubt that internal processes within the ice masses acted with external factors such as changes in global ocean circulation (THC) and fluctuations in concentration of greenhouse gases to produce possible explanations for short-term stadial and interstadial fluctuations. The timing of these events, such as the Loch Lomond Stadial and the Little Ice Age, can be correlated around the world.

Loch Lomond Stadial (the Younger Dryas event)

Ice sheets began retreating about 18,000 years ago, with rapid deglaciation by 15,000 years ago (late glacial interstadial) with similar temperatures as today leading to widespread distribution of deciduous woodlands across Europe. However, around 12,500 years ago the temperatures plunged downwards and, by 11,500 years ago, glacial conditions occurred with temperatures 6–7 °C lower. Glaciers re-advanced in many parts of the world including the formation of ice caps in the Scottish Highlands (Figure 4.4), from which cirque and valley glaciers flowed outwards, with smaller areas of cirque glaciers in the Lake District and North Wales.

Greenland ice core data suggested a very rapid temperature rise of 7 °C after the event (perhaps in only 50 years) with a corresponding rapid rise in sea level. The timing would seem to be inconsistent with orbital forcing, as neither solar forcing nor volcanic eruptions could lead to a fluctuation of such magnitude. One possibility was that the Loch Lomond Stadial was triggered when drainage of the huge proglacial Lake Agassiz disrupted the THC, thus cutting off the poleward heat transport from the Gulf Stream.

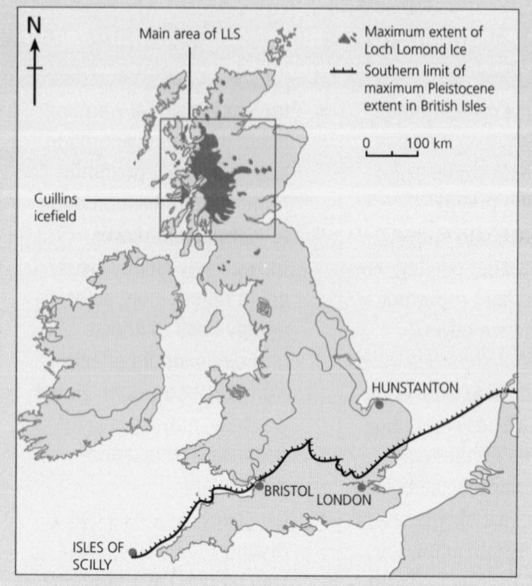

Figure 4.4 The Loch Lomond re-advance in the UK glaciation context

The Little Ice Age – the longest glacial oscillation in historical times

Proxy records from historical documents and paintings add increasing detail to our knowledge of past climate. Between 1350 and 1900, conditions were slightly colder – perhaps on average by between 1.0 °C and 2.0 °C – than at present over much of the globe. Between 1550 and 1750 there was a low trough of very cold conditions, known as the Little Ice Age, which occurred globally. There were many impacts:

- The widespread abandonment of upland farms in Scandinavia and Iceland.
- Many glaciers in Europe re-advanced down valleys; the Little Ice Age was a period of predominantly positive net mass balance leaving prominent terminal moraines from which the glaciers subsequently retreated, but often at different dates/times around the world.
- Arctic Sea ice spread further south with polar bears seen frequently in Iceland.
- Rivers in the UK and lowland Europe, and New York harbour, froze over.
- Curling developed as a national sport in Scotland as there were so many frozen lakes and rivers.

Some researchers argue that the Little Ice Age could have developed into a new stadial but that this was prevented by the onset of the Industrial Revolution, fired by coal. The release of carbon dioxide triggered climate warming, which dramatically halted the cold period.

4.2 Present and past distribution of ice cover

Definition and importance of the cryosphere

The cryosphere consists of ice sheets and glaciers, together with sea ice, lake ice, ground ice (permafrost) and snow cover. Mass and energy are constantly exchanged between the cryosphere and other major components of Earth systems: the hydrosphere, lithosphere, atmosphere and biosphere. Glaciers are very visible and sensitive barometers of climate change because they constantly grow/advance and shrink/retreat in response to changes in temperature and precipitation.

Types of ice mass and glacial environments

As can be seen from Table 4.2, ice masses can be classified by their morphological characteristics, size and location.

Some glaciers are land based and their base is at or above sea level, for example where the Mer de Glace flows from Mont Blanc in the Alps or glaciers flowing from the Rocky Mountains in North America.

Other glaciers are marine based, w[illegible] below sea level, for example the Wes[illegible] Sheet, which has a base of up to 2000 [illegible] actually frozen to the seabed.

A further morphological contrast is provide[illegible] degree of constraint. Unconstrained glaciers [illegible] be the larger glacial forms, such as ice sheets an[illegible] caps (see Table 4.2), which are so thick and extensive that they 'submerge' the landscape. They are drained by outflowing ice streams and outlet glaciers. Valley glaciers can sometimes only 'escape' across pre-existing cols by the process of diffluence (flowing away) as their snouts become 'trapped'.

The most significant classification is the thermal regime of a glacier as this has a major impact on glacier movement, the operation of glacial processes and subsequent landforms produced.

Key term

Cryosphere: The parts of the Earth's crust and atmosphere subject to temperatures below 0 °C for at least part of each year.

Table 4.2 Different types of ice mass

Type of ice mass	Description	Average size (sq km)	Degree of constraint	Example
Ice sheet	Complete submergence of regional topography; forms a gently sloping dome of ice several kilometres thick in the centre	10–100,000	Unconstrained	Greenland Antarctica
Ice cap	Smaller version of ice sheet occupying upland areas; outlet glaciers and ice sheets drain both ice sheets and ice caps	3–10,000	Unconstrained	Vatnajökull (Iceland)
Ice field	Ice covering an upland area, but not thick enough to bury topography; many do not extend beyond highland source	10–10,000	Unconstrained	Patagonia (Chile) Columbia (Canada)
Valley glacier	Glacier confined between valley walls and terminating in a narrow tongue; forms from ice caps/sheets or cirques; may terminate in sea as a tidewater glacier	3–1500	Constrained	Aletsch Glacier (Switzerland) Athabasca (Canada)
Piedmont glacier	Valley glacier which extends beyond the end of a mountain valley into a flatter area and spreads out like a fan	3–1000	Constrained	Malaspina (Alaska)
Cirque glacier	Smaller glacier occupying a hollow on the mountain side – carves out a corrie or cirque; smaller version is known as a niche glacier	0.5–8	Constrained	Hodges Glacier Grytviken (South Georgia)
Ice shelf	Large area of floating glacier ice extending from the coast where several glaciers have reached the sea and coalesce	10–100,000	Unconstrained	Ronne and Ross Ice Shelf (Antarctica)

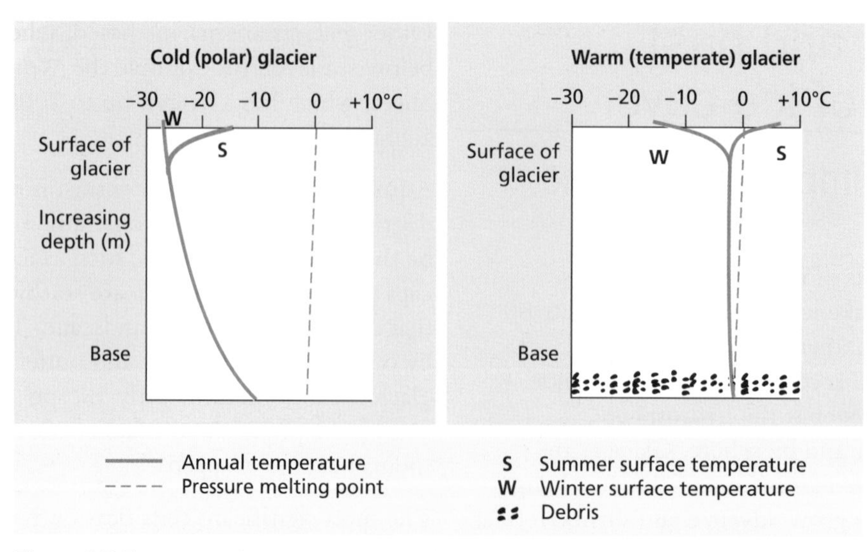

Figure 4.5 The contrasting temperature profiles of temperate and polar glaciers

Glaciers have traditionally been divided into:

- **Warm based** (temperate or wet) glaciers, which occur in high altitude areas outside the polar region, such as in the Alps and sub-Arctic areas. The temperature of the surface layer fluctuates above and below melting point, depending on the time of year, whereas the temperature of the rest of the ice, extending downwards to the base, is close to melting point. Because of increased pressure of overlying ice, water exists as a liquid at temperatures below 0 °C, causing the basal ice to melt continuously. The effects of pressure, geothermal energy and percolation of meltwater all contribute to prevent the glacier freezing to its bed. The glacier has lots of debris in its basal layers, and significant subglacial depositional features.
- **Cold based** (polar) glaciers, which occur in high latitudes, particularly in Antarctica and Greenland. As the average temperature of the ice is usually well below 0 °C, as a result of the extreme surface temperature (as low as −20–30 °C), the accumulation of heat from geothermal sources is not great enough to raise the temperature at the base of the glacier to 0 °C, even though the ice may be up to 500 m thick. There is relatively little surface melt in the very short and cool polar summer, so little meltwater percolates downwards. The glacier is permanently frozen to its bed, so there is no debris-rich basal layer.

Figure 4.5 summarises the contrasting temperature profiles of the two types and the key temperature controls within a glacier.

Figure 4.6 Hummocky moraines in Svalbard

A further subdivision is the hybrid *polythermal glacier*, whereby the underneath is warm (wet) based and the margin cold based. Many large glaciers are cold based in their upper regions and warm based lower down, when they extend into warmer climate zones – this is a common occurrence in Svalbard, Norway.

Surging glaciers or ice streams may occur within warm based, cold based or polythermal glaciers, and may have rates of flow of up to 100 m per day (examples include the Greenland outlet glaciers, which average 30 m per day) with huge amounts of calving (ice breaking off at the edges).

Present and past distribution of ice cover

At present glacial ice covers over ten per cent of the Earth's land area, and 75 per cent of the world's freshwater is locked up in this ice cover, about 1.8 per cent of all the water (fresh, brackish or salt) on Earth.

Table 4.3 shows that for the present day:

- About 85 per cent of all current glacier ice is contained in Antarctica (shared between the West and East Antarctic Ice Sheets).
- The Greenland Ice Sheet is the second largest accumulation of glacier ice, with nearly eleven per cent of the Earth's total ice cover.
- The remaining ice cover is distributed among ice caps such as Vatnajökull (Iceland) and northern Canada and Alaska, highland ice fields and many smaller glaciers in high altitude areas (Himalayas, Rockies, Cascades, Andes, European Alps, etc.).

Table 4.3 Estimates of ice cover present and past

Region	Present area (estimated, 10^6 km^2)	Past area (estimated, 10^6 km^2)
Antarctica	13.50	14.50
Greenland	1.80	2.35
Arctic Basin	0.24	16.00
Alaska	0.05	
Rest of North America	0.03	
Andes	0.03	0.88
European Alps	0.004	0.04
Scandinavia	0.004	6.60
Asia	0.12	3.90
Africa	0.0001	0.0003
Australasia	0.001	0.7
Britain	0.0	0.34
Total	**15.78**	**45.31**

Key term

Environmental lapse rate: The rate of atmospheric temperature decreases with altitude at a given time and location.

- There are even glaciers above 4000 m in Ecuador in the High Andes, Mount Kilimanjaro in Tanzania and in Indonesia in Equatorial regions!

A number of factors influence the distribution of ice cover. The two most important factors are *latitude* (for polar ice masses) and *altitude* for Alpine glaciers. In high latitudes, the Sun's rays hit the ground at a lower angle so the solar energy received has to heat a larger area. High altitudes are impacted by the environmental lapse rate (ELR) whereby temperature declines by 1 °C for every 100 m.

Other more locally significant factors include *aspect*, which can determine the amount of snow falling and settling. In mountainous areas, aspect and *relief* combine to affect the distribution of cirque glaciers. In the northern hemisphere, north- and east-facing slopes are both more sheltered and shadier, and thus more conducive to snow accumulation.

As can be seen from Table 4.3 and Figure 4.7, which compare the distribution of present-day and late Pleistocene ice sheets and glaciers in both the northern and southern hemispheres, the following differences emerge:

- Ice cover at the Pleistocene maximum was more than three times greater than the present day.
- The Antarctica and Greenland ice sheets only covered a slightly greater area than they do today.
- The major extensions were two ice sheets in North America (Laurentide and Cordilleran) and the Scandinavian Ice Sheet in Europe – these all grew to thicknesses of 3000–4000 m and transformed the landscape of North America and Europe respectively.
- Other significant extensions included all of southern South America, South Island New Zealand, Siberia and the Himalayas.

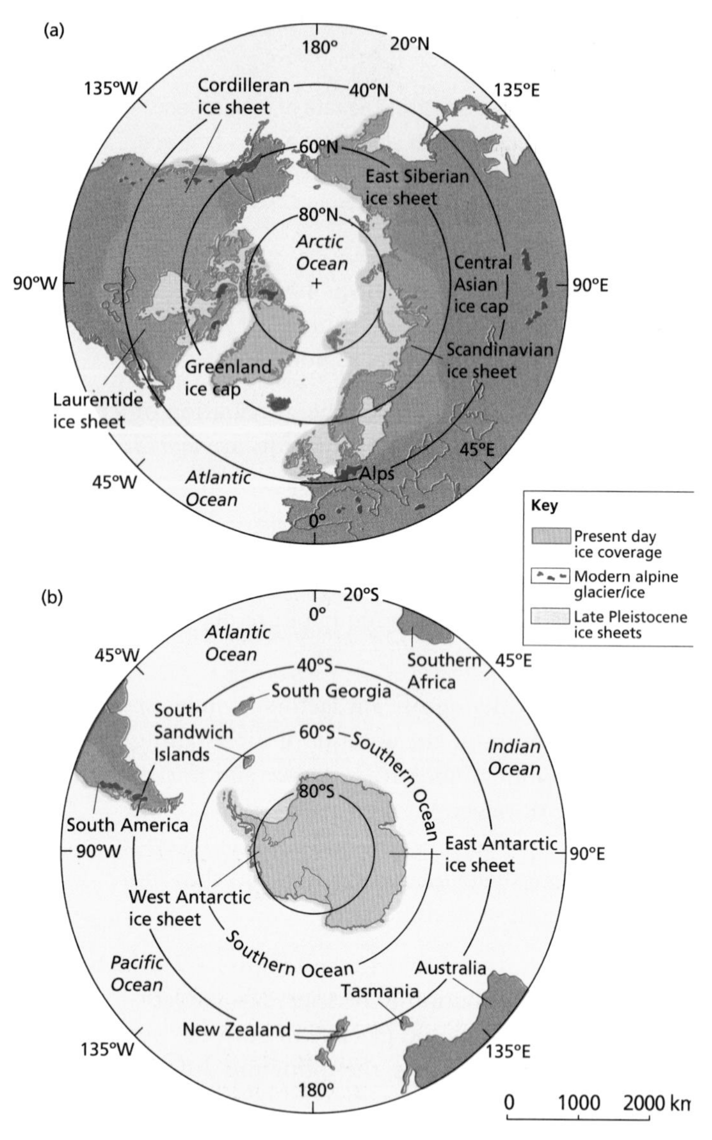

Figure 4.7 Past and present distribution of ice cover

Skills focus: Past and present ice coverage

1. Use maps of North America and Europe to reconstruct the past distribution of ice and compare it to the present day coverage.
2. Refer to Table 4.3 and calculate the percentage contribution of the following ice masses to total global ice coverage during the late Pleistocene maximum: Antarctica, Scandinavian Ice Sheet and the British Ice Sheet to assess their significance.

4.3 Periglacial processes and their distinctive landforms and landscapes

Periglacial environments

The term periglacial has traditionally been used to refer to the climate conditions and landscape that characterised the areas near the margins of glacier ice during the Pleistocene, or temporally before the onset of glacier conditions. However, the term is now more widely used to include all non-glacial cold climate areas with a high range of different high latitude and high altitude environments which may or may not contain glaciers.

Periglacial climates typically have many of the following characteristics:

- intense frosts during winter and on any snow-free ground in summer
- highest average annual temperatures usually range from 1 °C to −4 °C
- daily temperature below 0 °C for at least nine months, and below −10 °C for at least six months per year
- temperatures rarely rise above 18 °C, even in summer
- low precipitation: typically under 600 mm per year (<100 mm in winter and <500 mm in summer)
- temperatures fluctuating through frequent cycles of freezing and thawing to cause interstitial ice (ice within cracks) to melt (Figure 4.9).

These climate conditions give rise to a variety of processes, collectively known as periglacial. These processes combine to produce distinctive landscapes, containing landforms that are unique to periglacial areas. Some of these processes, such as frost action, occur elsewhere quite widely, although with less intensity than in periglacial areas.

Distribution past and present

Around 20 per cent of the Earth's land surface experiences periglacial conditions (Figure 4.8), largely in the northern hemisphere, with extensive areas in Siberia, northern Scandinavia, northern Canada and Alaska.

The distribution of periglacial environments in the Pleistocene glacial periods was very widespread, with an estimated 33 per cent of the world experiencing these conditions, at much lower latitudes than today. For

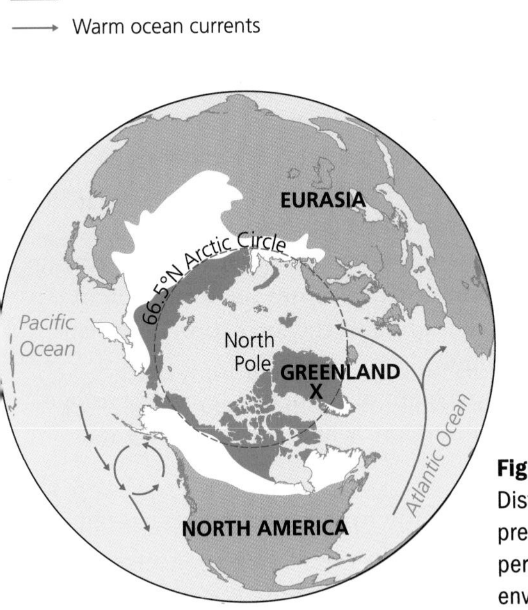

Figure 4.8 Distribution of present-day periglacial environments

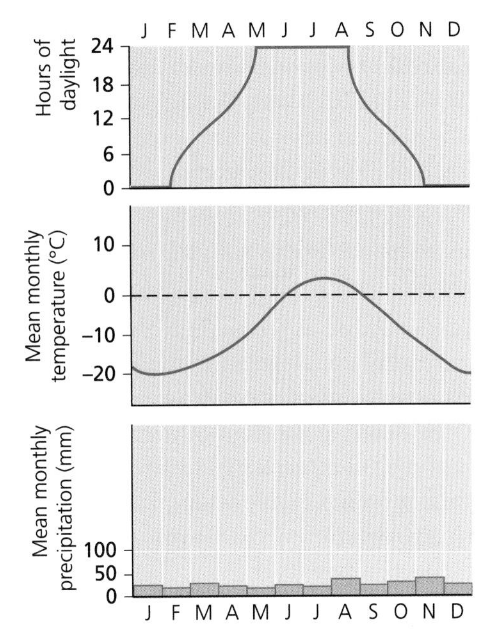

Figure 4.9 Climate of periglacial area X on Figure 4.8

example, periglacial conditions extended across Europe as far south as southern France, northern Italy and the Balkans. Interestingly, relict (surviving) landform evidence shows that southern Britain was not covered by ice but experienced periglacial conditions.

Key term

Active layer: The top layer of soil in permafrost environments that thaws during summer and freezes during winter.

Key concept: Permafrost

Permafrost is often loosely defined as 'permanently frozen ground', but technically the term refers to soil and rock that remains frozen as long as temperatures do not exceed 0 °C in the summer months for at least two consecutive years.

- **Continuous permafrost** forms in the coldest areas of the world where mean annual air temperatures are below −6 °C. It can extend downwards for hundreds of metres.
- **Discontinuous permafrost** is more fragmented and thinner.
- **Sporadic permafrost** occurs at the margins of periglacial environments and is usually very fragmented and only a few metres thick; it often occurs on shady hillsides or beneath peat.

In summer, the energy balance is positive, which causes overlying snow and ice to melt away to produce a seasonally unfrozen zone above the permafrost called the active layer, which varies from a few centimetres to as deep as 3.0 m.

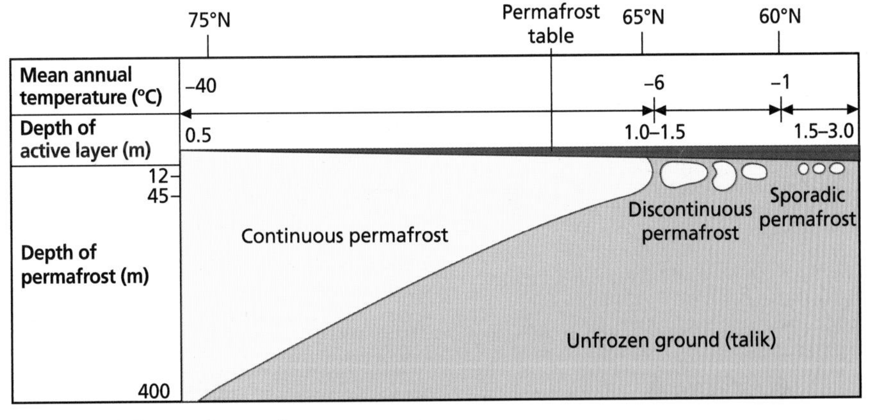

Figure 4.10 Types of permafrost

Up to 25 per cent of the Earth's surface is currently experiencing permafrost conditions (almost 50 per cent of Canada and 80 per cent of Alaska). While periglacial environments usually contain permafrost, sometimes areas of periglacial activity involving intense frost action do exist outside the permafrost zone.

A number of factors influence the distribution and character of permafrost:

- Climate is the main control, as temperature and the amount of moisture available determine the presence or absence, depth and extent of permafrost.
- On a local scale, the depth and extent of permafrost is influenced by a number of interrelated factors:
  - Proximity to water bodies is important; lakes are relatively warm so remain unfrozen throughout the year with a deep active layer.
  - Slope angle and orientation influence the amount of solar radiation, and therefore melting, freeze–thaw and wind.
  - Character of ground surface (different rock and soil types) can determine the degree and depth of permafrost; for example, dark compact rocks absorb a greater amount of solar radiation.
  - Vegetation cover can insulate the ground from temperature extremes.
  - Snow cover can slow the freezing process in winter and, in spring, delay the thaw and development of the active layer.

In conclusion, the depth of permafrost formation is affected by the energy balance at the surface, the thermal characteristics of subsurface material, and geothermal heat flow from below.

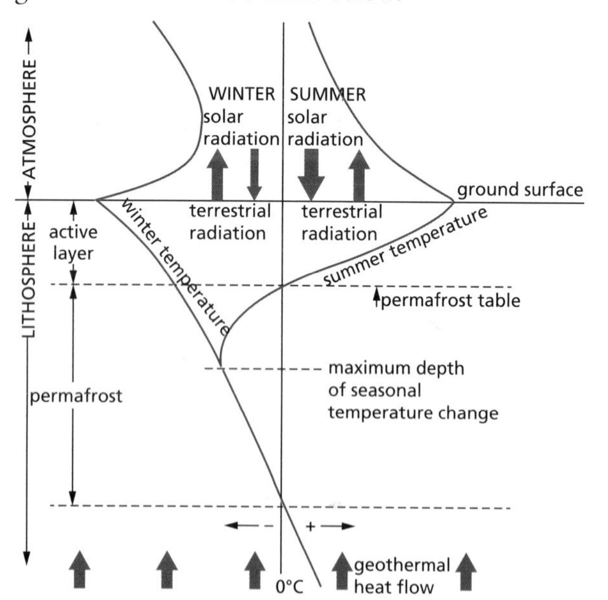

Figure 4.11 Factors affecting permafrost distribution

Periglacial processes and landforms

Cold climate environments develop distinctive geomorphology because of four basic processes.

- The nine per cent expansion of water on freezing; this causes frost shattering, which forms block fields and screes.
- The contraction and cracking of rapidly freezing soils in which ice wedges form, as well as frost heaving, which creates patterned ground.
- The migration of sub surface water to the 'freezing front' by suction, which causes the formation of segregated ice leading to the formation of ice lens, palsas and pingos.
- The mass movement of the active layer downslope largely by solifluction, which leads to lobes and terraces.

Of these processes only frost shattering occurs outside periglacial areas; the other three processes are associated with permafrost, and melting and movements within the active layer.

As Figure 4.12 shows, other processes such as wind and fluvial action are prevalent, but these are not exclusive to periglacial areas.

Ground ice features

Ice wedge polygons

Major ground ice features include networks of ice wedge polygons which are unique to periglacial areas. The process of frost cracking creates areas of irregular polygons, 5 to 30 m across, usually found on valley floors. When the active layer thaws, ice wedges can begin to form as water flows down into the cracks; it subsequently freezes and contracts, which means the ice wedge can build up over time. Larger ice wedges are usually a tapering shape 1 to 2 m wide and up to 10 m deep, extending down into the permafrost, taking over 100 years to form. Figure 4.13 shows the formation of ice wedges and polygons.

Patterned ground

Patterned ground is the general term for a range of features including circles, nets, polygons, steps and stripes. These features are unique to periglacial areas and are formed by a series of movements resulting from frost action. Frost push propels the stones upwards, while frost heave causes the stones to migrate outwards to form circles, which provides the basis for each of the patterns. The up doming of the circle created by heave

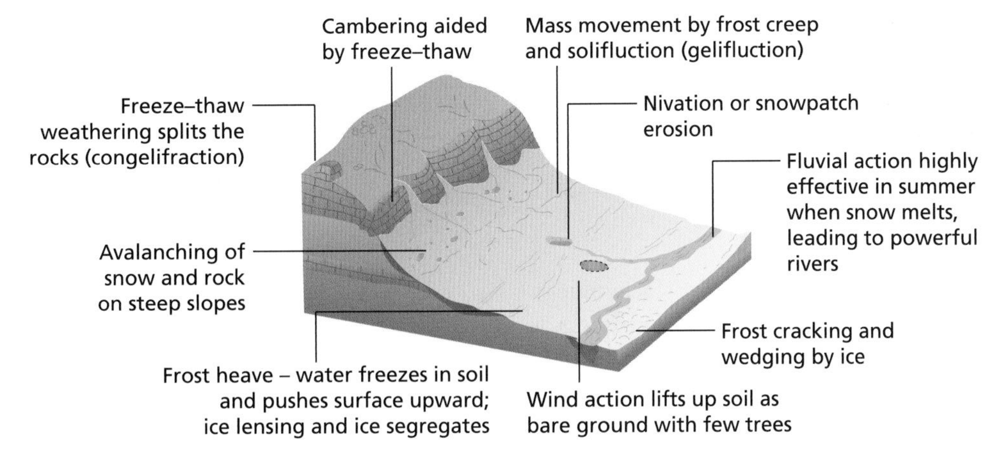

Figure 4.12 Processes forming periglacial landscapes

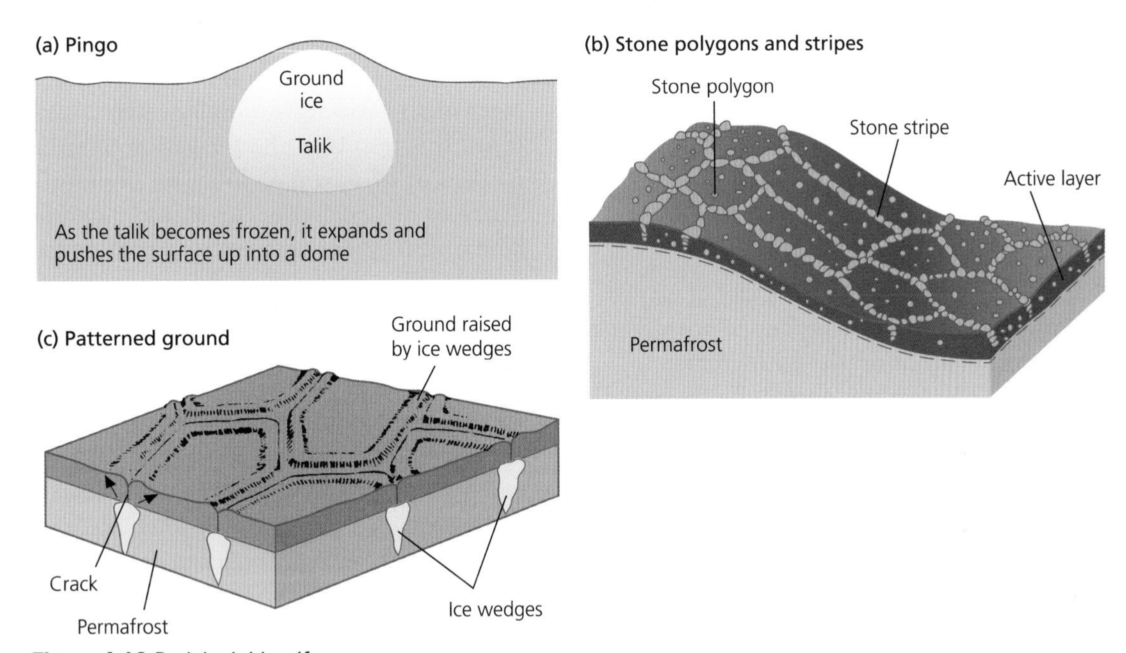

Figure 4.13 Periglacial landforms

means that larger stones roll outwards as a result of gravity, while finer sediments remain central, as shown in Figure 4.14. As a result of mass movement, stone polygons are elongated into stone nets and stripes, with a clear relationship between slope angle and the type of patterned ground (see Figure 4.13). Once slope gradients go beyond 30° patterned ground features no longer form and rock avalanches may occur.

Pingos

Pingos are another unique periglacial feature. They are ice core mounds 30 to 70 m in height and 100 to 500 m in diameter. The mounds can be either conical or elongated. The growth of an ice core forces up the overlying sediments, causing dilation cracks. Once the ice core is exposed at the surface it melts, causing the top of the pingo to collapse forming a crater, which may be filled with meltwater and sediments.

Figure 4.14 Stone circles in Svalbard, Norway

Two types of pingo occur:

- Open system (hydraulic pingos, or East Greenland type) are found in the discontinuous zone of permafrost or valley floors. Freely available groundwater is drawn towards the expanding ice core, so the pingo grows from *below* the ground.
- Closed system (hydrostatic pingos, or Mackenzie Delta type), which are associated with low-lying flat areas and only form in zones of continuous permafrost. They form from the *downward* growth of permafrost, often after a small lake is gradually enclosed with sediments. The loss of the insulating influence of the lake allows permafrost to advance, trapping the body of water and putting it under hydrostatic pressure and, ultimately, freezing it to push up the earth above it.

Smaller features similar to pingos, known as palsas, occur within peat beds.

The role of frost shattering

Freeze–thaw weathering puts pressure on any cracks in rocks and shatters them. While the process is not unique to periglacial environments, it occurs with greater severity within them. The features created by freeze–thaw include:

- **Block fields** (also referred to as felsenmeer, or 'rock seas') are accumulations of angular, frost-shattered rock, which pile up on flat plateau surfaces. They form *in situ*, created by frost heaving of jointed bedrock and freeze–thaw weathering.
- **Tors**, which 'crown' hill tops, stand out from block fields as they form where more resistant areas of rock occur, for example, less well jointed rock.
- **Scree or talus slopes** are formed when rock fragments fall and accumulate on the lower slopes or base of cliffs. The larger the material that makes up the slope, the steeper its angle of rest tends to be. Some research suggests slope is more a reflection of the rock type, length of slope and fragment shape, with shale/slates 'packing' together.
- **Pro-talus ramparts** are created if a patch of snow has settled at the base of a cliff. When rocks fall, as they are shattered by frost action, the snow patch acts as a buffer. The rocks settle at the base of the snow patch, leaving a rampart of boulders when the snow melts.
- **Rock glaciers** form when large amounts of frost-shattered rock mixes with ice. On the surface rock glaciers look like streams/fans or angular rocks, but they are conjoined with interstitial ice below and move slowly like glaciers, at rates of up to 1 m a year.

The role of mass movement

Frost creep and solifluction are the most important mass movement processes acting on slopes in periglacial environments.

- **Frost creep** is a very slow form of mass movement; material moves downslope by just a few centimetres per year, even on steeper slopes.
- Solifluction occurs in regions underlain by permafrost. During the summer months the active layer melts forming a mobile water-saturated layer; this results in the formation of either stone-banked or turf-banked lobes on slopes of 10–20°. Terraces or benches occur on more gentle slopes. The resulting deposits collect in the bottom of periglacial valleys and are known as head or coombe rock. Analysis of clasts (the stones in them) shows downslope orientation, and both angular and sub-angular shapes.
- **Asymmetric valleys** occur in periglacial environments – differential rates of solifluction and frost creep lead to one side of the valley being significantly steeper than the other. For example, in the northern hemisphere, south-facing slopes are more exposed to the Sun (insolation) and thaw more frequently, thus increasing soil moisture and promoting mass movement, leading to a less-steep slope.

The role of snow

The localised process of *nivation* occurs when both weathering and erosion take place around and beneath a snow patch. It is a common process in periglacial areas and leads to nivation hollows, which form at the base of a slope (these can initiate the formation of cirques).

Key term

Solifluction: The gradual movement of soil saturated with melt water down a slope over a permanently frozen soil in tundra regions.

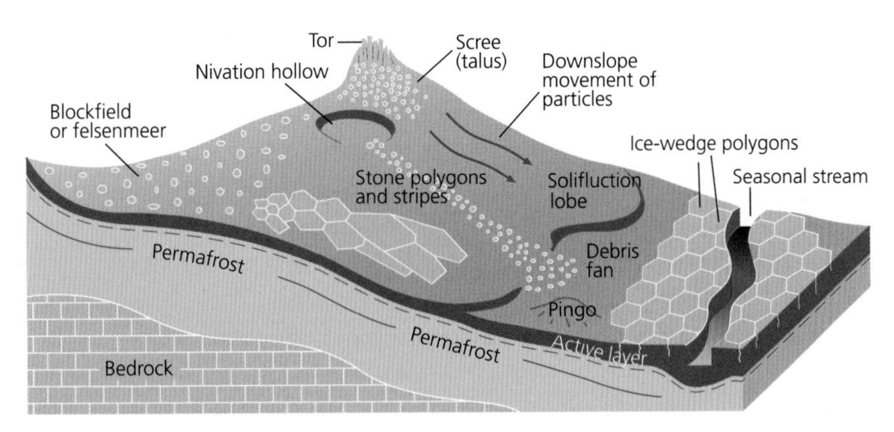

Figure 4.15 Periglacial landforms: the relationship of these features to the slope catena

The role of wind and meltwater rivers

Many periglacial areas are characterised by extreme aridity because most of the water is frozen and not available for plant growth (a physiological drought). The absence of vegetation provides abundant opportunities for wind action. In the Pleistocene ice ages, deposits of fine silt-sized sediment – formed on the extensive outwash plains (sandurs) from the great European and North American ice sheets – were blown southwards and deposited as loess over large areas of Europe and North America, forming soils of high agricultural potential.

A similar process is occurring today as winds blow fine material from the Gobi Desert to the loess plateau in northern China. The formation of loess is, therefore, not a process unique to periglacial areas.

Water erosion in periglacial areas is highly seasonal, occurring mainly in spring and early summer when surface snow and ice and the active layer melt leading to short periods of very high meltwater stream discharge. The drainage pattern near the margins of glaciers is typically braided (sometimes called anastomosing) because of the high amount of debris being carried by meltwater streams. Again, this is *not* a unique process as this type of drainage pattern is associated with any streams with variable discharge regimes, which carry large amounts of load.

Periglacial environments contain some unique landforms and some that can be found more widely. It is the assembly of these landforms within a tundra slope catena, and the occurrence of tundra ecosystems and soils, which creates the distinctive landscapes characteristic of periglacial environments.

Relict periglacial features

Periglacial features can form distinctive relict forms when the climate warms. In paraglacial conditions, just after the rapid melting of permafrost, a thermo-karst landscape can occur containing large areas of surface depressions and irregularly shaped lakes, so called because the depressions reminded scientists of the sink holes found in limestone (karst) landscapes.

In areas such as the UK, however, it is only comparatively recently that many 'mystery' features have been attributed to the periglacial conditions experienced during the last ice age, often by surveying areas of present-day periglaciation (the principle of uniformitarianism).

Table 4.4 lists some of the features attributed to periglaciation found largely in southern parts of Great Britain, an area beyond the extent of maximum glaciation. Other features are found in highland areas above the level of late-stage valley glaciations, including the Glyders in North Wales where, because of the height, some features are semi-active.

Key terms

Loess: A wind-blown deposit of fine-grained silt or clay in glacial conditions.

Catena: A connected series of related features, such features formed by periglacial processes which change down a slope.

Paraglacial: Rapidly changing landscapes which were once periglacial or glacial, but are moving towards not-glacial conditions.

Table 4.4 Possible relict periglacial landforms found in Great Britain

Landform	Location of possible examples
Ice wedge polygons – evidence of infilled wedges known as 'pipes'	Banks of the River Till, near Etal Northumberland
Patterned ground – some examples are still active but relict examples are found also	Tinto Hills in the Southern Uplands, Scotland (still active) On the Glyders, North Wales
Pingos found as circular depressions, almost like volcanic craters	Vale of Llanberis, North Wales North York Moors
Block fields are currently found on the highest summits above 1000 m	Cairngorms, Scotland Summit of Glyders and Snowdon, North Wales
Tors, relict residual features often surrounded by 'clitter' (loose boulders)	The Stiperstones, Shropshire Gritstone tors in the Pennines Summit of Glyders, North Wales
Screes, both active and relict	Slopes of Wastwater in the Lake District Slopes of the Glyders, North Wales
Solifluction terraces and lobes	College Valley, Northumberland
Solifluction deposits known as head or coombe rock	Scratchy Bottom near Durdle Door, Dorset Berwyn Mountain Valleys, Wales
Asymmetric valleys	Examples in Surrey and the River Exe, Devon
Dry valleys that may have been formed by meltwater flowing over permafrost in periglacial periods	Devils Dyke near Brighton White Horse Vale, Berkshire
Nivation hollows	Base of the Berkshire Downs
Loess deposits (brickearth)	Parts of East Anglia

Even though these features provide a very incomplete record, periglacial landforms are vital when mapping the extent of the ice cover as they provide evidence of areas that were not actually covered by ice, as they were beyond the margins.

Fieldwork opportunity

Create transects at 5 or 10 m intervals from the base to the top of scree deposits to carry out surveys of the size and shape of a sample of 25 scree deposits, measuring the angle of slope and the amount and type of vegetation coverage using sample quadrats.

Review questions

1 Use the National Snow and Ice Data Center (NSIDC) website to research data on the Greenland Ice Cores and explain how they are useful in constructing the chronology of Pleistocene ice-house–greenhouse cycles: http://nsidc.org.

2 Figure 4.16 shows different types of glacial environments. Research locations where you can find examples of the features shown in environments 1 to 6 outside of the UK.

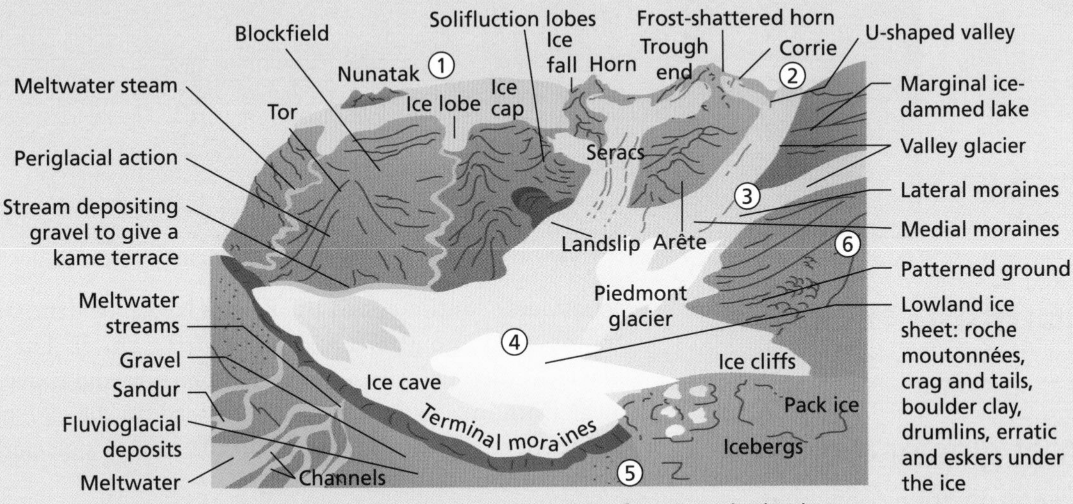

1. Ice sheet overall – knock and lochan scenery, e.g. Isle of Lewis, Sutherland
2. Highland corrie glaciation – late-stage re-advance, e.g. high peaks of the Cairngorms and North Western Highlands, Isle of Skye
3. Valley glaciation, e.g. valleys in the Cairngorms, Lake District and Snowdonia
4. Lowland glaciation from ice sheet deposition, e.g. Norfolk, south Shropshire and Cheshire, northeast Northumberland and the Vale of York
5. Proglacial environments (in front of the ice sheets), e.g. parts of the Midlands and the north coast of Devon
6. Periglacial environments in mountainous areas above the ice, e.g. North York Moors, or adjacent to the ice, e.g. much of southern Britain, south of a line from Bristol to London

Figure 4.16 Types of glacial environments

3 Use Figure 4.8 to describe and explain the distribution of present-day permafrost areas.

4 Explain why the climate shown in Figure 4.9 leads to permafrost conditions.

5 Study Figure 4.16. Make a list of the features you can see and briefly state how they were formed. Put a U beside them if you consider them unique to periglacial areas.

Further research

Research the Glaciers Online website to compare images of polar, temperate, polythermal and surging glaciers: www.swisseduc.ch/glaciers

Take a look at the website of the British Antarctic Survey (Natural Environment Research Council). The content is research related, with good background information and links: www.bas.ac.uk

Research the data available at the World Glacier Monitoring Service: www.geo.unizh.ch/wgms

Find out more about the Arctic and Antarctic at the Scott Polar Research Institute: www.spri.cam.ac.uk

5 The glacier system

What processes operate within glacier systems?

By the end of this chapter you should:

- understand why glacier mass balance varies and how this affects the dynamics of a glacier as a system
- be able to explain how the variety of glacier systems affects glacier movement and the variations in its rates
- know how processes operate in the glacier landform system and the variety of glacial landforms that occur at macro-, meso- and micro-scales
- understand how the assemblages of glacial landforms can be used to provide evidence for the extent of ice coverage in the Pleistocene.

5.1 The operation of glacier systems

The formation of glacier ice

Glacier ice is formed primarily from compacted snow, with smaller contributions from other forms of precipitation, such as hail or sleet, which freeze directly on top of or inside the glacier.

As Figure 5.1 shows, the first stage of formation of an ice mass is the accumulation of a permanent snowfield, either at high altitudes in mountainous areas or high latitude polar regions. There must be low numbers of positive degree days (when temperatures are above 0 °C) so that the snowfield can survive the summer melt and gradually increase each winter.

The lower layers of this granular snow (density 0.1 g cm^{-3}) become increasingly compressed to form névé or firn. Pressure-induced melting and refreezing of water filling gaps between individual ice crystals is the cause of this increasing density. As the snowfield increases in size, the process continues and the deeper layers, or firn, are transformed into glacier ice (density 0.9 g cm^{-3}). The glacier is then deformed by further pressure and moves away from the centre and flows outward (in the case of an ice sheet/cap) or downward (in the case of a glacier) by extrusion flow.

Figure 5.1 The formation of glacier ice

Timescale for the process

- The transformation from snowflake to firn can be very quick in more temperate areas (a few days) but much slower in polar areas (over a decade).
- The final stage from firn to glacial ice may take as little as 25 years but up to 150 years in polar areas such as Greenland.
- Overall rates of transformation from snow to ice can be, on average, as little as 100 years in some temperate areas, but can take up to 4000 years in Antarctica.

Inputs and outputs of the glacier system

To understand how glaciers behave it is helpful to view them as systems with inputs and outputs, and interactions with other systems, such as the atmosphere, oceans, rivers and landscape. The glacier system is driven by inputs of

Key terms

Névé (or firn): Crystalline or granular snow, especially on the upper part of a glacier, where it has not yet been compressed into ice.

Extrusion flow: The theory that glacier ice flows faster at depth.

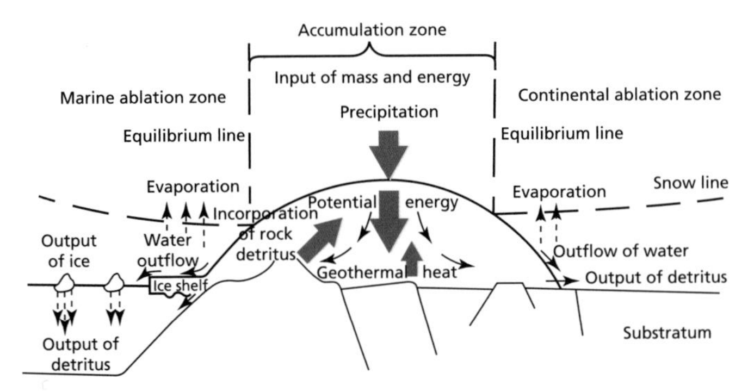

Figure 5.2 How mass and energy fluxes work in an idealised glacier system

energy from the Sun (which evaporates water from the oceans to create air masses, which can produce snowfall). Mass enters the systems in the form of snowfall and rock debris. As this mass generally occupies an elevated position in the Earth's gravitational field, this mass has potential energy, which is expended as the glacier flows downslope. The energy expended is used to warm or melt ice and then must be dissipated from the system in the form of heat and water. As this is going on, potential energy is turned into work, transferring ice and rock from the highlands towards lower levels and the oceans.

Glacier mass balance

Mass balance is defined as the gains and losses of the ice store in the glacier system (see Figure 5.3). *Accumulation* results from direct snowfall and other precipitation, ice, blown snow and avalanching from slopes above the glacier surface. The snow and ice are then transferred down valley by glacier movement until they reach lower areas where they are lost to the system either by melting, evaporation (sublimation) or the break away of ice blocks and icebergs (calving), processes collectively known as *ablation*. At the same time, there is an input and output of rock debris. Rock debris supplies come from weathering and erosion of

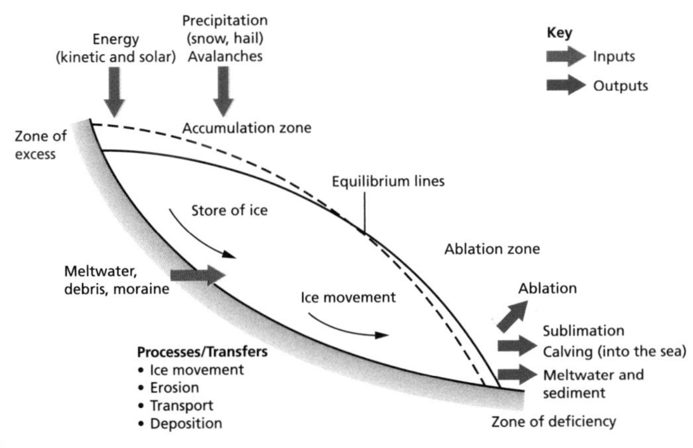

Figure 5.3 Glacier mass balance as a system

slopes above the glacier; it is transported and eventually deposited as a further glacier output. As Figure 5.3 shows, you get more accumulation than ablation in the upper part of the glacier, and more ablation than accumulation in the lower part of the glacier. The place where accumulation and ablation balance each other out is known as the glacier's equilibrium point.

> **Key term**
>
> Equilibrium point: Where losses from ablation are balanced by gains from accumulation in a glacier.

Glacier systems are dynamic; the ratios of inputs to outputs vary continually over both short- and longer-term timescales. The adjustment of the glacier system to changes is reflected by variations in the mass balance.

The mass balance year runs from autumn to autumn when ice masses are generally at their smallest volume. Figure 5.4, a model of the annual mass balance of a typical glacier, shows that theoretically:

- If accumulation exceeds ablation, a usual situation in the winter period when there are very few positive degree days, the glacier increases in mass, i.e. a *positive regime* in the glacial budget.
- Conversely in summer, when there is more ablation than accumulation (rising temperatures) the glacier has a *negative regime* in the glacial budget. If the accumulation in winter is equalled by the ablation in summer, then the annual net balance is zero and the glacier is likely to be at a still-stand. Even within the time span of the annual budget, the regimes have some visual impact on the size of glacier mass.
- A positive regime or mass balance causes the glacier to grow and therefore advance at the snout.
- A negative regime or mass balance causes the glacier to thin/shrink/downwaste and therefore the position of the snout begins to retreat.

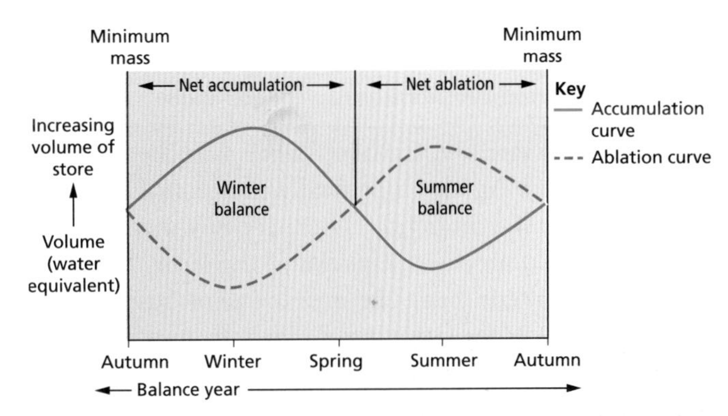

Figure 5.4 Annual mass balance of a typical glacier

Longer term the situation is more complex. The annual net balance can be calculated for each year and then, by looking at readings over longer periods of time (usually over a decade), trends can be summarised. It is these longer-term trends that determine the 'health' of the glacier and whether it will significantly advance or retreat, and if thinning/retreating contributes to increased concerns over global rises in sea level. Currently it is estimated that nearly 75 per cent of the world's ice masses are experiencing 'rising trends' in their net negative balances, almost certainly as a result of short-term climate change (the average global increase in temperatures is 0.6 °C in the last decade) and are thus thinning, melting and retreating.

In the past much of the evidence of longer-term fluctuations in the mass balance of glaciers was carried out using old photographs and maps, augmented by field surveys of the lateral and end margins (snout): see the box on Fluctuating glaciers in South Georgia.

In recent decades the interpretation of aerial photographs taken over a period of time has also been used to look at the 'health' of glaciers – to see evidence of thinning and retreat (a feature of a 'sick' glacier) – as it is possible to identify the extent of zones of accumulation and ablation.

In the last decades polar-orbiting satellites have revolutionised the study of ice sheets and glaciers, especially in remote locations. These satellites provide regular and comprehensive observations from space. Three main techniques are used:

- **Altimetry** provides repeated measurements of ice sheet surface elevation, which can be used to determine whether the ice sheet is thickening/thinning or expanding/shrinking.
- **Gravimetry** measures the gravitational attraction of the ice sheet from which changes in ice sheet mass can be inferred.
- **Mass budget method** compares the amount of snow accumulation on the ice sheet (looking for five years of fresh white snow) and the amount of meltwater or iceberg calving density leaving the ice sheet.

In 2010 a new satellite, CryoSat-2, was launched by the European Space Agency specifically to monitor the ice sheet situation in Antarctica. There are now five years of very detailed measurements that, long term, are of growing global concern. The measurements suggest that the recent rate of ice loss is approximately double the twenty-year average, across measurements from both the west and east Antarctic regions.

The Greenland Ice Sheet

The Greenland Ice Sheet is one of the world's two remaining ice sheets. It currently covers an area of 1.7 million km^2. As it contains more than 2.5 million km^3 of stored ice, it has a huge potential impact on other earth systems including the atmosphere and oceans. In the centre of the ice sheet the ice is over 3 km thick, so its weight isostatically depresses the earth's crust by about 1 km in depth.

In response to recent climate warming (very marked in the Arctic area) a number of changes are occurring to the mass balance of the ice store (figures indicate change per year in water km^3):

- Accumulation from snowfall in the central area: +520 km
- Ablation by melting and edges: –290
- Ablation by calving icebergs: –200
- Ablation by sublimation: –60
- Total ablation: –550
- **Mass balance: –30**

Cumulative study of research findings suggests that overall the Greenland Ice Sheet has a negative mass balance with an accelerating rate of increase. There is huge uncertainty as to how the mass balance of the huge ice store could change in the future largely because of feedback loops. Positive feedback loops amplify the speed of any processes whereas negative feedback loops diminish their impact (see page 91).

An example of positive feedback could occur if as a result of anthropogenic processes the Greenland Ice Sheet melted catastrophically, for example, snow melt on the edge of the ice sheet would lead to widespread occurrence of bare ground. This reduces the albedo of the surface (of snow cover) so accelerating warming of the land. Moreover, increased melting could lead to the release of huge quantities of methane, a naturally occurring greenhouse gas stored in tundra zones, accelerating ablation. If the Greenland Ice Sheet was to melt completely the height of the land would be so low that surface temperatures could be much warmer although isostatic recovery would counteract this and the amount of sea level rise. So the negative mass balance accelerating would have a huge impact on sea levels with a possible rise of 6 m globally. However, the rapid melting of the ice sheet would upset the thermohaline circulation cutting off warm water currents from the Gulf Stream from reaching countries such as the UK. This negative feedback occurs as the absence of warm currents diminishes the impacts of climate warming.

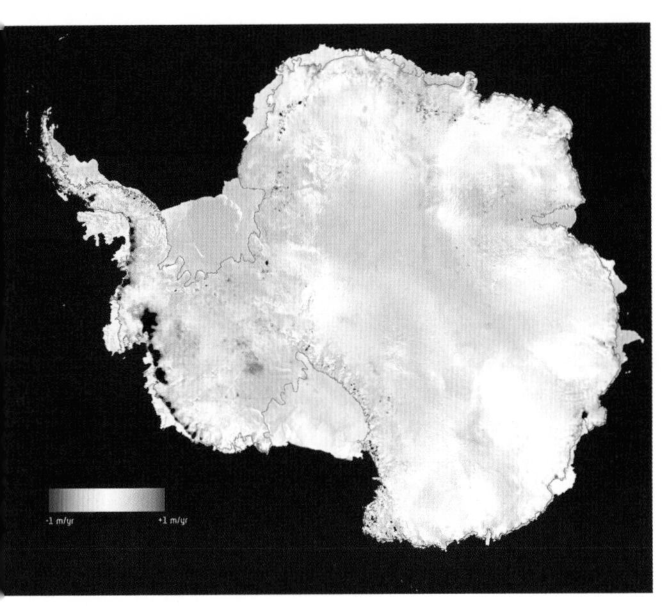

Figure 5.5 The latest melting situation in Antarctica

Similar research is also taking place in other remote regions, such as Greenland, Svalbard and northern Canada (Baffin and Ellesmere Islands).

The USGS Benchmark Glacier Research project measures changes in the mass balance for four benchmark glaciers: Gulkana and Wolverine in Alaska, South Cascade in Washington State and, most recently, Sperry in Montana (since 2005). These four research sites were unified into a single project with common strategies for field work analysis and research methodology to enable comparison between the glaciers and to measure each glacier's exposure to climate change. This project again provides long-term records of mass balance trends at a continental scale.

Key term

Benchmark glacier: A designated glacier in which the accumulation and ablation are measured annually using standardised techniques to monitor the impacts of climate change. USGS, for example, studies five such glaciers currently with more chosen for the future to reflect a variety of locational scenarios.

Fluctuating glaciers in South Georgia

The extent of the ice in South Georgia has fluctuated over time. Today just over 50 per cent of the island is covered in permanent snow and ice, largely on the windward south coast, which is more extensively glaciated and exposed to cold blasts from Antarctica.

Geological evidence has shown that during the coldest periods of the Pleistocene Ice Age, an ice cap covered all the lower ground and even extended offshore to the edge of the continental shelf, with only the very highest mountains standing above the surface of the ice (as nunataks).

- Following the last glacial maximum, around 22,000 years ago, glaciers retreated within the fjords where they halted around 12,000 years ago, forming very well-defined morainic ridges.
- Further thinning and retreat followed, but then a re-advance occurred between 3600 and 1100 years ago, and again during the late-nineteenth century.
- Although there has been a very marked general retreat since then, confirmed by field research and historic photographs, there was a brief re-advance between 1925 and 1935.
- Today there is a period of very marked retreat, creating inlets, lagoons and beaches where previously none existed.

At Gold Harbour (Figure 5.6) the lowest section of the Bertrab Glacier has disappeared entirely, exposing a large rock step which, 30 years ago, was covered by a spectacular icefall.

Figure 5.6 Features at Gold Harbour, South Georgia

Skills focus: Investigating cumulative mass balances

Figure 5.7 shows single season and longer-term cycles of the four benchmark glaciers. While it is possible to identify the general trends in annual mass balance and calculate the mean rate for five- or ten-year periods, a more useful measure is to draw a graph of cumulative mass change for each glacier so that the trends can be analysed more effectively.

Each glacier is measured relative to the start of mass balance measurements for that site (for the Sperry Glacier this was 2005).

Analysing the trends in seasonal cycles

1 Identify the glacier that has the least range in seasonal cycles.
2 Identify the glacier that starts with a positive net balance (when winter snowfall accumulation exceeds summer ablation).
3 Identify one year in which all four glaciers showed a negative net balance.
4 By visually looking at the graphs, which glacier do you think had the greatest decrease in cumulative balance?

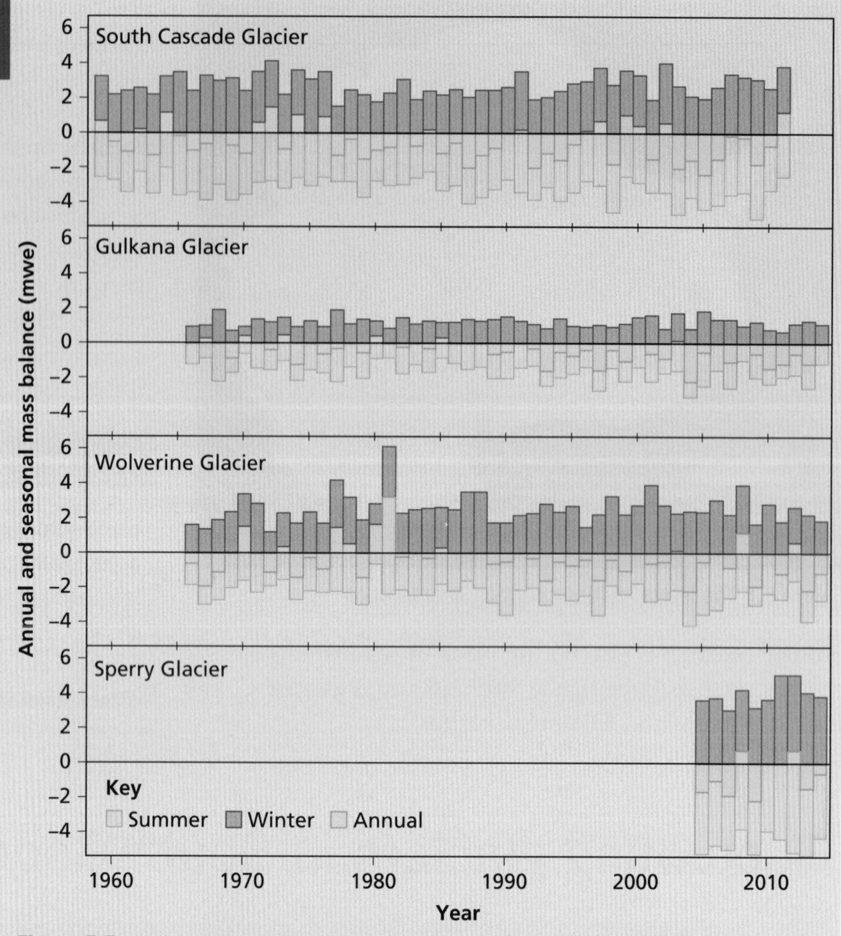

Figure 5.7 Mass balance measurements for the four benchmark glaciers

Cumulative annual balance

1 Copy and complete Table 5.1.
2 Calculate both the mean and total winter accumulation and summer ablation amounts, and comment on your results.
3 Draw an annotated graph to show trends in cumulative annual balance using the *x*-axis for cumulative annual balance amounts and *y*-axis for years 2005–14. Comment on your results.

Table 5.1 Mass balance measurements for the Sperry Glacier in Montana, USA (Source: www.usgs.gov/climate_landuse/clu_rd/glacierstudies/results.asp)

Year	Winter accumulation	Summer ablation	Annual net balance	Cumulative annual balance
2005	+3.6	−5.6	−1.9	
2006	+3.7	−4.9	−1.2	−3.1
2007	+3.2	−5.1	−1.9	−5.0
2008	+4.1	−3.7	−0.4	−4.6
2009	+3.4	−5.3		
2010	+3.5	−3.8		
2011	+3.7	−5.0		
2012	+5.2	−5.2		
2013	+5.2	−5.8		
2014	+3.7	−4.3		

All figures are in MTE (metres of water equivalent)

5.2 The processes that cause glacier movement and variations in rates

Glacier movement

The fundamental cause of ice movement is gravity. Ice moves downslope from higher altitudes to lower areas either on land or at sea level. As the ice mass builds up over time in the accumulation zone, the weight of the snow and ice exerts an increasing downslope force due to gravity (known as shear stress). Shear stress increases as the slope angle increases and, once the shear stress is great enough to overcome the resisting forces of ice strength and friction, the glacier ice pulls away and moves downward away from the zone of accumulation. The momentum of the ice's movement towards the ablation zone prevents further build-up, thereby maintaining the glacier at a state of dynamic equilibrium with the slope angle. This forward movement of glacial ice towards the margins/snout, occurs regardless of whether the glacier as a whole is advancing or retreating.

Thus the speed of glacier movement forward depends on the degree of imbalance, or the gradient, between the zone of accumulation and the zone of ablation.

- Warm, wet-based glaciers in temperate maritime climates experience greater snowfall in winter and more rapid ablation in summer; therefore the imbalance between accumulation and ablation zones is greater, so the glacier ice must move downslope more rapidly to maintain the equilibrium with the slope angle.
- In cold-based, polar glaciers the slower rates of accumulation, and especially ablation, result in a smaller gradient of equilibrium and slow ice movement. Figure 5.8 demonstrates glacier equilibrium.

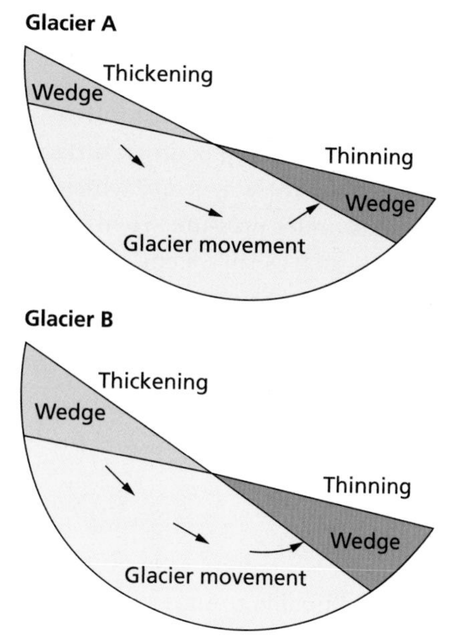

Figure 5.8 Glacier equilibrium

Further contrasts in movement rates occur because of the contrasts in the nature of the substrate (base) on which the glacier rests and its own base. This determines the relative importance of the three processes which facilitate glacier movement: basal sliding, internal deformation and subglacial bed deformation.

Basal sliding

This relates to the presence of meltwater beneath a glacier. This type of ice movement applies to warm-based glaciers; it cannot occur where a glacier is frozen to its bed. The meltwater acts as a lubricant reducing friction with both the entrained debris and with the underlying bedrock (this is known as slippage). It can account for up to 75 per cent of glacier movement in warm-based glaciers.

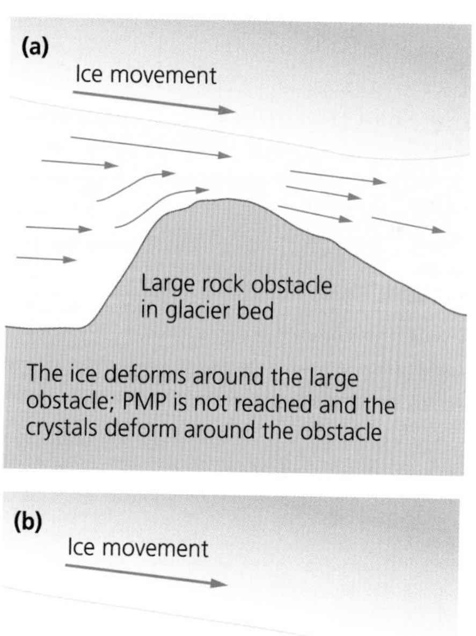

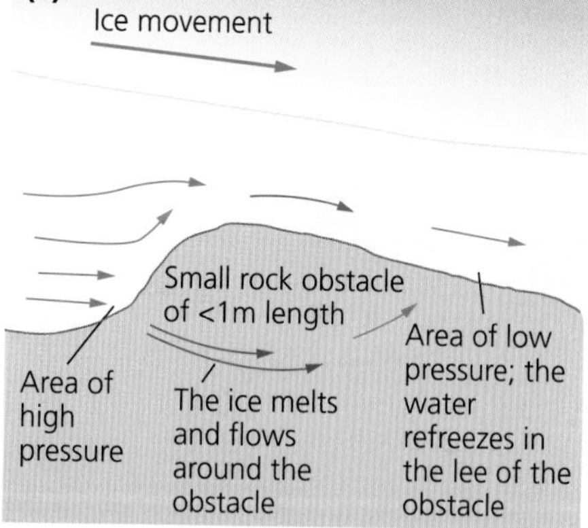

Figure 5.9 The processes that contribute to basal sliding

Two specific processes enable glaciers to slide over their beds:

- **enhanced basal creep**, whereby basal ice deforms around irregularities on the underlying bedrock surface
- **regelation creep**, sometimes known as *slip*, which occurs as basal ice deforms under pressure when encountering obstructions such as rock steps. As the glacier moves over the obstruction the pressure on the basal ice will increase up glacier, leading it to reforming in a plastic state as a result of melting under this pressure. Once the glacier has flowed over the obstruction the pressure is lowered and the meltwater refreezes.

Internal deformation

Cold-based, polar glaciers are unable to move by basal sliding as their basal temperature is below the pressure melting point. They therefore move by internal deformation, which has two main elements:

- **intergranular flow**, when individual ice crystals deform and move in relation to each other
- **laminar flow**, when there is movement of individual layers within the glacier.

The deformation of ice in response to stress is known as ice creep and is a result of the increased ice thickness and/or the surface slope angle. Where ice creep cannot respond quickly enough to the stress, ice faulting occurs, which manifests itself in a variety of crevasse types at the surface.

When the slope gradient is increased, there is acceleration of ice and extensional flow. Such conditions can occur in the zone of accumulation and can result in an ice fall. Near the ablation zone, where there is usually a reduction of slope angle, the ice decelerates and there is compressional flow, which leads to a whole series of thrust faults in the ice, with closed-up crevasses.

Transverse crevasses cut across the glacier at approximately right angles to the direction of glacier flow. These can be very deep and wide, and result from ice faulting at depth within the ice mass. Changes in the width of the valley can also lead to ice fracturing, for example forming longitudinal crevasses that are orientated parallel to the flow direction of the ice, as the ice masses spread out laterally in a less-constrained environment.

Radial crevasses can form in a splayed pattern at the snout of the glacier, where ice spreads out in a broad lobe.

Marginal crevasses form near the sides of a glacier as a result of differential movement within the glacier as friction on the sides of the valley slows ice movement relative to ice near the middle of a glacier.

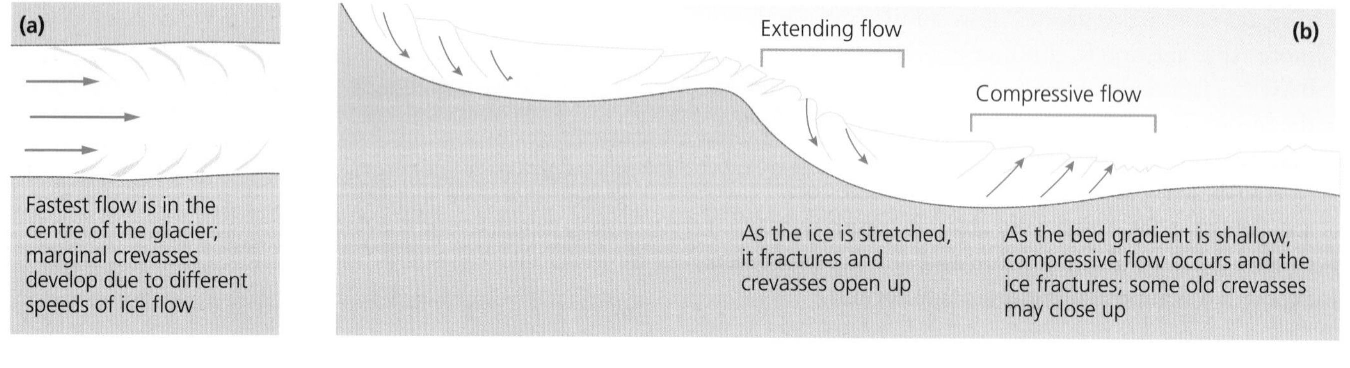

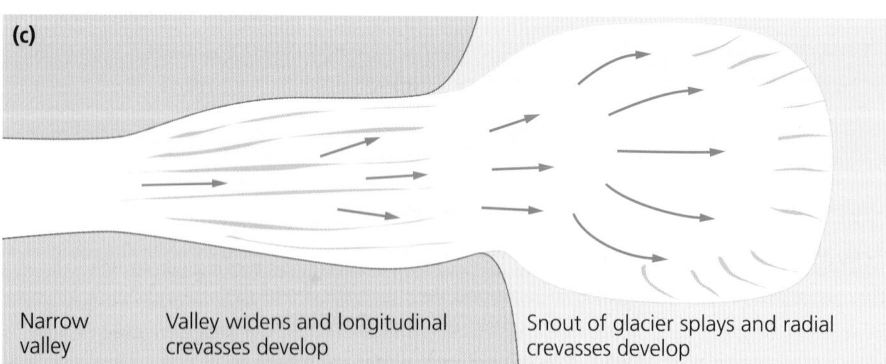

Figure 5.10 The types of crevasses prevalent in a glacier or ice mass

Subglacial bed deformation

This occurs locally when a glacier moves over relatively weak or unconsolidated rock, and the sediment itself can deform under the weight of the glacier, moving the ice 'on top' of it along with it. Locally this process can account for up to 90 per cent of the forward motion of glacier ice, often in polythermal outlet glaciers as in Iceland.

Skills focus: Glacier movement

Table 5.2 Contribution of different types of flow to movement of glaciers (Source: *Glaciation and Periglaciation Advanced Topic Master*, J. Knight, Philip Allan)

Glacier	Country	Basal sliding (%)	Internal deformation (%)	Ice thickness (m)
Aletsch	Switzerland	50	50	137
Tuyuk Su	Kazakhstan	65	35	52
Salmon	Canada	45	55	495
Upper Athabasca	Canada	75	25	322
Lower Athabasca	Canada	10	90	209
Blue	USA	9	91	26
Vesl-Skautbreen	Norway	9	91	50
Meserve	Antarctica	0	100	80

1 Research the glaciers shown in Table 5.2 and try to classify them by thermal regime (warm based, cold based, polythermal).
2 What relationships can you find between thermal regime and types of flow?

Using either a scatter graph (with regression lines) or a statistical correlation technique, assess the strength of the relationship between ice thickness and the percentage of movement caused by the two processes. Comment on your results.

Velocity of glacier ice

The overall velocity of a glacier comes from a combination of the processes described above. Warm-based glaciers have a greater overall velocity of ice movement than cold-based glaciers because of the addition of basal sliding to internal deformation and flow, which affect both types. Even greater velocities are reached when a warm-based glacier moves over deformable sediment.

Observations of glaciers across the world have shown that great variations in the total velocity of glacier ice occur, with most glaciers having velocities between 3 m and 300 m per year. A number of factors have an impact on the rate of movement:

- altitude, which affects the temperatures and precipitation inputs
- slope, which can be directly related to flow – steeper slopes lead to faster speeds
- lithology, which can affect basal processes and the possibility of subglacial bed deformation
- size, which can affect the rapidity of response
- mass balance, which affects the equilibrium of the glacier and also whether it is advancing or retreating.

Table 5.3 shows the rates of movement of a variety of glacier types.

The table shows that large outlet glaciers, also known as ice streams, have very fast velocities as they move vast quantities of ice outwards from ice caps and ice sheets, under great pressure.

The highest velocities of all occur during a glacier surge. Periodically glaciers collapse when the mass and slope angle of the ice builds up to a critical level within the accumulation zone. At the time of a surge – a rare

Table 5.3 How fast does ice move? Data from a variety of glacier types

Glacier	Type	Velocity (m/day)	Velocity (m/yr)	Category of movement
Jakobshavn Glacier, Greenland	Tide water, outlet glacier	19.0	Approx. 700	Fast
Columbia Glacier, Alaska	Outlet glacier	20.0	Approx. 720	Fast
Austdalsbreen, Norway	Standard warm based	0.15	Approx. 60	Normal
Saskatchewan Glacier, Canada	Standard warm based	0.35	Approx. 120	Normal
Lewis Glacier, Kenya	Corrie glacier	0.01	Approx. 3	Slow
Variegated Glacier, Alaska	Surging glacier	0.60–65.0	Variable, up to 2–3 km	Surge type

Key term

Glacial surge: This occurs where flow instabilities result in dramatic increases in glacier velocity.

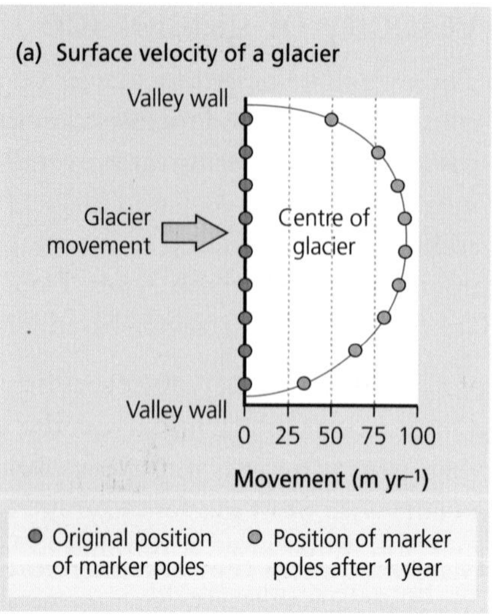

Figure 5.11 Measuring glacier velocities

event as only four per cent of all the world's glaciers are prone to surging – the ice races forward at velocities between 10 and 100 times the normal velocity.

Glaciologists have carried out research into glacier velocities around the world. On the surface they use a series of stakes, so they can survey the position of these stakes on an annual basis (Figure 5.11). It is far more difficult to measure the rates of glacier movement at various depths. In the 1940s scientist Max Perutz developed a series of flexible poles in order to do this and noted that movement was actually fastest just below the surface. In both cases the effect of friction of the glacier bottom and valley sides can be seen.

5.3 The glacier landform system

The movement of the ice allows the ice sheet or glacier (ice masses) to pick up debris and erode at its base and sides, as well as to transport and modify the materials it is carrying. The more rapid this movement is, the more likely the glacier is to transform the landscape. Conversely, stagnant ice, a frequent 'state' of lowland ice sheets, is more likely to 'protect the landscape' and only reshape it by dumping huge amounts of debris.

A combination of both direct ice action and indirect impacts – such as the formation of fluvio glacial features by meltwater, disturbance of pre-existing drainage systems, and complex ice induced sea level changes – shapes glacial landscapes.

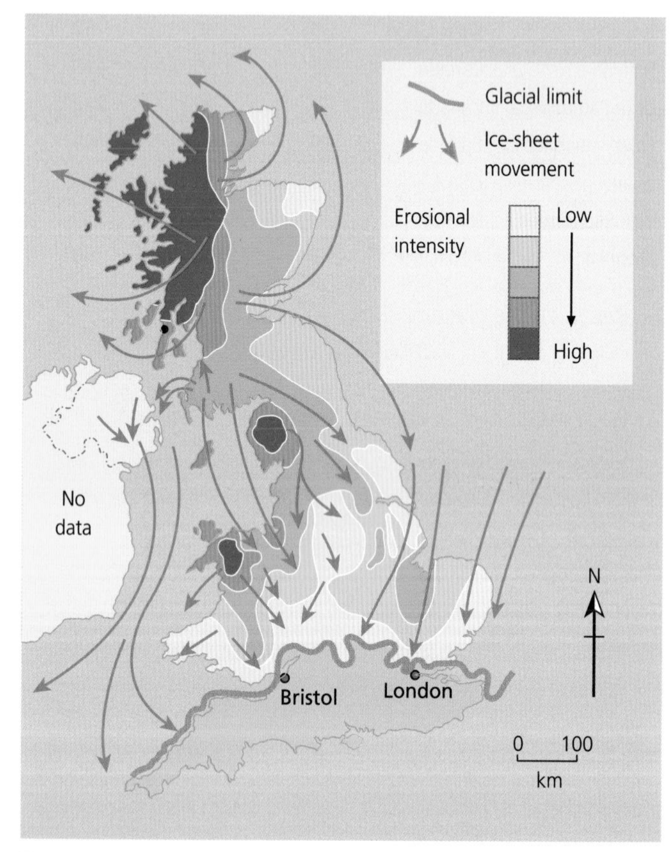

Figure 5.12 How erosional intensity varied across the UK in the Devensian glaciation

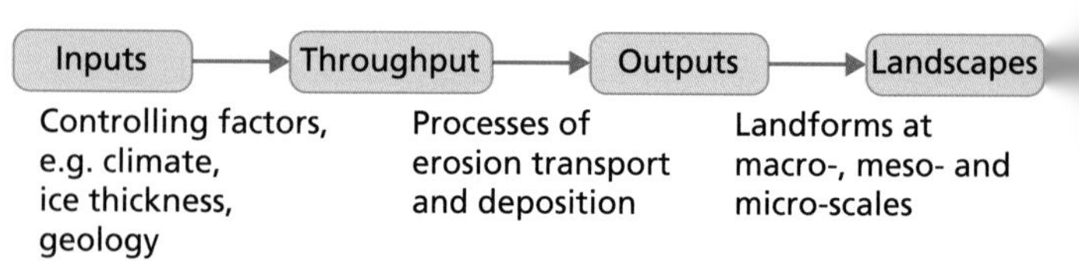

Figure 5.13 How the glacier landform system works

Glacial processes – a brief guide

Glacial erosion

Glacial erosion is the removal of material by ice and meltwater and involves a combination of several processes:

- **Abrasion** by individual clasts (stones), which leads to micro-features such as striations and chatter marks. Additionally rock flour (grade sizes under 0.1 mm in diameter) polishes the underlying rocks by 'sand paper' action.
- **Plucking** is often referred to as glacial quarrying. Quarrying is a two-stage process with the initial widening of the joints by fracture and the subsequent entrainment of any loosened material. The importance of plucking as a process is clearly very dependent on rock type and the incidence of pre-existing joints.
- **Fracture and traction** occur as a result of the crushing effect of the weight of moving ice passing over the rock and variations in pressures lead to freezing and thawing of the meltwater (basal melting), which aids the plucking process.
- **Dilation** occurs as overlying material is moved, causing fractures in the rock parallel to erosion surfaces as the bedrock adjusts to the unloading.
- **Meltwater erosion** can be both mechanical (similar to fluvial erosion, except that the water is under hydrostatic pressure) and chemical, whereby glacial meltwater can dissolve minerals and carry away the solutes, especially in limestone rocks.

Glacial debris entrainment

Entrainment is the incorporation of debris on to or into the glacier from supraglacial or subglacial sources.

- Supraglacial sources include material falling from hillsides being washed or blown on to the glacier from the surrounding land, plus atmospheric fall-out such as volcanic ash (a common feature on Icelandic glaciers).
- Subglacial sources include material eroded from the glacier bed and valley walls, material frozen to the base from subglacial streams, as well as englacial material that has worked its way down through the glacier or ice sheet.

Glacial sediment transportation

For ice sheets, most debris is transported subglacially via the basal layer. For valley glaciers there is more transport by englacial and supraglacial debris as a result of more frequent ice contact at their lateral margins. As pebbles (clasts) are transported they come into contact with each other and are ground down (comminuted) by a process similar to attrition in rivers.

Key terms

Supraglacial: Debris transported on the surface of the glacier.

Subglacial: Debris transported beneath the glacier.

Englacial: Debris transported inside the glacier.

Glacial sediment transport therefore occurs horizontally and vertically through glaciers, by the movement of the ice itself, meltwater transporting sediment through the glacier drainage system or by glacial deformation of subglacial and proglacial sediments.

Glacial deposition

Glacial deposition occurs when material is released from the ice at the margin or the base of a glacier. Deposition may occur directly on the ground (ice contact) or sediments may be released into meltwater. Deposition mechanisms include: release of debris by melting or sublimation of the surrounding ice, lodgement of debris by friction against the bed, deposition of material from meltwater, and disturbance and remodelling of previously deposited sediments.

Scales of glacial landforms

The imprint of glaciation can therefore be seen through glacial landforms at several scales, which are the result of both erosion and deposition. These scales are summarised in Table 5.4.

Note that this is only a guide; landforms often come in many different sizes.

Table 5.4 Selected glacial landforms classified by scale

Macro-scale	Ice sheet eroded knock and lochan landscapes, cirques, arêtes and pyramidal peaks, glacial troughs, ribbon lakes, till plains, terminal moraines, sandurs
Meso-scale	Crag and tail, roches moutonnées, drumlins, kames, eskers and kame terraces, kettle holes
Micro-scale	Features such as striations, glacial grooves and chatter marks, erratics

Key concept: Glacier process environments

Peter Knight identified how assemblages of landforms develop in glacial process environments at various locations within an ice mass as different parts of the glacier are associated with specific sets of geomorphic processes. Commonly identified glacial process environments include:

- subglacial geomorphology (beneath the glacier)
- glacier margin geomorphology, either lateral (at the sides of the glacier) or terminal (at the end of the glacier)
- proglacial geomorphology and meltwater landscape geomorphology
- paraglacial landscapes where, after glacial retreat, surface features have to rapidly adjust to their new post-glacial environments. Deglaciation causes instability (leading to massive landslipping) and rapid erosion lasting until a new equilibrium is established between any surface features and the post-glacial process environment
- periglacial process geomorphology in permafrost areas adjacent to ice cover.

Figure 5.14 Marginal zones of glaciers in Svalbard

Again, the rapid melting of permafrost can lead to a transitional paraglacial stage before an equilibrium occurs and post-glacial modification takes place.

While characteristic assemblages of features can be recognised, a further dimension of difference is added by contrasts between the ice mass types, especially between contrasting thermal regimes.

Becoming an ice detective

The distinctive assemblages of landforms left behind after glaciation are very useful when trying to reconstruct the exact position and extent of the ice cover – this is known as inversion modelling. It is not an exact science and morphological mapping of areas of past glaciation has to be combined with an analysis of any deposits (known as fabric analysis).

Hutton's principle of uniformitarianism is the key to understanding and identifying the features found in relict landscapes resulting from past glaciation. The principle states that 'the present is the key to the past'. By looking at present-day environments such as Svalbard, Greenland or northern Canada, glaciologists can see features actually in the process of formation. For example, ice-cored round hills (pingos) in the Mackenzie Delta in northern Canada were finally linked to some mysterious rounded hills with collapsed craters in the North York Moors, now also identified as possible former pingos.

It is also worth remembering the principle of equifinality, which states that a particular feature can be formed in a number of ways, when trying to explain the formation of drumlins (see page 80).

A further complication is that most glaciated regions are polycyclic/polygenetic; that is, they are the product of several episodes of glaciation and may have been modified under periglacial, paraglacial, interglacial or post-glacial conditions. Glacial landscapes have been considerably modified by subaerial weathering, mass movement and fluvial action even since the Loch Lomond Stadial (page 48).

Figure 5.15 shows a student's map of the Nant-y-Llyn Valley in the Berwyn Mountains, Wales. It is comparatively easy to chart the direction of ice flow from the corrie southward, but harder to work out the extent and height to which the glacial trough was filled with ice. On some U-shaped valley sides it is possible to identify a trimline – below this line you can see evidence of glacial abrasion, such as striations and polished rock surfaces, whereas above it only block fields and scree deposits occur, both of which are evidence of periglaciation.

Finding the extent of the ice cover during the last glaciation (the Loch Lomond Stadial) involved a trip down valley to look for a possible terminal moraine. It is always worth using a 'drift' geology map to help in the identification of superficial deposits such as moraine.

Fieldwork opportunity

Mapping and analysing the features of a glaciated valley forms a very manageable group fieldwork project. In some cases, as in the College Valley in the Cheviots, Northumberland, there is even the need to resolve whether the valley was actually glaciated at all during the Loch Lomond Stadial.

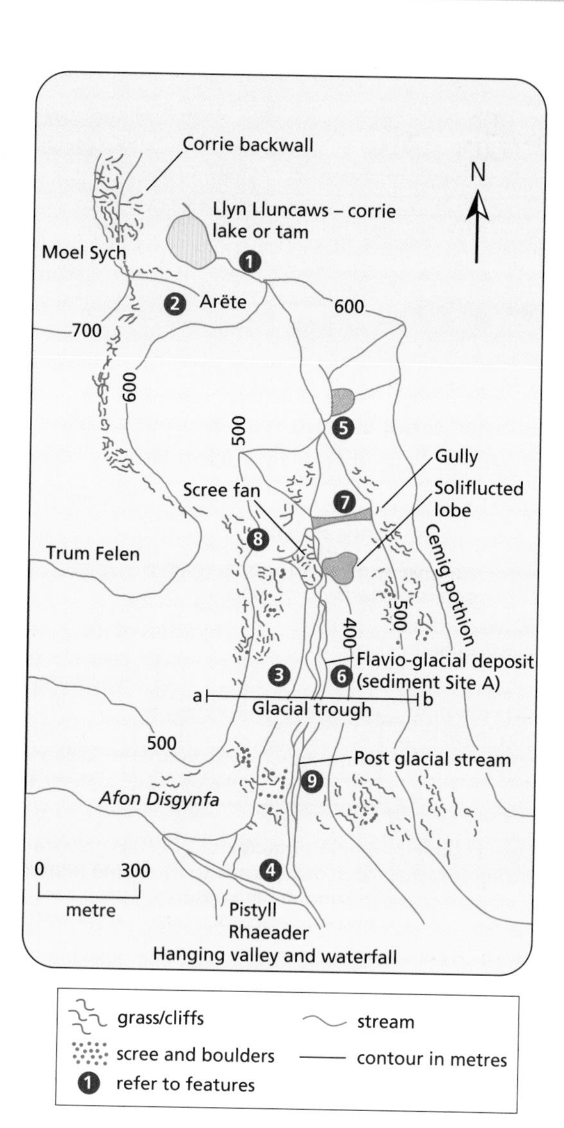

Figure 5.15 Student map of Nant-y-Llyn Valley, Wales

Review questions

1. Use a search engine to add further examples of glaciers and their speed of flow (to make 12 in total) to Table 5.3 (be sure to convert any measurements to common units). Use a statistical test, such as Student's t-test, to compare rates of glacier movement.
2. For each of the glacier process environments detailed on page 70, find images on the internet that illustrate the environment and its characteristic landforms (Peter Knight's Glacier Pages are a useful starting point: www.petergknight.com/glaciers/). Annotate your images with information about the key features.
3. Study Table 5.4 and test out the scale classification for selected features. How realistic is it to use macro-, meso- and micro-scales when writing about glacial landforms?
4. Research Hutton's concept of uniformitarianism and explain briefly how it can be used when studying periglacial and glacial environments.
5. Draw a simple field sketch to show how you could identify a trimline.

Further research

Explore further glacier terminology using the Glaciers Online photo glossary: www.swisseduc.ch/glaciers/glossary/index-en.html

Find further terms in the National Snow and Ice Data Center's Cryosphere Glossary: http://nsidc.org/glaciers/glossary

Find out more about Antarctic glaciers: www.antarcticglaciers.org

Glacial landforms and landscapes

How do glacial processes contribute to the formation of glacial landforms and landscapes?

By the end of this chapter you should:

- understand the relative importance of the main processes of erosion in forming glacially eroded landscapes at a macro-scale in glaciated highlands and ice-scoured lowlands
- know and understand how the macro-, meso- and micro-scale features are formed in glaciated environments
- understand why there is such a variety of glacial depositional features and how they can provide information on ice movement and provenance
- understand the importance of glacial meltwater within the glacial system.

6.1 The distinctive landforms of glacial erosion

Factors affecting glacial erosion

There is great variation in the intensity of glacial erosion, with perhaps the most important single factor determining the efficiency of glacial erosion being the glacier itself. Its size, which determines ice thickness, and its thermal regime, also determine the importance and intensity of the erosional processes. A general rule is that all types of glacial erosion operate far more effectively when glacier ice is warm based (meltwater and abundant debris facilitate abrasion) and regelation (whereby water melts under pressure and freezes again when the pressure is reduced), one of the key elements of the plucking process, can also occur.

Other important factors influencing the rate of glacier erosion include the glacier's velocity across the bed, the ice thickness and, hence, the power of the glacier to cause shattering. Also important are the quantity and shape of the rock debris. Subaerial processes of freeze–thaw (congelifraction) combine with extensive mass movement from scree slopes to supply the tools for glacial erosion. The bedrock characteristics, such as density of jointing and hardness, are also significant in influencing compressional and extensional flow (see page 66).

Figure 6.1 summarises the key factors influencing abrasion and plucking rates – in many cases there is an overlap.

Essentially erosion rates are more intense when the glaciers are warm based, thick and fast moving and the bedrock relatively weak, often due to dense jointing. On the other hand, erosion rates are much slower where glaciers are cold based and the rock relatively resistant.

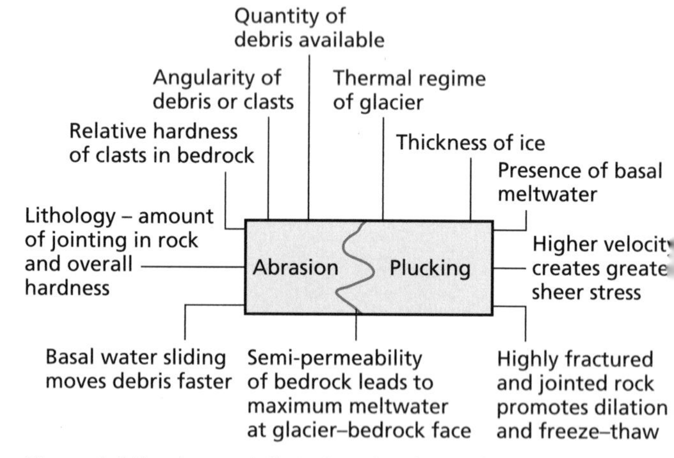

Figure 6.1 Key factors influencing abrasion and plucking rates

Features and landforms of glacial erosion

As Figure 5.13 (page 68) showed, the landforms which are the output of the glacier landform system result from the interaction of the processes and characteristics of the landscape experiencing glaciation (bedrock type, structure and topography) operating through time. However, the processes do not operate at a constant rate during time, and the landforms have to continually adjust, especially after glacial retreat in the short paraglacial period and then post-glacially when the landforms shaped by glaciation are reshaped by water, weathering and mass movement.

Key term

Topography: The shape and relief of the land.

A further complication is that most present-day landscapes resulting from glacials are polycyclic/polygenetic: the product of many successive advances of glacier ice because of the alternating ice-house–greenhouse conditions during the Quaternary period. As the last glacial period ended very recently in terms of geological time (the Loch Lomond Stadial, which ended around 11,500 years ago), the mountain areas of the UK provide clear examples of glacial erosional landforms. The Cairngorms and North West Highlands of Scotland, the Lake District and North Wales are all excellent areas for a glaciated highland case study.

Many different features and landforms are produced by glacial erosion, and they can be classified in different ways based on scale/size range, relative altitude or even the dominant erosion process that formed them (plucking or abrasion).

Table 6.1 provides a checklist of landforms associated with glaciated highland erosion, classified by scale.

Macro-scale features

These are around 1 km or greater in size and form the major elements in a glaciated highland landscape. They also contain many of the meso- and micro-scale erosional features, as well as depositional landforms.

Cirques (corries, cwms) are armchair/bowl-shaped depressions usually found at relatively high altitudes. The initial stage of formation is for snow to accumulate in a sheltered mountain side location. In the northern hemisphere, cirques most commonly form on the northeastern side of mountains, in the lee from prevailing westerly winds, and in shadier sites protected from insolation.

Once a sheltered area has accumulated snow, nivation or snow patch erosion begins, enlarging the hollow by a combination of freeze–thaw weathering to loosen the rock, and in summer melt water from melting snow transports the rock debris away, thus enlarging the hollow. Once a nivation hollow (a periglacial feature) is established, positive feedback occurs: the enlarged

Table 6.1 Checklist of landforms associated with glaciated highland erosion

	Landform	Appearance	Formation
Macro	Cirque (or corrie)	An armchair-shaped hollow on a hillside above a glacial valley	A pre-glacial hollow is enlarged by plucking and abrasion as ice moves in a rotational manner under gravity
	Arête	A narrow ridge between two cirques	As two cirques are enlarged back to back, the ridge between them becomes increasingly narrowed
	Pyramidal peak	A sharp, pointed hilltop	As three or more cirques are enlarged the hilltop between them becomes increasingly sharp and pointed
	Trough (or U-shaped valley)	A steep-sided, flat-floored, straight valley	A pre-glacial river valley is widened and deepened by erosion from an advancing glacier
	Truncated spur	A steep and possibly rocky section of the side of a trough	The pre-glacial interlocking spurs of the river valley are eroded by the much more powerful glacier
	Hanging valley	A small tributary valley high above the floor of the trough, often with a waterfall	Tributary glaciers with small amounts of ice did not erode their valley floor as deeply as the main glacier and so are left at a higher altitude
Meso	Roche moutonnée	Asymmetrical, bare rock outcrop with a gently sloping side facing up-valley	As ice crosses a resistant rock outcrop, the increased pressure causes melting and basal sliding and the up-valley side is smoothed by abrasion. On the leeward side pressure is reduced, refreezing occurs and plucking takes place, causing a steep, jagged slope
Micro	Striations	Grooves on exposed rocks	Abrasion by debris embedded in the base of the glacier as it passed over bare rock. They can indicate the direction of ice movement

hollow traps additional snow and gradually enlarges to provide a site for glacial ice formation.

Figure 6.2a shows how the processes of plucking and abrasion combine to develop the cirque. The glacier ice may expand in area and move down valley during a glacial period. Figure 6.2b shows how the cirque can be modified post-glacially with the formation of a small lake known as a *tarn* (an example is Red Tarn on the northeast face of Helvellyn in the Lake District).

The erosion of cirque headwalls backwards into the slopes behind can result in the formation of an *arête* (a steep, knife-like ridge produced from the intersection of two cirque headwalls) on either side of a slope divide. If three or more cirques interact back-to-back around the flanks of a mountain, a steep pointed peak is produced called a horn or a *pyramidal peak* because of its slope (for example the Matternhorn in the Alps).

Glacial troughs

When glacier ice moves through mountain valleys, it straightens, widens and deepens them, changing the valley from V shaped to U shaped. More accurately, these glacial troughs are described as parabolic in shape. Glacial troughs can be of varied length, from Nant Ffrancon in Snowdonia, which is around 5 km long, to spectacular features such as Yosemite Valley in Yosemite National Park California.

Along their lengths (long profile) many glacial troughs have a stepped profile, reflecting differential erosion as a result of both irregularities in the underlying bedrock and variations in intensity of erosion. For example, where several cirque glaciers meet at the head of a valley the enlarged glacier erodes very deeply to form a trough end to the valley. After deglaciation, successive rock basins down a glacial trough are separated by *riegels* or rock steps. Longer and deeper basins may contain linear lakes

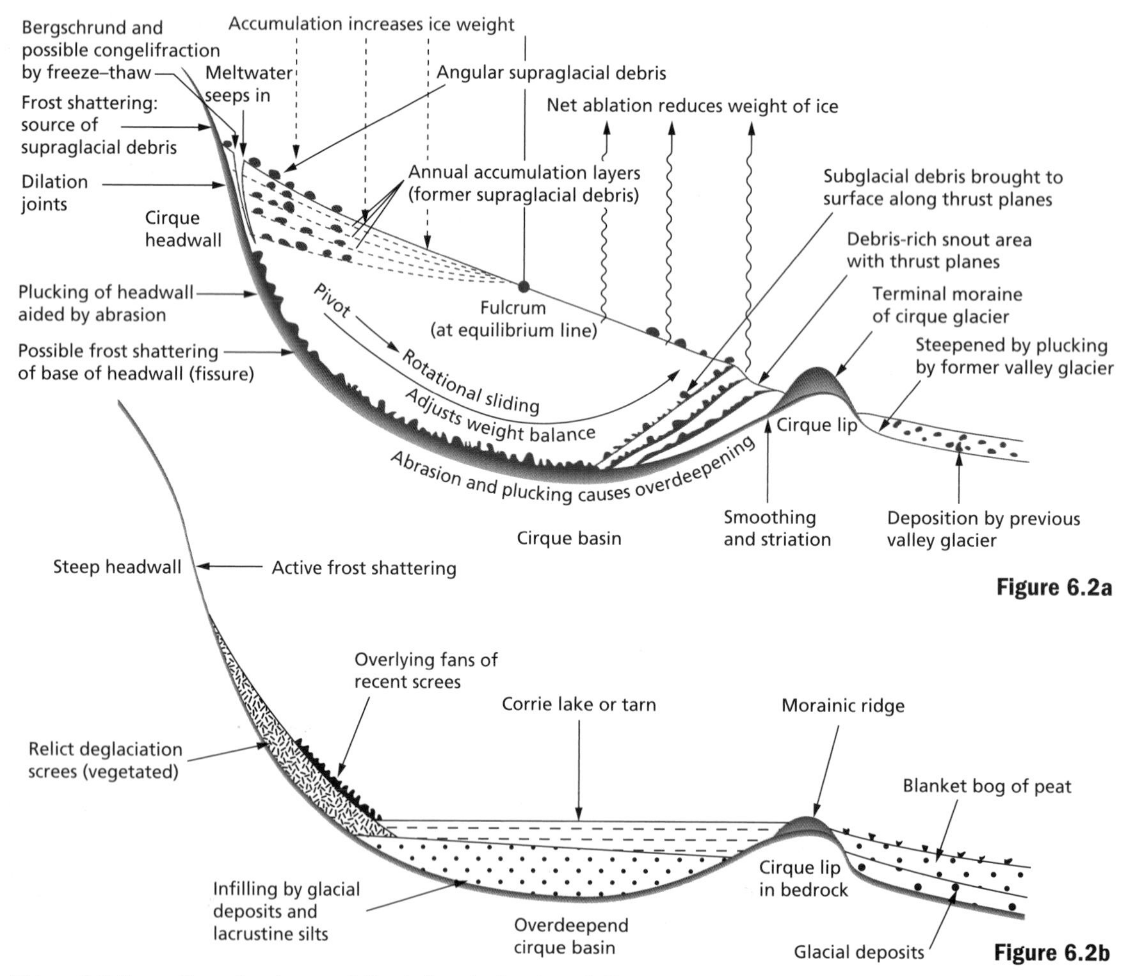

Figure 6.2 Formation of a cirque - 6.2a during the ice Age; 6.2b present day with post-glacial modification (Source: David Holmes, *Geography Factsheets 99*, Curriculum Press)

Fieldwork opportunity

The cirque skills focus could also be extended to be a fieldwork project with collection of primary data. Visit between ten and fifteen cirques, for example in the Glyders in Wales. For each cirque carry out measurements of size, length, breadth (by pacing), height of lip (use map or GPS device), height and angle of back wall (use clinometer) and orientation (use compass). Also photograph and map key features such a density of jointing in rock type, incidence of micro features and fabric analysis of any moraine. You could also record any post-glacial modifications. These field measurements might be useful when assessing reasons for any anomalies, for example in length to height ratios. You will also need copies of drift/solid geology maps and any field guides for the area.

termed *ribbon or finger lakes*. Post-glacial weathering and mass movement has led to infill of glacial troughs, which are now commonly occupied by misfit streams.

With relative sea level rises at the end of the last glacial period, many coastal glacial troughs were flooded by the sea to form sea *lochs* (Scotland) or *fjords* (Norway).

Hanging valleys occur where a small side tributary glacier meets a larger main valley glacier. During the glacial phase the surface ice elevation of the tributary and main valley glaciers is the same but, because the rate of erosion beneath the main valley glacier is much greater, once the glaciers have retreated the tributary valley can be left hanging hundreds of metres above, often with a waterfall plunging from the hanging valley to the main valley below (for example Pistyll Rhaeadr in the Berwyn Mountains).

Many glacial troughs show truncated spurs, marked by very steep, almost vertical, side walls where original interlocking spurs have been cut away or truncated by glacial erosion, because of the inflexibility of glaciers moving down the valley (for example Lauterbrunnen Valley in Switzerland or Yosemite in California).

Figure 6.3 summarises the nature and location of these macro-scale features within a glaciated upland area.

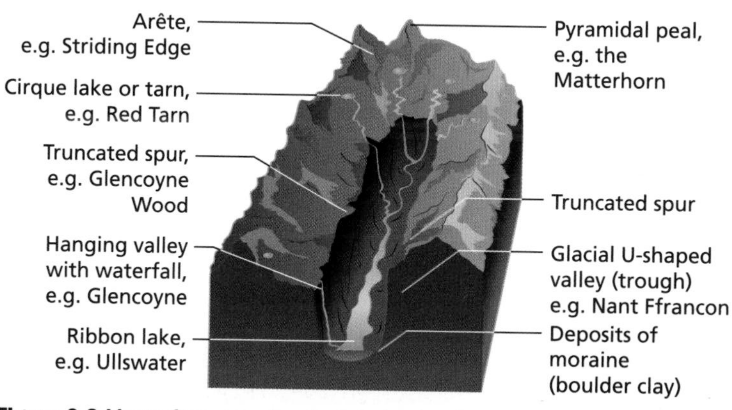

Figure 6.3 Macro features of a glaciated upland area

Meso-scale features

Meso-scale features are largely found within macro features, for example the *whalebacks* and *roches moutonnées* found on the floor of the Yosemite glacial trough. These intermediate-scale landforms can range from around 10 m to 1 km in length. Streamlined bedrock features such as whalebacks are the most common, where a glacier moves over a resistant rock knoll, so abrading it.

Roches moutonnées are stoss and lee features; abrasion smooths the up-glacier, stoss-side of a bedrock knoll, while glacial plucking makes the down-glacier, lee-side rugged and rough, thus producing an asymmetric landform. Figure 6.4 shows how they are formed beneath the ice.

Average-size examples, such as those found in the Cairngorms in Scotland, are around 300 m long and about 30 m in height.

Micro-scale features

Micro features of glacial erosion are those that are a few metres in size or less. They include *striations*, which are scratches on hard bedrock caused by debris being dragged across the surface during abrasion, almost like chisel marks. They tend to be parallel to the direction of ice movement, with the deepest part of the scratch at the initial point of impact, and are therefore useful for tracking the direction of past glacier movement.

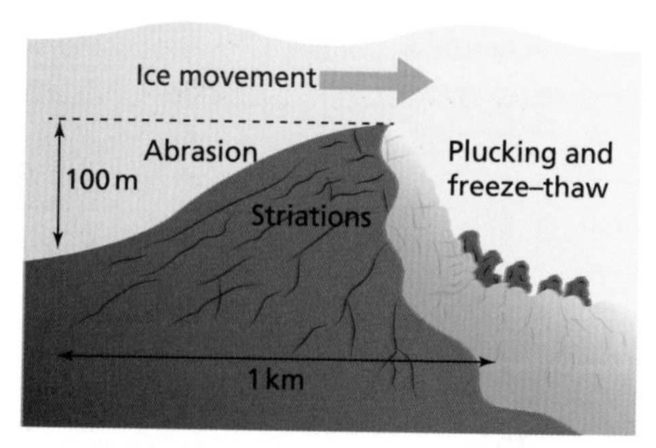

Figure 6.4 Formation of a roche moutonnée

Skills focus: Cirque analysis in Snowdonia

Figure 6.5 shows the glacial geomorphology of the Glyders and Nant Ffrancon Valley in Snowdonia, North Wales. Fifteen cirques, marked 1 to 15, are shown on the map with accompanying data on their dimensions, elevation and orientation (aspect).

A number of statistical and graphical analyses can be carried out:

- the relationship between cirque size and elevation of basin
- the relationship between cirque size and height of the back wall (use rock wall height).
- the orientation (aspect) of the cirques, which can be shown using a rose diagram; this can then be related to height of basin or cirque size using a frequency dispersion diagram or histogram.

In theory you would expect the most common aspect to be northeast facing, and that these would be the larger cirques. Equally, these cirques might develop at a lower height. Use your analysis to test out these hypotheses.

Extension of study

This study could be supplemented by a 1:25,000 large-scale map and a solid geology map. As in the Glyders, structure and lithology often play a key role in influencing the aspect and size of corries. You could extend your study

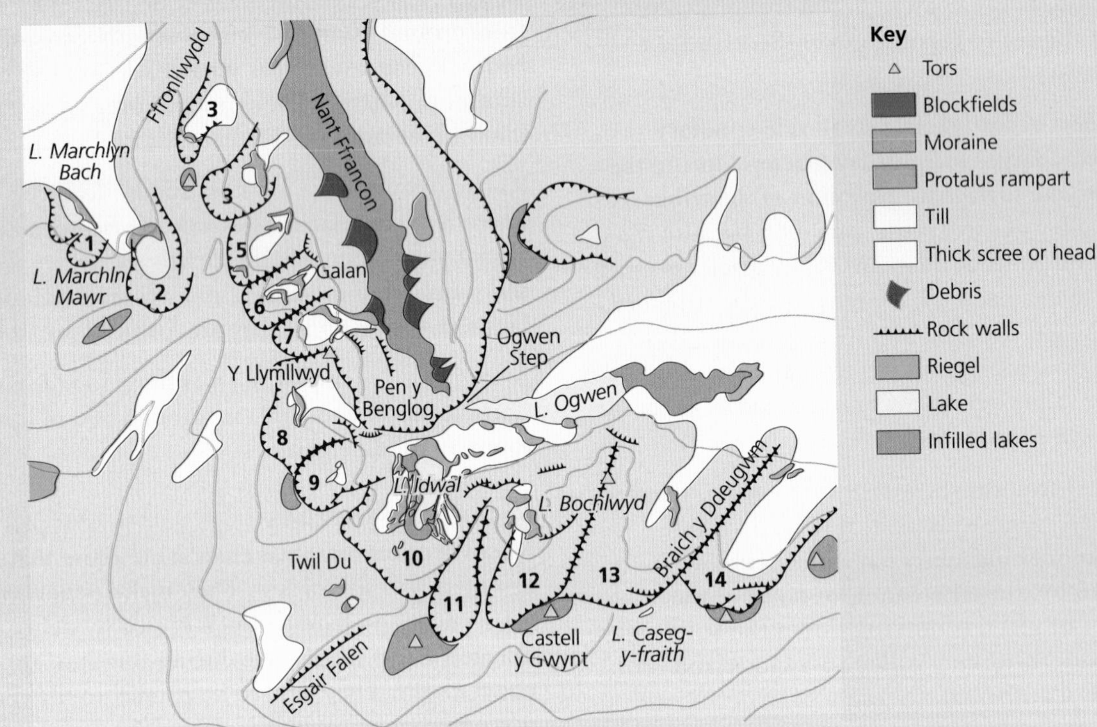

Glyder corries	Elevation (m) basin	Orientation 0° is true north	Area km² excoriated	Maximum rockwall height (m)	Length/breath ratio
G1	585	330	0.26	190	0.65
G2	670	340	0.66	190	1.64
G3	570	30	0.19	130	1.73
G4	600	60	0.47	200	1.65
G5	500	45	0.56	170	1.40
G6	500	55	0.47	170	1.95
G7	520	75	0.38	170	1.66
G8	585	50	0.60	220	1.28
G9	700	80	0.55	230	1.52
G10	420	50	1.37	420	1.94
G11	835	10	0.49	230	2.09
G12	800	15	1.09	230	1.74
G13	600	10	1.01	220	1.37
G14	500	35	1.05	180	1.47
G15	600	45	0.75	240	1.62

Figure 6.5 Glacial geomorphology of the Glyders and Nant Ffrancon Valley

(see Figure 6.6) to several groups of cirques in order to test a range of hypotheses; as in the Glyders there is comparatively little variation in the cirque aspect due to the impact of geological controls. Figure 6.7 shows how cirque orientation is related to altitude (elevation). Analyse the pattern for the three other groups: Snowdon, Carnedd and Hebog.

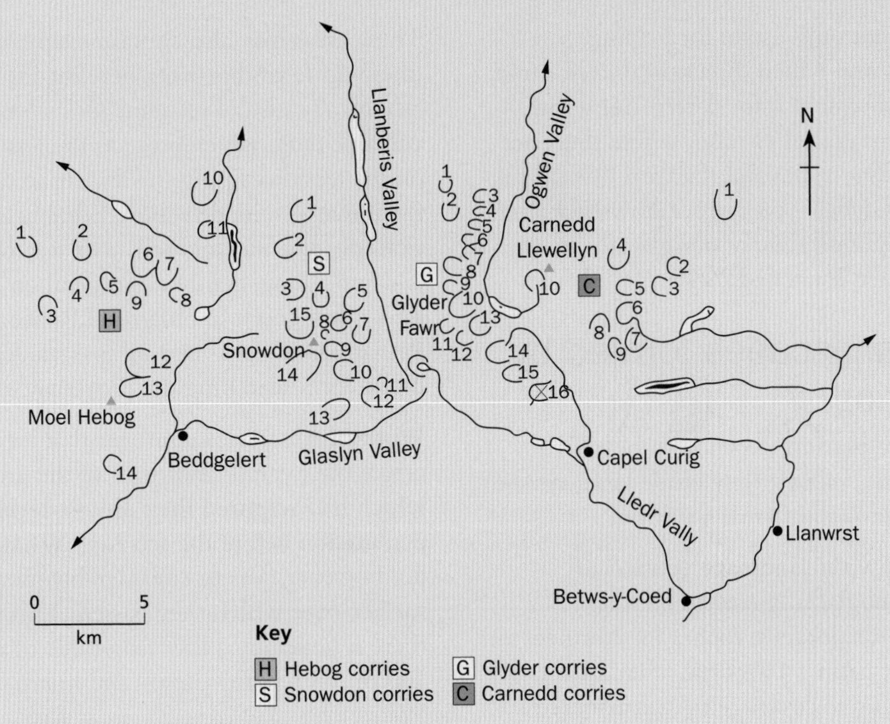

Figure 6.6 Cirque location within Snowdonia

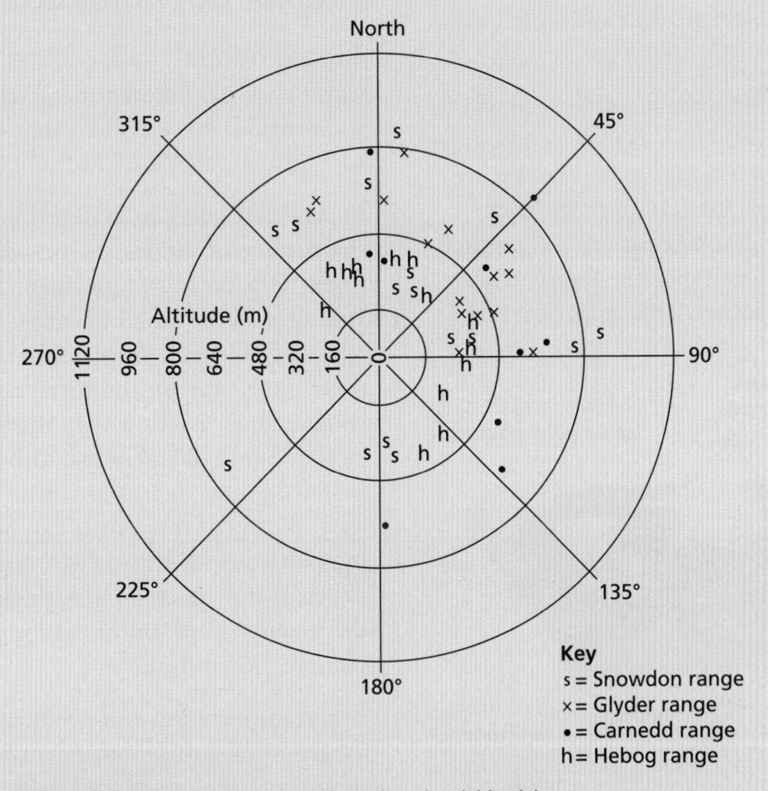

Figure 6.7 Cirque orientation against elevation (altitude)

Chatter marks are irregular chips and fractures in the rock, whereas *crescentic gouges* have a more regular pattern and are usually concave up-glacier. Look out for these micro features on abraded surfaces.

Micro features are not only useful for helping glaciologists understand which direction the ice came from (its provenance) but also for determining the maximum altitude of glacial erosion, where there are no micro features (that is, beyond the trim line on the valley side). In the Glyders you can see how *block fields*, *screes* and *tors*, clearly indicative of periglacial activity, supersede abraded, ice-scratched rocks.

Ice-eroded landscapes formed by glacial scouring

When ice sheets and glaciers expand out beyond constrained mountain valleys, they erode large areas of lower relief by the process of areal scouring.

Figure 6.8 shows how the landscape consists of extensive tracts of subglacially eroded bedrock, composed of many whalebacks, roches moutonnées and over-deepened rock basins. This type of landscape is associated with extensive coverage by warm-based ice, quite slow moving, which differentially eroded the hard bedrock. The structure of the underlying rock therefore has a major impact on the orientation and scale of the erosional landforms.

In North West Scotland (Sutherland and the Isle of Lewis) the landscape is called *knock and lochan* topography because the higher areas of resistant rock (knocks) are interspersed with numerous small lakes in the rock basin (lochans). A chaotic drainage pattern has resulted, often where patches of residual moraine interfere with the drainage. Other areas where landscapes of areal scour occur include Central Finland Lake Belt and the Canadian Shield, both areas of ancient resistant igneous and metamorphic rocks where differential erosion is controlled by the density of jointing.

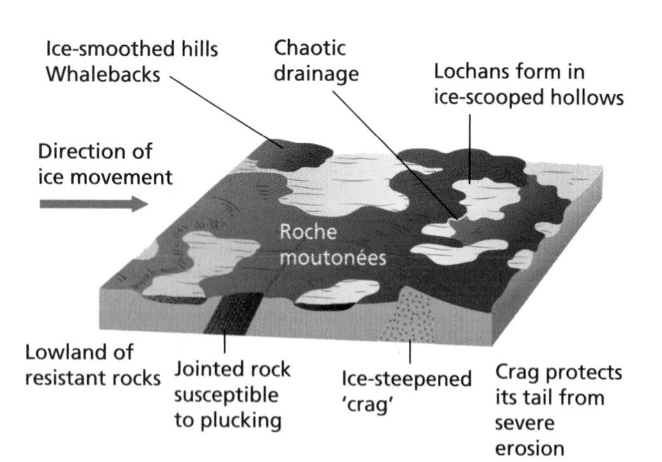

Figure 6.8 Lowland glaciation erosion

These areas may also contain examples of *crag and tail* (meso-scale landforms) where the glacier ice is forced around a large and resistant rock obstacle, such as a volcanic plug, which protects the less-resistant material on the leeside causing the feature to taper into a tail in the down-glacier direction. Edinburgh Castle is a famous example of a crag and tail, with a steep up-glacier stoss end and a long, gently sloping tail on the leeside that runs for 1.4 km down to Holyrood Palace and the Scottish Parliament. Micro- and meso-scale features are again useful for researching the provenance of the ice.

Work in Antarctica confirms that ice sheets did not create the overall landforms of the great shield areas, which had acquired their almost-level surfaces by denudation before the Ice Age. What these ice sheets did do, however, was to considerably modify the underlying surface over which they passed. This is confirmed as there is generally a low amplitude of relief (less than 100 m) with many meso- and micro-scale features.

Glacial depositional landforms

Glacial debris

Once rock has entered the glacial system and is being transported, it is classified into three kinds of debris: supraglacial, englacial and subglacial.

Supraglacial refers to material being transported along the surface of the ice. If rock debris from surrounding slopes falls on to the glacier in the accumulation zone it will become buried by new snow and become *englacial*. If it falls on the ablation zone area, it will most likely stay on the surface till it reaches the snout, which is often very dirty from accumulated debris. Supraglacial debris also can become englacial by falling into deep tranverse crevasses opened up by extensional flow, or it can be carried downwards by meltwater in warm (wet) based glaciers. *Subglacial* debris is transported beneath the ice all the way to the glacier snout, but it can be thrust upwards during compressional flow.

The transportation and eventual deposition of debris by glaciers is just as significant as glacial erosion in modifying the pre-glacial landscape, in this case by covering it over.

The presence of large boulders known as glacial *erratics*, so called because they are of a different rock type to the bedrock they 'sit' on, testifies to the sheer scale of the ability of glaciers (especially the vast ice sheets) to transport enormous quantities and weight of rock debris over great distances. Huge erratic – boulders weighing up to 16,000 tonnes – were carried over 300 km from the Canadian Rockies to the plains of Alberta by the Cordilleran Ice Sheet. Some are actually dumped as perched boulders, for example the Bowder Stone in Borrowdale in the Lake District. If the erratic is made of a rock source of a distinctive geology from a restricted location – for example Ailsa Craig granite from West Scotland – you can precisely map the direction of movement of the glaciers (see Figure 6.9). Erratics from Scandinavia have been found in the boulder clay of the Northumberland, Durham and Yorkshire coasts of North East England, confirming the presence of the continental ice sheets from Scandinavia.

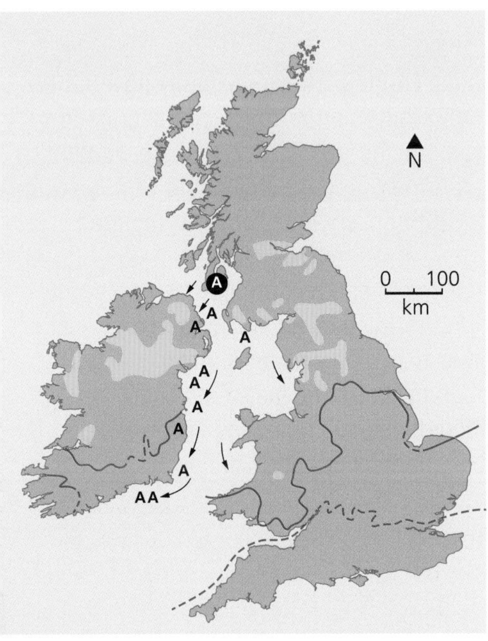

Figure 6.9 A map of ice age deposition in the UK shows the distribution of Ailsa Craig erratics

The processes of glacial deposition

The main processes by which glaciers deposit material are:

- **Lodgement:** this process occurs beneath the ice mass when subglacial debris that was being transported becomes 'lodged' or stuck on the glacier bed. Lodgement occurs when the friction between the subglacial debris and the bed becomes greater than the drag of the ice moving over it. It is commonly associated with glaciers carrying huge loads of debris and where the glacier is very slow moving, if not static.
- **Ablation:** this process refers to debris being dumped as the glacier melts and thaws. It can include supraglacial and englacial material, as well as subglacial material.
- **Deformation** is a less-common process associated with weak underlying bedrock, whereby these sediments are defined by the movement of the glacier.
- **Flow** occurs if high meltwater content causes the glacial debris to creep/slide or flow during deposition.

All of these processes produce till or boulder clay of different compositions, enabling scientists to analyse the types of depositional process. Lodgement till has relatively rounded clasts because of the grinding that occurs at the ice bed interface, not within a matrix of clay or silt-size particles (rock flour).

Ablation till consists of more angular clasts as they are not ground down, and also the matrix is of larger-sized material and less compact.

6.2 Glacial deposition, landforms and landscapes

Landforms of glacial deposition

The term moraine is used to refer to an accumulation of glacial debris, whether it is dumped by an active glacier or left behind as a deposit after glacial retreat. There are two broad categories:

- moraines formed beneath the glacier (subglacial)
- moraines formed along the margins of a glacier (ice-marginal).

Subglacially formed moraines

These moraines are composed primarily of lodgement till as they are formed from glacial debris beneath the glacier. Till plains of ground moraine are extensive flat areas that cover pre-existing topography, often to depths of 50 m. In some places beneath active glaciers, lodgement till is moulded into streamlined mounds called *drumlins* that have their long axis orientated parallel to the direction of ice movement.

Drumlins vary widely in size, usually ranging from 10 to 50 m high and between 200 and 2000 m long. The steeper, blunt end of the drumlin (stoss end) is the up-glacier side, whereas the gently sloping, tapered end occurs down-glacier. Drumlins usually occur in 'swarms' forming what is often called a 'basket of eggs' topography. They often occur regularly spaced, with a length-to-width ratio never more than 50, and are typically found in lowland areas in relative close proximity to upland centres of ice dispersal. Excellent examples of drumlin swarms occur in Northern Ireland, the Ribble Valley (Lancashire), the Cheshire Plain, North Shropshire and the Eden Valley, Cumbria (see Figure 6.9).

Drumlins are an example of equifinality, in that a number of mechanisms have been proposed for their formation. Not all drumlins are necessarily formed in the same way; some drumlins have a rock core, which also needs an explanation, whereas others do not.

- The Boulton-Menzies theory suggests that a drumlin is formed by deposition in the lee of a slowly moving obstacle in the deforming layer. The obstacle of bedrock, or thermally frozen material, forms the core of the drumlin and ground moraine is plastered round it.
- The Shaw theory suggests that all drumlins, even rock core drumlins, were formed by subglacial meltwater in flood causing irregularities to form in the river bed which were subsequently moulded into drumlins and streamlined by the advancing ice.

It is only recently that time-lapse geophysical surveys have been carried out subglacially. These actually show a drumlin forming from deforming sediments beneath the Rutford Ice Stream in West Antarctica, which helps to reveal more about their formation.

In some areas lodgement till remoulded into streamlined *flutes*, with a length-to-width ratio in excess of 30. These long, narrow features are usually less than 3 m in height and less than 100 m long.

Skills focus: Investigating drumlin swarms

- Drumlin morphometry can be investigated using a 1:25,000 or even 1:10,000 map or a land-sat image of a drumlin field.
- Make a tracing of a drumlin field and number each drumlin identified. You can record length, width and height, and the orientation of the long axis and the direction the stoss end faces in order to gain a picture of the morphometry.
- Calculate the drumlin length to drumlin width ratio so that you can calculate the elongation ratio. It has been suggested that all ratios should be below 50 for true drumlins. The faster the ice was flowing, the greater the elongation ratios. Drumlin swarms usually have little variation in their elongation ratio, so two sets need to be compared. It may be possible to deduce whether the swarm was formed under near-stable or non-stable fast-flowing conditions, such as those associated with surging.
- The relationships found can be displayed using a variety of graphical techniques, such as frequency graphs or scatter graphs, as well as rose diagrams for drumlin orientation.
- The spacing of drumlins can be measured using the nearest-neighbour technique to assess the degree of regularity and/or randomness of clustering.

Fieldwork opportunity

The Skills Focus work on drumlins can be extended into a fieldwork investigation. All the dimensions – height, length, width and angle of stoss end – can be measured in the field. Additionally, fabric analysis of any exposures can be carried out, and soil samples can be taken to assess the texture.

Figure 6.10 shows the dimensions that can be measured into order to analyse drumlin morphometry.

Ice-marginal moraines

Table 6.2 summarises the range of marginal moraines that commonly occur and suggests how they are formed. Most are linear features and often form within close proximity in either lateral or terminal margin process environments.

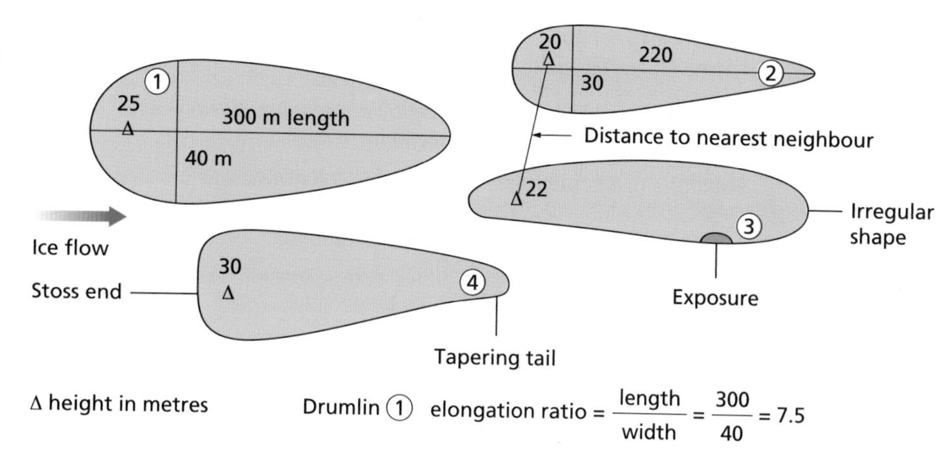

Figure 6.10 Features of drumlins

Table 6.2 Range of marginal moraines that commonly occur

Landform	Appearance	Formation
Lateral moraine (linear)	A ridge of moraine along the edge of the valley floor	Exposed rock on the valley side is weathered and fragments fall down on to the edge of the glacier. This is then carried along the valley and deposited when the ice melts. Parallel to ice flow.
Medial moraine (linear)	A ridge of moraine down the middle of the valley floor	When two valley glaciers converge, two lateral moraines combine to form a medial moraine. Material is carried and deposited when melting occurs. Parallel to ice flow.
Terminal or end moraine (linear)	A ridge of moraine extending across the valley at the furthest point the glacier reached	Advancing ice carries moraine forward and deposits it at the point of maximum advance when it retreats. The up-valley (ice contact) side is generally steeper than the other side as the advancing ice rose over the debris. Transverse to ice flow.
Recessional moraine (linear)	A series of ridges running across the valley behind the terminal moraine	Each recessional moraine, and there may be many, represents a still-stand during ice retreat. They are good indicators of the cycle of advance and retreat that many glaciers experience. Transverse to ice flow.
Push moraine (linear)	A ridge of moraine with stones tilted upwards	Any morainic material at the glacier snout will be pushed forward during advance. The faster the velocity of advance, the steeper the angle of tilt or stones. Transverse to ice flow.
Hummocky or disintegration moraine (non-linear)	Chaotic jumble of till mounds	Originally considered a product of ice stagnation and dropped from a debris-rich glacier, now associated with active glacial retreat. Limited orientation.

These glacial depositional landforms are complex to interpret as they are frequently interspersed with features of fluvio-glacial deposition. Moreover, as glaciers expand and retreat, or still-stand, they often rework older glacial deposits into new forms, adding to the complexity of depositional landforms.

As well as creating distinctive landscapes in both lowland areas such as eastern Denmark or in the floors of glaciated valleys, glacial depositional landforms are particularly useful in helping glaciologists to understand not only the extent of ice cover but also which direction the ice came from (its provenance).

Figure 6.11 How lateral moraines join to form medial moraine in Svalbard

Table 6.3 A quick guide to provenance

	Feature	Use	Scale	Visual guide
Ice erosion	Roche moutonnée	Up-glacier abrasion, down-glacier plucking	Meso	
	Striations (ice scratches)	Parallel to ice movement Deepest at initial point of impact	Micro	
	Crag and tail	Sloss end up-glacier, tapering tail down-glacier	Macro	Tail Crag
Ice deposition	Drumlins	Sloss end up-glacier, elongated tapering end down-glacier	Meso	
	Till fabric for boulder clay	Analysis of stones may show orientation parallel to ice advance; till may contain clasts from diagnostic geology, e.g. chalky boulder clay	Micro	
	Erratics	Rocks may have distinctive geology so can be sourced	Micro	
	Terminal moraines	Demarcate glacier snout or ice sheet front; behind is the glacial area, in front proglacial	Macro–meso	Terminal moraine Till plain Sandur
	Sandur	Sorted fluvio-glacial deposits, largest near ice front	Macro	Terminal moraine

In some cases it is the orientation of the feature and in others it is the contrasting up-glacier and down-glacier shape or the actual debris that yields the clues. The overall geography of the assemblage of features is also very important, in particular behind and in front of any terminal morainic ridges, as these mark glacial snouts or ice sheet edges.

Table 6.3 lists some of the glacial features that help 'ice detectives' and explains how they could be used. Often it is a case of piecing a range of evidence together.

6.3 Glacier hydrology – the role of meltwater

Meltwater from glaciers plays a vital role in the processes of erosion, entrainment and transport, as well as deposition. It is indirectly involved in the processes of glacial abrasion and plucking, but above all it plays a vital role in glacier movement by basal sliding and subglacial bed formation (see pages 65–67). Meltwater beneath a glacier is also responsible for erosion; because of its fast speed and power it can scour and groove the underlying rock. There are two main sources of meltwater from glaciers: surface melting and basal melting.

Surface melting contributes most of the supply and peaks in late summer; it is the only source of meltwater for cold basal glaciers. Supraglacial surface streams form running along the top of the ice, especially in the ablation zone. These supraglacial channels are often very fast flowing and may plunge down into the ice either through a crevasse or, more commonly, via a moulin (a cylindrical, vertical tunnel rather like a pothole in limestone landscapes), becoming englacial streams.

Figure 6.12 shows a diagram of meltwater movement through a warm-based temperate glacier. As meltwater

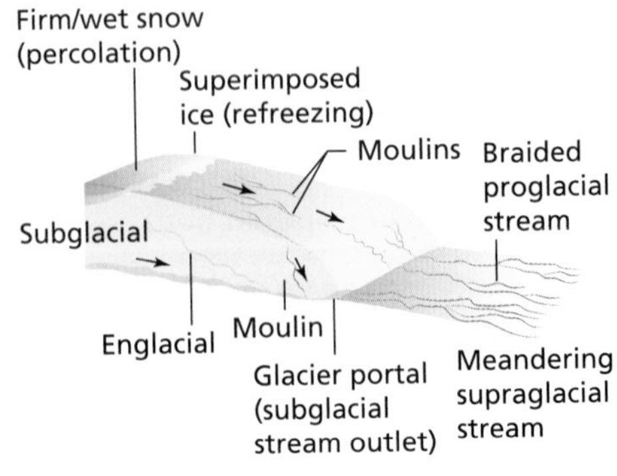

Figure 6.12 Meltwater movement through a warm-based temperate glacier

Figure 6.13 Portal at glacial snout Svalbard

moves through a glacier it may refreeze or contribute to further melting and reach the subglacial supply, depending on the temperature of the ice inside the glacier.

Basal melting occurs if the temperature of the ice at the base of a glacier is at pressure melting point (in a warm-based glacier). The basal meltwater flows under hydrostatic pressure beneath the glacier and can excavate subglacial tunnels by cutting through the bedrock. The meltwater streams eventually emerge from subglacial tunnels at the glacier snout via portals (caves).

Processes of fluvio-glacial erosion and deposition

It is important to remember that away from the glacier, outlet streams behave similarly to normal streams, although their discharge, sediment loads and the lack of vegetation lead to some variations in operation. Within the glacier, however, fluvio-glacial streams operate very differently because of the high pressure and velocity of flow. This causes the erosion of underlying bedrock by abrasion, cavitation and chemical means beneath the glacier ice, and can also lead to intense erosion by meltwater streams as they exit the glacier snout. The ablation rates are very high during deglaciation, and many of the meltwater streams have very high discharges leading to powerful erosion.

When meltwater deposits material subglacially, englacially and supraglacially, the material is referred to as an *ice-contact fluvio-glacial deposit*. Where the fluvio-glacial material is deposited at or beyond the ice margin, by streams coming out of the snout, it is known as *outwash* or proglacial.

Characteristics of fluvio-glacial deposits

In comparison to glacial deposits (tills), fluvio-glacial deposits tend to be:

- generally smaller than glacial till as meltwater streams; although having seasonally high discharge, they still have less energy than large valley glaciers so they generally carry finer material
- generally smoother and rounder through water contact and attrition
- sorted horizontally, especially in the case of outwash deposition, with the largest material found up-valley or nearer the glacier snout, and progressively finer material down-valley, due to the sequential nature of deposition mechanisms
- stratified vertically with distinctive layers that reflect either seasonal or annual sediment accumulation.

In contrast, glacial till is classed as a diamicton, being angular, poorly sorted and non-stratified.

A further distinction can be made between ice-contact fluvio-glacial deposits and outwash deposits. Outwash deposits experience more attrition, causing clasts to become more rounded, and the material is better sorted horizontally.

Three main zones of outwash deposition extend from the front of the glacier, and the characteristics change through these zones, as shown in Table 6.4.

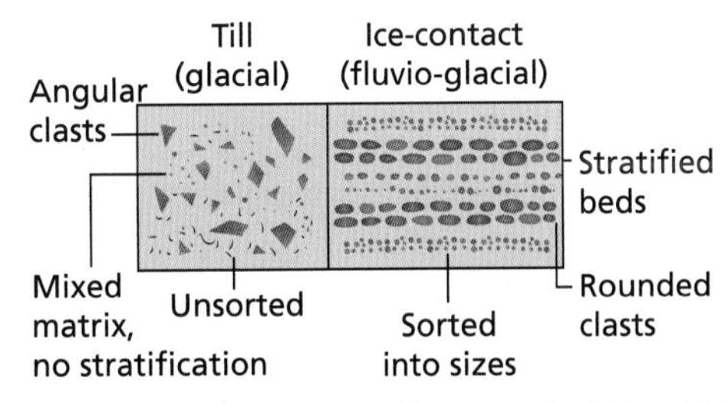

Figure 6.14 Differences between glacial and fluvio-glacial deposits

Table 6.4 Sorting of outwash deposits

Zone	Description
Proximal zone	Immediately in front of the glacier, close to the snout. Meltwater has greatest power here so the outwash deposit contains a large particle size. Outwash may be interbedded with layers of till as some glacial deposition may occur. Outwash may occur in alluvial fan forms.
Medial zone	Further from ice margin, meltwater streams tend to anastomose and form braided channels because of high daily and seasonal variability of meltwater discharge. Particle size is less coarse than in the proximal zone and clasts are more rounded.
Distal zone	Furthest from ice margin. Drainage pattern resembles normal drainage system, with meanders on a broad flood plain. Outwash is well sorted and characterised by smaller and even more rounded particles.

Note that varved deposits form in meltwater lakes along or beyond a glacier margin (see page 87).

Fieldwork opportunity

Fieldwork in glacial landscapes is often centred around the examination and analysis of sedimentary deposits. You can find out about the origin of the deposit, its mode of formation and possibly the type of glacial environment it was deposited in.

- Locate and photograph your chosen site(s). A local geology map, aerial photograph and field guide should help you locate possible survey sites.
- Use a stone board to measure the size of a chosen sample of 50 stones, to be sampled randomly or from various layers. Try to measure both the long and middle axis to get an idea of size. Use a sieve for the analysis of any fine-grained matrix.
- Angularity and roundness can be measured quantitatively using Power's index (matching your stones to a picture to see whether they are in one of six categories from very angular to well rounded) or the Cailleux Index (which involves an index of roundness) (Figure 6.15).
- Orientation is used to refer to the three-dimensional position of the sediment, as the positioning and direction of the clasts allows you to make inferences about the nature and direction of the depositional process. Use a compass along the long axis and present your results as a rose or star diagram.
- Numerical analysis: calculate the mean, mode, median and standard deviation of all data sets to enable reliable comparisons.
- Sorting (which indicates the presence of water) can be gauged by developing a simple visual chart and scale as shown below (Figure 6.15).
- The type of geology, especially of very large deposits such as erratics, can be photographed and matched up using a geology map to try to assess the provenance of the ice that has transported the deposits.

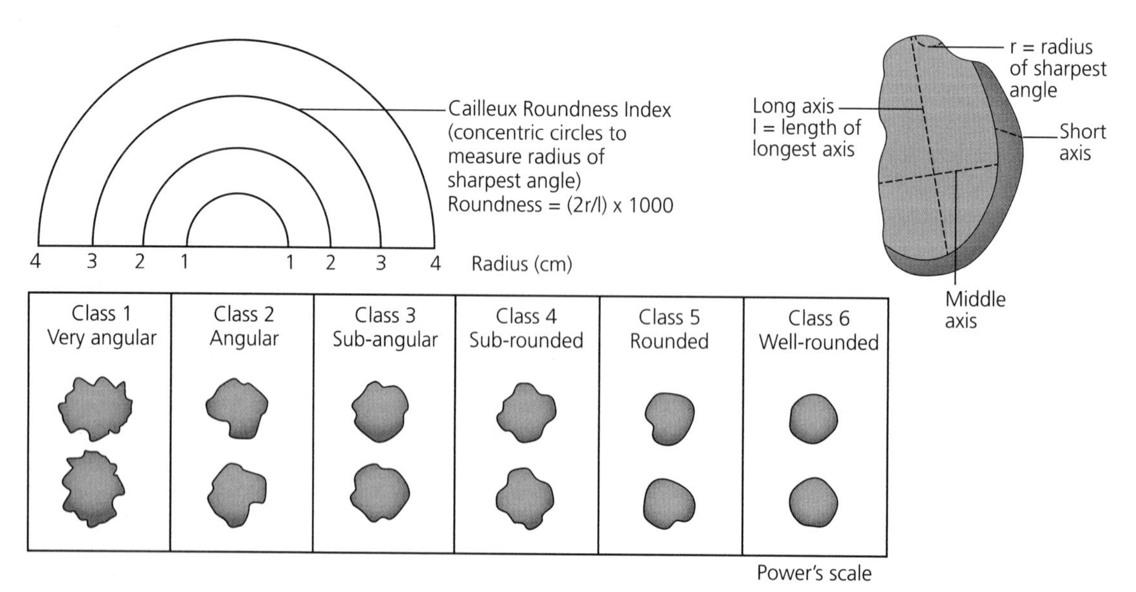

Figure 6.15 Stone board and sorting scale estimation

Skills focus: Sediment analysis

As well as fieldwork, investigation of data from sediment analysis can be analysed. A number of suggestions include:

- Student's t-test to analyse changes in sediment size and shape, with data from the three zones of an outwash plain
- central tendency analysis of both glacial and fluvio-glacial deposits
- comparison of the size, shape, orientation and degree of sorting of clasts from a variety of deposits.

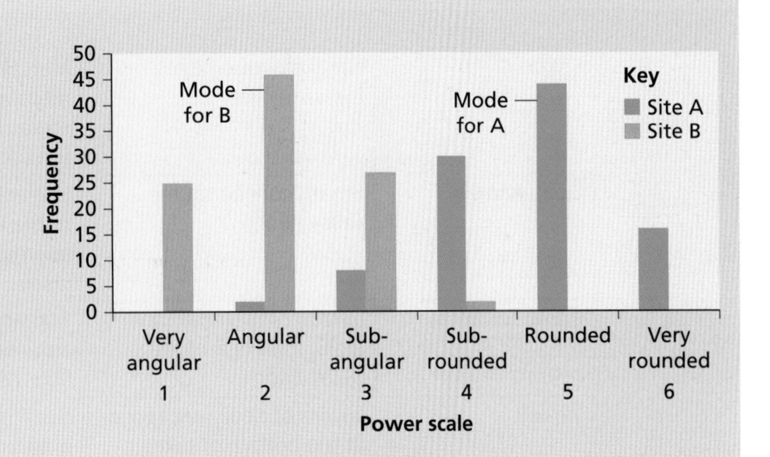

Exercise

Study the following data. You have a choice of four types of deposit to choose from, and the results from two sites:

- **Site A** is located on a river cliff exposure. Stones have some sorting.
- **Site B** is positioned on the eastern valley side, stones are poorly sorted.

Describe the shape and orientation of the stones at sites A and B (Figure 6.16).

Suggest which category of deposit they could be from the four types listed in Table 6.5. Explain your reasons.

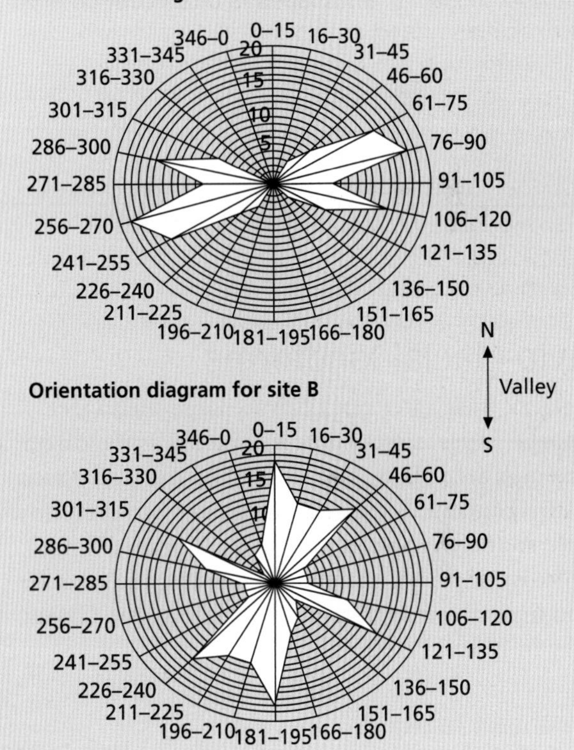

Figure 6.16 Deposits A and B

Table 6.5 Deposit and depositional features

Deposit	Position in valley	Orientation	Shape	Degree of sorting
Fluvial	Post-glacial stream in bottom of valley	Downstream/valley	Mostly well rounded	Well sorted
Fluvio-glacial	Bottom	Some downstream	Course material dominant Variable, some rounded	Some sorting
Moraine/till	Bottom and side in sheets	No orientation apparent in this location	Angular	No sorting apparent Mixture of sizes
Soliflucted 'head'	Valley sides in crops or lobes	Downslope resulting from mass movement	Angular clasts	Poor sorting, long axes aligned parallel to direction of flow

Table 6.6 Landforms of fluvio-glacial deposition – a checklist

	Landform	Appearance	Formation
Ice contact	Eskers	Long, sinuous ridges on the valley floor	Material is deposited in subglacial tunnels as the supply of meltwater decreases at the end of the glacial period. Subglacial streams may carry huge amounts of debris under pressure in the confined tunnels in the base of the ice.
	Delta kames	Small mounds on the valley floor	Englacial streams emerging at the snout of the glacier fall to the valley floor, lose energy and deposit their load, or supraglacial streams deposit material on entering ice marginal lakes.
	Kame terraces	Ridges of material running along the edge of the valley floor	Supraglacial streams on the edge of the glacier pick up and carry lateral moraine, which is then deposited on the valley floor as the glacier retreats.
Proglacial	Varves	Layers of sediment found at the bottom of lakes	Sediment carried by meltwater streams is deposited on entering a lake as energy is lost. In summer, when large amounts of meltwater are available, the sediment is coarse and plentiful, leading to a wide band of sediment of relatively large material. In winter, with little meltwater present, sediment is limited in amount and size, so the bands are thin and fine.
	Outwash (sandur)	A flat expanse of sediment in the proglacial area	As meltwater streams gradually lose energy on entering lowland areas, they deposit their material. The largest material is deposited nearest the snout and the finest further away.
	Kettle holes	Small circular lakes in outwash plains	During ice retreat blocks of dead ice become detached. Sediment builds up around them and, when they eventually melt, a small hollow is formed in which water accumulates to form a lake.

Fluvio-glacial landforms

Subglacial meltwater can excavate large meltwater channels. These can cut across contours as the direction of the meltwater flow is controlled by the hydrostatic pressure gradient. Subglacial meltwater can even flow uphill, so these channels can have a 'humped' long profile. Examples of these meltwater channels can be found in many parts of the UK including the Gwaun Valley in North Pembrokeshire.

Ice-contact features

Eskers result from subglacial meltwater deposition. They are sinuous ridges of relatively coarse sand and gravel deposited by meltwater flowing through tunnels, sometimes englacially, but normally subglacially. Eskers came in a variety of sizes. A small esker can be found in the UK at Wark on the River Tweed; it is about 1 km long, 40 m wide and about 20 m high.

Eskers are thought to occur when a subglacial or englacial channel becomes obstructed, leading to deposition of material upstream from the blockage.

The ice needs to be stagnant for englacial eskers to form, otherwise the material would be reworked by glacier ice movement.

Another possible mode of formation is where a delta of fluvio-glacial material extends outward, perpendicular to the ice margin, taking on an elongated form under conditions of rapid ice retreat.

Kames are generally steep-sided conical hills although they come in a variety of shapes and sizes. They are formed by deposition of material in ice in either surface depressions or crevasses, or alternatively as deltas along the sides of a glacier between the ice margin and the hillside. They show some evidence of stratification, although the bedding can be disturbed by subsidence as the ice melts away.

Kame terraces form relatively continuous bench-like features along the valley side, when a gap or a lake between the valley side and the ice margin is filled with fluvio-glacial deposits.

Some regions of extensive kame deposits also contain kettle holes (see page 87) and can be described as areas of kame and kettle topography. This type of topography is usually developed when there is a large amount of fluvio-glacial material deposited over the surface of stagnant ice, which melts *in situ*.

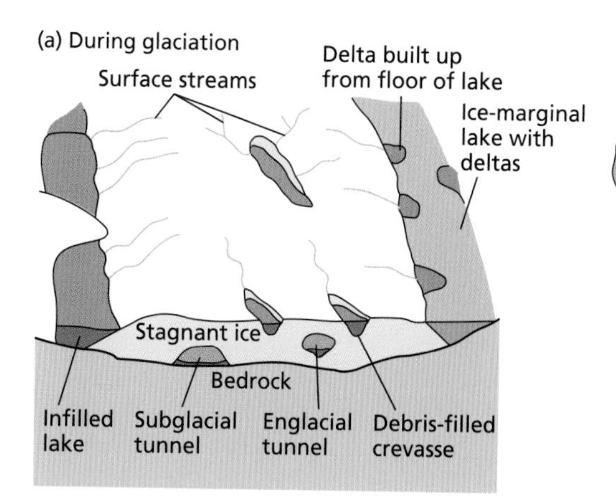

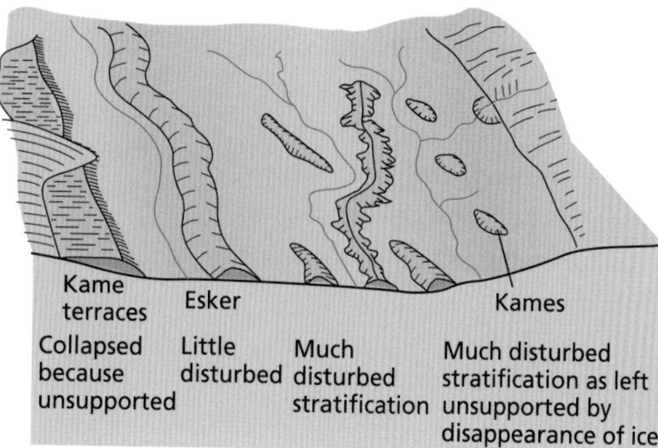

Figure 6.17 Formation of fluvio-glacial features (Source: David Holmes, *Geography Factsheets 99*, Curriculum Press)

Proglacial features

When an englacial or subglacial stream exits the snout of a glacier there is a rapid decrease in water pressure and velocity, causing the deposition of coarse fluvio-glacial material as an outwash fan. Outwash fans merge to become part of a debris-rich, anastomosing braided drainage system. As the discharge of meltwater decreases with deglaciation, the broad expanse of fluvio-glacial material that was deposited and spread out by the braided river systems is left behind as an *outwash plain* or *sandur*. An outwash plain is a gently sloping surface made up of rounded, sorted and stratified sands and gravels, with the particle size becoming progressively smaller away from the ice front.

Outwash plains may contain *kettle holes* where any surviving blocks of ice during deglaciation were buried by outwash material. After the ice melted, the ground above it subsided leading to the formation of a depression, which subsequently filled with water. As these kettle holes are only fed by rainwater, many of the smaller ones are gradually colonised by vegetation (hydroseres) and subsequently dry up. In the Ellesmere area of north Shropshire there is an area with many kettle holes varying from 400 m to 1.5 km in size. They are also very common in the Alpine Foreland area of southwest Germany near Wolfegg.

Proglacial lakes (also known as ice-margin lakes) are formed along the front of glaciers and ice sheets where meltwater from the glacier becomes impounded within a depression blocked by glacier ice and bounded by high ground.

These lakes are ephemeral (temporary) features; depending on the rapidity of deglaciation the proglacial lake can empty completely, usually via a pre-existing col, or could stabilise at a lower level if the ice margin had not completely disappeared.

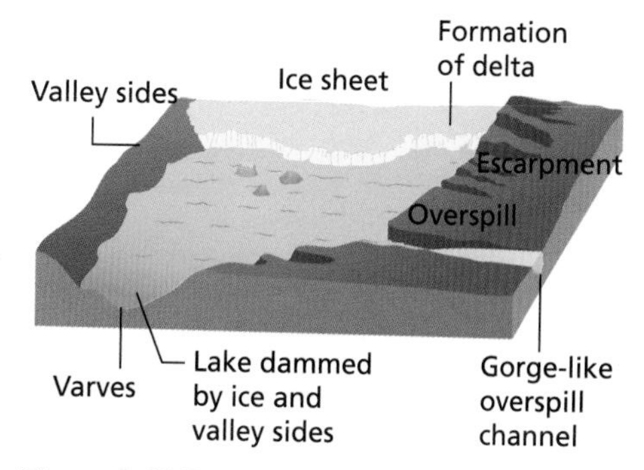

Figure 6.18 The formation of a proglacial lake

The dimensions of former proglacial lakes can be inferred from both erosional and depositional forms. The parallel 'roads' of Glen Roy in the Scottish Highlands mark the former shoreline of a proglacial lake formed during the Loch Lomond Stadial. Strandlines marking the shore of the proglacial lake may occur if the water level was stationary for a relatively long time. It may also be possible to find former lake deltas, where meltwater streams deposited outwash as they entered the lake. If the water was relatively calm in the lake, *varved* deposits form. They are characterised by alternate bands of relatively coarse grained sand, reflecting the rapid ice melt in summer at the base of the layer, overlaid by fine, dark-coloured silt or clay that came out of suspension when the lake's surface (and the streams that fed it) were frozen in winter. These annual bands of sediment reflect the seasonal variation in discharge from the glacier.

Proglacial lakes were a very common feature with many forming in the English Midlands, such as Lake Harrison or Lake Lapworth. Many proglacial lakes, such as those formed along the margins of the Laurentide Ice Sheet in North America, were enormous; at its maximum Lake Agassiz covered an area of around 300,000 km^2.

Overflow channels, also known as meltwater spillways, are formed when proglacial lakes overflow their confines. These channels are an open V shape, often gorge-like and sinuous, as they were caused by intense fluvial erosion along an outflow path. In the present day, many of these channels are now dry or contain only a very small (misfit) stream. These overflow channels can lead to diversions of preglacial drainage systems.

Figure 6.19 shows a very well-known example of a series of proglacial lakes and overflow channels in East Yorkshire.

The Channeled Scablands in northwestern USA contain huge numbers of stream-less gorges known as coulees that formed during periods of catastrophic flooding from proglacial Lake Missoula, which was dammed by a lobe of the Cordilleran Ice Sheet.

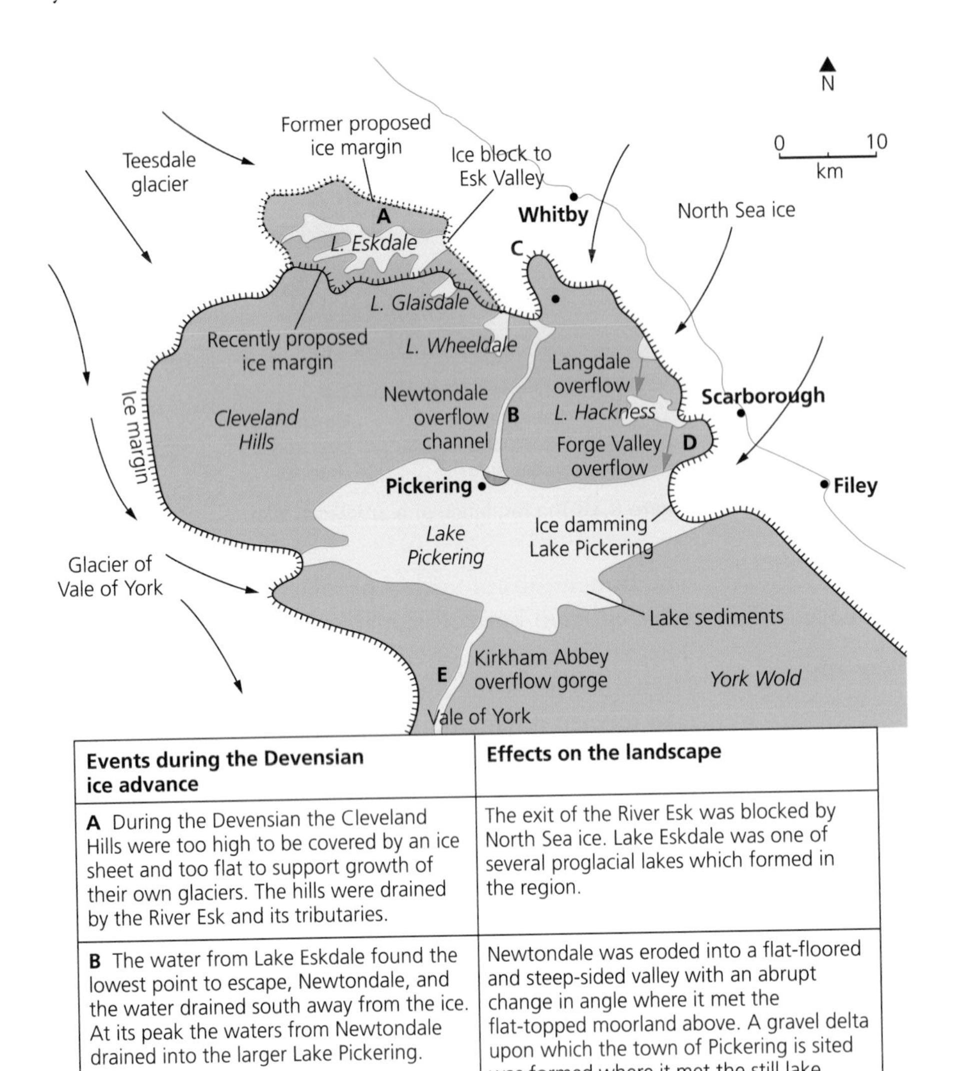

Events during the Devensian ice advance	Effects on the landscape
A During the Devensian the Cleveland Hills were too high to be covered by an ice sheet and too flat to support growth of their own glaciers. The hills were drained by the River Esk and its tributaries.	The exit of the River Esk was blocked by North Sea ice. Lake Eskdale was one of several proglacial lakes which formed in the region.
B The water from Lake Eskdale found the lowest point to escape, Newtondale, and the water drained south away from the ice. At its peak the waters from Newtondale drained into the larger Lake Pickering.	Newtondale was eroded into a flat-floored and steep-sided valley with an abrupt change in angle where it met the flat-topped moorland above. A gravel delta upon which the town of Pickering is sited was formed where it met the still lake waters.
C The levels of Lake Eskdale fell as the North Sea ice retreated eastwards.	The River Esk was able to resume its easterly course, which it still follows, reaching the sea at Whitby.
D The waters of the River Derwent, which used to reach the sea north of Scarborough, were also blocked by ice.	This lake overflowed through the Forge Valley channel in the larger Lake Pickering, also blocked by North Sea ice.
E When the ice retreated the River Derwent did not resume its preglacial course.	Lake Pickering overflowed south through a sinuous gorge near to Kirkham Abbey into the even larger Lake Fenland. The River Derwent continues to follow this course to this day.

Figure 6.19 Proglacial lakes in East Yorkshire (Source: *Process and Landform*, Peter Comfort and Alan Clowes, Oliver and Boyd)

Review questions

1 Define 'equifinality'. Choose three examples of depositional features (glacial or fluvio-glacial) and explain how they illustrate this concept.

2 Explain how you could distinguish between the following features found at an ice marginal process environment: lateral moraines, solifluction terraces and kame terraces.

3 Study Figure 6.20, which summarises the effects of the Pleistocene in Europe. Describe and explain the distribution of the following three environments: glacial deposition, fluvio-glacial sands and gravel, and loess deposition.

   Research examples of these environments in Europe to describe their main features, for example, within Denmark or the North European plain.

4 Using examples, explain how cirques and glacial troughs can be modified by post-glacial activity.

5 Research two case studies of glacial highland scenery (include a map, photographs and descriptions of the key features). One should be a relict area, such as North Wales, the Lake District or Cairngorms in the Scottish Highlands. The other should be an active area such as the French Alps or part of the Swiss Alps. You should try to obtain maps and field guides for your two chosen areas.

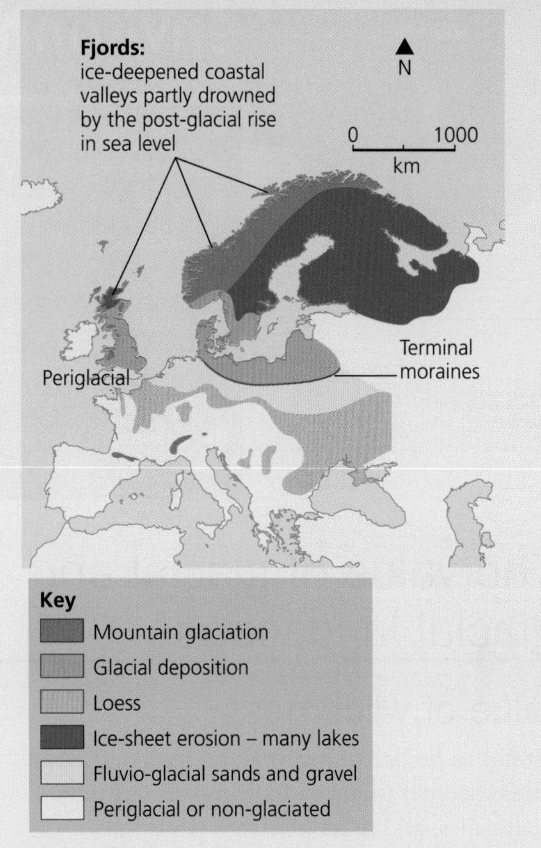

Figure 6.20 The effects of the Pleistocene in Europe

Further research

Olafus Ingolfsson is a professor of glacial and quaternary geology at the University of Iceland; his website has some useful material and links: www3.hi.is/~oi/index.htm

Explore glacial landforms and landscapes further on the Glaciers Online page: www.swisseduc.ch/glaciers

Human activity and glacial environments

How are glaciated landscapes used and managed today?

By the end of this chapter you should:

- appreciate that glacial and periglacial landscapes have a range of intrinsic cultural, economic and environmental values
- understand that these fragile landscapes are facing increasing threats from both natural hazards, human exploitation and global warming
- know how these challenges can be managed using a spectrum of approaches from 'business as usual' through to total protection
- appreciate that there are many different players involved and that there is no one solution that fits the diversity of these glacial and periglacial environments.

7.1 The value of glacial and periglacial landscapes

The value of wilderness

A contrast has to be drawn between active glacial and periglacial landscapes found in high (polar) latitudes and high altitudes, and relict glaciated landscapes. Whereas nearly all of the former can be classified as true wilderness as they are remote, possess a harsh physical environment and carry little or no population, except for small groups of indigenous people or groups exploiting their resources short term, the latter relict areas are often more densely populated as they provide many more opportunities for economic development and employment, for example farming, forestry and tourism.

It is precisely the qualities of wilderness which provide opportunities for spiritual refreshment and enjoyment for many travellers and explorers who wish to experience pristine, almost wholly natural environments and almost always untouched by humans. Wildernesses have inspired an enormous range of people to write, paint and communicate their feelings and experiences – for instance wilderness poets such as Robert W Service from Alaska, polar explorer John Muir or legendary broadcaster and naturalist David Attenborough.

Figure 7.1 develops the concept of the wilderness continuum and shows how high-quality wildernesses such as Antarctica or the Arctic, both barely modified by human activity, contrast with lower wilderness quality areas, which merge into quite heavily settled rural areas. For example, the relict environments of the summits of Snowdon or the Cairngorms, both reached by mountain railways, only have *elements* of pristine quality; they have lost their innate remoteness to large quantities of tourists.

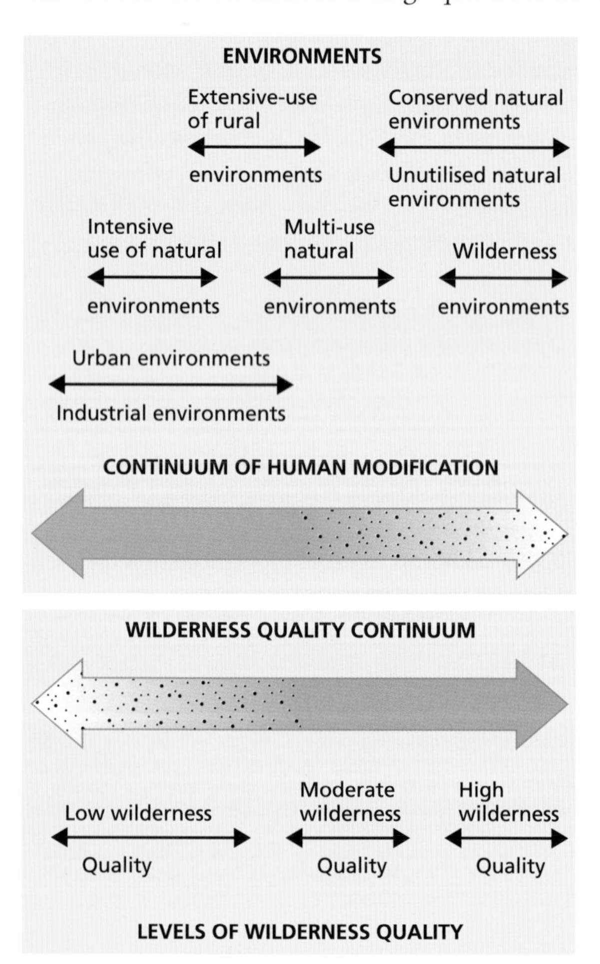

Figure 7.1 The wilderness continuum

Environmentalists and ecologists would point to wilderness areas, such as the polar environments, being of inherent value for scientific reasons. They argue that areas of a truly pristine nature are very useful for research for a number of reasons:

- the need to maintain a gene pool of wild organisms to ensure the maintenance of genetic variety (global seed bank in Svalbard)
- the need to retain wilderness so that animal communities can remain in their natural environment, for example providing sanctuary for the migratory bird and animal communities of the Alaskan North Slope
- to use wholly natural communities that still exist as control systems for comparison with exploited, mismanaged systems elsewhere.

Scientific research

Both Arctic and Antarctic polar environments have become 'living labs' for scientists. They have both similarities and significant differences in their research programmes.

Economic value of glacial environments

Millions of people, who live far from any mountains, benefit from a range of goods and services provided by glacial environments, both relict and active. These include supplies of pure mountain water, timber (usually coniferous woodland), hydroelectric power for their homes, as well as the opportunity to enjoy beautiful mountain environments such as the Alps or Himalayas for recreation and leisure.

About eight per cent of the world's people live in polar and mountainous and upland regions, many of which have been glaciated in the past or still contain some glaciers, albeit mostly diminishing in volume.

Farming

Within mountainous regions in developing nations, there may be limited transport links and access to essential supplies and markets may be poor; employment opportunities are also limited. In countries such as Nepal, Bolivia (Altiplano), Ethiopia (Bale Mountains), the highlands are largely inhabited by indigenous communities who gain their living from subsistence farming. In Bolivia 70 per cent of the population lives in the High Andes, growing crops such as potatoes, quinoa and beans to feed themselves, as well as rearing llamas and alpacas, yet they earn only 30 per cent of the country's GDP. Almost all of the 60 per cent of Bolivia's population living below the poverty line are indigenous Indians living in the Altiplano of the High Andes. The development of internet and mobile phones, leap frogging old technology of landlines and cables, has revolutionised their lives, providing many opportunities for cottage industries such as weaving and knitting co-operatives, and ecotourism.

In Alpine areas in developed countries, the agriculture in upland glaciated regions is primarily pastoral because of the above-average precipitation, rugged terrain with steep slopes and stony, shallow soils, which together make cultivation difficult. In the truly Alpine areas

Antarctica

Antarctica was designated a continent of peace and science as a result of the 1959 Antarctic Treaty and subsequent protocols. In the International Polar Year of 2007–09, the United Nations Environment Programme reports that 60 countries and more than 10,000 researchers took part in scientific research there costing over $1.5 billion.

Some of the planned or ongoing research programmes being carried out by international Antarctic scientists include:

- understanding global change – past, present and future – by looking at ice cores
- researching 'life on the edge' to explore how ecosystems cope with intensely harsh conditions
- investigating subglacial Lake Vostock and other lakes beneath the ice sheet
- developing sustainable food webs in the Southern Ocean ecosystem; in particular, looking at changing impacts on the various trophic levels such as the decline in krill, a key component of the Antarctic marine food web
- studying the Earth's upper atmosphere and its links to the lower atmosphere and the Earth's climate – taking advantage of the unpolluted atmosphere above the continent
- understanding how the Antarctic ice is melting, in terms of scale and pace, using satellites and field data.

(Source: www.bas.ac.uk)

Key term

Silviculture: The planting of trees for commercial forestry.

transhumance is a traditional pattern of livestock management; the farming system takes advantage of the seasonal climate cycle – in summer animals are grazed at high altitudes on Alpine meadows, which become free from snow and provide high-quality grass (the summer settlements are known as *sacters*) – at the same time the grass in the valley bottom can be made into hay for winter feed. In the snowy winters the animals are brought down and housed near the farm houses and are fed on hay or graze the lower pastures.

Forestry (silviculture)

Depending on farm prices and the degree of government support systems (such as EU rural payments), hill farming is an increasing struggle in many regions and uplands are now increasingly used for forestry (silviculture). In the UK this is carried out by the Forestry Commission and private investors, with the main type of tree being non-native, quick growing conifers, such as Sitka spruce, grown for softwood timber, wood pulp and even paper. Conifers tolerate harsh climates and acidic soils that would not be suitable for other land uses.

Mining and quarrying

Glacial erosion plays an important role in removing regolith (loose overlying soil) and vegetation to expose economically valuable rocks. In many active or relict areas there are mines and quarries of mineral deposits and ores, as well as rocks such as slates as many of the glaciated mountains are made from igneous and metamorphic rocks. In lowland areas, outwash deposits from the Pleistocene Ice Sheets provide a very important source of sand and gravel for the building industry, pre-sorted by meltwater into sands and gravels to be sold as aggregates, making them very useful for making concrete.

Hydroelectricity

Hydroelectric power (HEP) is a major use of water derived from glaciers. Both Norway and New Zealand derive over 90 per cent of their electricity from this source. In most cases either a natural ribbon lake or a dam and reservoir in a glaciated valley provide the HEP. Switzerland has over 500 HEP stations, producing some 70 per cent of its electricity. Clearly HEP is a renewable 'green' source, although there are issues with both the reliability of water supply and environmental concerns over damming of rivers. In mountain settlements in developing nations, such as Nepal or Bolivia, micro-hydros can revolutionise the quality of life in many villages.

Figure 7.2 A hydroelectric power station

Tourism

The tourism industry has seen tremendous growth in recent decades, which has brought many economic benefits to mountain regions, with visitors attracted to the spectacular scenery of both present-day and relict glaciated landscapes. A huge range of year-round, outdoor activities are possible in Alpine landscapes – hill walking, climbing, mountaineering and skiing – which has led to whole regions capitalising on their tourist potential.

Glaciated regions are increasingly visited for the glaciers themselves, which puts pressure on some very fragile landscapes.

Table 7.1 shows some of the developing areas for glacial tourism – note that it is a worldwide activity.

This table demonstrates how long-haul travel and modern communications has brought mass tourism not only to traditional areas, such as the Swiss Alps and the Rockies, but also to remote polar regions in the Arctic (Alaska, Greenland, Iceland and Svalbard) and Antarctic (South Georgia and the Antarctic peninsula). These areas have become increasingly popular, especially for expedition ship cruising, which has all sorts of implications for their environments (see page 97).

The economic value of any mountain tourism requires careful management of the benefits so that these are not

able 7.1 Developing areas for glacial tourism

Region	Tourist activities
Aletsch Glacier, Bernese Oberland, Swiss Alps	Jungfrau railway to Jungfraujoch (3454 m); access by road tunnel; all year round tourism
Mer de Glace, Chamonix, French Alps	Cable car to Aiguille du Midi for glacier viewing and hiking; visit to ice cave beneath glacier
Columbia Icefield and Athabasca Glacier, Alberta, Canada	Brewster snow coach tour and glacier walk; access by icefield parkway and ice centre
Franz Josef Glacier and Fox Glacier, Fjordland National Park, New Zealand	Guided walks, heli-rides and heli-skiing on glaciers
Everest base camp and ascent, Himalayas, Nepal	Trekking to Everest base camp (Kleenex trail); huge pressure on Everest itself as permitted numbers are allowed to rise and concentrated during a relatively short season
Torres del Paine National Park, Chile/Argentina	Glacier viewing platform, glacier hikes
Southwest Iceland	Ski and snowcat rides on the glacier

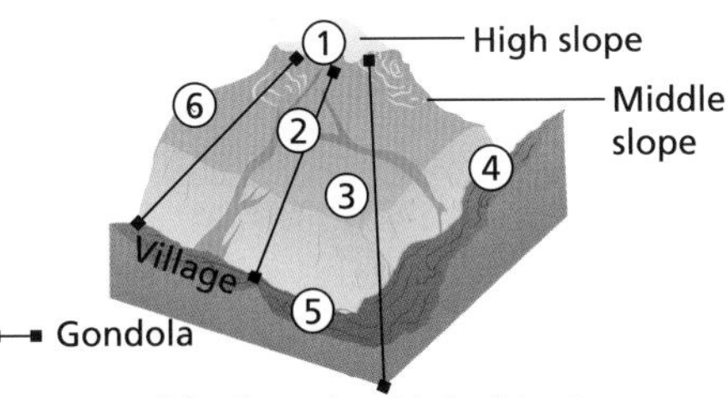

1 Skiers will be forced to high altitude resorts as lower resorts lose snow cover. This may increase the environmental pressures on a smaller number of higher resorts.
2 Removal of vegetation for chair lifts and gondolas may lead to an increase in erosion and increase the risk of avalanches.
3 Snow cover on middle and lower slopes is less certain – many banks are now cutting off funding for the lower resorts, which may face economic ruin.
4 More gondolas and chair lists will be needed to take skiers from the lower resorts to upper slopes, e.g. at Mayrhofen, Austria, where gondolas capable of holding 160 take people to the higher slopes.
5 Resorts in the lower parts of the valleys could face economic ruin if they cannot a) produce more snow or b) get skiers to upper slopes.
6 Increased usage of environmentally damaging artificial snow.

Figure 7.3 The increase in damage from ski resorts as a result of climate warming

outweighed by environmental costs to the scenery and to the culture of the local people. It's a very fine balance. So much depends on the fragility of the landscape and the nature and intensity of the economic activities.

Ecological and environmental value

Both glacial landscapes and periglacial landscapes make a very significant contribution to the world's life support systems. Approximately 75 per cent of all the freshwater of the world is locked up in ice – so glaciers contain nearly two per cent of all water overall. Glacially eroded valleys in many mountainous regions form natural hollows for water collection, forming ribbon lakes, or can be dammed for reservoirs, for example in the English Lake District.

Glaciers are especially valuable as a source of water for irrigation as they produce most water in late spring and summer, which is often the hot, dry season when other sources dry up. For example, in the USA the Arapaho Glacier currently produces about 260 million gallons of drinking water per year for the city of Boulder, Colorado, as well as irrigating huge areas of crops such as fruit and grapes. Glaciers have a very cool and beautiful image and this is used to promote sales of bottled water.

Being composed of freshwater, icebergs also offer a potential water resource, but there are numerous complications even in their potential use.

The extensive areas of permafrost and tundra peat are major areas for soil carbon storage. Currently the Arctic lands remain a weak carbon sink, meaning that more carbon is added each year than is lost. However, as permafrost melts as a result of positive feedback within the climate warming process, ancient carbon is being released, especially via methane emissions, thus upsetting the balance in the system.

Tundra vegetation occurs in periglacial areas that are not ice covered at present, both in high latitudes (Arctic

Key term

Carbon sink: A natural or artificial reservoir that absorbs more carbon than it releases, leading to carbon accumulation.

Key terms

Biodiversity: A measure of the variety of organisms present at a particular location.

Primary productivity: The rate at which energy is converted by photosynthesis; it has a major influence on the level of biodiversity.

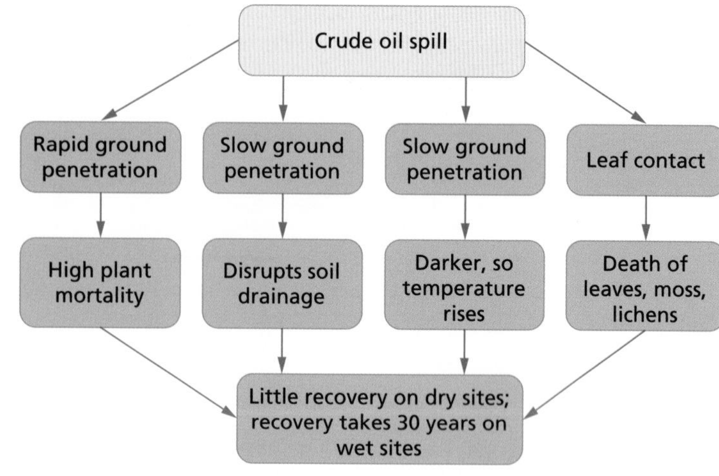

Figure 7.4 Impact of a crude oil spill on the Arctic tundra in Alaska

ecosystems) and high altitude (alpine ecosystems), covering around 8 million km^2 of the Earth's surface.

The presence of permafrost in most areas in the soil increases their fragility. In the lower Arctic latitudes (around 70 to 75 °N) there is a continuous cover of ground vegetation, with sedges and mosses in the wetter hollows and scattered dwarf trees (elder and birches) on the lower ridges. Elsewhere heaths, grasses and rapidly flowering plants flourish. At higher latitudes (75 to 80 °N) and higher altitudes, polar desert conditions prevail, but a small range of plants survive in favourable sheltered locations such as the purple saxifrage and arctic poppy.

Tundra plants have to adapt to low temperatures, drying winds and snow blasts in winter blizzards. The system has low biodiversity because of low primary productivity and the low nutrient content of the Arctic soils, leading to very weak nutrient cycling in dynamic equilibrium with the underlying permafrost, so any melting will have dramatic effects.

Human activities can have dramatic impacts, both directly, such as vegetation being removed during the building of roads and infrastructure, and indirectly through pollution. Toxic chemicals/acid rain from coal mines (Svalbard) and metal smelting in the Kola Peninsula (Russia) lead to the contamination and melting of the permafrost and the formation of thermokarst conditions. A growing threat is the potential for oil spills as Arctic oilfields are developed on the Alaskan North Slope and Russian Siberia, controversially in areas of outstanding ecological value, such as the Arctic National Wildlife Refuge (ANWR) where permission has just been given to drill (2015) amid huge controversy as it affects the lives of the native peoples as well as the environment and ecology.

In 2005 the Millennium Ecosystems Assessment (MEA) identified the value of polar ecosystems and posed questions about what humans would lose if they were irreparably degraded.

Table 7.2 summarises the value of these ecosystems using the MEA framework.

Table 7.2 Value of polar ecosystems

Ecosystem services	Examples	Importance
Provisioning (goods)	Fisheries	Importance of Southern Ocean and Arctic fisheries for local and international use
	Wild foods	Seasonal hunting of seals for indigenous tribes; former hunting of whales (now banned except for scientific purposes)
	Genetic resources	Important genetic resources exist on land and sea even though productivity is low; many species are endemic and some may not have been discovered yet
	Fresh water	Huge stores of fresh water in ice caps and glaciers
Regulating	Carbon sequestration	Large stores of carbon in peat and Arctic soils
	Climate regulation	Warming could release large volumes of methane and carbon dioxide
		Arctic and Antarctic play crucial roles in regulating the Earth's climate by working to cool the planet (via ocean currents and atmosphere movement)
Cultural	Aesthetic value	Largely unspoilt pristine wildernesses to be enjoyed by travellers
	Spiritual value	Some Arctic indigenous people practise animism, giving the landscape ecosystems a unique cultural and spiritual significance

7.2 The threats to glaciated upland landscapes

Natural hazards in glacial environments – rising risks

Both current glaciated and relict upland regions are hazardous because of the high incidence of avalanches, rock falls, debris slides and flooding. These hazards have the capacity to develop into disasters because of the rising human vulnerability in these areas, resulting from increasing population and development as well as the growing popularity of outdoor sports and adventure tourism, which put more people at risk.

Avalanches

An avalanche risk exists when shear stress exceeds shear strength of a mass of snow located on a slope. The shear strength of a snow pack is related to its density and temperature.

Snow avalanches result from two different types of snow pack failure:

- **loose snow** acts rather like dry sand; a small amount of snow slips out of place and starts to move down slope
- **slab avalanches** occur when a strongly cohesive layer of snow breaks away from a weaker underlying layer. A run of higher temperatures followed by refreezing creates ice crusts, which provide a source of instability. Slabs can be as large as 100,000 m^3 and can bring down 100 times the initial volume of snow and cause huge danger.

Most avalanches start off with a gliding motion then rapidly accelerate, especially on steep slopes in excess of 30°. Three types of avalanche motion commonly occur:

- **powder** avalanches (the most hazardous)
- **dry flow** avalanches
- **wet flow** avalanches, which occur mainly in spring

While avalanches tend to follow well-known tracks and can often be predicted, they are nevertheless a significant hazard, usually killing around 200 people per year with most of these deaths in the Alps or the Rockies.

Figure 7.5 shows there is a large variety of ways by which avalanche hazards could be reduced.

In 1970 the Peruvian towns of Yungay and Ranrahirea were destroyed by an earthquake-induced ice and rock avalanche from Mount Huascarán. It travelled 16 km down valley as a muddy flood, killing over 18,000 people. The 2015 Nepalese earthquake set off many ice and rock avalanches that killed some members of expeditions at Everest Base Camp.

Lahars

Some of the most-destructive volcanic hazard events are caused by lahars. The second-deadliest eruption recorded in historic times resulted from lahars generated by the 1985 eruption of Nevado del Ruiz, Colombia (Figure 7.6). Volcanic activity caused large-scale glacier melting, producing a huge lahar. It rushed down the Lagunillas Valley overwhelming the town of Armero, 50 km downstream. With a mudflow deposit 3 to 8 m deep, it killed more than 23,000 people almost instantly.

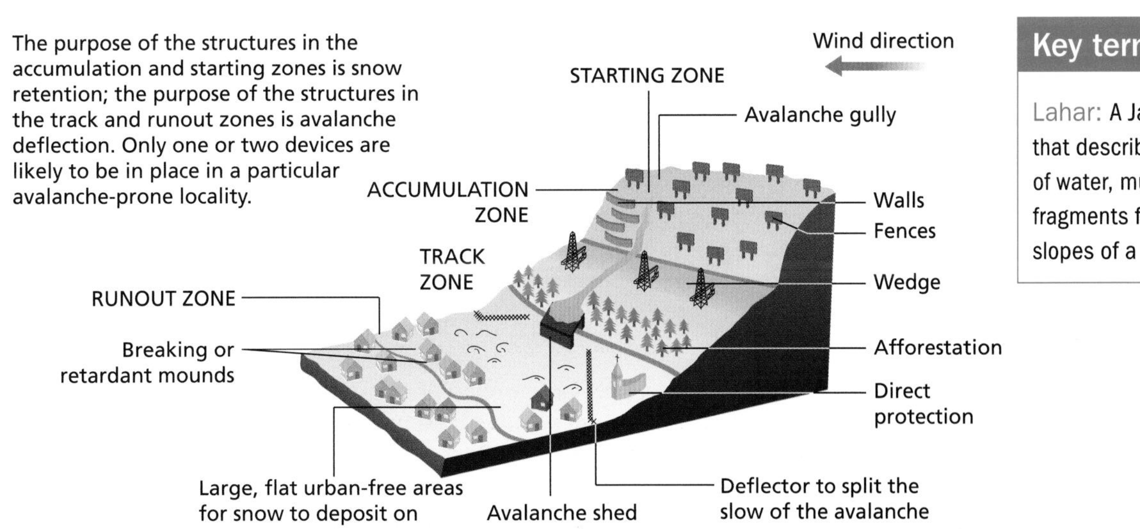

Figure 7.5 Measures to reduce the impact of avalanches

Key term

Lahar: A Javanese word that describes a mixture of water, mud and rock fragments flowing down the slopes of a volcano.

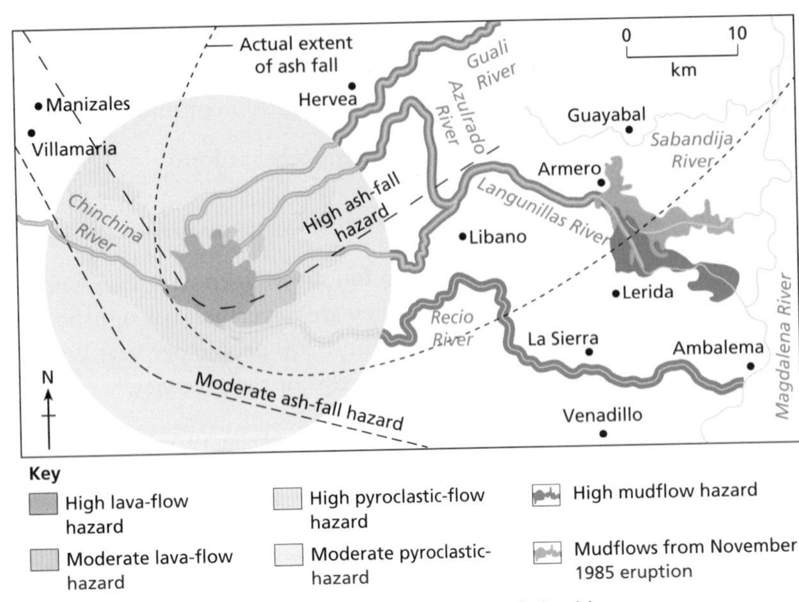

Figure 7.6 The mudflow risk from Nevada del Ruiz, Colombia

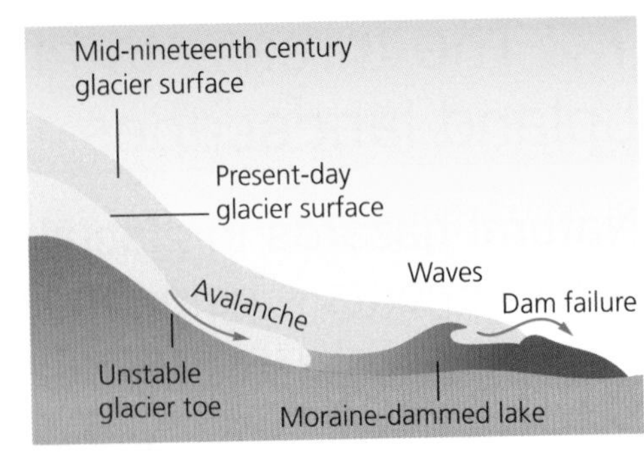

Figure 7.7 Schematic diagram of the relationship between glacier thinning and retreat and the formation and destruction of moraine-dammed lakes

Glacial outburst floods

A glacial outburst flood is also known by the Icelandic term jökulhlaup. It is a powerful flood caused by the sudden discharge of a subglacial or ice moraine dammed lake. There is potential for an outburst flood whenever meltwater collects behind an ice or moraine obstruction. The sudden catastrophic release can be triggered in six main ways:

- increased flotation of ice as water levels rise
- overflow and melting of an ice dam – common in climate warming
- breakdown of an ice dam because of tectonic activity
- irreversible overtopping of a moraine dam by large tsunami-style waves triggered by a snow/ice avalanche or landslide into a lake
- failure of moraine dam by slow melt of ice within it or removal of fine sediment from the moraine by underwater 'piping'
- enlargement of pre-existing tunnels beneath an ice dam because of increased water pressure.

Note that outburst floods may be cyclical in nature as the ice dam and lake may reform following a flood.

These very large floods are a huge threat to people and property in inhabited mountain valleys around the world. They may destroy property tens or even hundreds of kilometres from their source, especially in areas with a long history of settlement, for example the Andes, European Alps and Himalayas.

Periglacial areas also experience widespread ground subsidence when ground ice melts, which means that special engineering designs need to be developed when these areas, largely in northern Canada, Alaska and Siberia, are exploited for their resources.

Figure 7.8 A glacier outburst

Glacial outburst floods in Iceland

These are particularly frequent in Iceland because of volcanic activity that generates both meltwater beneath glaciers and acts as a trigger for ice instability and the sudden release of meltwater. Heat from the Grímsvötn volcano beneath the Vatnajökull ice cap melts the ice and creates a subglacial lake within its cauldron. When this lake reaches a critical size it forces its way through subglacial tunnels and the lake drains catastrophically within hours, with total discharges ranging between 0.5 and 3.5 km³ approximately every three to six years. In 1996, an eruption of Grímsvötn actually broke through the ice, sending up a huge ash column 10 km high and melting a large quantity of ice. On 5 November the meltwater burst through the glacier with a massive peak discharge (the largest since 1930) causing damage to infrastructure of around US$15 million.

Threats from human activity

Human activities can also degrade and damage fragile glacial and periglacial environments and ecosystems.

While continuous settlement in relict glaciated areas has traditionally been agriculturally inspired, within true polar environments this is not a feasible option. In polar lands, settlements tend to be nucleated and surrounded by vast areas of nothingness. These settlements tend to be built for resource exploitation by outsiders: for whaling, sealing and fishing, or for mining (initially for gold or copper or uranium, and more latterly for oil).

There are inevitably issues from pollution and toxic waste from these often hastily built urban areas; also, the exploitation often leads to conflicts with the way of life of native peoples. Regular contact from outside has progressively reduced the chances for the survival of the traditional culture of groups such as the Sami (Lapland) or Inuit (Greenland).

There are many other threats from human activities, especially in the more remote and fragile polar wilderness areas. These include the widespread impacts of tourism.

Polar tourism

Polar tourism is one frequently cited example of a threat to glacial environments. Romanticised, over the centuries, travel to the polar areas – for so long the stuff of Amundsen and Shackleton 'daring dos' – now increasingly represents an expensive leisure activity suitable for all, the only real barrier being cost. In the late Victorian era cruise travel to the Arctic, for example North Cape and Svalbard, became the province of many wealthy adventure-seeking travellers. Today it is the Antarctic which, as the last great wilderness, represents the journey of a lifetime for around 40,000 tourists a year.

Figure 7.9 Tourism in North Cape

There has been an explosion of polar tourism. Annual figures for the increasingly accessible Arctic, where tourism has long since been relied upon by local communities, have doubled since the early 1990s, from 1 million tourists to over 2 million in 2014. Factors such as climate warming have lengthened the summer season, especially for ships with ice-strengthened hulls, and winter activities such as snowmobiling and husky sledging, as well as viewing the Northern Lights, have also been developed. Most visitors arrive by ship, sailing up the Norwegian coast to the North Cape where the recently built centre receives over 1 million visitors per year. Over 50 cruise ships visited Svalbard in 2015. The limits are currently only controlled by the limited number of feeder flights. The main issue is the huge increase in the number of landing sites (now nearly 200), which could spread the damage to uninhabited pristine areas.

Since 1990, Iceland, Greenland and northeastern Canada have also enjoyed very strong growth rates, especially from cruise ship tourism; with Greenland up 400 per cent since 2004. Alaska too receives over a million passengers. For middle-class Americans, 'doing Alaska' by car or cruise ship is a rite of passage. For this reason numbers have had to be controlled at pressure points such as Denali National Park and Glacier Bay National Park.

In contrast, because of its greater remoteness and therefore higher costs (in 2015 a standard trip cost around £12,000 to £14,000 for 20 days) tourism was comparatively low key in Antarctica and has only increased dramatically since 1990. The protocols

Figure 7.10 Antarctic expedition

and environmental protection measures adopted in 1996, and subsequently added to the Antarctic treaty, combined with voluntary policing by the International Association of Antarctica Tour Operators (IAATO), have meant that there is a framework in place to manage Antarctic tourism. While numbers are only currently at 40,000 per year, the season is very concentrated and there are concerns about a number of potential issues:

- the development of land-based tourism, for example in the Patriot Hills area
- the spread of tourism to areas beyond the Antarctic peninsula, where most cruise and expedition ships are currently concentrated
- the development of 'fly-sail', where helicopters bring tourists to ships; these tourists often come from Southeast Asia, a rapidly growing market
- the increasingly large ships (up to 800 passengers) being used
- the fact that IAATO is only a voluntary organisation and one or two companies, usually Russian, are not members.

Table 7.3 summarises the potential impact of tourism on Antarctica's environment

While tourism is seen as a legitimate polar activity, it does need much more regulation. There is a need for:

- limits in tourist numbers
- mandatory safety codes for tourist vessels in Antarctic waters
- ratification of laws to prevent ships of more than 500 passengers from landing
- limiting the numbers of passengers going ashore to a particular site to a maximum of 80 to 100 at any one time, with one registered guide for every twenty people.

There was heightened urgency with the *MS Explorer* sinking in the Bransfield Straits in 2007 near the South Shetland Islands and the grounding of two further ships in 2008, which led to localised pollution from engine oil. This was followed by tourists becoming stranded in the ice shelf in an Australian research/expedition ship in 2014. However, compared to many other threats, if tourists are educated in responsible tourism, the impact may not be as great as suspected. Surveys from the heavily visited Port Lockroy, a former research base, suggest that tourists have had minimal impact on the Gentoo penguin colonies.

Figure 7.11 Port Lockroy, Antarctica

Table 7.3 Potential impact of tourism on Antarctica's environment

Potential impacts	Part of the environment at risk	Ways to minimise impact
Disturbance of wildlife	Breeding birds, hauled out seals	Impose minimum approach distances to wildlife Educate visitors on behaving responsibly
Litter, waste, fuel spills	Damages land-based ecosystems Marine wildlife, particularly seals and birds, becoming entangled in rubbish or coated in fuel	Ensure ship operation conforms to international maritime standards Ensure ships are ice-strengthened and have modern ice-navigation equipment Limit size of tourist vessels entering Antarctic waters Ensure ship has an oil spill response plan in case of an accident Oil spill equipment available and crew trained in clean-up techniques
Environmental degradation (e.g. trampling)	Fragile moss mats	Limit numbers going ashore Avoid sensitive areas Brief tourists before arrival
Removing historic artefacts, fossils, bones	Historic sites, fossils	Tell tourists not to collect souvenirs Brief tourists before arrival
Disruption of important scientific research	Research stations, field study sites	Allow only a few tourist visits per season Brief tourists before arrival Guide tourists around station

Human degradation of the landscape and ecosystems

Table 7.4 attempts to quantify the impact of activities such as walking on a range of commonly occurring ecosystems in glaciated environments.

Table 7.4 The effects of people walking on mountain ecosystems

	Plant community type	Number of people who could walk across it in a week before the plant cover is under threat (halved)
Increasing fragility ↑	Tundra 'flower' meadows	25 (visitors are encouraged to spread out but avoid delicate plants)
	Cushion mosses and moss mats	50 (can be very delicate; polygonal patterns under threat)
	Heathland	100
	Wet meadows	250 (some issues in permafrost areas)
↓ Increasing biodiversity	Mountain 'prairie' grasses or tundra tussock grasses	400 (generally resistant, but habitat of rare ptarmigan)

There seems to be a threshold beyond which the vegetation begins to suffer and the downward spiral of wear shown in Figure 7.12 sets in. Other types of tourism, such as pony trekking or mountain biking, are even more damaging than walking.

Inevitably glaciated highland areas suffer from soil erosion, especially when slopes are exposed, for

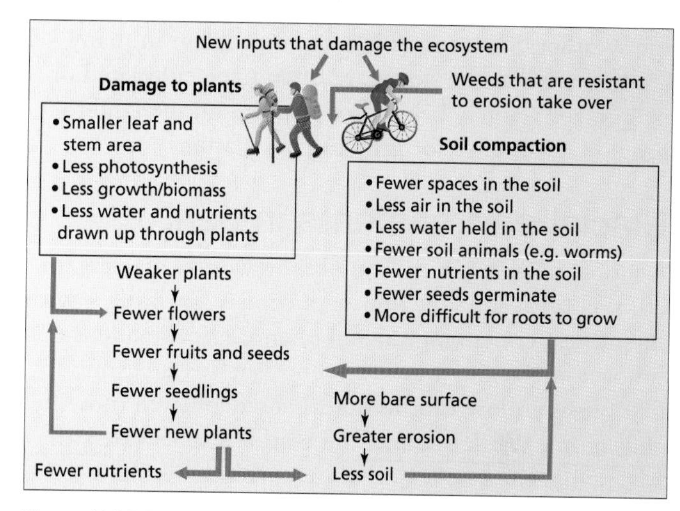

Figure 7.12 The effects of wear on an ecosystem

Fieldwork opportunity

A very useful fieldwork investigation is to examine the impact of outdoor recreation and the impacts of trampling on the landscape and ecosystems within a designated area by doing a footpath and track survey.

- Develop a visual scale for the degree of footpath erosion and carry out a visual and photographic survey of all designated paths within a defined area (use a choropleth mapping system).
- Develop a sampling strategy to carry out footpath erosion surveys, for example comparing a severely eroded stretch with a minimally eroded stretch at particular locations. Measure the width of the footpath and depth of erosion at regular intervals (rather like doing river cross-sections) to calculate the amount of erosion.
- At each of your chosen sites measure the angle of slope, carry out a texture test for a soil sample, a compaction survey (to measure the degree of trampling) and also use quadrants to carry out a vegetation survey, including height of vegetation, type of vegetation and percentage of bare ground at various sites across your erosion transects. Calculate an index of diversity for the vegetation.
- Carry out activity surveys on the paths and surrounding countryside in your chosen area to find out the pedestrian densities and whether any paths are used for highly damaging activities such as moto-scrambling, mountain biking or pony trekking.
- If there are screes, it is also worth investigating the impact of scree running on the stability of the scree slope.
- You can then develop a number of hypotheses to test, such as:
  - the link between slope and footpath erosion levels
  - the link between stoniness/soil texture and footpath erosion
  - the link between visitor activity and degree of erosion
  - the impact of visitors on different vegetation types.
- A further line of enquiry is to research any management plans and evaluate their success and suggest modifications. Alternatively, develop a management plan to improve the situation in an unmanaged area.

Key term

Phenology: The study of the timing of natural events and phenomena, such as the first day snowdrops appear, in relation to climate.

example by clear-cut tree felling or other examples of deforestation, as this exposes the fragile ecosystems to the weather. Soil erosion is a major problem in many Andean areas as the slopes are often over-cultivated or overgrazed, largely because of pressure on the land to provide subsistence for growing populations.

Glacial environments in peril

A global snapshot of the state of the world's glaciers (in 2015) clearly shows the impact of climate warming – with supporting climatological and phenological evidence (impact on various weather events, such as first snows, first pussy willow catkins out, first day of lawn mowing and so on). While some of the world's glaciers are still advancing, the vast majority are currently retreating. Here are some scary facts:

- On the eastern slopes of the Rocky Mountains, all the glaciers have lost between 25 and 75 per cent of their mass since 1850.
- In 1949 in Tajikistan, Central Asia, glaciers covered 18,500 km^2 of land; in 2012 the covered only 11,000 km^2, a 38 per cent decrease.
- Around 95 per cent of Himalayan glaciers are in rapid retreat; for example, the Khumbu Glacier (one of the highest in the world at the base of Everest) has retreated over 5 km since 1953.
- Areas in Peru and Bolivia covered by glaciers shrank by 25 per cent over the last 30 years.

There are very few areas where glaciers are expanding. One example is that of the maritime glaciers in Scandinavia; here the elements of changing precipitation have contributed to a more positive mass balance.

Data from the satellite surveys of the Greenland Ice Sheet shows a huge decrease in the ice-covered area, and new data from West Antarctica shows it is beginning to follow trends in East Antarctica with a massive loss in shelf ice.

Recent and regular surveys confirm that this melt and retreat is happening at an ever-increasing (exponential) rate as positive feedback is amplifying the process (see page 47).

less ice → loss of albedo → reduced reflection → more atmospheric warming → more melting

The destabilising effect of climate warming can be looked at in two very significant contexts – changes to the hydrological cycle and changes in sea level.

Changes to the hydrological cycle

This will have serious consequences for millions of people.

Mountainous areas and diminishing water supply

In mountainous areas such as the Andes and Himalayas, glacial meltwater feeds rivers; changes in discharge will have knock-on effects on sediment yield and water quality.

Rivers in Asia, such as the Mekong, Yangtze, Brahmaputra, Ganges and Hwange Ho, are all fed by Himalayan glacial meltwater. The loss of a steady supply has huge implications for the population powerhouses of India and China (together containing over a third of the world's population), both emerging as superpowers with almost insatiable demands for water for development of both their people's quality of life and their economies.

Western China's semi-desert area contains 350 million farmers dependent on water supplied from the glaciers of the Tibetan plateau, an area experiencing high amounts of glacial thinning. Water shortages could affect 538 million people – some 42 per cent of China's people – hence the development of massive hard-engineering solutions for water security such as dams, and the South–North water transfer scheme.

In India the reduction of glacial meltwater flowing into the Ganges-Brahmaputra system is likely to result in at least 500 million people facing water shortages, with nearly 40 per cent of India's irrigated (post-Green Revolution) land being affected.

The High Andes

Runoff from 'glacierised' basins is an important element of water budgets in the High Andes, assuring sufficient supplies of drinking water, HEP, ecosystem viability and integrity, as well as year-round flows for an increasingly intensive irrigation-based agriculture. Any changes induced by Andean glacier retreat will therefore have both economic and social consequences and will require adaptation measures.

Andean 'tropical glaciers' have declined by around sixteen per cent overall since 1970, but many smaller glaciers have disappeared completely, for example Cotacachi in Ecuador or, almost disappeared (82 per cent gone) Chacaltaya in Bolivia. While the rapid melting initially led to an unsustainable net increase in hydrological run off, now there is the issue of loss of biodiversity and declines in agriculture and tourism, with many almost water-less streams.

As glaciers cease to act as runoff regulators, seasonal water and HEP power supplies will be affected. Bolivian urban centres such as La Paz, El Alto (total population 2.5 million) rely on glacier meltwater for about 40 per cent of their drinking water supply. In Quito in Ecuador, the situation could be even worse, with glacial meltwater currently contributing 50 per cent of the water supply for its 2 million people. Changing hydrological conditions will affect water costs and the ability of the two urban areas to maintain vibrant economies.

Figure 7.13 The diminished size of Andean glaciers in Bolivia

Various scenarios have been modelled for the discharges of the Andean streams with and without melting, and there is no doubt that additional streams will need diverting for both urban areas from a wider area, and soon. Glacier retreat will also affect HEP generation (50 per cent of total energy in Ecuador and 80 per cent in Peru comes from HEP).

There are economic consequences of this reduction – including the extra costs of electricity as well as the possible need for rationing supply. There are cultural, almost spiritual, costs also for the Quechuan and Aymaran Indian people, who have long-revered the highest snow-covered Andean peaks as religious icons.

Rapid glacier retreat will disrupt the water cycle in a whole series of individual glacier-dependent basins, thus having a local impact on agricultural communities, but at the same time there will be national impacts as countries have to change their energy mix for generating electricity. For example, Peru may have to invest in additional thermal-based power stations or gradually develop alternative energy sources, as well as building additional reservoirs for drinking water, all at higher costs to its economy even though its own population has contributed little to the causes of climate change.

(Source: Summary of World Bank Report, source EOS Vol 818 No 25 June 2007)

7.3 The management of glaciated landscapes

There are a number of possible approaches to the management of cold environments, as shown in Figure 7.14.

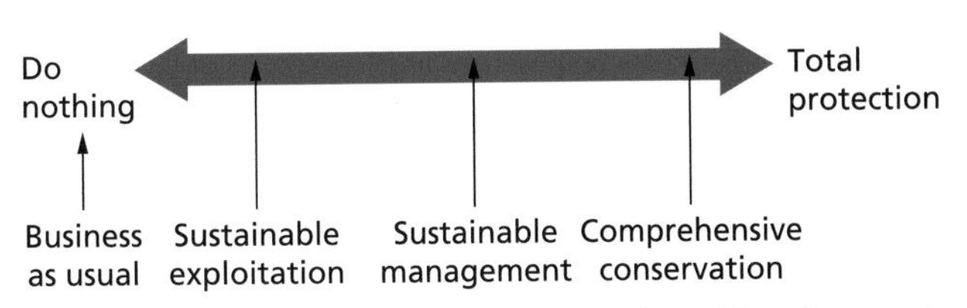

Figure 7.14 The spectrum of management for cold environments

- **Do nothing** lies at one end of the spectrum. It allows multiple economic uses to flourish. The ethos would be to allow cold environments to be exploited for whatever resources are in demand and profitable. This approach might be supported by governments at local or national level for revenues, or by some local people, for example chambers of commerce or trade unions for employment potential, or by developers such as industrialists and globalised TNCs, for example energy and mining companies.
- **Business as usual** is a very similar approach, leaving the area as it currently stands, but this might include aspects of pre-existing sustainability such as

self-regulation on environmental issues. All TNCs have pre-existing environmental policies as part of their mission statements. With the exception of conservationists, most players are content with the status quo.

- **Sustainable exploitation** can be regarded as a middle way; it targets development for profit but with the insistence on mandatory environmental regulation, for instance waste disposal. It can be channelled to provide distinctive benefits for the community, for example the development of fishing for local communities or sustainable hunting. In theory it takes into account the vested interests of many players at a variety of scales, but it relies on considerable compromise for it to be successful.
- **Sustainable management** attempts to develop an area in a way that uses resources for the benefit of the existing community without destroying the environment but, at the same time, conserving resources for future generations. The four facets of the sustainability quadrant or the three facets of the stool of sustainability (Figure 7.15) are very difficult to achieve, especially in cold environments.

There are tensional forces between the need to conserve fragile, vulnerable environments yet at the same time to exploit vital resources for the economic well-being of future generations. This is very clearly shown by the controversy over Alaskan oil where there is a clash between environmentalists, local indigenous peoples, state and national governments, and oil companies. Many NGOs, such as the World Wide Fund for Nature (WWF), favour sustainable development as an approach as ultimately it could be a good way to conserve the landscape and support indigenous communities.

- **Comprehensive conservation** aims to protect and conserve glacial and periglacial environments as wilderness, especially where still in a pristine condition. Only carefully regulated ecotourism or organic eco-farming is likely to be favoured by environmentalists and those allowed to practice and enjoy it (local businesses and tourists). Exploitative activities such as mining would not be permitted. Governments might be ambivalent towards it as, in the short term, less income might be earned.
- **Total protection** is an approach really only favoured by conservationists and some traditionalists among local people, as this does not permit access to the pristine environment at all, except perhaps for scientific monitoring and research purposes. It therefore does not allow local people to earn revenue from it, or tourists to enjoy it.

Which strategy is appropriate depends on the area and the interplay of the views of involved players. In some areas there are immediate crises, for others time to plan ahead. For most areas there are a number of alternative strategies; often these are appropriate only for certain parts of a large area. Zoning is often a very useful middle way, with the highest-value wilderness areas fully protected, possibly surrounded by areas where sustainable activity is permitted – for example, the buffer zone within a biosphere reserve, and designated areas which are targeted for economic development.

Figure 7.15a

Futurity	'Greenness'
Pro-poor	Bottom-up community involvement

Figure 7.15b

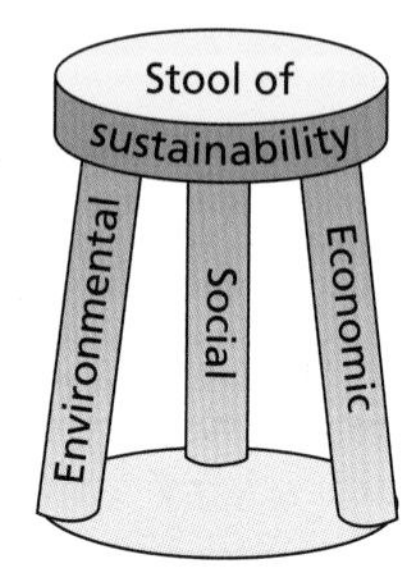

Figure 7.15 Assessing sustainability – 7.15a Quadrant; 7.15b Stool

Legislative frameworks to protect and conserve glaciated landscapes

Legislative frameworks can be developed at a number of scales and can aid the protection and conservation of glaciated landscapes considerably, *provided* they are closely policed and monitored. Mandatory legislation has 'teeth' and is likely to be far more successful in protection and conservation than frameworks and agreements (hard versus soft strategies).

International scale

Antarctica has had a unique system of international governance since the passing of the Antarctic Treaty in 1959 for all areas south of 60°S. The ATS established Antarctica as a continent of peace and science, and set all pre-existing territorial claims of seven countries to one side. Around sixty countries have now signed (including

Yosemite National Park – loved to death?

Yosemite National Park covers an area of 302,687 km^2 on the western slopes of the Sierra Nevada in California. As it encompasses a height range from 648 m to nearly 4,000 m, it has very high levels of biodiversity with giant sequoia stands as well as alpine meadows and the wetlands of the Merced valley. The scenery can best be described as 'awesome' with sheer sided granite domes over which spectacular waterfalls form combined with 'textbook' glaciated features such as troughs and hanging valleys. Yosemite finally became a National Park in 1916 after many years of partial Federal Administration, originally to protect its environment from exploitation from activities such as logging. World Heritage status for both natural and cultural categories was awarded in 1984.

Yosemite is the archetypal location to study the issues of conservation and protection versus the impacts of tourism and recreational use. The main problem is that of managing the 5 million visitors (2016), 90 per cent of whom are concentrated in just 6 per cent of the park's area in the main Yosemite valley. The classic visitor profile is that of day trippers who arrive by car and explore the main valley where the facilities are concentrated, as well as viewing from honeypot sites. There is accommodation for 16,000 overnighters. Only a small minority of visitors are long stay, active hikers who go up country to Wilderness areas which are all permit controlled to avoid over capacity and subsequent ecological damage.

Visitor numbers have doubled since the 1980s and a number of management problems have occurred. These include:

- The degradation of the natural vegetation with habitat fragmentation especially in the main Yosemite valley.
- The invasion by alien plant species.
- The frequent occurrence of wildfires, some started by people.
- Traffic congestion from visitors resulting in atmospheric pollution, for example, along Glacier Point Road.
- A major brown bear problem – they raid the garbage bins on the campsites
- Overcrowding by cars and 'selfie taking' tourists at the most beautiful views overlooking the waterfalls, with eroded trails.

As these problems have escalated, this has resulted in many attempts over the last 40 years to develop management plans. This has proved very difficult as there are many conflicts of interest, with a spectrum of opinions from Environmental groups such as the Sierra Club who favour protectionist strategies through to commercial enterprises at Curry Village who favour unfettered tourism access.

The general management principles to guide the planning and management of the Park in the future are:

A Retain the unique, priceless beauty of the whole park.

B Allow natural processes to prevail for the maintenance of healthy valley ecosystems and also to protect native Indian settlements.

C Promote visitor understanding and enjoyment by developing education programmes via high quality visitor services and facilities.

D Reduce traffic congestion and overcrowding by developing car parks outside the park with visitors transported by fleets of electric shuttle buses to park vistas, so cutting the number of private cars. Quotas and limits on day visitor numbers especially in peak periods would also support this.

Whilst there is general agreement on these underlying principles, some of the proposed strategies such as cutting the number of overnight beds and car parking facilities and putting the main visitor centre not only outside the main valley, but even outside the park, have proved extremely controversial. Many of the problems which result from over use for recreation and tourism can be resolved to an extent, but the underlying one of 'visitor management' is opposed by many groups as it could entail plans to control access for recreation and tourism to some sites at certain times. Visitors can be resistant to plans to disperse them away from the iconic vistas and they soon make their feelings known on social media.

all those with bases on Antarctica), representing some 88 per cent of the world's peoples. Since 1959 more than 250 recommendations and four separate international agreements have been adopted to form the Antarctic Treaty System (ATS):

- 1964 Agreed Measures for the Conservation of Antarctic Flora and Fauna (AMCAFF)
- 1972 Convention for Conservation of Antarctic Seals (CCAS)
- 1982 Convention for Conservation of Antarctic Marine Living Resources (CCAMLR)
- 1998 Protocol on Environmental Protection to the Antarctic Treaty (EP)

Now over 50 years old, the Treaty is recognised as one of the most successful international agreements, but there is the need to go a stage further towards Greenpeace's World Park concept.

However, Antarctica is a long way ahead compared to other areas and at least by being proactive and forward looking it is to an extent 'future proof'. The other large polar area, the Arctic, is in a far less healthy situation. While there are many similarities with Antarctica, the physiography is very different. It is essentially an oceanic area surrounded by powerful countries (including the superpowers USA and Russia). Moreover whereas 97 per cent of Antarctica is covered by ice, in the Arctic there are extensive areas of terrestrial tundra vegetation, as well as valuable marine ecosystems. As a result, over 4 million people live in the Arctic, a third of these being indigenous groups whose traditional way of life and cultural values must be protected and conserved.

Eight countries currently have territory and territorial waters within the Arctic Circle. These countries work together through the intergovernmental Arctic Council, which was set up by the 1996 Ottawa Declaration as a forum to promote co-operation among the Arctic states. Indigenous communities are consulted on key issues, such as sustainable development and environmental protection. It was strengthened in 2003 by the Polar Code which, in theory, enables nations to enforce stricter environmental regulations in Arctic areas.

The UN manages territorial disputes using vehicles such as the UN Convention for the Law of the Sea (UNCLOS), but recently increased geopolitical tensions have occurred as the melting sea ice has led to greater accessibility via the opening of new sea routes, and the feasible exploitation of more mineral deposits. Many would argue that the Arctic Council and its framework could be strengthened into a treaty-based organisation with regulatory powers in order to directly manage issues such as territorial claims, fishing rights and quotas, biodiversity conservation and shipping.

Other examples of international co-operation include the *Alpine Convention* and the *Svalbard Treaty*.

National scale

At a national scale there are frameworks for the development of a whole hierarchy of conservation areas ranging from National Parks through to areas of special scientific interest, with varying rules and regulations for each country concerning permitted activities and access.

In the Arctic, for example, there are probably more protected areas than in any other region, with over fifteen per cent of the area protected because of the relatively few competing land uses there (compared to Alpine areas). All eight countries have some parts of their Arctic territories protected – for example 56 per cent of Alaska (USA) has some kind of protected status.

Within Alaska there are many National Parks, with federally owned land. Some, like the Arctic National Wildlife Refuge (ANWR), are vast wildernesses with a range of ecosystems from boreal forest in the south and treeless tundra in the north, with many animal species such as wolves, caribou, bears and snow geese. However, in the continued search for oil – 'black gold' – to reinforce US energy security, permission has now been granted to drill in part of the ANWR. So much for the protection status!

Finland has an alternative system. Around 33 per cent of its Arctic region is protected with a system of National Parks for public access, but also a number of strict Nature Reserves with very limited public areas. An example is Kevo in Finnish Lapland, which has only a few marked trails, with entry by permit only.

Global systems for conservation

Individual species are protected by global strategies such as Convention on International Trade in Endangered Species (CITES) signed in 1973 – many Arctic species are on the list, including polar bears and walruses.

Conservationists would challenge their effectiveness, citing the examples of whales and elephants (ivory trade). Whales also are protected but limited numbers are allowed to be caught for scientific purposes, usually by Russia and Japan, leading to global protests from organisations such as Greenpeace.

The International Union for Conservation of Nature (IUCN) has published a Red List for endangered species, with polar bears currently classified as 'vulnerable'. However, so far the US refuses its accepted designation as this would mean acknowledging climate warming as the main reason for vulnerability.

UNESCO administers a global system of World Heritage Sites, with site being listed because of their ecological or cultural importance. Wrangel Island (owned by Russia) was designated a World Heritage Ecological Site in 2004 as it has a very high level of biodiversity (including 23 endemics, large numbers of polar bear breeding dens, feeding grounds for grey whales, the world's largest population of Pacific walrus and is a nesting ground for over 100 migratory species of bird). Some of the explorer's huts in the Antarctic are designated cultural sites as they contain historic artefacts from expeditions such as those of Scott.

The real issue is that without overarching legislation, conservation and protection is spatially piecemeal. Also, as these global interventions are restricted to specifics, they are reliant on national government priorities, which frequently change.

Much of the legislation is without real 'teeth' (except in Antarctica) and rarely gives protection from growing tourist activities (the SMART project is only a loose agreement to carry out sustainable tourism in the Arctic). National economic 'needs' mean that activities such as mining and drilling are allowed to override environmental protection in some cases. Moreover, not even the ATS can give protection against climate warming and trans-boundary pollution, which are considered by many decision makers to be the most pressing issues.

Futures

Climate warming is described as a context hazard area with worldwide, far-reaching environmental consequences. In spite of the efforts to model its impact using emissions scenarios by numerous scientists, usually working for the Intergovernmental Panel on Climate Change (IPCC), it is very difficult to forecast its likely environmental and socio-economic impact.

Most scientists now consider that rising levels of greenhouse gases from human activities are the key driver of the scale and pace of any changes in climate. The problem is that changes are amplified by positive feedback loops, such as the melting of the glaciers, ice caps and ice sheets, which lower the albedo, or the melting of permafrost which releases large quantities of methane, a very powerful greenhouse gas.

As a result, active glacier environments face multiple stresses as global changes on the physical environment impact on the human environment in a multitude of ways.

Changes would occur on a number of fronts, often as an indirect outcome from climate warming. A number of questions can be asked:

- **Would the world's climates become more varied and more hazardous?** Certainly changing weather patterns are forecast, with more extreme events leading to higher incidences of weather hazards and disasters brought about by changes in atmospheric circulation.
- **What will happen to our life support systems of water and carbon cycles?** Certainly water insecurity will increase for many people and communities for a variety of reasons and, with the melting of permafrost and the loss of masses of methane, there will be losses in the carbon stores and the potential for instability within the carbon cycle.
- **Which of the glaciated and periglacial environments are most vulnerable to climate warming? In particular how vulnerable are relict environments compared to active environments, or upland areas compared to lowland areas?** This is a very complex issue. While active landscapes are likely to experience more dramatic impacts of melting from climate warming than relict environments, upland areas are more exposed to the impacts of temperature changes than lowland areas.
- **What will happen to sea levels?** Certainly they will rise – various scenarios predict by up to 0.5 to 0.75 m by the end of the twenty-first century – putting at least 100 million people's lives at risk. All the US cities from Miami to Boston, large cities in Asia, Bangladesh and also many delta areas and coral reef islands will be flooded or beset with saltwater incursions, affecting people's subsistence, should the major ice sheets melt.
- **What will happen to wildlife?** Iconic images of polar bears abound with no pack ice to feed off; their lives will be threatened as there will be complex changes in food webs, especially in polar areas. The loss of krill, perhaps linked to decreasing sea ice cover, will have consequences for the whole Antarctic food web. Note that it's not all decimation – many groups of animals such as seals extend their habitats by migration.
- **What will happen to the oceanic circulation?** The Doomsday scenario is for the total collapse of the thermohaline circulations and complete disruption of ocean currents. Recent modelling suggests a 20 to 50 per cent collapse as ice sheets melt and icebergs cool the water, partially blocking the passage of currents such as the Gulf Stream.
- **What will happen to people's lifestyles, especially traditional lifestyles of indigenous people whose hunting habits will be affected as migration patterns and numbers of animals**

and fish change? Diseases and pests are spreading pole-wards with major implications for biodiversity.

With the threat of these changes and the multiplication of stressors, there is general agreement that drastic action must be taken to manage the impacts of climate change in order to avoid reaching the tipping point (the point at which climate changes occur irreversibly and at an increasing rate), originally seen as a certain level of greenhouse gas emissions (450 ppm carbon dioxide) or as a threshold of global temperature rise of 2 °C. The visible manifestation would be the melting of the Greenland Ice Cap, the melting of the East and parts of the West Antarctica Ice Sheet, and partial collapse of the global ocean current system.

The two-pronged strategy of *mitigating* against the impacts and *adapting* to them has to take place at all scales. Most scientists say that even if effective mitigation could occur, by 'stopping' greenhouse gas pollution, there is so much delay in the system that adaptation is likely to be necessary too, especially as an interim measure.

As Table 7.5 shows, every little helps towards improving the global situation as local strategy can often be replicated more widely.

Think local ⟶ Act global

Act local ⟵ Think global

So, will current ice-covered areas become relict areas as we head towards a really warm interstadial, or could the situation change and the Earth suddenly swing into a new stadial, or even a new ice age event, with the glaciers returning to their maximum extent?

Table 7.5 Examples of strategies available for mitigation

Individual	Local	National	Global and international
Lifestyle and consumption changes	Local government strategies for planning, recycling, transport, micro-energy, green homes, etc.	General policies and legislation, e.g. on energy mix, transport, taxation	International agreements, e.g. Kyoto; co-operation to control emissions and agree targets for carbon dioxide emissions

Review questions

1 Research the writings of a range of famous people on the spiritual value of wilderness and explain what the writers feel about it.

2 Use the British Antarctic Survey (BAS) website to write a report on current scientific research in Antarctica. Can we justify the billions spent globally?

3 Research the value of 'micro-hydros' and explain how this form of energy has revolutionised the quality of life in rural areas in the developing world such as Nepal or Bolivia.

4 Use an internet search engine to research the Alpine Convention. Evaluate its strength as a strategy for conservation and protection.

5 Explain why it is so difficult for the countries of the world to develop a common strategy for mitigating global warming. You can research the Climate Conferences of Johannesburg and Copenhagen and the recent 2015 Paris Conference, which has offered more hope and commitment, to support your answer.

6 Research the latest IPCC report on the likely impacts of climate change.
   a Explain why there are a number of scenarios for each impact.
   b What are the direct and indirect consequences for glaciated and periglaciated environments?

Further research

On the PBS Glacier Hazards from Space page you can view a series of glacier-related disasters and risks imaged by a NASA satellite: www.pbs.org/wgbh/nova/earth/glaciers-satellite.html

Visit Extreme Ice Survey, which illustrates visually the impact of climate change on glaciers: http://extremeicesurvey.org

Take a look at an interesting blog 'from a glacier's perspective': http://glacierchange.wordpress.com

Exam-style questions

AS questions

Study Table 1 below. It shows the orientation of striations (ice scratches) on the surface of an area of rocks eroded by an ice sheet in Labrador, Canada.

Table 1

Orientation (°)	Frequency	Orientation (°)	Frequency
350	2	30	2
357	3	32	3
360	4	35	5
12	3	40	6
15	3	42	9
18	3	44	11
20	3	45	12
25	4	48	10
26	4	50	6
28	4	52	3

1 Name the glacial process which caused/formed the striations. [1]

2 Which of the following best describes the modal direction for the orientation of the striations? Tick the correct answer. [1]
   a NW ↔ SE
   b NE ↔ SW
   c N ↔ S
   d W ↔ E

3 Explain the importance of **either** glacial landforms **or** glacial deposits when 'working out' the provenance of the ice. [6]

4 Assess the importance of glacial landscapes in providing a resource for humans. [12]

5 Study Figure 1 below. It shows part of a stone board a student uses to record the differences between glacial and fluvial-glacial deposits.
   There are two methods, Powers' scale and the Cailleux Index. Explain **one** strength and **one** weakness of each method. [6]

6 You have carried out field research investigating how glacial landscapes have changed since the last ice age.
   Assess the value of the primary data you collected in helping you form conclusions to your research question. [9]

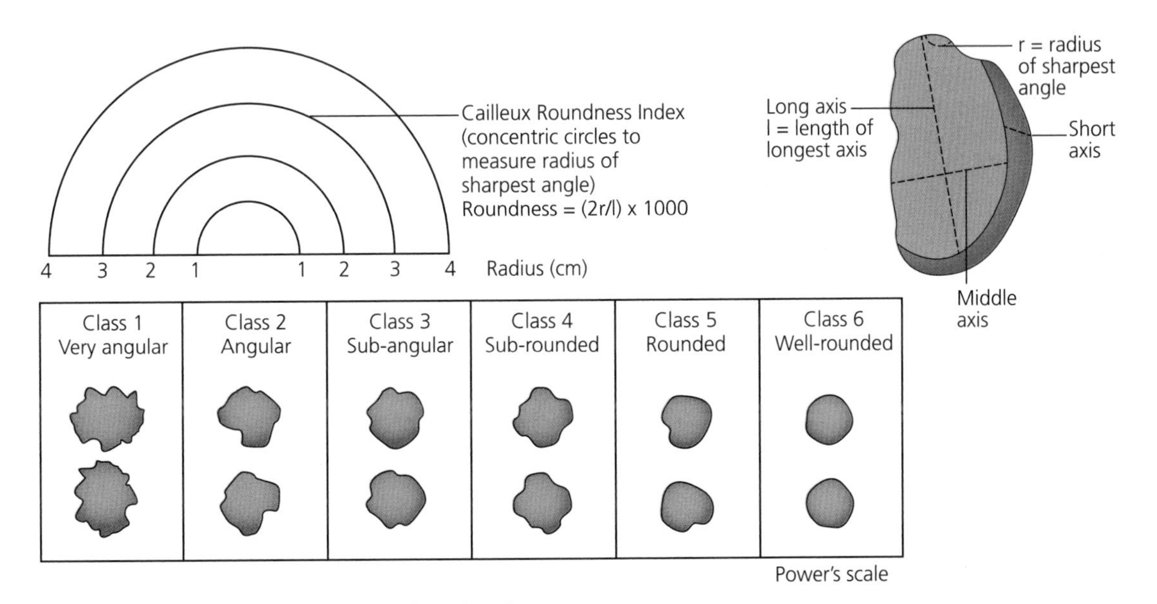

Figure 1: Stone board and sorting scale estimation

A level questions

1 Study Figure 2 below. Explain the causes of the variations in glacier mass balance at annual and longer-term scales. [6]

2 Study Figure 2 below. Explain why some glaciers move at faster rates than others. [6]

3 Explain how glacial meltwater plays a significant role in contributing to glacial landscapes. [8]

4 Evaluate the extent to which the actions of people are the main threat facing the survival of both active and relict glaciated and periglacial landscapes. [20]

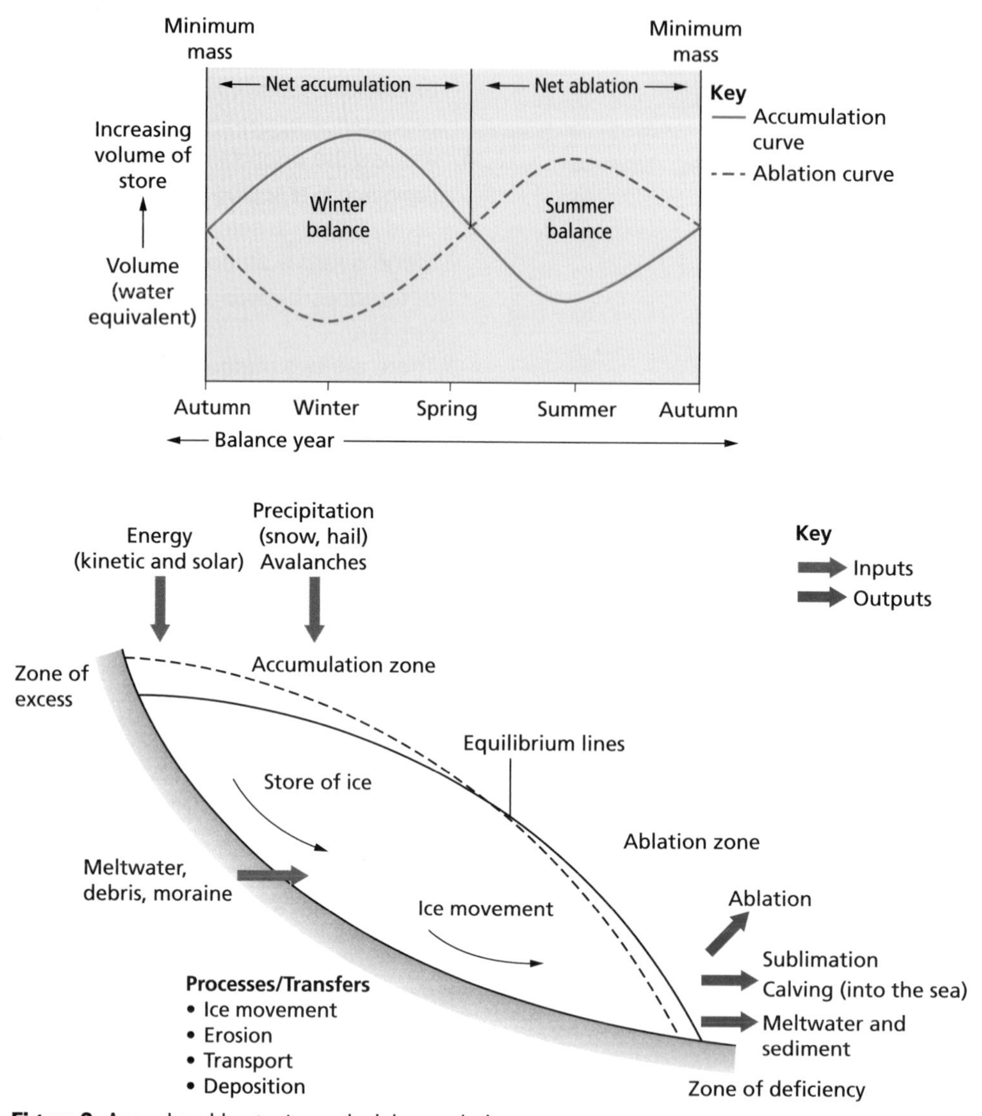

Figure 2: Annual and longer-term glacial mass balances

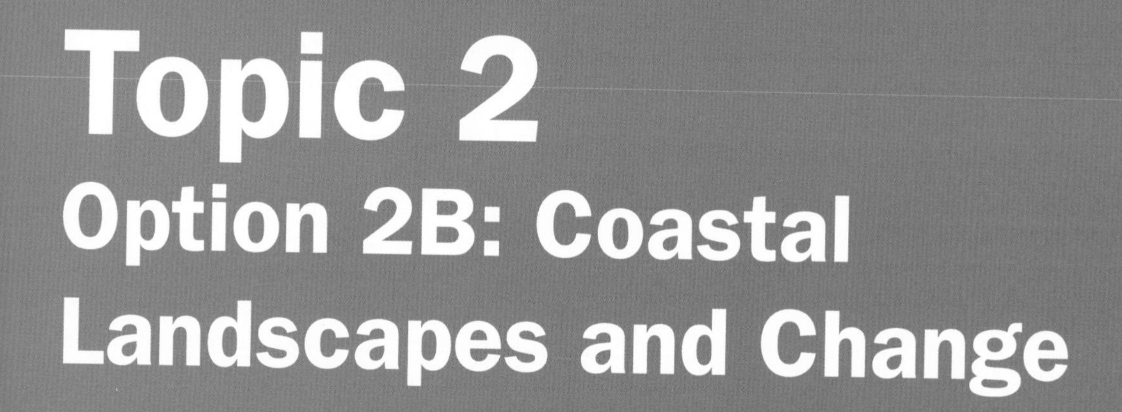

Topic 2

Option 2B: Coastal Landscapes and Change

8 Coastal processes

Why are coastal landscapes different and what processes cause these differences?

By the end of this chapter you should:

- understand the variety of coastal landscapes and why they differ
- understand the importance of underlying geology in generating different coastal landscapes and features
- understand the influence of geology and other factors in affecting rates of coastal retreat.

8.1 The coastal zone

Very large numbers of people live near the coast. Globally over 1 billion people live on coasts that are at risk from flooding, and roughly half of the world's population live within 200 km of the coast. They live in a dynamic environment because:

- Coasts represent a boundary zone where land and sea meet, and where both marine and terrestrial processes operate and interact.
- Coasts experience extreme events, including tropical cyclones, storms and tsunamis, which although rare can cause significant, rapid change.
- Human development on coasts is very varied (ports and transport, industrial locations, residential and tourism land uses) and constantly changing.

The littoral zone

The littoral zone can be divided into a number of subzones as shown in Figure 8.1. The backshore zone, above high tide level, is only affected by waves during exceptionally high tides (often called spring tides) and during major storms. Wave processes are normally confined to the foreshore between the high and low tide marks. Shallow water areas close to land are termed the nearshore. This zone is often one of intense human activity, such as fishing and leisure, but also forms part of the physical system of the coastline through transfers of sediment by currents close to the shore.

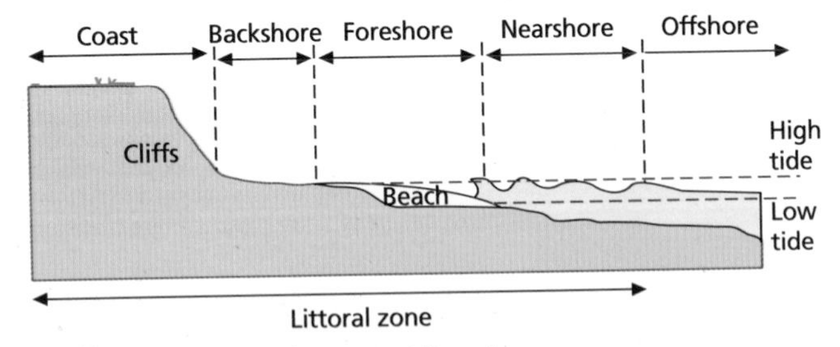

Figure 8.1 The littoral zone and its subzones

All coastlines can be divided into similar zones, but not all coastlines have similar landscapes. A distinction can be made between two main types of coast:

- Rocky (or cliffed) coastlines, which have cliffs varying in height from a few metres to hundreds of metres; cliffs are formed from rock but the hardness of the rock is very variable.
- Coastal plains; the land gradually slopes towards the sea across an area of deposited sediment, with sand dunes and mud flats being the most common examples. These coasts are sometimes referred to as alluvial coasts.

Cliffs create a clear, sometimes spectacular, distinction between land and sea, whereas coastal plains often have a blurred boundary (Figure 8.2).

Coasts can be classified in a variety of different ways according to different physical features and processes, as shown in Table 8.1.

Rocky coastlines

About 1000 km of the UK coastline consists of cliffs, but the relief of these is variable. In some locations there are very high cliffs, including:

- Hangman's Cliffs in Devon, which rise 318 m above the Bristol Channel
- Boulby Cliff in North Yorkshire, which is 203 m above the North Sea
- Conachair on Hirta in the Outer Hebrides has a sea cliff of 427 m, the highest in the UK.

Key term

Littoral zone: The wider coastal zone including adjacent land areas and shallow parts of the sea just offshore.

Cliffed coast

Chalk cliffs at Flamborough Head in Yorkshire.

- The transition from land to sea is abrupt.
- At low tide the foreshore zone is exposed as a rocky platform (wave-cut platform).
- The cliffs here are vertical, but cliff angles can be much lower.

Sandy coastline

Sand dunes fringe many coastal plains, as seen here in Belgium.

- At high tide the sandy beach is inundated, but the vegetated dunes are not.
- Dune vegetation plays a crucial role in stabilising the coast and preventing erosion.

Estuarine coastline

Estuaries are found at the mouths of rivers, such as here at Lymington in Hampshire.

- Extensive mud flats, cut by channels, are exposed at low tide but inundated at high tide.
- Closer to the backshore the mud flats are vegetated, forming a salt marsh.
- This type of coastline gradually transitions from land to sea.

Figure 8.2 Contrasting coastlines

Table 8.1 A classification of coastal environments

Formation processes	**Primary coasts** are dominated by land-based processes such as deposition at the coast from rivers or new coastal land formed from lava flows. **Secondary coasts** are dominated by marine erosion or deposition processes.
Relative sea level change	**Emergent coasts** where the coasts are rising relative to sea level, for example due to tectonic uplift. **Submergent coasts** are being flooded by the sea, either due to rising sea levels and/or subsiding land.
Tidal range	Tidal range varies hugely on coastlines, meaning coasts can be: • **Microtidal coasts** (tidal range of 0–2 m) • **Mesotidal coasts** (tidal range of 2–4 m) • **Macrotidal coast** (tidal range greater than 4 m)
Wave energy	**Low energy** sheltered coasts with limited fetch and low wind speeds resulting in small waves. **High energy** exposed coasts, facing prevailing winds with long wave fetches resulting in powerful waves.

These high-relief cliffs are composed of relatively hard rock. In the case of Conachair, the rock type is granite. There are two main cliff profile types (Figure 8.3). Where marine erosion by wave action dominates, cliffs tend to be steep, unvegetated and there is little in the way of rock debris at the base of the cliff, because it is quickly broken up by erosion and carried away by waves. Cliffs which are not actively eroded at their base by waves have shallower, curved profiles and lower relief. On these cliffs the sub-aerial processes of surface runoff erosion and mass movement (landslides) slowly move rock and sediment downslope, but the limited amount of marine wave erosion means it is not removed.

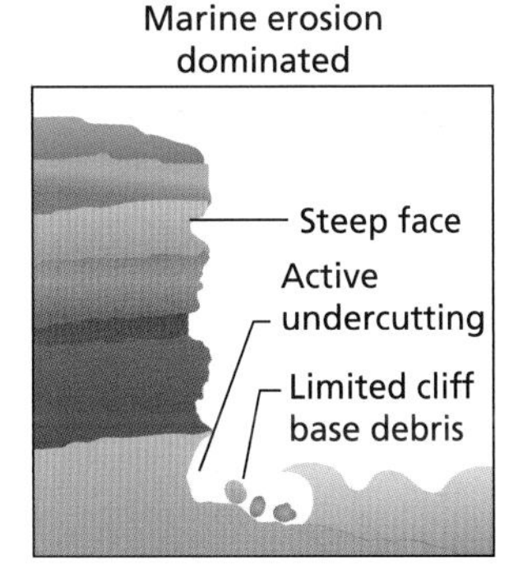

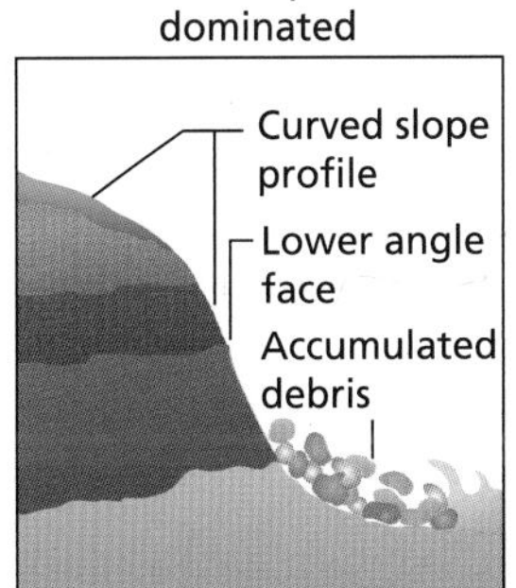

Figure 8.3 Marine and sub-aerial cliff profiles

Key term

Cliff profile: The height and angle of a cliff face as well as its features, such as wave-cut notches or changes in slope angle.

Key concept: Sub-aerial processes

It is tempting to think that cliffs are caused by the action of waves pounding at their base. Wave erosion is important but is not the only process acting on cliffs. Other processes are termed sub-aerial and include:

- **weathering:** the chemical, biological and mechanical breakdown of rocks into smaller fragments and new minerals *in situ*
- **mass movement:** landslides, slumps and rock falls, all of which move material downslope under the influence of gravity
- **surface runoff:** water, usually during heavy rain, flowing down the cliff face and causing erosion of it.

Sub-aerial processes are likely to be more important when cliffs are made of less-resistant rocks such as shale, clay or mudstone.

Erosion resistance

Rocks forming cliffs vary in terms of their resistance to erosion and weathering. Resistance refers to the 'hardness' of rock and this is influenced by:

- how reactive minerals in the rock are when exposed to chemical weathering; calcite (found in limestone) can be weathered by solution, whereas quartz (found in sandstone) is not subject to chemical weathering
- whether rocks are clastic (sedimentary rocks like sandstone, made of cemented sediment particles) or crystalline (igneous and metamorphic rocks, made of interlocking crystals); the latter are more erosion resistant
- the degree to which rocks have cracks, fractures and fissures, which are weaknesses exploited by the forces of weathering and erosion.

For sedimentary rocks, the cement holding sediment clasts together is important. If the cement (usually minerals such as iron oxide, calcite or quartz) is weak, then the rock will be weak. The weakest rocks are sedimentary rocks, which are barely cemented at all. These are termed 'unconsolidated' and include gravels, boulder clay and sands. As can be seen in Table 8.2, coastal recession rates are very variable, reflecting different degrees of resistance to erosion.

Table 8.2 Typical coastal recession rates for five different rock types

Rock type (lithology)	Coastal recession rates (cm per year)
Granite (igneous)	0.1–0.3
Limestone (sedimentary)	1–2
Chalk (sedimentary)	1–100
Sandstone (sedimentary)	10–100
Boulder clay (unconsolidated)	100–1000

Coastal plains

Coastal plains are low-lying, low-relief areas close to the coast. Many coastal plains contain wetlands and marshes because they are only just above sea level and poorly drained due to the flatness of the landscape. Coastal plains form in one of two ways:

- Some, such as the Atlantic coastal plain in the USA, are a result of a fall in sea level exposing the seabed of what was once a shallow continental shelf sea.
- Deposition of sediment from the land, brought down to the coast by river systems, can cause coastal accretion where the coastline gradually moves seaward, such as in a river delta.

In many locations coastal plains are maintained in a state of dynamic equilibrium by the balancing forces of:

- deposition of sediment from river systems inland and deposition of sediment from offshore and longshore sources
- erosion by marine action at the coast.

Overall, coastal plains are often a low-energy environment, usually lacking very large and powerful waves except on rare occasions such as during hurricanes.

Key terms

Coastal accretion: The deposition of sediment at the coast and the seaward growth of the coastline, creating new land. It often involves sediment deposition being stabilised by vegetation.

Dynamic equilibrium: The balanced state of a system when inputs and outputs balance over time. If one element of the system changes because of an outside influence, the internal equilibrium of the system is upset and other components of the system change. By a process of feedback, the system adjusts to the change and the equilibrium is regained.

8.2 Geological structure and landscape

Key concept: Geological structure

This refers to the arrangement of rocks in three dimensions. There are three key elements to geological structure:

- **Strata**: the different layers of rock within an area and how they relate to each other.
- **Deformation**: the degree to which rock units have been deformed (tilted or folded) by tectonic activity.
- **Faulting**: the presence of major fractures that have moved rocks from their original positions.

All of these aspects of geological structure influence coastal landscapes and the development of landforms, often as much as the specific lithologies (rock types) at the coast.

Figure 8.4 illustrates several components of geological structure.

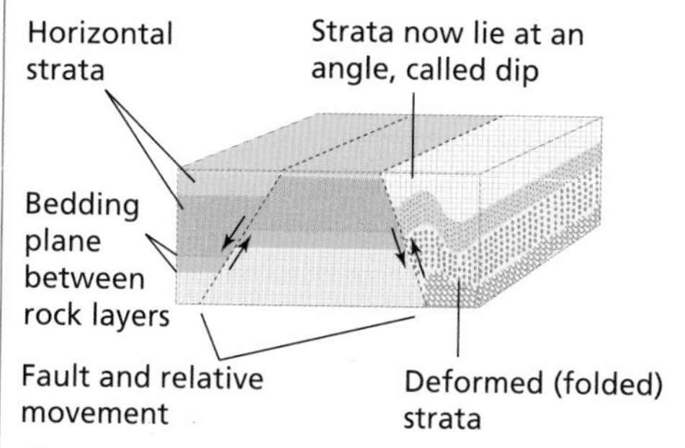

Figure 8.4 Block diagram showing key elements of geological structure

Concordant and discordant coasts

Geological structure produces two dominant types of coast:

- **Concordant**, or Pacific coasts, are generated when rock strata run parallel to the coastline.
- **Discordant**, or Atlantic coasts, form when different rock strata intersect the coast at an angle, so geology varies along the coastline.

Discordant coastlines are dominated by headlands and bays. Less-resistant rocks are eroded to form bays whereas more resistant geology remains as headlands protruding into the sea. Figure 8.5 shows the example of the West Cork coast in Ireland. Here rock strata meet the coast at 90 degrees in parallel bands. Weak rocks have been eroded, creating elongated, narrow bays, whereas more-resistant rocks form headlands.

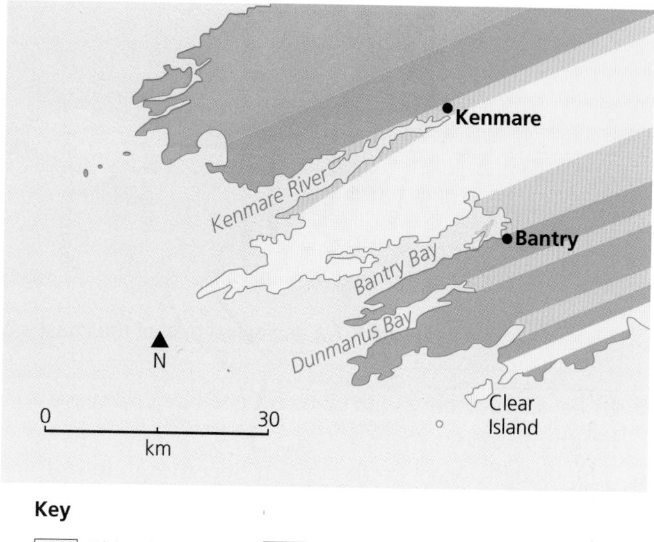

Figure 8.5 The discordant coast of West Cork, Ireland

Especially resistant areas remain as detached islands, such as Clear Island. This coastline also illustrates the complexity of coasts. The long, narrow bays are actually rias (drowned river valleys). In the past rivers eroded the softer rocks to form valleys, which were subsequently drowned by rising sea levels at the end of the last ice age. Since then, marine erosion has continued to erode the softer bays.

On coasts with headlands and bays, the headlands are eroded more than the bays. This tends to reduce the difference between headlands and bays and 'smooth' the coastline over time. This happens because of the effect such coasts have on wave crests (see Figure 8.6).

- In deep water wave crests are parallel.
- As waves approach the shallower water offshore of a headland they slow down and wave height increases.

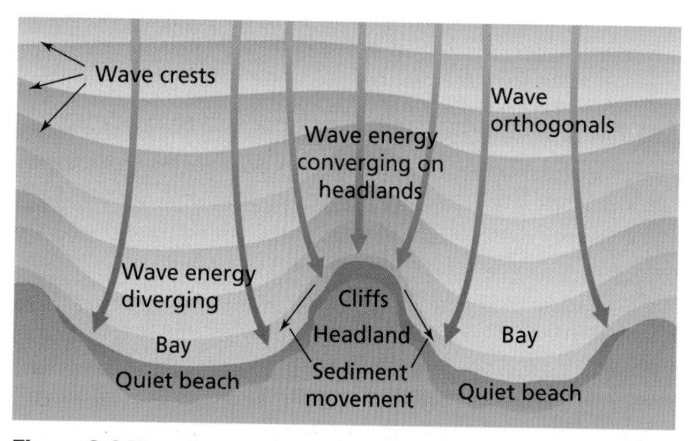

Figure 8.6 The effect of headland and bays on waves

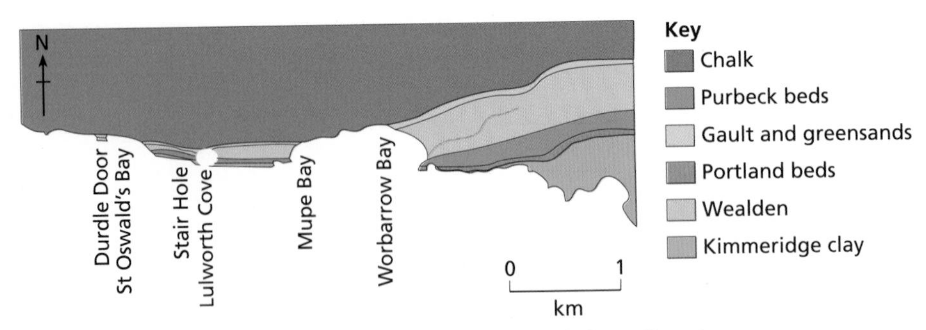

Figure 8.7 A geological map of the coast around Lulworth Cove, Dorset

- In bays, wave crests curve to fill the bay and wave height decreases.
- The straight wave crests refract, becoming curved, spreading out in bays and concentrating on headlands.
- The overall effect of wave refractions is to concentrate powerful waves at headlands (meaning greater erosion) and create lower, diverging wave crests in bays, reducing erosion.

Concordant coasts are more complex in nature. The classic example is the coast around Lulworth in Dorset. Geological strata run parallel to the coast (see Figure 8.7) but vary in terms of their resistance to the sea:

- The hard Portland limestone and fairly resistant Purbeck beds protect much softer rocks landward (the Wealden and Gault beds).
- At Lulworth Cove and Worbarrow Bay, marine erosion has broken through the resistant beds, and then rapidly eroded the wide coves behind.
- At the back of these coves is the resistant chalk, which prevents erosion further inland.

The coast of Dalmatia, Croatia, in the Adriatic Sea is another example of a concordant coastline. Like the West Cork coast in Ireland it was drowned by sea level rise during the Holocene. The geology of Dalmatia is limestone. It has been folded by tectonic activity into a series of anticlines and synclines that trend parallel to the modern coastline (see Figure 8.8). This underlying structure of upstanding anticlines and lower syncline basins (which would have been eroded by rivers in the past) has been drowned by rising sea levels to create a concordant coastline of long, narrow islands arranged in lines offshore.

Haff coastlines are another type of concordant coast found on the southern fringes of the Baltic Sea. Long sediment ridges topped by sand dunes run parallel to the coast just offshore, creating lagoons (haffs) between the ridges and the shore.

Key term

Holocene: The geological epoch that began about 12,000 years ago at the end of the last Pleistocene ice age. Its early stages were marked by large sea level rises of about 35 m and a warming interglacial climate.

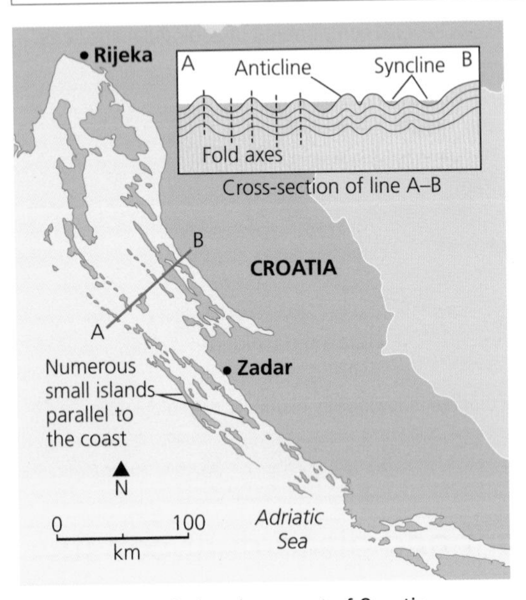

Figure 8.8 The Dalmatian coast of Croatia

Cliff profiles

Cliff profiles are influenced by several different aspects of geology. Two characteristics are dominant:

1. The resistance erosion of the rock.
2. The dip of rock strata in relation to the coastline.

Dip, simply meaning the angle of rock strata in relation to the horizontal, is important. Dip is a tectonic feature. Sedimentary rocks are formed in horizontal layers but can be tilted by plate tectonic forces. When this is exposed on a cliffed coastline it has a dramatic effect on cliff profiles, as shown in Figure 8.9.

Horizontal dip
Vertical or near vertical profile with notches reflecting strata that are more easily eroded.

Seaward dip, high angle
Sloping, low angle profile with one rock layer facing the sea; vulnerable to rock slides down the dip slope.

Seaward dip, low angle
Profile may exceed 90° producing areas of overhanging rock; very vulnerable to rock falls.

Landward dip
Steep profiles of 70–80° producing a very stable cliff with reduced rock falls.

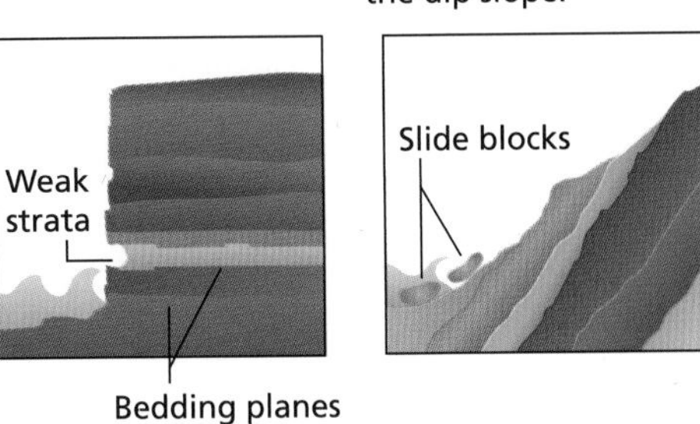

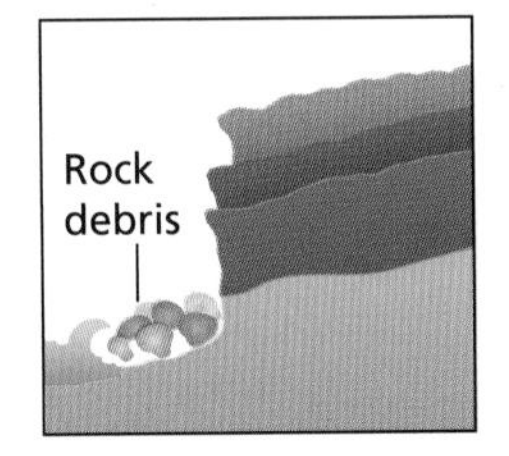

Figure 8.9 The influence of dip on cliff profiles

> **Key term**
>
> Faults: Major fractures in rocks produced by tectonic forces and involving the displacement of rocks on either side of the fault line.

In addition to the dip of strata, other geological features influence cliff profiles and rates of erosion (see Figure 8.10).

- Faults represent major weaknesses within rock layers. Either side of a fault line, rocks are often heavily fractured and broken and these weaknesses are exploited by marine erosion.
- **Joints** occur in most rocks, often in regular patterns, dividing rock strata up into blocks with a regular shape.
- **Fissures** are much smaller cracks in rocks, often only a few centimetres or millimetres long, but they represent weaknesses that erosion can exploit.

The location of microfeatures found within cliffs, such as caves and wave-cut notches, are often controlled by the location of faults and/or strata which have a particularly high density of joints and fissures.

Regular patterns of jointing in rocks are often a result of sedimentary rocks being folded by tectonic forces. Folding occurs due to crustal compression. When horizontal strata are 'squeezed' they can be folded into a series of anticlines and synclines. The folding is usually accompanied by the development of joints and fissures. There are two basic types of folds: anticlines and synclines. Anticline folds are described as convex up (A-shaped) while synclines are concave down (U-shaped). The two types of fold usually occur together.

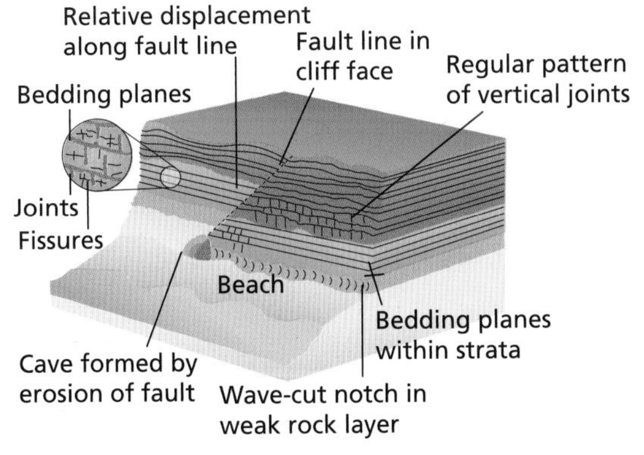

Figure 8.10 The influence of geological structure on cliff features

Northumberland Heritage Coast

Folding can produce some interesting coastal features. Figure 8.11 shows Cullernose Point on the Northumberland Heritage Coast. In the background is the point itself, which is a headland made of highly resistant dolerite (an igneous rock similar to basalt).

- The cliff on the left in the foreground is made of sedimentary limestone; micro-features like horizontal bedding planes, vertical joints and small caves can be seen within the layers of limestone.
- To the right on the foreshore, and beneath the limestone cliff, are a series of folds.
- There is a prominent anticline on the left, plus an anticline and syncline on the right with waves breaking over it.
- Erosion has removed relatively soft rock strata, exposing a hard layer of grey limestone and numerous folds.

Figure 8.11 Cullernose point on the Northumberland Heritage Coast

Skills focus: Drawing a fieldsketch

Fieldsketching is an important geographical skill. Picking up a pencil to draw what you see might seem a little old fashioned in an age of digital images, but a fieldsketch has a different purpose to a photograph. Fieldsketches:

- focus on identifying key features in a landscape, rather than trying to record everything
- are annotated with labels to explain observed features
- help you to remember what you saw
- provide an opportunity to apply key terminology to a real place.

Fieldsketches are not art; they are neat, outline drawings, so no artistic skills are needed. There are seven easy steps to drawing a good fieldsketch:

1. Decide on the view you want to record, and draw a frame.
2. Divide the frame up into four, this will make sketching in features much easier.
3. Draw in the skyline/horizon as a starting point and any major features in the foreground; this will divide up your sketch into a small number of large areas.
4. Add in the detail, but do not try to record everything.
5. Important minor features can be added in using close-up sketches.
6. Annotate the key features.
7. Add a title and location.

Figure 8.12 is a fieldsketch of the view in Figure 8.11 that illustrates the process of completing a fieldsketch.

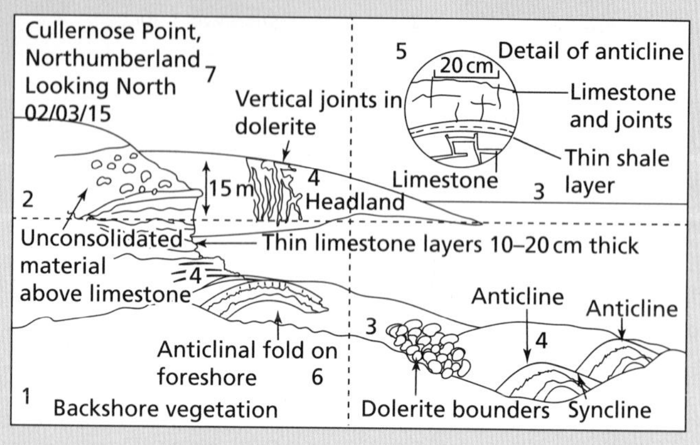

Figure 8.12 Fieldsketch of Cullernose Point, Northumberland

8.3 Rates of erosion

A crucial element of coastal management is understanding rates of erosion, or rates of recession. The latter term means how fast a coastline is moving inland. This is influenced by many factors but the key one is lithology, or rock type. The three major rock types – igneous, sedimentary and metamorphic – erode at different rates.

There are exceptions to the general order of resistance to erosion. Some limestones, which are actually crystalline sedimentary rocks, are erosion resistant, as are ancient sandstones that have been compressed and compacted over hundreds of millions of years. Recently erupted volcanic lava flow rocks and layers of volcanic ash or tephra (called volcani-clastic rocks) tend to be weak and easily eroded. The weakest coastal material is unconsolidated sediment.

Many cliffed coastlines are not made of only one rock type. They are composite cliffs with different rock layers, possibly from different geological periods. This produces

Table 8.3 Rates of erosion for different rock types

Rock type	Examples	Erosion rate and explanation
Igneous	Granite Basalt Dolerite	VERY SLOW • Igneous rocks are crystalline; the interlocking crystals make for strong, hard erosion-resistant rock. • Igneous rocks such as granite often have few joints, so there are limited weaknesses that erosion can exploit.
Metamorphic	Slate Schist Marble	SLOW • Crystalline metamorphic rocks are resistant to erosion. • Many metamorphic rocks exhibit a feature called foliation, where crystal are all orientated in one direction, which produces weaknesses. • Metamorphic rocks are often folded and heavily fractured, which are weaknesses that erosion can exploit.
Sedimentary	Sandstone Limestone Shale	MODERATE to FAST • Most sedimentary rocks are clastic and erode faster than crystalline igneous and sedimentary rocks. • The age of sedimentary rocks is important; geologically young rocks tend to be weaker. • Rocks with many bedding places and fractures, such as shale, are often most vulnerable to erosion.

Key terms

Unconsolidated sediment: Material such as sand, gravel, clay and silt that has not been compacted and cemented to become sedimentary rock (it has not undergone the process of lithification) and so is loose and easily eroded.

Pore water pressure: The pressure water experiences at a particular point below the water table due to the weight of water above it.

complex cliff profiles. Generally, the resistance of a cliff will be strongly influenced by its weakest rock type. Cliff profiles can also be influenced by the permeability of strata:

- Permeable rocks allow water to flow through them, and include many sandstones and limestones.
- Impermeable rocks do not allow groundwater flow and include clays, mudstones and most igneous and metamorphic rocks.

Permeability is important because groundwater flow through rock layers can weaken rocks by removing the cement that binds sediment in the rock together. It can also create high pore water pressure within cliffs, which affects their stability. Water emerging from below ground on to a cliff face at a spring can run down the cliff face and cause surface runoff erosion, weakening the cliff. Figure 8.13 shows a complex cliff profile with numerous changes in slope angle. This is a result of differences in the erosion resistance of rock including unconsolidated material on the cliff top. Groundwater flow and pore water pressure cause slumping in the unconsolidated sediments on the cliff top.

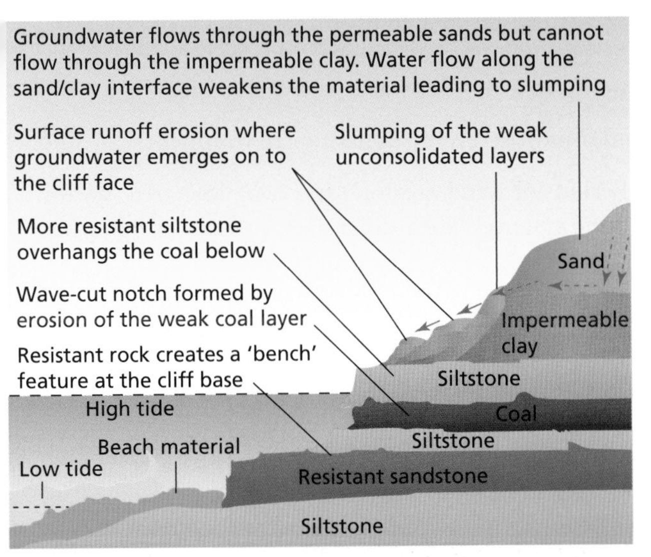

Figure 8.13 Complex cliff profile reflecting different alternating rock strata

Skills focus: Using maps and satellite images

Satellite images can be useful as a starting point for coastal analysis. Satellite images show locations that have rocky shorelines versus areas where low-lying sand dunes and salt marshes dominate. A better understanding of a coastline can often be gained by combining a satellite image with another map, especially:

- Ordnance Survey (OS) maps, which show height as well as coastal features such as cliffs and caves
- geological maps showing solid bedrock geology and unconsolidated material (sometimes called 'drift geology').

An online geological map of the UK published by the British Geological Survey (BGS) can be found at http://mapapps.bgs.ac.uk/geologyofbritain/home.html. Figure 8.14 shows a BGS solid geology map and a satellite image of the area around Strumble Head, just west of Fishguard in Wales. This is an area of ancient Ordovician rocks, both igneous and sedimentary. Using these two maps together it is possible to begin to see the relationship between geology and coastal landscape. Think about these questions:

1. Is the geological structure of this coastline concordant or discordant?
2. Is there a relationship between the numerous islands just offshore and the bedrock geology?
3. What landform evidence is there that the sedimentary rocks are more easily eroded than the igneous rocks?

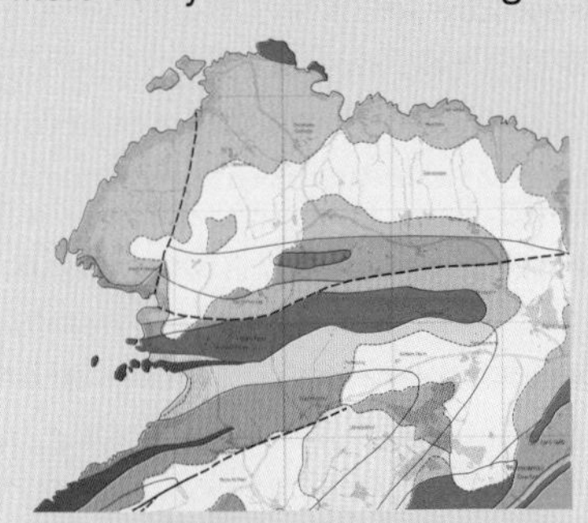

Figure 8.14 The area around Strumble Head, Wales (8.14a Contains British Geological Survey materials © NERC 2016; 8.14b Google and the Google logo are registered trademarks of Google Inc., used with permission. Imagery ©2016 Getmapping plc, Data SIO NOAA, US Navy, NGA, GEBCO, Map data ©2016 Google)

Coastal vegetation

Rocks and sediment play the most important role in influencing the shape of the coastal landscape, but vegetation is also important. Many coastlines are protected from erosion of unconsolidated sediment by the stabilising influence of plants. This includes:

- coastal sand dunes
- coastal salt marshes, found in many river estuaries
- coastal mangrove swamps, which are found on tropical coastlines.

Vegetation stabilises sediment in a number of ways:

- The roots of plants bind sediment particles together making them harder to erode.
- When submerged, plants growing in sediment provide a protective layer so the surface of the sediment is not directly exposed to moving water and therefore erosion.
- Plants protect sediment from wind erosion by reducing wind speed at the surface due to friction with the vegetation.

Many plants that grow in coastal environments are *halophytes* or *xerophytes*:

- Halophytes can tolerate salt water, either around their roots, being submerged in salt water (at high tide) or salt spray from the sea.
- Xerophytes can tolerate very dry conditions, such as those found on coastal sand dunes where the sandy soil retains very little water due to drainage.

The combination of salt water, strong winds, tide and waves makes the coast an extreme environment for plants, so only specially adapted plants can grow here. Maintaining, and in some cases expanding, sand dunes, salt marshes and mangroves has become a key goal in coastal management because of the importance of these environments in reducing the twin risks of coastal erosion and coastal flooding.

Plant succession

Succession refers to the changing structure of a plant community over time as an area of initially bare sediment is colonised by plants. The process is especially important on coasts because of its role in coastal accretion.

On a coast where there is a supply of sediment and deposition takes place, certain very specialised plants will begin to grow in the bare sand or mud. These are called pioneer species, and they begin the first stage of plant succession. Each step in plant succession is called a seral stage. The end result of plant succession is called a climatic climax community – essentially a new ecosystem area. In the case of coastlines:

- a sand dune ecosystem, called a *psammosere*
- a salt marsh ecosystem, called a *halosere*.

On coastal dunes the succession begins with the colonisation of embryo dunes by pioneer species (see Figure 8.15). Embryo dune pioneer plants:

- stabilise the mobile sand with their root systems
- reduce wind speeds at the sand surface, allowing more sand to be deposited
- add dead organic matter to the sand, beginning the process of soil formation.

Embryo dunes alter the environmental conditions from harsh, salty, mobile sand to an environment that other plants can tolerate. New plant species therefore colonise the embryo dunes creating a fore dune. Further environmental modification continues as a soil develops, nutrients and water become more available and plants that need better conditions can colonise the area. In Figure 8.15, the grey dune area was once an embryo dune but, due to plant colonisation and succession, the dunes have grown upwards and out to sea.

Sand dunes are a very dynamic environment:

- Periods of wind erosion can create low areas within dune systems called dune slacks; erosion stops when

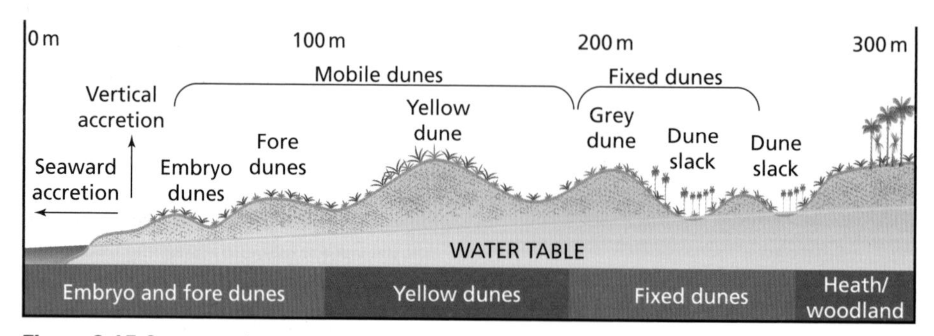

Figure 8.15 Cross-section across sand dunes showing plant succession

Table 8.4 Stages in salt marsh succession

Seral stage	Algal stage	Pioneer stage	Establishment stage	Stabilisation	Climax vegetation
Plants	Blue-green algae Wrack (algae) Gut weed	Glasswort Cord grass	Salt marsh grass Sea aster	Sea thrift Scurvy grass Sea lavender	Rush Sedge Red fescue grass
Processes	Grow on and within bare mud, binding it together.	Roots begin to stabilise the mud, allowing further mud accretion.	A continuous carpet of vegetation is established and the salt marsh height increases.	This area of the salt marsh is rarely submerged.	Developed soil profile and only submerged once or twice each year.

damp sand at the water table is exposed as this cannot be easily eroded by wind.

- Embryo and fore dunes are prone to wind and wave erosion, especially during major storms, but as long as the supply of sediment to the coast resumes, new embryo dunes will form and the dune front will stabilise.

Sand dune succession relies on specialised plants. Marram grass is one of the best examples. Its tough, long, flexible waxy leaves can cope with being 'sand-blasted' in gale-force winds and are designed to limit water loss through transpiration. Marram grass has roots of up to 3 m long that can tap water far below the dune surface. It can tolerate temperatures of up to 60 °C and grows up to 1 m per year, so can keep pace with deposition of wind-blown sand.

A similar process of plant succession happens on bare mud deposited in estuaries at the mouths of rivers, which is exposed at low tide but submerged at high tide. Estuaries are ideal for the development of salt marshes because:

- they are sheltered from strong waves, so sediment (mud and silt) can be deposited
- rivers transport a supply of sediment to the river mouth, which may be added to by sediment flowing into the estuary at high tide.

The stages in salt marsh succession are shown in Table 8.4.

Figure 8.16 shows an area of salt marsh in the Eden Estuary of Fife, Scotland. In the foreground is bare mud, which is host to species of algae. Beyond this are the pioneer plants, which are the first to colonise the mudflats. Beyond this, taller plant species from later seral stages can be seen. The salt marsh gradually slopes upwards toward the shore, because plants trap sediment on incoming tides, which is deposited. This slowly increases the height of the salt marsh, so flooding is less frequent.

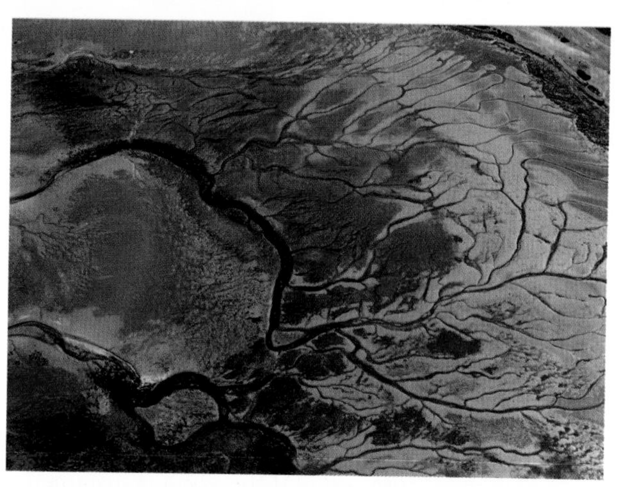

Figure 8.16 Salt marsh in the Eden Estuary of Fife, Scotland

Review questions

1. Briefly explain the different ways of classifying coasts.
2. Explain how geological structure influences the coastal landscape.
3. Explain how lithology (rock type) can influence rates of erosion at the coast.
4. How can the permeability of rocks influence the stability of cliffs?
5. Outline the value of ecosystems, such as salt marshes and sand dunes, that are found on some coastlines.

Further research

Explore visual panoramas of the East Riding of Yorkshire coast: www.eastriding.gov.uk/coastalexplorer/ipix/1.html

Find out about the geology of the coast near where you live, or other parts of the UK, using the BGS map viewer: www.bgs.ac.uk/discoveringGeology/geologyOfBritain/viewer.html

Use Google Earth to navigate along coastlines in different parts of the world: www.google.co.uk/intl/en_uk/earth

Explore the complex geology of the Jurassic Coast: www.southampton.ac.uk/~imw/index.htm#List-of-Webpages

9 Landforms and landscapes

How do characteristic coastal landforms contribute to coastal landscapes?

By the end of this chapter you should:

- understand how wave types and erosion processes generate erosional landforms that together create distinctive landscapes
- understand the processes of sediment transport and deposition and how these generate depositional landforms
- recognise the influence of sub-aerial processes that help shape and modify landforms, especially slope profiles and processes.

9.1 Marine processes and waves

Waves are, of course, a crucial element on any coast as they directly influence erosion, transport and deposition, and shape the coastline. Waves are caused by friction between wind and water, with some energy from the wind being transferred into the water. The force of wind blowing on the surface of water generates ripples, which grow into waves if the wind is sustained.

In open water out at sea, waves are simply energy moving through water. The water itself only moves up and down, not horizontally. There is some circular or orbital water particle motion within the wave, but no net forward water particle motion. Wave size depends on a number of factors:

- the strength of the wind
- the duration the wind blows for
- water depth
- wave fetch.

The UK's largest waves are generally experienced in Cornwall. This is no surprise, because the prevailing wind is from the southwest and the fetch from Florida to Cornwall is over 4000 km. Large waves generated by a sustained southwesterly wind have a great distance over which to grow.

It is possible to stand at the coast on a windless day and still see large waves breaking on the beach. These are swell waves. The wind that caused the waves has dropped but the remaining waves gradually make their way onshore.

Waves are affected by water depth. As waves approach a shoreline the water shallows and the shape of waves changes significantly, as shown in Figure 9.1. At a water depth of about half the wave length, the internal orbital motion of water within the wave touches the seabed. This creates friction between the wave and the seabed, and this slows down the wave. As waves approach the shore, wave length decreases and wave height increases, so waves 'bunch' together.

Key term

Fetch: The uninterrupted distance across water over which a wind blows, and therefore the distance waves have to grow in size.

Key concept: Tides

Waves and tides are frequently confused. Tides are formed by the gravitational pull of the Moon acting on water on the Earth's surface. The gravitational pull of the Sun, and the Earth's rotation, make a minor contribution to tides. Put simply, tides are produced as a bulge of water rotates around locations in the ocean called *amphidromic* points. As the bulge passes a location on the coast, high tide occurs – twice a day with roughly a twelve-hour interval.

In a sea with no waves, water at the coast would gradually rise and fall. This happens with very little force. In many locations the tidal range is also very small, perhaps 1–2 m. This begs the question: are tides important in terms of coastal processes? The answer is yes, in some circumstances:

- Some locations have large tidal ranges – the height difference between low and hide tide – which can produce waves capable of erosion, called tidal bores. Examples include the Bay of Fundy in Canada and the Severn Estuary in the UK, both of which can experience tidal ranges of over 15 m.
- When storms occur at the same time as spring tides (the highest tides), storm waves will act on parts of the coastline that do not normally experience them and, in addition, high waves on top of a spring tide dramatically increases the risk of coastal flooding.

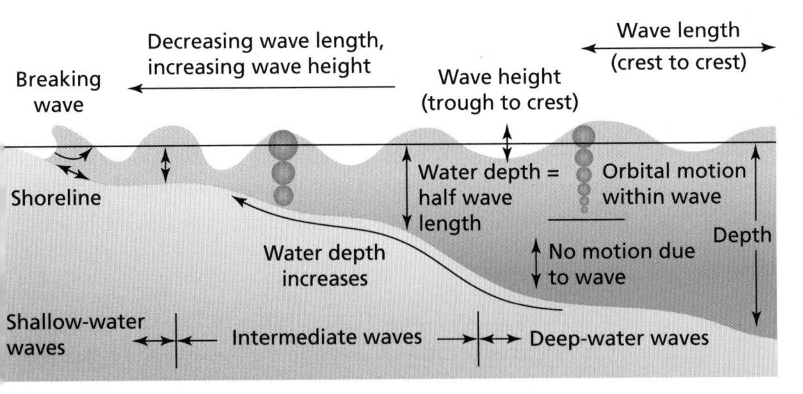

Figure 9.1 Changes in waves approaching a shoreline

Waves break in shallow water because the crest of the wave begins to move forward much faster than the wave trough. The trough experiences significant friction with the sediment and rock of the shore. Eventually the wave crest outruns the trough and the wave topples forward – a breaker.

Constructive and destructive waves

Waves breaking on a shoreline do not all have the same shape. The basic difference between constructive and destructive waves is shown in Figure 9.2.

- **Constructive** waves have a low wave height (less than 1 m) and long wave length (up to 100 m). They are also called spilling or surging waves. They are gentle, flat waves with a strong swash but weak backwash. The strong swash pushes sediment up the beach, depositing it as a ridge of sediment (berm) at the top of the beach. The relatively gentle beach profile with steep berm at the top of the beach means that most of the backwash percolates into the beach, rather than running across the surface.
- **Destructive** waves have a wave height of over 1 m and a wave length of around 20 m. They are common during storms and are also called plunging waves. These waves have a strong backwash that erodes beach material and carries it offshore, creating an offshore ridge or bar.

Key terms

Swash: The flow of water up a beach as a wave breaks.

Backwash: When water runs back down the beach to meet the next incoming wave.

Depending on conditions, beaches experience both constructive and destructive waves over the course of time, and this can mean significant changes to

Skills focus: Wave frequency

Measuring breaking waves can be quite a challenge, but wave frequency is something that can be measured as part of fieldwork. All that is required is a stopwatch, a simple recording sheet and time. Wave frequency can be measured in several ways:

- Time the gap, in seconds, between the arrival of one wave crest and the next at some fixed point on the beach, for example the end of a groyne or a sea wall.
- Time the gap between one wave breaking and the next (this is harder with constructive/swell waves, which break more gradually).

Measuring over ten minutes would provide a suitable number of timed 'gaps', which could then be averaged to give the mean wave period. Constructive waves have a frequency of six to nine per minute, whereas destructive waves have a frequency of about eleven to fifteen per minute.

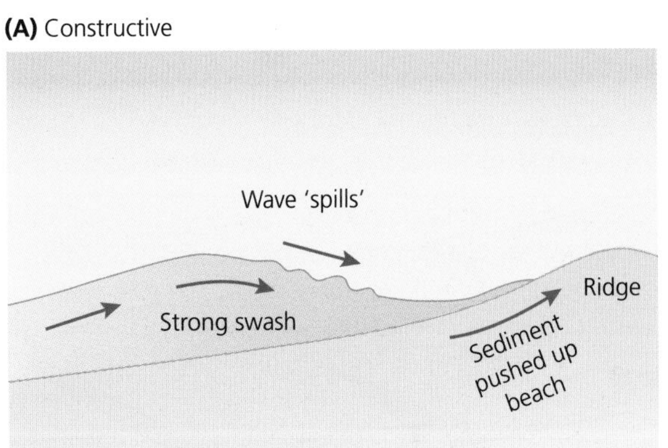

Figure 9.2 Constructive and destructive waves

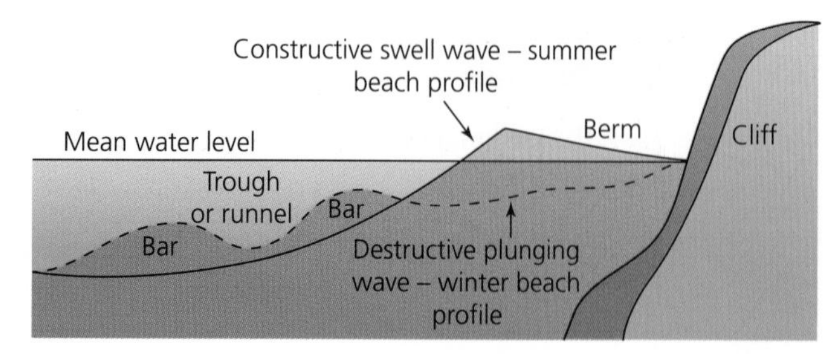

Figure 9.3 Changes between summer and winter beach profiles

beach morphology. Figure 9.3 shows how UK beach profiles generally change between summer and winter, when constructive and destructive waves dominate respectively. Even during the course of a day a beach profile may change, as plunging storm waves give way to swell waves as a storm passes. Over longer periods of time beach profiles may change because:

- sediment supply from rivers is reduced, for instance due to the construction of dams on rivers that traps sediment upstream
- of interference in sediment supply along the coast, often a result of coastal management in one place having an effect on processes further along the coast
- of changes to climate; for instance, if global warming made the UK climate on average stormier, then destructive waves and 'winter' beach profiles would become more common.

Beaches themselves have many landforms, but as wave conditions change these landforms also change. Figure 9.4 illustrates the relationship between beach morphology and beach sediment size.

Key term

Beach morphology: The shape of a beach, including its width and slope (the beach profile) and features such as berms, ridges and runnels. It also includes the type of sediment (shingle, sand, mud) found at different locations on the beach.

- Storm beaches result from high-energy deposition of very coarse sediment during the most severe storms.
- Berms, typically of shingle/gravel, result from summer swell wave deposition.
- The middle area of the beach is mostly sand, but the sand is coarser where berms/ridges have been deposited than in channels and runnels.
- Offshore ridges/bars are formed by destructive wave erosion and subsequent deposition of sand and shingle offshore.

Erosion processes

Waves cause erosion, but erosion is not a constant process. Most erosion occurs during a small number of large storms. Under normal conditions very little erosion takes place, even on very soft rock coastlines such as Holderness in the East Riding of Yorkshire. Most erosion will occur when:

- waves are at their largest, which is influenced by wind speed and fetch, meaning they have a lot of energy and can pick up sediment and hurl it at a cliff face
- waves approach the coast at 90° to the cliff face
- the *tide* is high, propelling waves higher up the cliff face

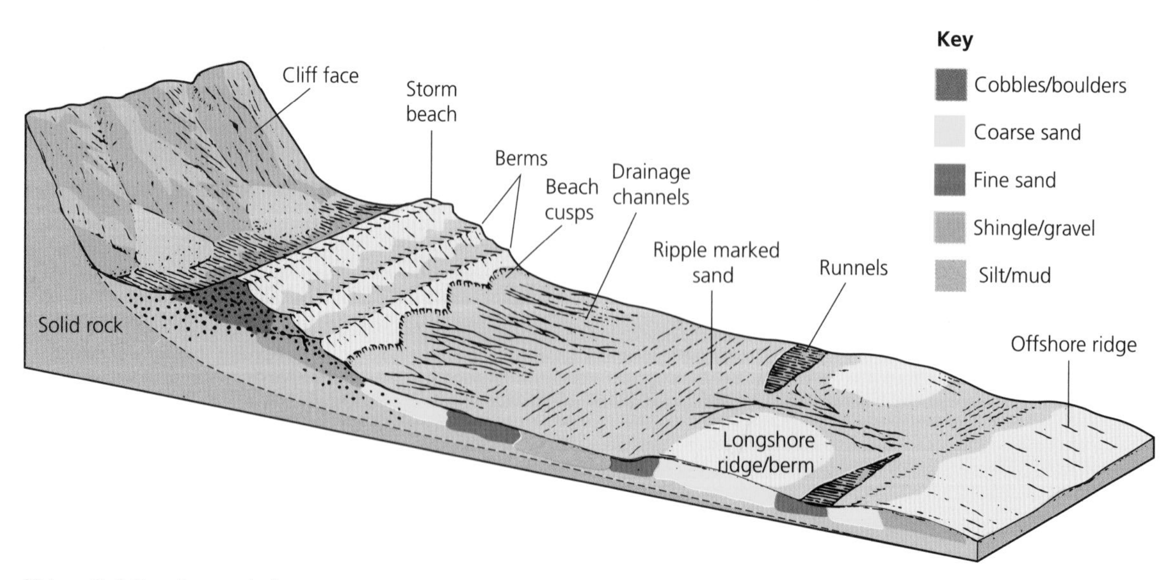

Figure 9.4 Beach morphology

- heavy rainfall, percolation of water through permeable strata and surface runoff weakens the cliff
- debris from previous erosion has had time to be removed from the cliff foot, as debris can protect against waves.

There are four main processes that cause erosion:

Table 9.1 The processes of erosion

Process	Explanation	Influence of lithology
Hydraulic action (wave quarrying)	Air trapped in cracks and fissures is compressed by the force of waves crashing against the cliff face. Pressure forces cracks open, meaning more air is trapped and greater force experienced in the next cycle of compression. This process dislodges blocks of rock from the cliff face.	Heavily jointed/fissured sedimentary rocks are vulnerable. In very hard igneous rocks (basalt, granite) hydraulic action on cooling cracks may be the only erosive process operating.
Abrasion (corrasion)	Sediment picked up by breaking waves is thrown against the cliff face. The sediment acts like a tool on the cliff, chiselling away at the surface and gradually wearing it down by removing small rock particles.	For abrasion to be effective, suitably loose sediment has to be available, for example shingle or pebbles. Softer sedimentary rocks are more vulnerable than hard igneous ones.
Attrition	This process acts on already eroded sediment. As sediment is moved around by waves, the numerous collisions between particles slowly chip fragments off the sediment. The net result is that sediment gets smaller and more rounded over time.	Softer rocks are very rapidly reduced in size by attrition.
Corrosion (solution)	Carbonate rocks (limestones) are vulnerable to solution by rainwater, spray from the sea and seawater. On tropical coastlines large rock overhangs, called visors, often indicate locations were limestone has been dissolved.	Mainly affects limestone, which is vulnerable to solution by weak acids.

Erosional landforms

There are many different types of landform produced by erosion, many of which are well known (arches, stacks, caves). The classic suite of coastal landforms (Figure 9.5) is most commonly found where the rocks are sedimentary (sandstone, chalk) with well-defined

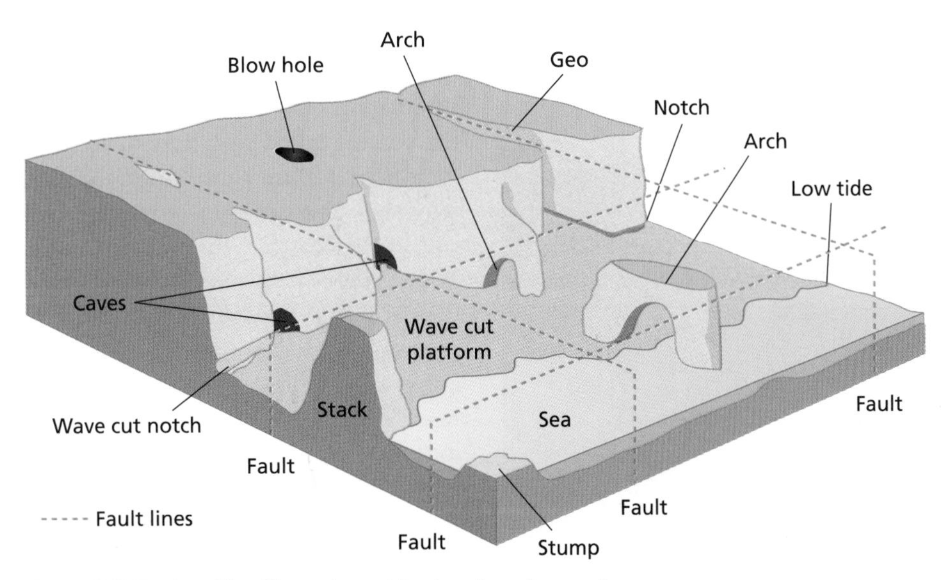

Figure 9.5 Erosional landforms in resistant sedimentary rocks

bedding planes and joints. Softer rocks (shales, clays) are not competent enough to form many of these erosive features.

The most fundamental process of landform formation on a coastline is the creation of a wave-cut notch. This is eroded at the base of a cliff by hydraulic action and abrasion. As the notch becomes deeper, the overhanging rock above becomes unstable and eventually collapses as a rockfall. Repeated cycles of notch cutting and collapse cause cliffs to recede inland. The former cliff position is shown by a horizontal rock platform visible at low tide called a wave-cut platform. Many other erosional landforms sit on or next to this platform.

The position of some landforms is influenced by structural geology. The location of fault lines (Figure 9.5) allows geos to form. These are narrow fault-guided gullies. Cave and blow hole formation is more common in areas weakened by the presence of a fault. Upstanding features on the wave-cut platform, such as stacks, may indicate more resistant areas of rock, which can be related to faults.

Stacks are one of the most impressive coastal landforms. The Old Man of Hoy is the UK's tallest stack at 137 m high. In 1750 the location was a headland but, by 1820, an arch had formed. Today only a stack remains. Future erosion will inevitably lead to its collapse.

Key term

Blow hole: Forms when a coastal cave turns upwards and breaks through the flat cliff top. Usually this is because of erosion of especially weak strata or the presence of a fault line.

Figure 9.6 The Old Man of Hoy eroded from red sandstone on Orkney

Figure 9.6 clearly shows the bedding planes of the sandstone strata and their slight seaward dip. On the right and at the base of the tall cliff is a vegetated slope of rock debris. This protects the base of the cliff at that location from erosion, whereas to the left bare rock and a small wave-cut notch can be seen. The Old Man of Hoy is towards the end of a sequence of landform formation called cave-arch-stack, shown in Figure 9.7. Caves eroded in headlands can break through the headland completely forming an arch, which is subsequently attacked by erosion and weathering; on its collapse a stack is left. The stack gradually erodes until only a stump remains.

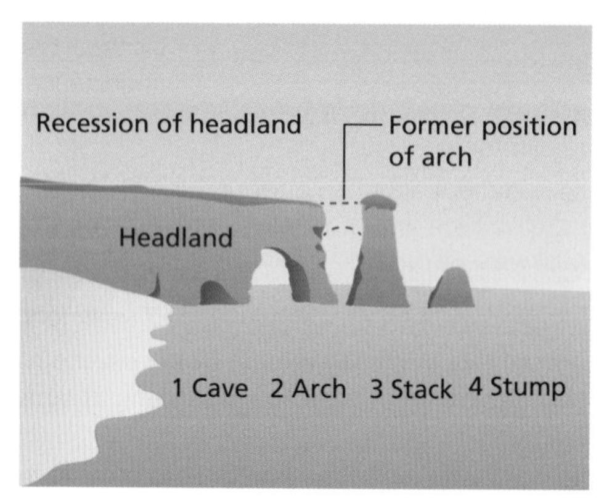

Figure 9.7 The cave-arch-stack sequence

9.2 Transport and deposition

Material eroded from cliffs is transported by the sea as sediment. Transport of sediment happens in four ways, outlined in Table 9.2.

Rarely does sediment simply move seaward and landward due to waves. Usually it is transported *along* coastlines by currents. The process of sediment transport along the coast is called longshore drift. Figure 9.8 shows a longshore current running parallel to the coast caused by waves approaching a coast at an angle. It has a crucial role in the coastal sediment cell and sediment transport. When wave crests approach the coast at an angle (rather than at 90° to the coast) the swash from the breaking waves, and the subsequent backwash, follow different angles up and down the beach in a zigzag pattern. The result is net sediment transport along the beach and a longshore flow of sediment. Wave direction is caused by wind direction, which of course changes frequently. However, on most coastlines there is a dominant prevailing wind so, over time, there is a dominant direction of longshore drift.

Table 9.2 Processes of sediment transport

Traction	Sediment rolls along, pushed by waves and currents	The sound of rolling pebbles and shingles can often be heard clearly at the beach	Pebbles, cobbles, boulders
Saltation	Sediment bounces along, either due to the force of water or wind	On a dry, windy day, a layer of saltating sand is often seen 2–10 cm above the beach surface	Sand-sized particles
Suspension	Sediment is carried in the water column	On soft-rock coasts such as Holderness, the sea is often muddy brown in colour due to suspended sediment	Silt and clay particles
Solution	Dissolved material is carried in the water as a solution	This type of sediment transport is of limited importance on coasts	Chemical compounds in solution

Key term

Currents: Flows of seawater in a particular direction driven by winds or differences in water density, salinity or temperature. Some are almost continuous, such as those that form the global thermohaline circulation, and others are more sporadic, such as longshore currents, while some last only for a few hours, such as rip currents.

Depositional landforms

Depositional landforms are found on many coasts, but they are most common on drift-aligned coasts. Coasts can be divided into two broad categories:

- Swash-aligned: wave crests approach parallel to the coast, so there is limited longshore movement of sediment.
- Drift-aligned: wave crests break at an angle to the coast, so there is consistent longshore drift and the generation of elongated depositional features.

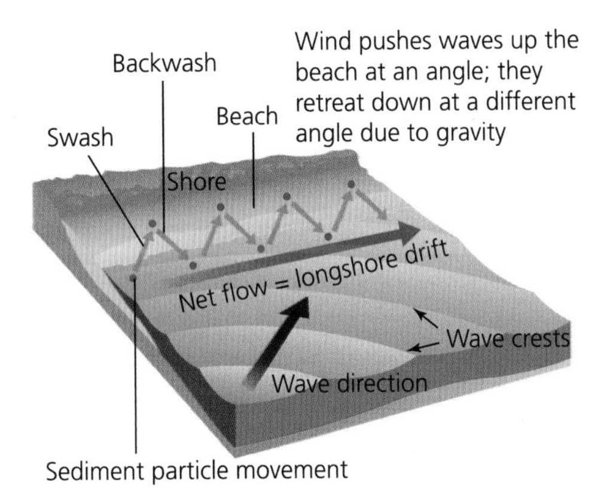

Figure 9.8 The process of longshore drift

Longshore drift is a key source of sediment for depositional landforms on coasts. Sediment transported down river systems to the coast or from offshore sources is also important. Sediment is deposited when the force transporting sediment drops. Deposition can occur in several ways:

- **Gravity settling** occurs when the energy of transporting water becomes too low to move sediment; large sediment will be deposited first followed by smaller sediment (pebbles, then sand, then silt).
- **Flocculation** is a depositional process that is important for very small particles, such as clay, which are so small they remain suspended in water. Clay particles clump together due to electrical or chemical attraction and become large enough to sink.

Figure 9.9 shows a number of depositional landforms.

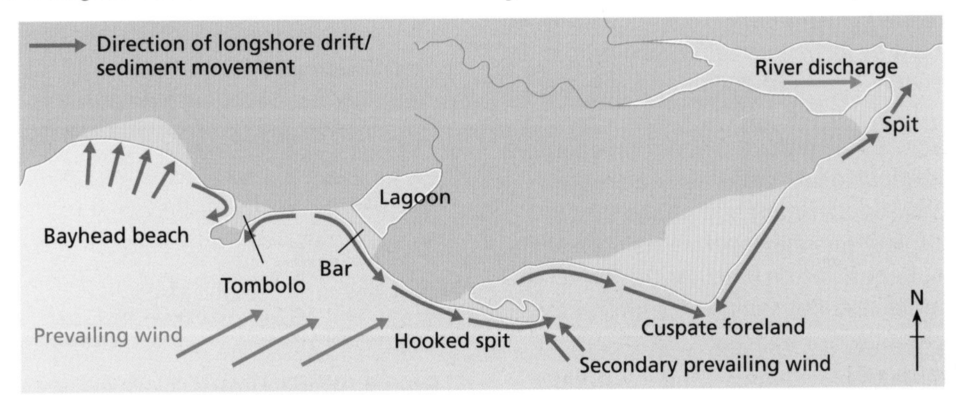

Figure 9.9 Depositional landforms

Table 9.3 Summary of depositional landform formation

Landform	Processes	Example
Spit	Sand or shingle beach ridge extending beyond a turn in the coastline, usually greater than 30°. At the turn, the longshore drift current spreads out and loses energy leading to deposition. The length of a spit is determined by the existence of secondary currents causing erosion, either the flow of a river or wave action which limits its length.	Spurn Head on the Holderness Coast, East Riding of Yorkshire
Bayhead beach	A swash-aligned feature, where waves break at 90° to the shoreline and move sediment into a bay, where a beach forms. Due to wave refraction, erosion is concentrated at headlands and the bay is an area of deposition.	Lulworth Cove, Dorset
Tombolo	A sand or shingle bar that attaches the coastline to an offshore island. Tombolos form due to wave refraction around an offshore island which creates an area of calm water and deposition between the island and the coast. Opposing longshore currents may play a role, in which case the depositional feature is similar to a spit.	St Ninian's tombolo, Shetland
Barrier beach/bar	A sand or shingle beach connecting two areas of land with a shallow water lagoon behind. These features occur when a spit grows so long that it extends across a bay, closing it off.	Chesil Beach, Dorset
Hooked/recurved spit	A spit whose end is curved landward, into a bay or inlet. The seaward (distal) end of the spit naturally curves landward into shallower water. The 'hook' may be made more pronounced by waves from a secondary direction to the prevailing wind.	Hurst Castle spit, Hampshire
Cuspate foreland	These are roughly triangular-shaped features extending out from a shoreline. There is some debate about their formation, but one hypothesis suggests the growth of two spits from opposing longshore drift directions.	Dungeness, Kent

Holderness Coast sediment cell

Coastlines operate as a physical system. Attempts to manage erosion on one small area of a coast are unlikely to be successful unless the location is understood in a wider context. This context is the sediment cell (Figure 9.10). Long stretches of coastline operate as sediment cells. There are eleven of these around the English and Welsh coast. The Holderness Coast is an example. In each cell are:

- Sources – where sediment is eroded from cliffs (Flamborough Head's chalk, Hornsea's boulder clay) or sand dunes; sources can be offshore bars. River systems (River Humber) are also an important sources.
- Transfer zones – places where sediment is moving along the coast by longshore drift and offshore currents; beaches, parts of dunes and salt marshes (Humber Estuary) perform this function.
- Sinks – locations where the dominant process is deposition; depositional landforms are created including spits (Spurn Head) and offshore bars.
- Some coastal features may operate as both sinks and sources, depending on whether the dominant process is erosion or deposition at a given time.

Sediment cells operate as a system with inputs, transfers and outputs of sediment. Under natural conditions the systems operate in a state of dynamic equilibrium with sediment inputs balancing outputs to sinks. For short periods of time – for instance during a major storm that erodes a spit like Spurn Head – the system's equilibrium might be disrupted but will tend to return to balance over time.

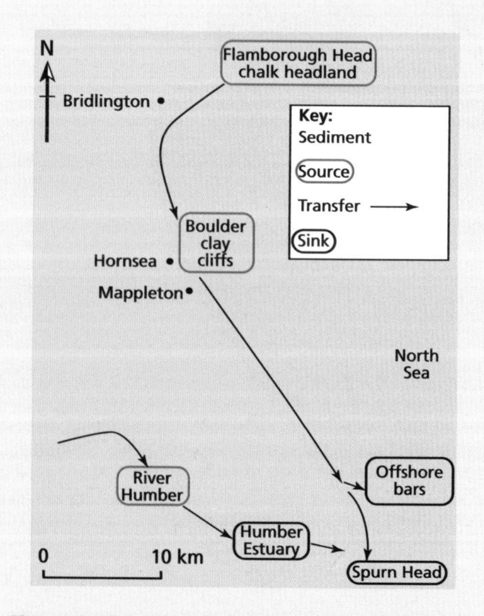

Figure 9.10 The Holderness Coast's sediment cell

Negative feedback mechanisms help maintain the balance by pushing the system back towards equilibrium:

- During a major erosion event a large amount of cliff collapse may occur, but the chalk or boulder clay at the base of the cliff will slow down erosion by protecting the cliff base from wave attack.
- Major erosion of Spurn Head spit could lead to increased deposition offshore, creating an offshore bar that reduces wave energy, allowing the spit time to recover.

Human intervention, in the form of coastal management, plus threats such as a sea-level rise as a result of global warming, represent risks to the long-term dynamic equilibrium of sediment cells like Holderness. These are examples of positive feedback which lead to disequilibrium in the coastal system:

- Increased storminess erode beach material faster, so recent debris never protects the cliff base temporarily and overall erosions is faster.
- Rising sea levels could increase the erosion of spits like Spurn Head, removing sediment faster than it can be replaced.

Mobility and stability

Depositional landforms, by their very nature, are made from unconsolidated sediment. This means they are vulnerable to change and are dynamic landforms. Change can result from major storms, especially when these occur during very high tides (spring tides). Huge volumes of sediment can be eroded during a single storm, leading to erosion and redeposition elsewhere. Some recent examples include:

- A major storm and tidal surge in December 2013 that breached Spurn Head spit on the Holderness Coast; this was the latest of many breaches over the last few centuries with the spit usually recovering. Rising sea levels due to global warming may present a longer-term threat to its stability.
- Chesil Beach was breached by a major storm in 1990; a storm in January 2014 dramatically altered the beach profile.

Depositional landforms depend on a continual supply of sediment and continuous deposition to balance any erosion that takes place. This is especially true of spits, as the seaward end of these landforms is constantly being eroded.

Vegetation plays a very important role in stabilising depositional landforms. Plant succession, in the form of salt marshes and sand dunes, bind the loose sediment together and encourage further deposition. Figure 9.11 shows a satellite view of Hurst Castle spit. Sand dunes stabilise the sediment immediately landward of the beach; behind the dunes is a large salt marsh formed in area of the calm water behind the spit. Depositional landforms are therefore vulnerable when their vegetation is damaged. This can happen due to overgrazing or trampling from tourism and leisure activities.

Figure 9.11 Hurst Castle spit (Google and the Google logo are registered trademarks of Google Inc., used with permission. Imagery ©2016 Google, Map data ©2016 Google)

9.3 Sub-aerial processes

The focus of coastal geography is often on dramatic events and landforms, such as the impact of major storms and cliff collapses. However, a number of underlying and largely unseen processes are important too. Weathering is one of these. Weathering is a sub-aerial process that affects all coastlines, although at very different rates. Weathering is often confused with erosion:

- **Weathering** is the *in-situ* breakdown of rocks by chemical, mechanical or biological agents. It does not involve any movement.
- **Erosion** is the breakdown of rock due the action of some external force (wind, waves, flowing water, ice flow) which then transports the eroded material to a new location.

There are three types of weathering:

- **Mechanical** weathering breaks down rocks due to the exertion of a physical force; it does not involve any chemical change.

- **Chemical** weathering involves a chemical reaction and the generation of new chemical compounds.
- **Biological** weathering often speeds up mechanical or chemical weathering through the action of plants, bacteria or animals.

As the name suggests, weathering is influenced by the weather or, more correctly, the climate conditions. Temperature and precipitation have a very strong influence of the type of weathering experienced as Figure 9.12 shows. Mechanical weathering dominates in cold climates, while chemical reactions speed up in hot, wet conditions.

The lithology of coastal rocks is also a major influence, as some rocks are more prone to certain types of weathering than others:

Weathering contributes to rates of coastal recession in a number of ways. Weathering weakens rocks, making them more vulnerable to erosion or mass movement processes. Some strata may be more vulnerable to weathering than others, contributing to the formation of wave-cut notches and therefore having an effect on overall cliff stability. Rates of weathering are very slow. Even in a hot, wet climate, the igneous rock basalt weathers at a rate of 1–2 mm every 1000 years.

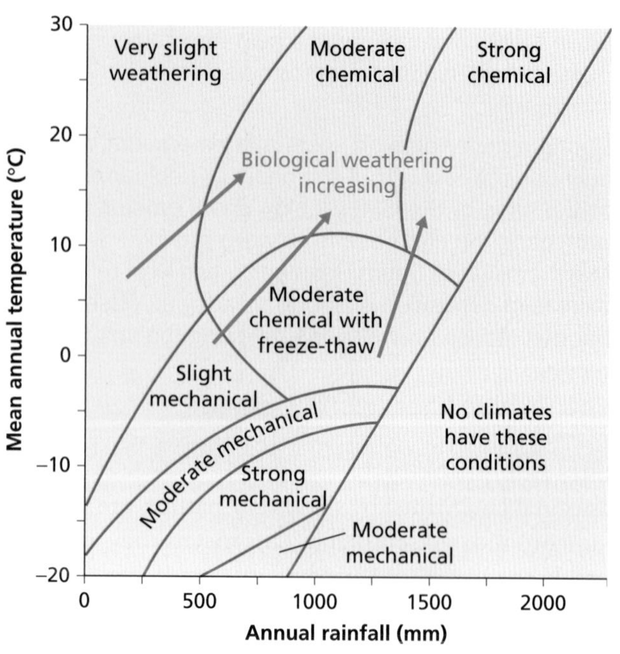

Figure 9.12 The relationship between climate and weathering

Key term

Mass movement: The downslope movement of rock and soil; it is an umbrella term for a wide range of specific movements including landslide, rockfall and rotational slide.

Table 9.4 Weathering processes

Type	Process	Explanation	Vulnerable rocks
Mechanical	Freeze–thaw	Water expands by 9% in volume when freezing, exerting a force within cracks and fissures; repeated cycles force cracks open and loosen rocks.	Any rocks with cracks and fissures, especially high up cliffs away from salt spray. Freezing is relatively uncommon on UK coasts, especially in the south, and salt spray can reduce its effect even further.
	Salt crystallisation	The growth of salt crystals in cracks and pore spaces can exert a breaking force, although less than for freeze–thaw.	Porous and fractured rocks, for example sandstone. The effect is greater in hotter, drier climates where evaporation and the precipitation of salt crystals in more pronounced.
Chemical	Carbonation	The slow dissolution of limestone due to rainfall (weak carbonic acid, pH 5.6) producing calcium bicarbonate in solution.	Limestone and other carbonate rocks.
	Hydrolysis	The breakdown of minerals to form new clay minerals, plus materials in solution, due to the effect of water and dissolved CO_2.	Igneous and metamorphic rocks containing feldspar and other silicate minerals.
	Oxidation	The addition of oxygen to minerals, especially iron compounds, which produces iron oxides and increases volume contributing to mechanical breakdown.	Sandstones, siltstones and shales often contain iron compounds that can be oxidised.
Biological	Plant roots	Trees and plants roots growing in cracks and fissures forcing rocks apart.	An important process on vegetated cliff tops that can contribute to rockfalls.
	Rock boring	Many species of clams and molluscs bore into rock and may also secrete chemicals that dissolve rocks.	Sedimentary rocks, especially carbonate rocks (limestone) located in the inter-tidal zone.

Table 9.5 Classification of mass movements

Fall	Scar Cliffs Rockfall Rockfall debris	Rockfalls, or blockfalls, are a rapid form of mass movement. On coasts blocks of rock can be dislodged by mechanical weathering or by hydraulic action erosion. Undercutting of cliffs by the creation of wave-cut notches can lead to large falls.
Topple		Geological structure influences topples. Where rock strata have a very steep seaward dip, undercutting by erosion will quickly lead to instability and blocks of material toppling seaward.
Translational slide		A very low angle seaward dip in strata will prevent falls. In this case, material will tend to slide down the dipslope towards the sea.
Rotational slide/slumping	Failure surface	Mass movements can occur along a curved failure surface. In the case of a rotational slide, huge masses of material can slowly rotate downslope over periods lasting from days to years. Water plays an important role in rotational slides.
Flow		Flows are common in weak rocks such as clay or unconsolidated sands. These materials can become saturated, lose their cohesion and flow downslope. Heavy rainfall combined with high tides can contribute to saturation.

Mass movement

Weathering and erosion are important on many coasts. On some coastlines mass movement is the dominant cause of cliff collapse. There are many different types of mass movement, and the role of water is very important in a number of types. Mass movements can be classified in a number of ways including how rapid the movement is and the type of material (solid rock, debris or soil):

Rockfalls on coastal cliffs can be dramatic and involve very large volumes of material. In April 2013 a large rockfall occurred in St Oswald's Bay, Dorset, in the cretaceous chalk cliffs. The mass movement (Figure 9.13) occurred overnight without warning. An 80–100 m long section of cliff collapsed, taking a section of the South Coast Coastal Path with it. Figure 9.13 shows:

- the landslide scar: an area of unvegetated fresh chalk with older vegetated cliff face to the left and right
- a fan-shaped talus scree slope of broken chalk blocks extending down the cliff and out into the sea
- the debris from the rockfall has caused the effective coastline to protrude further out to sea than normal.

Figure 9.13 The St Oswald's Bay rockfall, Dorset

A rockfall of this magnitude will take many years to be transported away by wave action. In the meantime the base of the cliff is protected by this debris. Undercutting by abrasion and hydraulic action was one cause of the St Oswald's Bay rockfall, but unusually wet weather probably contributed too, saturating the cliff from above and weakening it internally.

Rotational slides

Coastlines with weak sedimentary rocks, such as clays and shales, or unconsolidated material such as boulder clay, sands and gravels, are vulnerable to rotational slides. This type of mass movement is very difficult to manage because it often has multiple causes. It leads to failure deep within the cliff and can move very large volumes of cliff material. Rotational slides usually happen slowly so are very different to the sudden rockfalls like that experienced at St Oswald's Bay.

The most vulnerable situation is where permeable strata (chalk, sandstone) sit on top of impermeable strata (clay, mudstone). This arrangement of rocks is common along the south coast of England and can be found at Folkestone, Lyme Bay and Christchurch Bay. Rotational slides also occur in the glacial boulder clay of the Holderness Coast in the East Riding of Yorkshire and on the Norfolk Coast. Figure 9.14 illustrates the causes and features of rotational slides.

1 The bedding plane between the impermeable clay and permeable sand dips seaward, which promotes mass movement.
2 Cracks develop in the cliff top during dry weather as soil and sediment dry out; these later become routes that rainwater can take into the sand.
3 Heavy rain saturates the permeable sands, loading the cliff material.
4 Water percolates through the permeable sand but is forced to move along the sand/clay boundary as the clay is impermeable; this contributes to high pore water pressure in the sand and creates internal pressure within the cliff.
5 Toe erosion by marine processes (abrasion, hydraulic action) undercuts the cliff from below, adding to instability.
6 Curved failure surfaces develop in the sand, and the whole cliff begins to rotate about a pivot point (7).

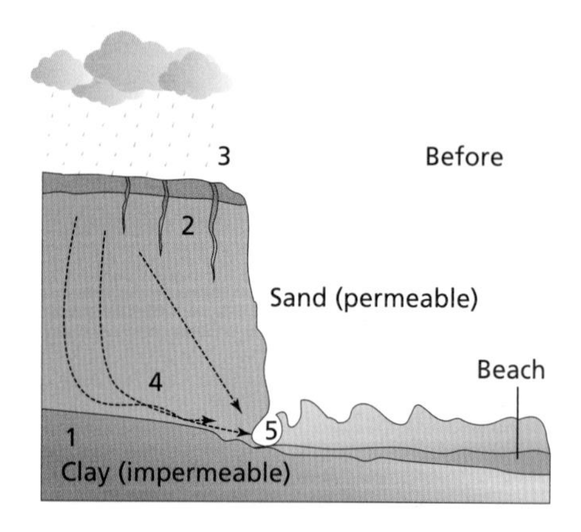

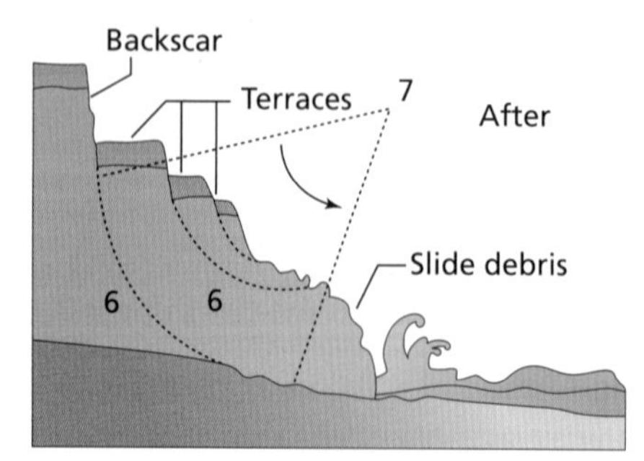

Figure 9.14 Rotational slides

Rotational slides produce a characteristic cliff profile. The rotational failure surface can be seen as a scar high on the cliff wall. Further down the cliff are a series of terraces or benches which represent the smaller failure surfaces of rotating blocks.

Review questions

1 Outline the factors that influence the size of breaking waves at the coast.
2 Describe how waves change as they approach a coastline.
3 Briefly explain how constructive and destructive waves influence beach morphology.
4 How does the presence of structural features, such as faults, influence coastal erosional landforms?
5 Explain how the sediment cell model helps us to understand the coastline as a system.
6 Outline the role that vegetation and succession play in stabilising some depositional landforms.
7 Using examples, discuss the importance of mass movements and weathering in influencing the stability of cliffs.

Further research

Explore the differences between waves, tide and currents: http://ocean.si.edu/ocean-news/currents-waves-and-tides-ocean-motion

Find out more about the variety of different beach types among the thousands of beaches in Australia: www.ozcoasts.gov.au/conceptual_mods/beaches/beach_intro.jsp

Use this website to explore the wide variety of coastal landforms around the coast of Britain and Ireland: www.geograph.org.uk

10 Coastal risks

How do coastal erosion and sea level change alter the physical characteristics of coastlines and increase risks?

By the end of this chapter you should:

- understand how sea level changes, both short and longer term, influence the physical geography of coastlines and increase risks for people
- understand why rapid coastal recession occurs in some locations, and the physical and human factors that influence this
- understand how coastal flooding is a risk on some coastlines, and the impact of global warming on coastal flood risk.

10.1 Sea level change

Sea levels change on a day-to-day basis because:

- high and low tides alter the local sea level every few hours (see Chapter 9)
- atmospheric air pressure has an influence on sea level – low air pressure causes a slight rise in sea level
- winds can 'push' water towards a coast, and wave height varies from day to day.

It is usual to refer to 'mean sea level', which is an averaged value taking into account the variations above. Over long timescales – thousands of years – sea level is influenced by a number of factors and can change very significantly. Sea level change is complex because it can involve a change in both land level (isostatic change) and water level (eustatic change).

The complexity arises from the fact that, on a given coastline, both isostatic and eustatic change can occur at the same time. Table 10.1 summarises the causes of sea level change.

Key terms

Isostatic change: A *local* rise or fall in land level.

Eustatic change: Involves a rise or fall in water level caused by a change in the volume of water. This is a *global* change, affecting all the world's connected seas and oceans.

Table 10.1 The complex causes of sea level change

Marine regression	Marine transgression
In both cases below the effect on a coastline is that former seabed is exposed as the sea level drops, producing an **emergent coast**.	In both cases below areas of land flood, so the coastline is 'drowned', producing a **submergent coast**.
Eustatic fall in sea level	**Eustatic rise in sea level**
During glacial periods, when ice sheets form on land in high latitudes, water evaporated from the sea is locked up on land as ice leading to a global fall in sea level.	At the end of a glacial period, melting ice sheets return water to the sea causing the sea level to rise globally. Global temperature increases cause the volume of ocean water to increase (thermal expansion) leading to sea level rise.
Isostatic fall in sea level	**Isostatic rise in sea level**
During the build up of land-based ice sheets, the colossal weight of the ice causes the Earth's crust to sag. When the ice sheets melt, the land surface slowly rebounds upward over thousands of years. This post-glacial adjustment slowly lifts the land surface out of the sea.	Land can 'sink' at the coast due to the deposition of sediment (accretion), especially in large river deltas where the weight of sediment deposition leads to very slow 'crustal sag' and delta subsidence.

Key terms

Accretion: This occurs when sediment is added to a landform, such as a river delta, by deposition. It can build up to form new land, allowing a delta to grow out to sea. It tends to be balanced by subsidence, caused by the weight of the newly deposited sediment.

Post-glacial isostatic adjustment: Refers to the uplift experienced by land following the removal of the weight of ice sheets. It is sometimes called post-glacial rebound or post-glacial re-adjustment.

Although the last glacial period ended around 12,000 years ago, the impact of the sea level rise that occurred then and post-glacial isostatic adjustment are very important to coastal development in northern Europe and North America. Figure 10.1 shows that, even today, isostatic adjustment is happening.

- Scotland is still rebounding upward, in some places by up to 1.5 mm per year.
- In contrast, England and Wales are subsiding at up to 1 mm per year.

The area which is rising corresponds roughly to the area covered by the last major ice sheet during the Devensian glacial period. In effect, Great Britain is slowly pivoting upward in the north and downward in the south. It is well known that the sea level is slowly rising today due to global warming. The effect of the isostatic adjustment is to compound the effect of this current sea level rise in the south, but cancel it out in the north.

Emergent coasts

The scale of isostatic and eustatic changes during and after the last glacial period is large. Global sea levels fell by up to 120 m as ice sheets grew. A similar rise in sea level happened very quickly at the end of the last ice age, over a period of a few thousand years. In North America and Northern Europe, the post-glacial isostatic adjustment has been more than 300 m. The two changes are linked but happen at very different rates:

- Post-glacial sea level rise was very rapid and would have 'drowned' many coastlines.
- Isostatic adjustment was very slow, meaning that in previously ice-covered areas newly drowned coastlines slowly emerged from the sea (and are still emerging today).

The effect can be seen in Figure 10.2. It shows part of the eastern Scottish coastline near Earlsferry in Fife. The modern-day beach and short cliff can be seen on the right. In the background are two flat platforms, each with a steep slope to the left. These are raised beaches (flat area) and fossil cliffs (slope). They were formed during the last 10,000 years as this part of Scotland has rebounded, gradually stranding cliff lines and beach platforms well above the current sea level. Each fossil cliff is approximately 10 m high. Figure 10.3 is from western Scotland, near Lendalfoot in Ayrshire. Here the A77 road runs along a raised beach, with the modern beach to the left. To the right is a heavily vegetated 40 m high steep slope. This is a fossil cliff, about 200 m inland of the present-day beach. The heavy vegetation on the fossil cliff shows that it is not actively eroding.

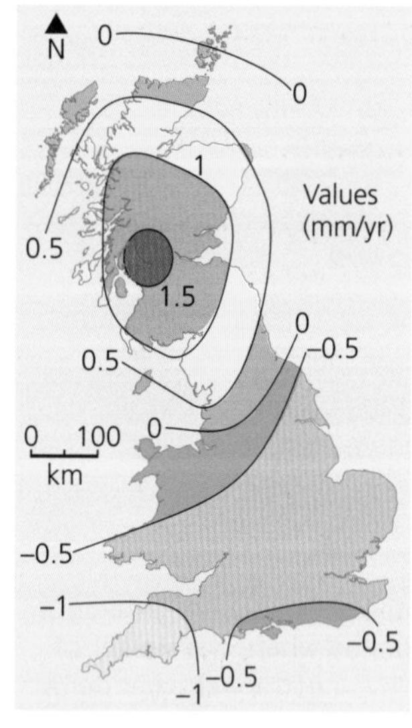

Figure 10.1 The annual rate of isostatic adjustment for Great Britain

Figure 10.2 Raised beaches and fossil cliffs in Fife, Scotland

Figure 10.3 Fossil cliff and raised beach in Ayrshire, Scotland

Key terms

Ria: A drowned river valley in an unglaciated area caused by sea level rises flooding the river valley, making it much wider than would be expected based on the river flowing into it.

Barrier islands: Offshore sediment bars, usually sand dune covered but, unlike spits, they are not attached to the coast. They are found between 500 m and 30 km offshore and can be tens of kilometres long.

Submergent coasts

On coastlines that were not affected by glacial ice cover, the post-glacial sea level rise has created submergent (drowned) coastlines. These are found along the south coast of England and the east coast of America. The most common coastal feature is a ria. Figure 10.4 shows the Kingsbridge Estuary near Salcombe in Devon. Different arms of the estuary, called creeks are between 500 m and 1000 m wide. No significant river runs into this estuary, however, only a number of small streams. These streams could not have eroded the wide valleys that form the creeks. The estuary is an example of a ria because sea level rise has pushed the sea inland, drowning valleys that were eroded by rivers millennia ago. Rias are economically important: they are sheltered ports, often extending miles inland.

Fjords are also examples of drowned coastlines and are found on the coasts of Norway, Canada, New Zealand and Chile. Fjords are drowned valleys, but they differ from rias in a number of important ways:

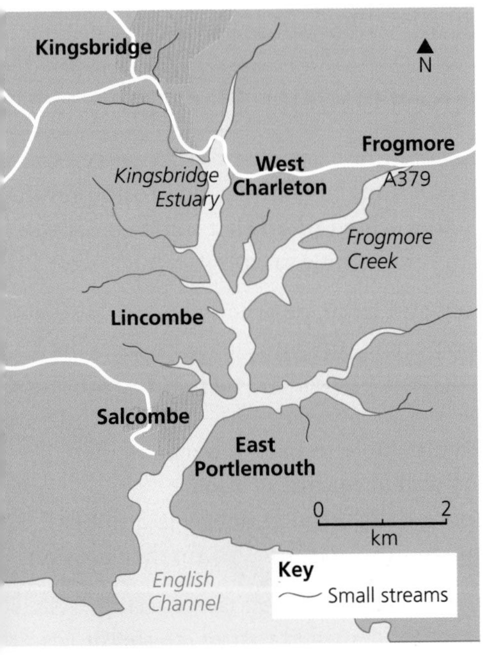

Figure 10.4 The Kingsbridge Estuary in Devon, an example of a ria

- In fjords the drowned valley is always a U-shaped, glacially eroded valley.
- The fjord itself is often deeper than the adjacent sea; some are over 1000 m deep.
- At the seaward end of the fjord there is usually a submerged 'lip' representing the former extent of the glacier (its terminal moraine, or the a rock sill representing the seaward extent of glacial erosion) that filled the valley.

In many areas where fjords are common, post-glacial isostatic adjustment is slowly raising the land out of the sea, shallowing the fjords by a few millimetres each year. This contrasts with many ria coastlines (for example southern England) which are still being slowly submerged.

The east coast of the USA is dominated by a complex coastline landform called a barrier island. Barrier islands can be found from Florida all the way north to Connecticut. There is ongoing debate about how barrier islands form. Most explanations focus on coastal submergence as a key process, as shown in Figure 10.5. Barrier islands may have formed initially as lines of coastal sand dunes attached to the shore. Subsequent sea level rises flooded the land behind the dunes forming a lagoon, but the dunes themselves were not eroded and so became islands. As the sea level continued to rise, the dune systems slowly moved landward.

Barrier islands are a natural form of coastal defence. They force ocean waves to break out at sea, protecting the true coastline landward of the lagoon. In the USA they form a natural protective barrier against hurricanes on the Gulf of Mexico coast and Eastern seaboard.

The formation of a Dalmatian-type coast was mentioned in Chapter 8. These are concordant

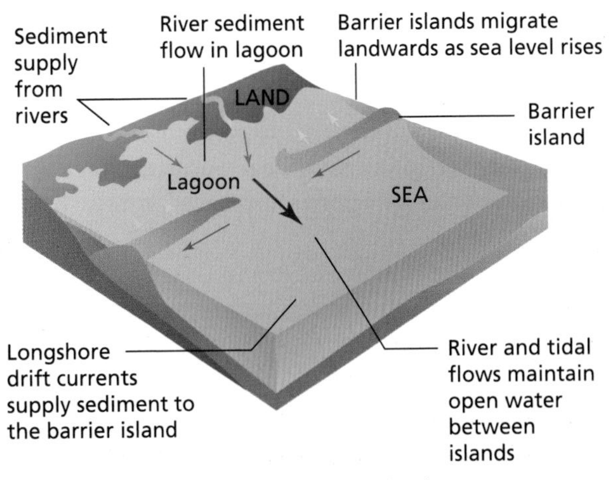

Figure 10.5 The formation of barrier islands

coasts which are strongly influenced by underlying structural geology. In the case of the Dalmatian Coast, limestone anticlines and synclines form a distinctive coastal landscape of parallel hilly islands (anticlines) and elongated bays (synclines). Post-glacial sea level rises submerged this area, creating elongated bays in places that were once low-lying valleys.

Contemporary sea level change

Sea levels are rising globally today and most scientists attribute this to the impact of global warming. The current rate of rise is about 2 mm per year. Figure 10.6 shows past and projected sea level data as collated by the UN Intergovernmental Panel on Climate Change (IPPC). It is noticeable that:

- sea levels are estimated to have been stable between 1800 and 1870 (although there is very little accurate, directly measured sea level data from this period)
- sea levels rose slowly between 1870 and 1940, but accelerated after that
- since 1980, sea level rise has been faster still
- over time, the minimum and maximum range of the instrumental record for sea level rise has narrowed as tide gauges and, more recently, satellite measurements have become more accurate.

Future projections cover a wide range. In 2007 the IPPC estimated increases in sea level of 20 to 50 cm by 2100 (Figure 10.6), amending this to 28 to 98 cm in 2013. Some scientists expect the global sea level to increase by over 100 cm by 2100.

Sea level is very difficult to predict because there are a number of different components that could affect the potential sea level rise:

- A major component is thermal expansion of the oceans as they warm due to global warming; the contribution of this component depends on how high global temperatures climb.
- The melting of mountain glaciers in the Alps, Himalayas and other mountain ranges will increase ocean water volume.
- The melting of major ice sheets (Greenland, Antarctica) could dramatically increase the global sea level, but there is significant uncertainty about when and by how much these ice sheets will melt.

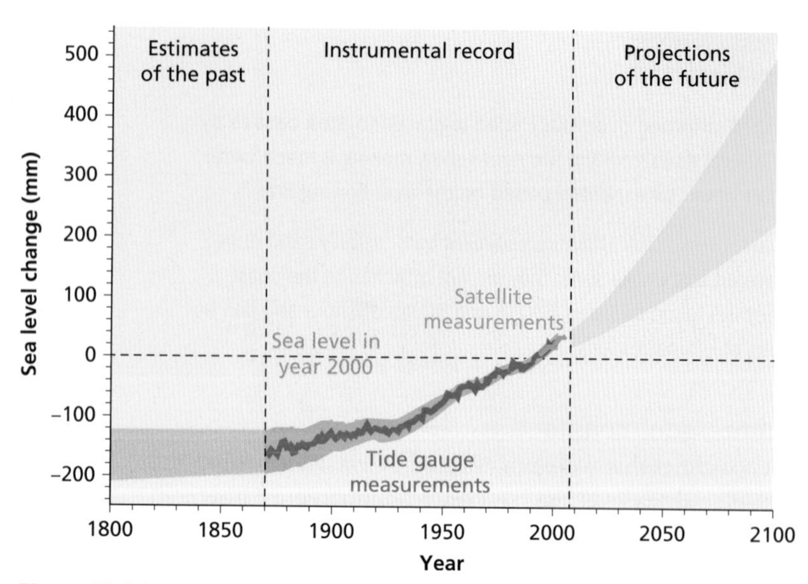

Figure 10.6 Past, present and future global sea levels

In some locations, sea level can change locally due to tectonic activity. A famous example is Turakirae Head near Wellington in New Zealand. Successive major earthquakes have repeatedly lifted the shoreline by several metres, five times in the last 7000 years. As a result of the earthquake that caused the 2004 Indian Ocean tsunami, the coastline of Aceh province on Sumatra dropped by 1 m, while some offshore islands were raised out of the sea by up to 2 m. The tectonic subsidence created rias, while in some locations coral reefs were raised up and subsequently eroded by wave action. This type of change is rare, localised and limited to tectonically active zones – but it can cause significant, instant change to coastlines.

10.2 The erosion threat

Rapidly eroding coastlines can threaten property and even people's lives. Fundamentally, rapid erosion results from physical factors which mean that some coasts recede faster than others, for example:

- long wave fetch and large, destructive ocean waves
- soft geology, especially unconsolidated sediments
- cliffs with structural weaknesses such as seaward rock dip and faults
- cliffs that are vulnerable to mass movement and weathering, as well as marine erosion
- strong longshore drift; eroded debris is quickly removed exposing the cliff base to further erosion.

In many places where rapid erosion is occurring, human actions can be identified that make the situation worse. Almost always this is because the coastal diment cell (see page 126) has been interfered with human actions.

emoving sediment from coastal sediment cells can ve severe consequences for erosion. A famous example curred at Hallsands in Devon in the 1890s. Offshore edging to provide construction material for a new naval ockyard removed over 1500 tonnes of sediment per day. he level of the beach at Hallsands began to fall as it was arved of sediment. By 1900, the village of Hallsands as under serious threat with no beach to dissipate wave ergy. Despite dredging stopping in 1902, in 1917 a ajor storm breached the village's sea defences and only ne house remained inhabitable. Today, Hallsands is an andoned 'ghost village' on the Devon coast.

ariations in erosion rate

rosion is not constant, even on coastlines that are roding very rapidly. It varies:

- in time, with peaks of erosion occurring in some seasons, and with some years having a lot of recession and some very little
- spatially, with some locations having much less erosion than others.

n the Holderness Coast in the East Riding of Yorkshire verage annual erosion is around 1.25 m per year but, as igure 10.7 shows, there are wide spatial variations in is rate, from 0 m per year to 6 m per year. This cannot e wholly explained by variations in geology as the ntire coastline (from Barmston to Kilnsea) consists of nconsolidated boulder clay. The variations are a result of:

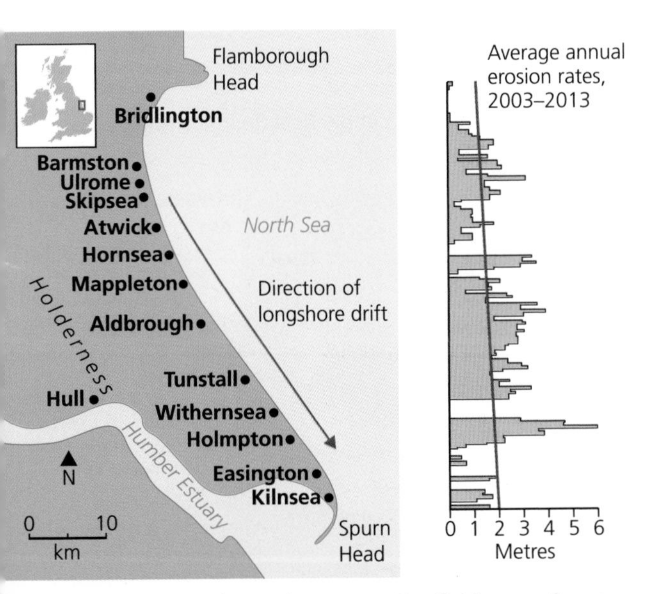

igure 10.7 Variations in erosion rate on the Holderness Coast, 2003–2013

- coastal defences in locations, such as Hornsea, Mappleton and Withernsea, which have stopped erosion
- starvation of sediment further south due to the construction of groynes and breakwaters that interrupt longshore drift and trap sediment, but lead to greater erosion just south of these defences
- sediment starvation, which means that erosion rates generally increase from north to south, as beaches further south are increasingly sediment starved
- variations in cliff height and the susceptibility to erosion of some areas of boulder clay compared to others
- mass movement susceptibility in some locations, as well as debris from previous erosion and mass movement protecting some parts of the cliff.

Erosion of Holderness also varies over time. Figure 10.8 shows erosion measured every six months at Hollym, just south of Withernsea, between 1999 and 2014. Erosion varies from 0 m to 11 m. Most erosion peaks occur when the biannual spring measurement is taken. Some reasons for these variations include:

- winter storms, which cause most erosion, especially when these coincide with a high spring tide
- storms are rare in the summer months so erosion rates measured in autumn are lower
- northeasterly storms cause most erosion because of the long wave fetch – 1500 km – from the north Norwegian coast, but these are rare.

In addition, the shape of the beach in front of the Holderness boulder clay cliffs can change and promote erosion. Ords are deep hollows on the beach running parallel to the cliff line at the base of the cliff. Ords seem to concentrate erosion in particular locations by allowing waves to directly attack the cliff with little energy dissipation on the beach. Ords migrate downdrift by about 500 m per year so the locations of most erosion change over time. Ord locations erode four times faster than locations without ords.

Key terms

Dredging: Involves scooping or sucking sediment up from the seabed or a river bed, usually for construction sand or gravel, or to deepen a channel so that large boats can navigate it.

Dissipation: The term used to describe how the energy of a wave is decreased by friction with beach material during the wave swash up the beach. A wide beach slows waves down and saps their energy so, when they break, most energy has gone.

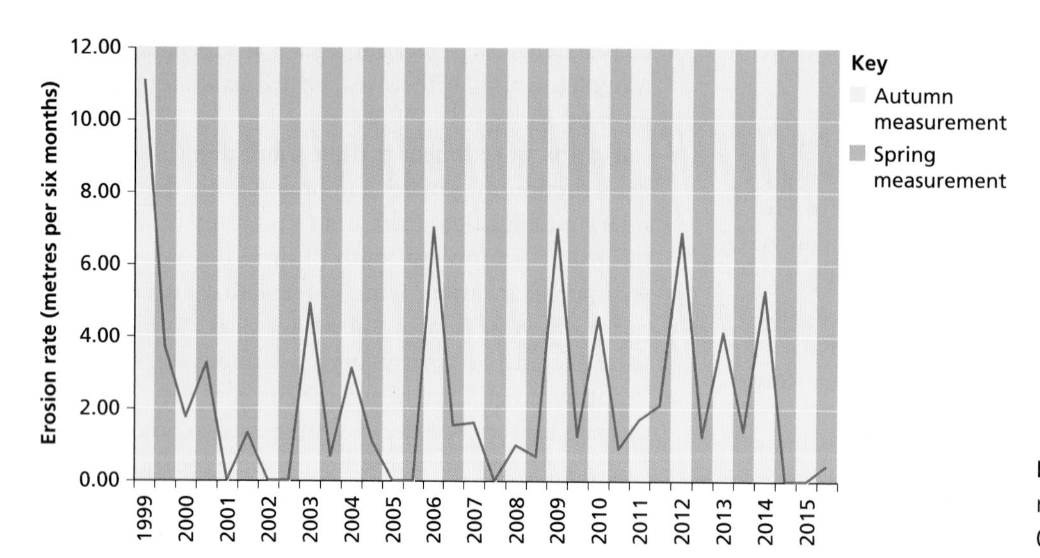

Figure 10.8 Variations in erosion rate at Hollym on the Holderness Coast, 1999–2014

Dams that cause erosion

A famous case study of coastal erosion is the Nile Delta. The delta is a depositional landform formed from sediment brought down the Nile by annual floods. Following construction of the Aswan High Dam on the River Nile in the 1960s, river discharge fell from about 35 billion m^3 metres per year to around 10 billion m^3. More significantly, sediment volume fell from 130 million tonnes to about 15 million tonnes. These changes were caused by:

- water withdrawals for industry, cities and farming from the reservoir behind the Aswan High dam (Lake Nasser)
- sediment being trapped by the reservoir and dam; water in the reservoir flows very slowly allowing sediment to be deposited.

The effect on the Nile Delta has been dramatic (Figure 10.9). Erosion rates at the Rosetta, where one of the main delta branches of the river meets the sea, jumped from 20 to 25 m per year to over 200 m per year as the delta was starved of sediment.

The River Nile is not an isolated case. The construction of the Akosombo Dam in Ghana in 1965 reduced the flow of sediment down the River Volta from some 70 million m^3 per year to less than 7 million m^3, with major impacts on longshore drift and coastal erosion in Ghana and even in neighbouring countries (Figure 10.10).

Both examples show how important river-transported sediment can be to maintaining the dynamic equilibrium between deposition and erosion at the coast. As a major sediment source, interference with the flow of sediment down rivers can lead to dramatic coastal change.

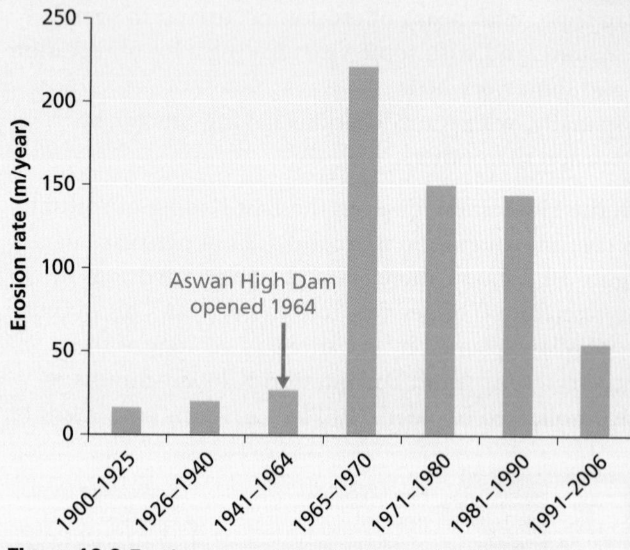

Figure 10.9 Erosion rates at the Rosetta in the Nile Delta, Egypt, 1900–2006

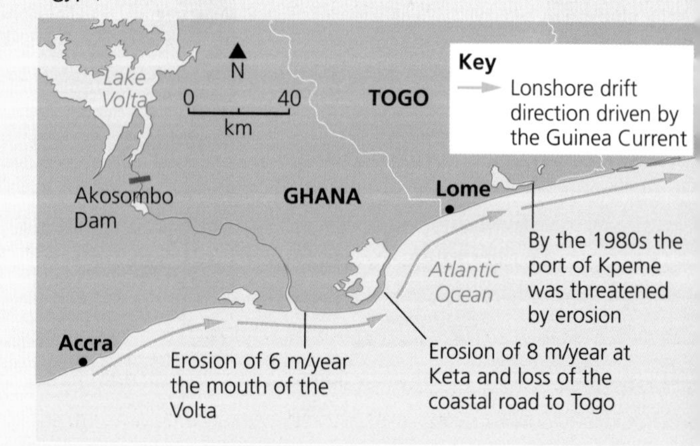

Figure 10.10 The Akosombo Dam, Ghana, and coastal erosion

Skills focus: Measuring erosion

Erosion can be measured in several different ways, at least over periods of months or years. On Yorkshire's Holderness Coast:

- In 1951 a series of fixed posts were set up inland of the coastline to act as fixed points to measure recession from, either using a tape measure or surveying equipment.
- In 1999 GPS (Global Positioning System) measurements replaced the posts; GPS uses a network of satellites to locate positions on the Earth's surface.
- In 2009 LIDAR (Light Detection And Ranging) replaced GPS measurements as an even more accurate method.

Satellite images can be used to measure erosion rates. The Earth's surface is continually being photographed from space and images from different years can be put together to create a 'time lapse' image of coastal recession.

Modern technology like this only extends so far back in time, however. For earlier periods, old maps can be useful. Ordnance Survey (OS) maps and Boundary Commission maps from the nineteenth century can be compared with modern OS maps, although there are difficulties:

- The scale of pre-1970 maps is normally in yards and miles, compared to modern metres and kilometres, so a conversion will be needed.
- Finding 'fixed points' that exist today and in the past can be difficult, but triangulation pillars, spot height locations, roads and especially churches can be useful.

Figure 10.11 shows two maps of Mappleton, a village on Holderness. The map on the left is the 1945–1948 'New Popular' OS edition whereas the one on the right is the modern 2015 OS map of the same area. The horizontal and vertical grid lines are in the same position for both maps.

Look closely at these two maps and notice that:

- The church at Great Cowden has disappeared by 2015, as has a track running north towards Mappleton.
- Some of the blue water courses between Great Cowden and Mappleton have disappeared by 2015.
- Mappleton itself is much closer to the coast in 2015 compared to the 1940s.

The total erosion here since 1990 is in the order of 250 m.

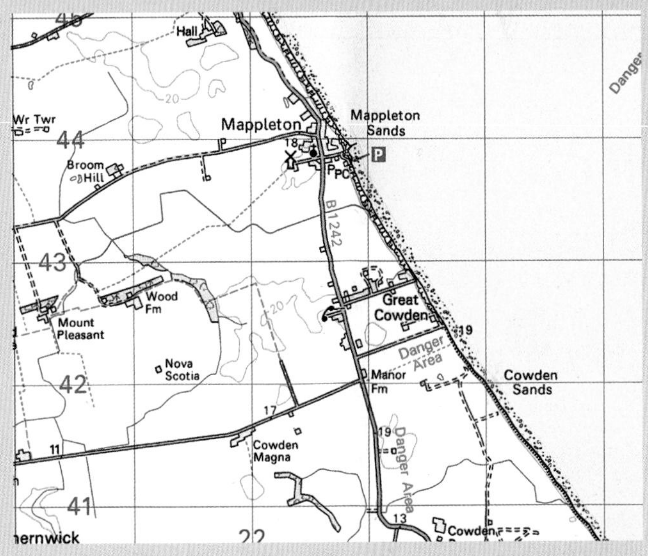

Figure 10.11 OS maps from the 1940s and 2015 for Mappleton (Source 10.11a: This work is based on data provided through www.VisionofBritain.org.uk and uses historical material which is copyright of the Great Britain Historical GIS Project and the University of Portsmouth)

10.3 Coastal flood risk

On many populated coastlines the threat of flooding outweighs the risk from erosion. This is because many people live on low-lying coasts that are only a few metres above sea level. These areas include:

- coastal plains, such as the east coast of the USA
- estuaries, such as those of the rivers Thames, Severn and Tees in the UK
- river deltas, such as the Nile and Mississippi.

It might seem odd that people live in areas at risk from coastal flooding but there are good economic reasons to inhabit these areas:

- Coastlines are popular with tourists, especially when access to beaches and the sea is easy.
- Deltas and estuaries are ideal locations for trade between upriver places and places along the coast or across the sea.
- Deltas are especially fertile and ideal for farming.

The mega-deltas of Asia are especially vulnerable to coastal flooding. These are the deltas of Asia's major rivers, each with a low-lying delta surface of more than 10,000 km². Many of these deltas are home to a very large city or megacity:

Table 10.2 Asia's mega-deltas and their populations

Delta	City	Population
Huang He- Hai	Tianjin	15 million
Yangtze	Shanghai	24 million
Pearl	Guangzhou	13 million
Chao Phraya	Bangkok	8.5 million
Ganges-Brahmaputra	Dhaka	14.3 million
Indus	Karachi	24 million

Other Asian mega-deltas, such as the Irrawaddy and Mekong, lack a major city but are home to millions of farmers. These deltas are vulnerable to rising sea levels in a number of ways, as Figure 10.12 shows. The dynamic equilibrium of sediment supply, deposition, accretion and erosion can be unbalanced easily in a delta environment, increasing risks of coastal flooding.

Sediment trapped by upstream dams, which reduce new deposition on delta surface

Start of delta deposition

Sea level rise increases flood risk and erosion

Megacity

Storm surges from hurricanes flood the delta more frequently

Delta subsidence is caused by weight of sediment deposition, which is balanced by new deposition on the delta surface

Groundwater extraction causes subsidence and accelerates sinking

River straightening for navigation means faster river flow and sediment being propelled too far offshore

Destruction of mangrove forests for wood and charcoal exposes the coast to erosion

Figure 10.12 Sea level rise risks in Asia's mega-deltas

Storm surges

Most coastal flooding, at least in the short term, is caused by storm surges. A storm surge is a short-term change in sea level caused by low air pressure. This can be:

- a depression (low-pressure weather system) in the mid latitudes, for example the UK

Islands at risk

There are many examples of islands that are at risk from rising sea levels. These include the Maldives in the Indian Ocean, and Tuvalu and Vanuatu in the Pacific. Many of the most at risk islands are coral atolls only 1 to 3 m above sea level.

The Maldives has a population of 340,000 people spread out across 1200 islands, although many islands are uninhabited. The highest point in the country is a mere 2.3 m above sea level. A sea level rise of 50 cm by 2100 would mean the Maldives losing 77 per cent of its land area. Areas that remained above sea level would become vulnerable to storm surges and erosion. Very small changes in sea level translate into major losses of land because of the country's unusual topography.

Figure 10.13 shows Malé, the capital city island, and surrounding islands. The Maldivians have gone to extraordinary lengths to create new inhabitable space and protect against rising sea levels. Malé is ringed by a 3 m high sea wall. Hulhumalé is a new artificial island built from coral and sediment dredged from the seabed between 1997 and 2002 at a cost of US$32 million. It is a full metre higher than Malé, which may come in useful in decades to come.

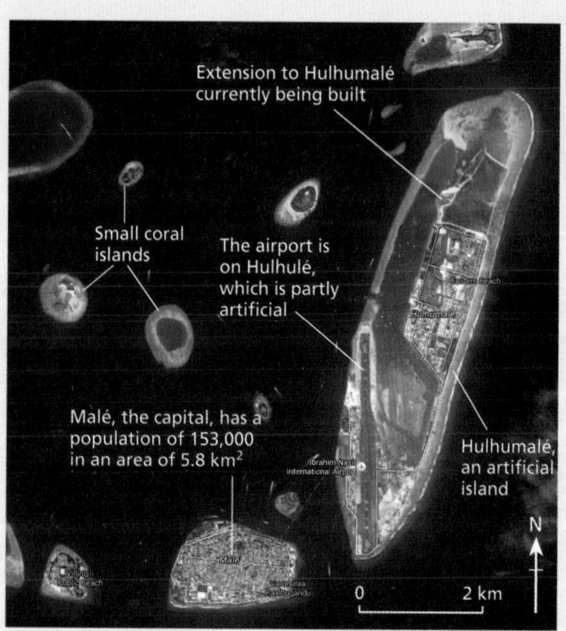

Figure 10.13 A satellite view of Malé and its surrounding islands in the Maldives (Google and the Google logo are registered trademarks of Google Inc., used with permission. Imagery ©2016 Data SIO NOAA, US Navy, NGA, GEBCO, Landsat, DigitalGlobe, CNES/Astrium, Map data ©2016 Google)

a tropical cyclone (hurricane, typhoon) in areas just north and south of the equator.

n both cases, a fall in air pressure of 1 millibar leads to 1 cm rise in local sea level. As air pressure drops, the weight of air pressing down on the sea surface drops, o the sea surface rises. This drop in air pressure would ause coastal flooding on its own but, during a severe epression or cyclone, it is made worse when:

- strong winds push waves onshore, so wave height increases the effective height of the sea even more
- high or spring tides occur at the same time as the storm, making the sea level even higher than normal.

Coastal topography can also have an effect. In the North Sea (see Figure 10.14) and in other locations such s the Bay of Bengal, the coastline narrows into a funnel hape. If a storm approaches from a particular direction, he storm surge can be funnelled into an increasingly arrow space between coastlines; as the sea shallows owards the coast the effect is severe coastal flooding. Storm surges can be very large. During Hurricane Katrina which struck New Orleans in the USA in 2005, some coastal areas experienced an 8 m storm surge (similar in height to a three-storey building).

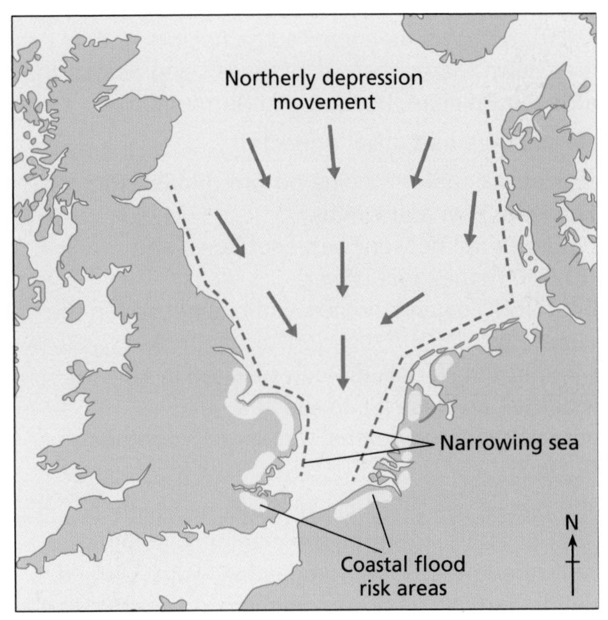

Figure 10.14 Coastal topography and storm surges

2013 North Sea storm surge

The worst storm surge in the North Sea on record occurred in January 1953. This historic event took the lives of 2500 people, mostly in the Netherlands but including 325 people in the UK. Sixty years later in December 2013 the unusual meteorological conditions repeated themselves.

Figure 10.15 shows a synoptic chart for 5 December 2013. A very deep depression named 'Cyclone Xaver' moved southeast down the North Sea:

- As the storm moved over Iceland it deepened dramatically, with very low air pressure (962 millibars) and rising storm surge potential.
- Winds of over 140 mph were recorded in mountainous areas of Scotland.
- Gale-force northerly winds drove the storm waves on to North Sea coasts, with a storm surge of 5.8 m recorded at Immingham in Lincolnshire.
- The surge corresponded with hide tide in many locations, making flooding even worse.

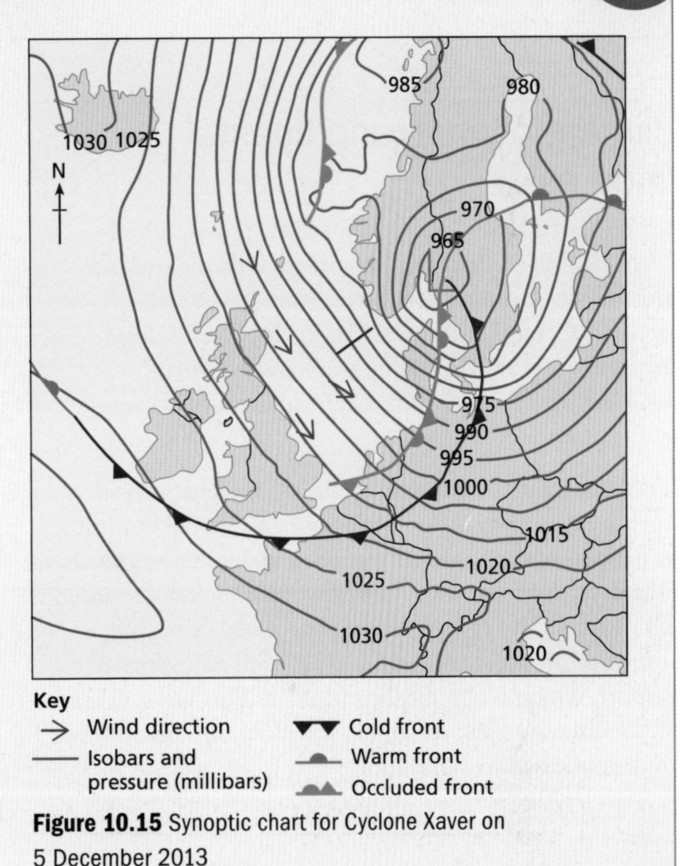

Figure 10.15 Synoptic chart for Cyclone Xaver on 5 December 2013

Figure 10.16 shows the storm surge heights and its progress down the east coast of the UK and across the North Sea in Belgium, Holland and Germany.

The storm surge had major impacts:

- Significant coastal flooding occurred in Boston, Hull, Skegness, Rhyl and Whitby.
- Scotland's rail network was shut down and 100,000 homes lost power.
- About 2500 coastal homes and businesses in the UK were flooded.
- At Hemsby in Norfolk erosion resulted in several properties collapsing into the sea.
- There were fifteen deaths across the countries affected.

In some places the 2013 storm surge was higher than that of 1953. However, the economic and human impacts were lower. Better forecasting, warnings and evacuation and, above all, much improved coastal flood defences limited the scale of the damage.

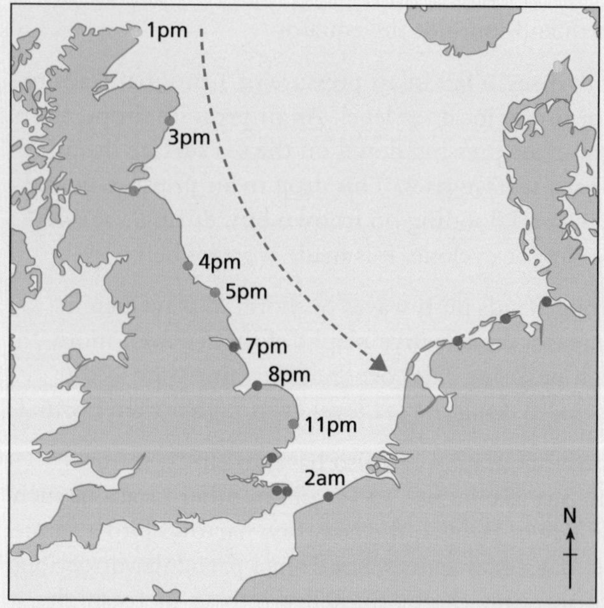

Figure 10.16 Storm surge heights and timings, 5 and 6 December 2013

Climate change and coastal flood risk

Sea level can have a major impact on coasts, whether it be long-term changes due to isostatic and eustatic movement or more short-term events such as tectonic activity and storm surges. It is well known that sea levels are rising due to global warming, but how significant is this rise for coastal flooding and erosion?

It is important to state that cyclones, depressions and storm surges have always happened and would continue to happen without global warming and rising sea levels. However, there is reason to believe that global warming will increase the risks to coasts. The global scientific authority on these risks is the IPCC 5th Assessment Report, which was published in 2014. This is a synthesis of scientific opinion on global warming. Its conclusions are summarised in Table 10.4.

Some predictions of the impacts of global warming are much more certain than others. Global data on trends in average wind speeds and wave heights is too poor to make accurate future forecasts, compared to data on sea level.

Even for sea level data, the magnitude of future sea level rise falls within a very large range, and the timing of this sea level rise is also not clear.

The IPCC 5th Assessment Report made another interesting statement about coastal changes which might be blamed on global warming:

> *'The impacts of climate change are difficult to tease apart from human-related drivers e.g. land-use change, coastal development, pollution.'*

This is interesting because it acknowledges that coasts are complex systems affected by many factors. Blaming increased coastal flooding or erosion on global warming misunderstands the interplay of factors that affect the level of risk on coasts.

Storm surges in Bangladesh

Locations that are in the path of tropical cyclones are especially vulnerable to storm surges. Compared to the depressions that affect the UK, tropical cyclones have lower air pressure and stronger winds, resulting in larger storm surge heights.

Bangladesh is especially vulnerable to the impacts of tropical cyclones for a number of reasons:

- Much of the country is a very low-lying river delta, only 1 to 3 m above sea level (Figure 10.17).
- Incoming storm surges often meet outflowing river discharge from the Ganges and Brahmaputra rivers, meaning river flooding and coastal flooding combine.
- Intense rainfall from tropical cyclones contributes to flooding.
- Almost all of the coastline consists of unconsolidated delta sediment, which is very susceptible to erosion.
- Deforestation of coastal mangrove forests has removed vegetation that once stabilised coastal swamps and dissipated wave energy during tropical cyclones.
- The triangular shape of the Bay of Bengal concentrates a cyclone storm surge as it moves north, increasing its height when it makes landfall.

Three major cyclones have struck Bangladesh since 1970, plus many more minor ones. Death tolls have fallen over time because of much-improved warnings, the construction of cyclone shelters and better aid response. However, vast areas are flooded by these storm surges forcing millions of people from their homes and farms in the densely populated coastal areas.

Cyclones in Bangladesh cause many metres of coastal erosion and can reshape whole stretches of coastline. River channels in the delta can shift dramatically, eroding farmland but creating new areas of deposition elsewhere.

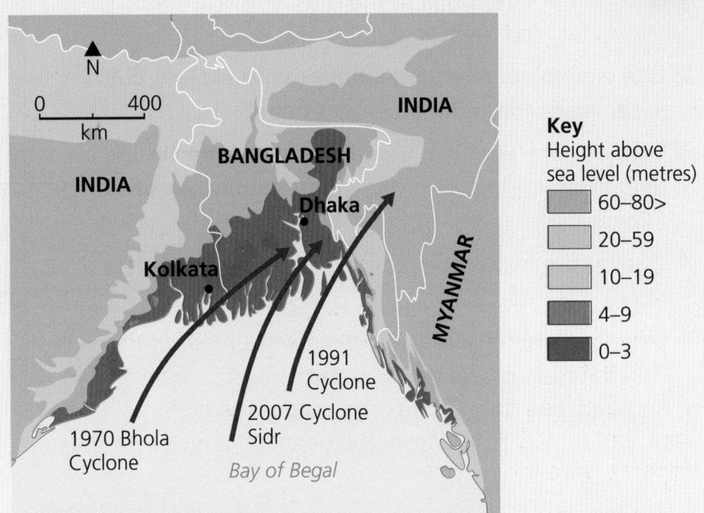

Figure 10.17 Height above sea level in Bangladesh, India and Myanmar

Table 10.3 A comparison of three major Bangladesh cyclones

	1970 Bhola Cyclone	1991 Cyclone	2007 Cyclone Sidr
Storm surge height	10 m	6 m	3 m
Maximum 1-minute sustained winds speed	205 kmph	250 kmph	260 kmph
Lowest air pressure	966 mb	918 mb	944 mb
Deaths	300,000–500,000	139,000	15,000
Economic losses	US$90 million	US$1.7 billion	US$1.7 billion

Table 10.4 A summary of the 2014 IPPC AR5 Report

Sea level	Sea level will rise by between 28 cm and 98 cm by 2100, with the most likely rise being about 55 cm by 2100
Delta flooding	The area of the world's major deltas at risk from coastal flooding is likely to increase by 50 per cent
Wind and waves	There is some evidence of increased winds speeds and larger waves
Coastal erosion	Erosion will generally increase due to the combined effects of changes to weather systems and sea level
Tropical cyclones	The frequency of tropical cyclones is likely to remain unchanged, but there could be more large storms
Storm surges	Storm surges linked to depressions are likely to become more common

High confidence – the evidence has a high degree of certainty; Medium confidence – the evidence has some certainty; Low confidence – the evidence is weak and uncertainty high

Review questions

1. Outline the difference in the meaning of the terms 'isostatic' and 'eustatic'.
2. Use examples of named landforms to explain the meaning of the term 'emergent coast'.
3. Suggest why coastlines in the north of the UK are usually emergent, whereas in the south they are submergent.
4. Outline the major components of rising sea levels today.
5. Using a named example, explain why erosion rates vary in time and space on a stretch of coastline.
6. Suggest why some coastlines are more threatened by rising sea levels and coastal flooding than others.
7. Discuss the significance of storm surges as a threat to coastal communities.

Further research

Explore the very varied coastline of Scotland, including emergent features: www.snh.gov.uk/about-scotlands-nature/habitats-and-ecosystems/coasts-and-seas/coastal-habitats

Explore the issue of sea level rise in terms of causes and consequences: http://climate.nasa.gov/vital-signs/sea-level

Investigate data sets about coastal erosion rates on the Holderness Coast: http://urbanrim.org.uk/Holderness.htm

Investigate the impact of one of Bangladesh's mostly deadly storm surges: www.weather.com/storms/hurricane/news/deadliest-cyclone-history-bangladesh-20130605#/1

11 Managing risks and conflicts

How can coastlines be managed to meet the needs of all players?

By the end of this chapter you should:

- understand the consequences for people and their property of coastal risks including erosion, flooding and rising sea levels
- understand the wide range of engineering approaches that can be used to reduce risk, including their advantages and disadvantages
- understand how decisions are made about coastal management and how this can lead to conflict and winners and losers.

11.1 Consequences for communities

The costs of rapid coastal recession caused by erosion can be classified into three broad categories:

- **Economic** costs include the loss of property in the form of homes, businesses and farmland; these are relatively easy to quantify.
- **Social** costs are the impacts on people, such as the costs of relocation and loss of jobs (which can be quantified) but also include impacts on health (such as stress and worry), for example, which are much harder to quantify.
- **Environmental** impacts include loss of coastal ecosystems and habitats; these are almost impossible to quantify financially but are likely to be small anyway because erosion is part of the natural coastal system to which ecosystems are adapted.

Losses from erosion are usually localised. This means that costs are very specific to particular locations, as shown in Figure 11.1. The UK government produces average land value estimates which provide some idea of the value 'lost' as land disappears into the sea as a result of erosion. In some places residential land values are likely to be high because coasts are desirable locations. Locations of known risk, however, will have lower values.

The loss of roads as a result of coastal erosion can be costly if new roads have to be built on a different route. The cost for a 100 m length of new road (two lanes) is about £150,000 to £250,000.

The destruction of a section of the South Devon Main Line railway at Dawlish due to erosion in February 2014 gives some indication of the costs of major infrastructure loss. The railway links Exeter and Plymouth to London and Bristol. Repairing the damage to reopen the railway cost £35 million. Estimates of the cost to businesses in the South West were put at between £60 million and £1.2 billion.

In most cases economic losses from erosion are localised and small. This is because:

- erosion is incremental, with a small number of properties affected over a long period of time

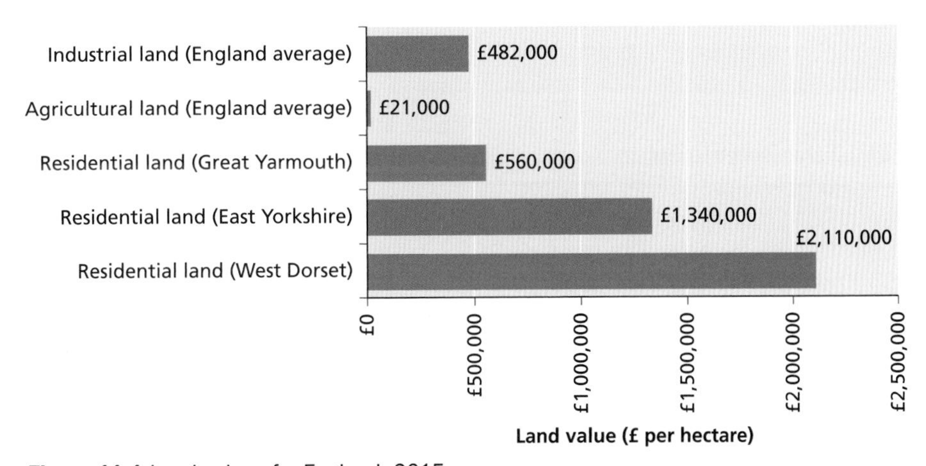

Figure 11.1 Land values for England, 2015

- property at risk loses its value long before it is destroyed by erosion, because potential buyers recognise the risk
- areas of high-density population, especially towns and villages, tend to be protected by coastal defences, thus reducing erosion risk.

When there are significant economic losses they tend to be the result of unexpected erosion events, often mass movement related. An example is the collapse of the Holbeck Hall Hotel in Scarborough in 1993. A major coastal rotational landslide of 1 million tonnes of glacial till caused cliff recession of 70 m and debris slumped 135 m beyond the original cliff foot. Heavy rainfall of 140 mm in the two months before June 1993 was a major contributing factor. The owners of the hotel sued Scarborough Borough Council for £2 million in compensation, claiming the council should have prevented the landslide, but lost the case in 2000.

In the UK property insurance for homes and businesses does not cover coastal erosion. This is because, in many cases, it is not a risk but a certainty. On the Holderness Coast in the East Riding of Yorkshire, which is currently eroding at 2 m per year, a house 20 m from the cliff edge will be lost to the sea in a few years time. For residents this means:

- falling property values, as the date of eventual loss approaches
- an inability to sell their property because the possibility of loss by erosion is too great
- the loss of their major asset, and the costs of getting a new home.

Figure 11.2 The coastal road at Aldbrough on the Holderness Coas

There are wider impacts too. Figure 11.2 shows Aldbrough on the Holderness Coast. Here the impacts of erosion include:

- increased costs to the owners of cliff-top caravan parks, who face moving caravans and relocating site roads and services
- loss of access as roads, paths and steps down to the beach disappear
- loss of amenity value as the coastline is visually scarred by collapsing roads, abandoned properties anc warning signs.

The consequences of coastal flooding

When coastal land is inundated as a result of storm surges there are immediate economic and social impacts. Table 11.1 compares the impacts of different storm surge events.

Compensation and help in the East Riding of Yorkshire

The coast of much of the East Riding of Yorkshire is rapidly eroding and properties are at risk. There is no national scheme in the UK to compensate property owners who lose their property to erosion, but there is some limited help.

Between 2010 and 2012 the Department for Environment, Food and Rural Affairs (DEFRA) provided East Riding Council with £1.2 million as one of fifteen UK 'Coastal Change Pathfinder' projects. The money was spent assisting 43 home owners with relocation and demolition expenses.

East Riding Council now offers help through its Coastal Change Fund:

- covering the cost of property demolition and site restoration
- offering up to £1000 in relocation expenses, such as removal vans and storage
- offering up to £200 in hardship expenses.

There are 'rollback' policies also. These give people who are losing homes to coastal erosion preferential treatment in terms of finding somewhere to build a new home. Land that would not normally be given planning permission for a new home will be approved in these exceptional circumstances.

East Riding Council expects up to 100 homes to be lost to the sea in the next few decades. Not a large number, but a huge upheaval for these residents with very little in the way of compensation.

able 11.1 The social and economic impacts of storm surges

Example	Cause	Economic costs	Social costs
Netherlands 1953 (North Sea flood)	Mid-latitude depression moving south through the North Sea generating a 5 m storm surge	Close to 10% of Dutch farmland flooded 40,000 buildings damaged and 10,000 destroyed	1800 deaths
UK 2013–14 winter storms	Coastal and other flooding caused by a succession of depressions and their storm surges	Damage of around £1 billion over the course of the winter.	17 deaths (from all causes)
USA 2012 Hurricane Sandy	Landfall of Hurricane Sandy in New Jersey and other US states with a storm surge up to 4 m.	Damage of US$70 billion. Six million people lost power and 350,000 homes in New Jersey were damaged or destroyed.	71 deaths
Philippines 2013 Typhoon Haiyan	One of the most powerful tropical storms ever with a 4–5 m storm surge.	Damages of around US$2 billion, centred on the city of Tacloban.	At least 6300 deaths and 30,000 injured

There are also long-term impacts once the immediate destruction from coastal flooding has been dealt with. In the Netherlands, the 1953 storm surge eventually led to one of the largest coastal engineering projects on the planet, the Delta Works (Deltawerken) in order to reduce the risks from future storm surges.

Key terms

Megaproject: A very expensive (over US$1 billion), technically difficult and usually long-term engineering project. Many megaprojects have multiple aims and often large environmental impacts.

Return period (or recurrence interval): Refers to the frequency of a flood of a particular magnitude. A 1:100 flood event will occur, on average, every 100 years (there is a one per cent chance of that flood occurring in a given year).

Deltawerken

The Deltawerken is an example of a hard-engineering **megaproject**. It was begun less than three weeks after the devastation of the 1953 North Sea floods. The aims of the project were to:

- reduce the risk of flooding in the low-lying Eastern Scheldt area of the Netherlands, where much land was below sea level
- shorten the length of the coastline exposed to the sea by 700 km
- control the flow of the Rhine, Maas and Scheldt rivers to reduce flood risk
- maintain safe access to the North Sea for shipping from important Dutch ports such as Rotterdam.

Flood defences were designed to protect from a 1:2000 year coastal flood and 1:250 year river flood **return periods**. The Deltawerken took place between 1958 and 1997. A series of dams and sluice gates were constructed between the islands that make up the Eastern Schelt area to control the flow of water (Figure 11.3). During a storm surge, these can be closed to shut the sea out. Embankments (called ring-dykes) were built to act as flood walls around the islands and along the coast. The entire plan cost over US$5 billion. However, the risk of rising sea levels due to global warming means that the Dutch will need to continue to raise and strengthen their flood defences. In 2008 a Dutch government report concluded that the Netherlands should assume a sea level rise of 1.3 m by 2100, meaning US$1.5 billion would need to be spent each year on new flood defences up to 2100.

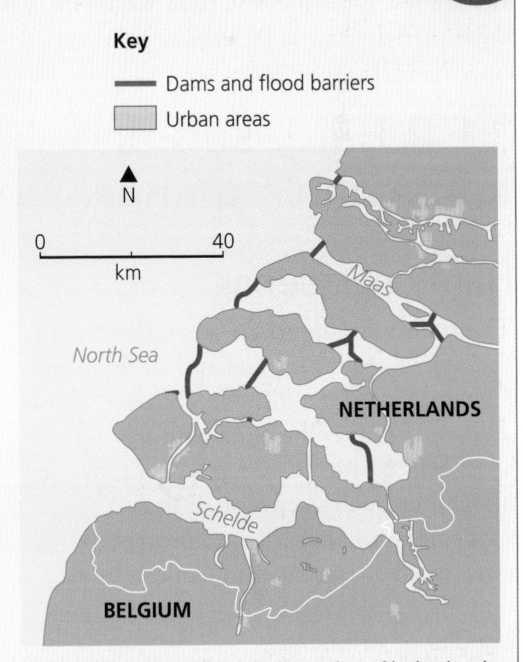

Figure 11.3 Map of the Deltawerken, Netherlands

Environmental refugees

In many parts of the world rising sea-levels caused by global warming will be problematic by 2100, but manageable at a cost. Rises of 40 to 100 cm will be managed by flood walls and sea defences in some places, but not everywhere.

Most at risk are islands such as the Maldives, Tuvalu and Barbados, which have particular risk factors:

- Tuvalu's highest point is 4.5 m above sea level, and most land is 1 to 2 m above sea level.
- Around 80 per cent of people in the Seychelles live and work on the coast.
- Many are fringed by coral reefs, which act as a natural coastal defence against erosion, but rising ocean temperatures due to global warming risks reef destruction due to coral bleaching.
- Water supply is limited and at risk from salt water incursion as sea levels rise and groundwater is overused.
- They have small and narrow economies based on tourism and fishing, which are easily disrupted.
- They have high population densities and very limited space, so there is no opportunity for relocation.

The worst case scenario for Tuvalu, and parts of the Maldives, is that some or all islands will have to be abandoned. The question then would be, where do these environmental refugees go?

Key term

Environmental refugees: Communities forced to abandon their homes due to natural processes including sudden ones, such as landslides, or gradual ones, such as erosion or rising sea levels.

11.2 Hard, soft and sustainable management

Hard engineering

The traditional approach when faced with the risk of coastal erosion and/or flooding is to encase the coastline in concrete, stone and steel. This is the hard-engineering approach. It aims to stop physical processes altogether (such as erosion or mass movement) or alter them to protect the coast (such as encouraging deposition to build larger beaches). This approach has a number of advantages:

- It is obvious to at-risk people that 'something is being done' to protect them.
- It can be a 'one-off' solution that could protect a stretch of coast for decades.

However, hard engineering has disadvantages:

- Costs are usually very high, and may involve ongoing maintenance costs.
- Even very carefully designed engineering solutions are prone to failure.
- Coastlines are made visually unattractive and the needs of coastal ecosystems are usually overlooked.
- Defences built in one place frequently have adverse affects further along the coast.

As Figure 11.4 shows, the costs of hard-engineering are very high. Exact costs depend on a number of factors such as:

- how extreme the location is in terms of tides and waves, which affect the difficulty of construction
- the size, strength and height of individual structures
- the degree to which the structures are made to look attractive or more 'natural'.

Table 11.2 summarises the key characteristics of hard-engineering approaches.

Of all the hard-engineering solutions, groynes are the most problematic. Sediment builds up behind

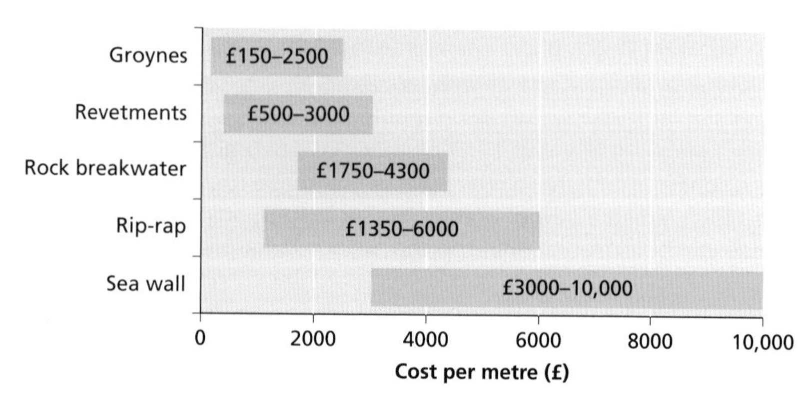

Data from the Environment Agency, 2015

Figure 11.4 Cost ranges (per metre) for hard-engineering coastal defences

able 11.2 Hard engineering coastal defences

Type	Construction and materials	Purpose	Impact on physical processes
Rip-rap (rock armour) 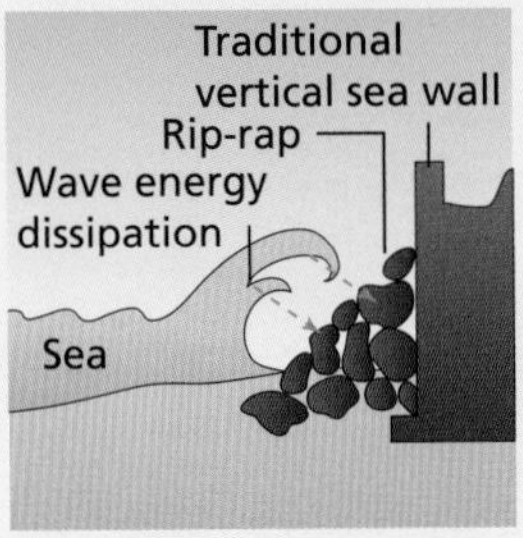	Large igneous or metamorphic rock boulders, weighing several tonnes	Break up and dissipate wave energy Often used at the base of sea walls to protect them from undercutting and scour	Reduced wave energy Sediment deposition between rocks May become vegetated over time
Rock breakwater 	Large igneous or metamorphic rock boulders, weighing several tonnes	Forces waves to break offshore rather than at the coast, reducing wave energy and erosive force	Deposition encouraged between breakwater and beach Can interfere with longshore drift
Sea wall 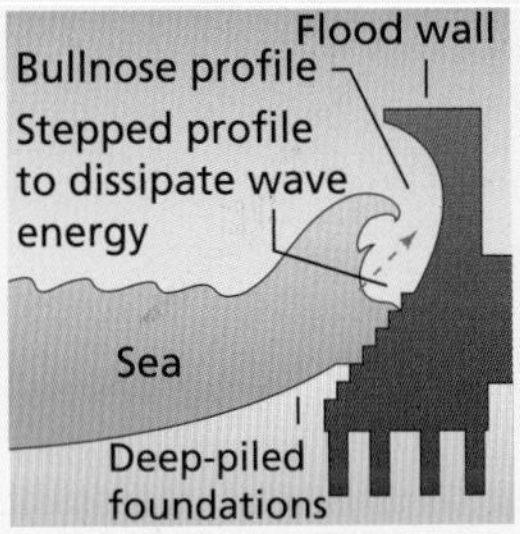	Concrete with steel reinforcement and deep-piled foundations; can have a stepped and/or 'bull nose' profile	A physical barrier against erosion They often also act as flood barriers Modern sea walls are designed to dissipate, not reflect, wave energy	Destruction of the natural cliff face and foreshore environment If reflective, can reduce beach volume
Revetments 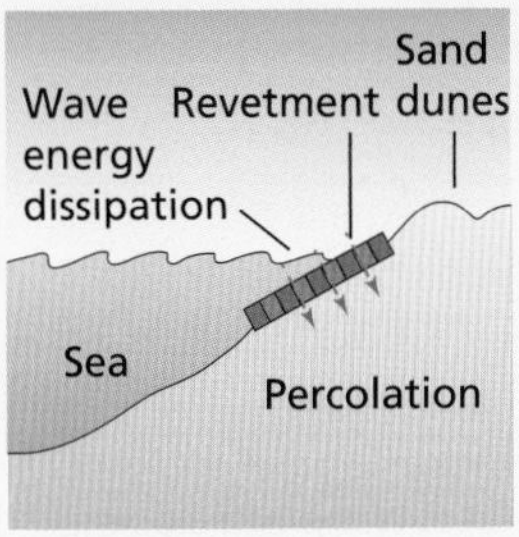	Stone, timber or interlocking concrete sloping structures, which are permeable	To absorb wave energy and reduce swash distance by encouraging infiltration Reduce erosion on dune faces and mud banks	Reduced wave power Can encourage deposition and may become vegetated
Groynes 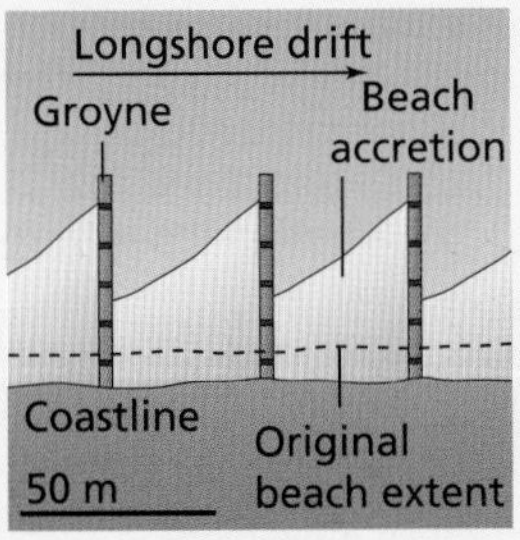	Vertical stone or timber 'fences' built at 90 degrees to the coast, spaced along the beach	To prevent longshore movement of sediment, and encourage deposition, building a wider, higher beach	Deposition and beach accretion Prevention of longshore drift, sediment starvation and increased erosion downdrift

the groyne on the updrift side, extending the beach seaward. Groynes interfere with sediment movement by longshore drift to such an extent that they 'starve' beaches downdrift of the sediment they need. This can lead to increased erosion elsewhere.

A larger beach is the best way of dissipating wave energy. As a wave breaks the swash runs up the beach and the wave's energy is lost due to a combination of the slope of the beach and water draining into the sediment.

Soft engineering

There is an alternative to hard-engineering, which attempts to work with natural processes to reduce coastal erosion and flood threat. Soft engineering is usually less obvious and intrusive at the coast, and may be cheaper in the long term.

Beach nourishment

This process involves artificially replenishing the sediment on a beach. It can range from a few truckloads of sediment to multi-million pound schemes. Beach nourishment is done for several reasons:

- to replace sediment lost by erosion
- to enlarge the beach, so that it dissipates wave energy and reduces wave erosion
- to increase the amenity value of beaches by adding fresh sand

Costs depend on the location of the sediment to be used (transport costs) and the volume needed. Costs of £2 million per kilometre of beach are typical. The benefit of the nourishment approach is that it works with natural processes. The material used (sand, gravel) is natural and the nourished beach naturally dissipates wave energy. However, there are issues:

- Ongoing costs are high; new sediment is required every few years as sand is continually removed by waves and currents.
- Care must be taken to find a sustainable source of sediment; often this is sediment dredged from offshore and pumped on to the beach (Figure 11.5) but this must not impact on the wider coastal sediment cell.

Figure 11.5 Beach nourishment using offshore sediment

Cliff stabilisation

In some locations it is the stability of rocks in cliffs that cause recession, rather than erosion by waves at the foot of a cliff. Engineering solutions can help stabilise cliffs prone to mass movement. This is often a costly process but less intrusive than large-scale cliff-foot defences (Figure 11.6).

Dune stabilisation

Sand dunes are a very effective coastal defence. Where they fringe the coast they not only absorb wave energy and prevent erosion but also protect low-lying areas from coastal flooding. Many sand dunes are unique and valuable habitats for flowering plants, insects and reptiles. Sand dunes are at risk from a number of processes:

- Major storm surges can breach dune systems causing rapid and significant erosion.
- Dunes are prone to overgrazing by animals, stripping them of vegetation and exposing bare sand, which is then eroded by wind.
- Human recreation and tourism can trample and kill vegetation, exposing the surface to erosion.

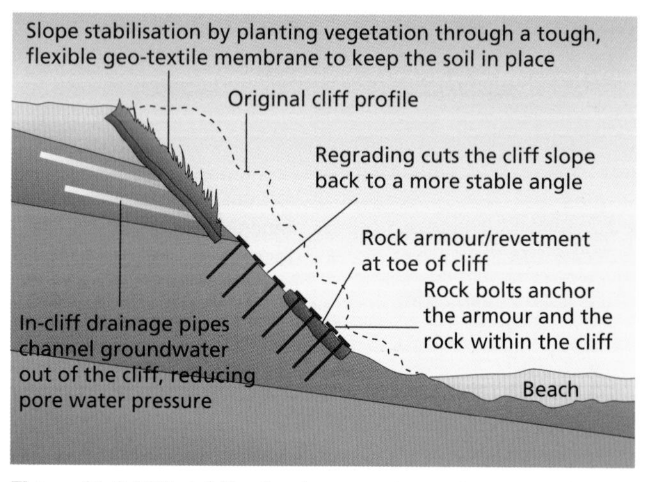

Figure 11.6 Cliff stabilisation to prevent mass movement

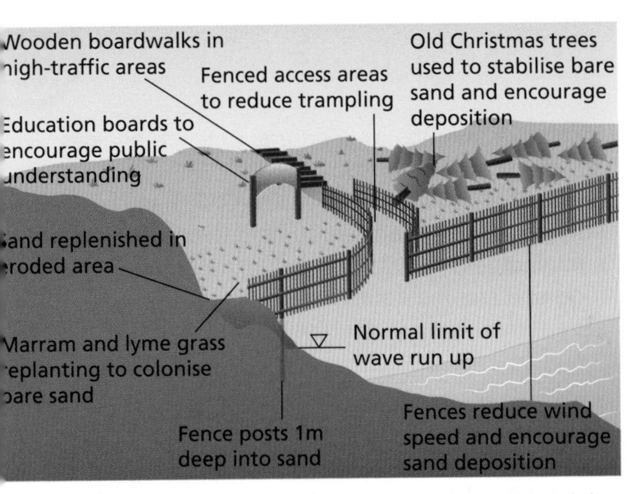

Figure 11.7 Techniques to stabilise and renew coastal sand dunes

Minor degradation of sand dunes can be managed using a number of low-cost, low-impact techniques (Figure 11.7). The costs of these methods are low compared to hard engineering. Dune fencing costs £400 to £2000 per 100 m, and replanting dunes about £1000 per 100 m. This means that working to maintain natural sand dunes can be very cost effective in the long term.

Sustainable coastal management

People at the coast have always lived with occasional major storms and erosion events, landslides and flooding. However, many coastal communities in the twenty-first century face the twin trends of:

- rising global sea levels, but uncertainty about the scale and timing of the rise
- increased frequency of storms and the possibility of increased erosion and flooding.

To cope with these threats, communities will need to adapt within the wider coastal zone to ensure the long-term well-being of people and the coastal environment. Figure 11.8 shows the different strands of sustainable coastal management.

Key term

Sustainable coastal management: Managing the wider coastal zone in terms of people and their economic livelihoods, social and cultural well-being, and safety from coastal hazards, as well as minimising environmental and ecological impacts.

Skills focus: Measuring coastal vegetation

Coastal ecosystems can be measured using fieldwork surveys to assess whether they are being degraded by erosion and human activities such as trampling, or recovering due to replanting and management. Surveys can also test whether the idealised plant succession can be found in a particular location. This is done using transect lines:

- The dune system or salt marsh ecosystem is sampled systematically along a transect line, for instance every 10 m.
- At each sample point the slope, height and percentage vegetation cover/bare soil and number of different plant species (species richness or diversity) present is measured.
- Vegetation cover is estimated using quadrants (50 × 50 cm or 1 × 1 m sampling grids).
- Soil samples, for pH and organic content testing, can also be taken.

It is often useful to use more than one transect across a coastal ecosystem, such as one in an area experiencing erosion versus one across a more pristine area. A statistical test, such as the chi-squared test, is useful to determine whether there are significant differences between one area versus another.

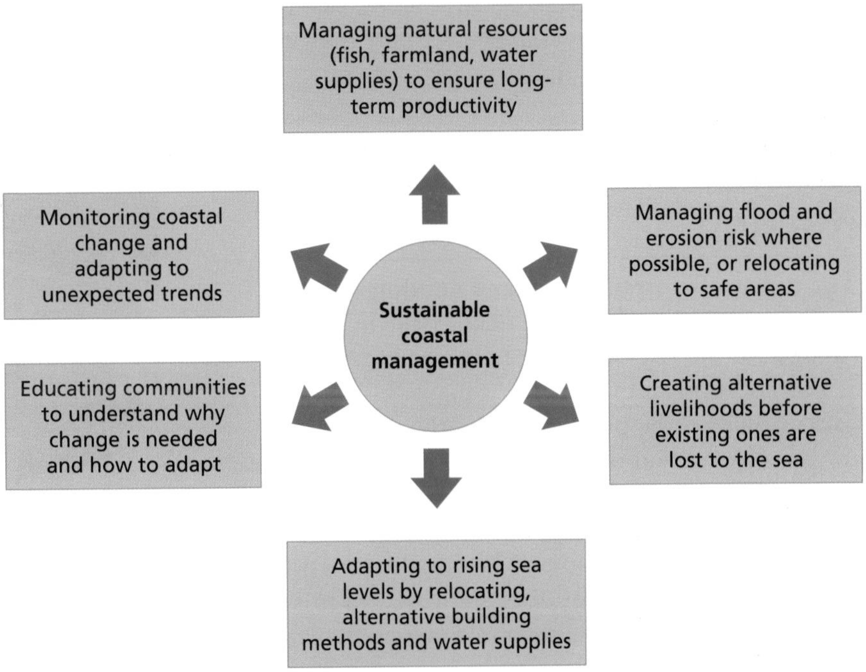

Sustainable coastal management

Managing natural resources (fish, farmland, water supplies) to ensure long-term productivity

Managing flood and erosion risk where possible, or relocating to safe areas

Creating alternative livelihoods before existing ones are lost to the sea

Adapting to rising sea levels by relocating, alternative building methods and water supplies

Educating communities to understand why change is needed and how to adapt

Monitoring coastal change and adapting to unexpected trends

Figure 11.8 Sustainable coastal management

Sustainable management in the Maldives

We saw in Chapter 10 how the Maldives is going to extraordinary lengths to battle rising sea levels, with the construction of artificial islands. However, this approach is unlikely to help the whole population. Rising sea levels threaten almost everyone in the country, and 97 per cent of the inhabited islands are experiencing coastal erosion. There is a risk that:

- money spent protecting the capital city Malé and creating new artificial islands such as Hulhumalé means that isolated islands are ignored
- sustainable management of traditional income sources (such as fishing) and resources (such as mangroves) is overlooked in favour of protecting urban and tourism development from coastal threats.

The Maldives has the potential for **conflict** to occur if coastal management focuses on some areas rather than others, and the needs of some groups over others. Some actions are being taken to help isolated, vulnerable coastal communities in the Maldives:

- The organisation Mangroves for the Future (MFF) is working with Maldivian communities to educate them on the importance of maintaining coastal mangrove swamps as a natural defence against coastal erosion and flooding.
- The Global Environment Facility (GEF) has provided small grants to islanders to help them develop sustainable and organic farming as an alternative food and income source to coral reef fish (threatened by both overfishing and global warming).

The Japanese government has funded mangrove nurseries on the Maldives so that damaged mangrove areas can be replanted (Figure 11.9).

Figure 11.9 Labour intensive mangrove forest restoration in the Seychelles

Key term

Conflict: In the context of coastal management, conflict means disagreement over how the coast should be protected from threats and which areas should be protected. Conflict often exists between different stakeholders, such as residents versus the local council.

11.3 Coastal decision making

The threat of coastal erosion and flooding is not new. In the UK, it became a major issue in Victorian times with the rise of seaside holiday resorts such as Scarborough and Cromer. Promenades, piers and parks needed to be protected from the sea. This led to the construction of sea walls and groynes (Figure 11.10). Almost always, construction was done in isolation with no consideration of how defences in one place might affect others. Today, the situation is different. Increasingly coasts are managed in a more holistic way using integrated coastal zone management (ICZM).

Figure 11.10 A postcard of Hunstanton, Norfolk, from the 1950s showing groynes and the sea wall

Integrated Coastal Zone Management

ICZM dates from the Rio Earth Summit in 1992 and has a number of key characteristics:

1 The entire coastal zone is managed, not just the narrow zone where breaking waves cause erosion or flooding. This includes all ecosystems, resources and human activity in the zone. It may include managing rivers that supply sediment to the coast. It involves management across political boundaries (for example, different council areas) to ensure a holistic approach.

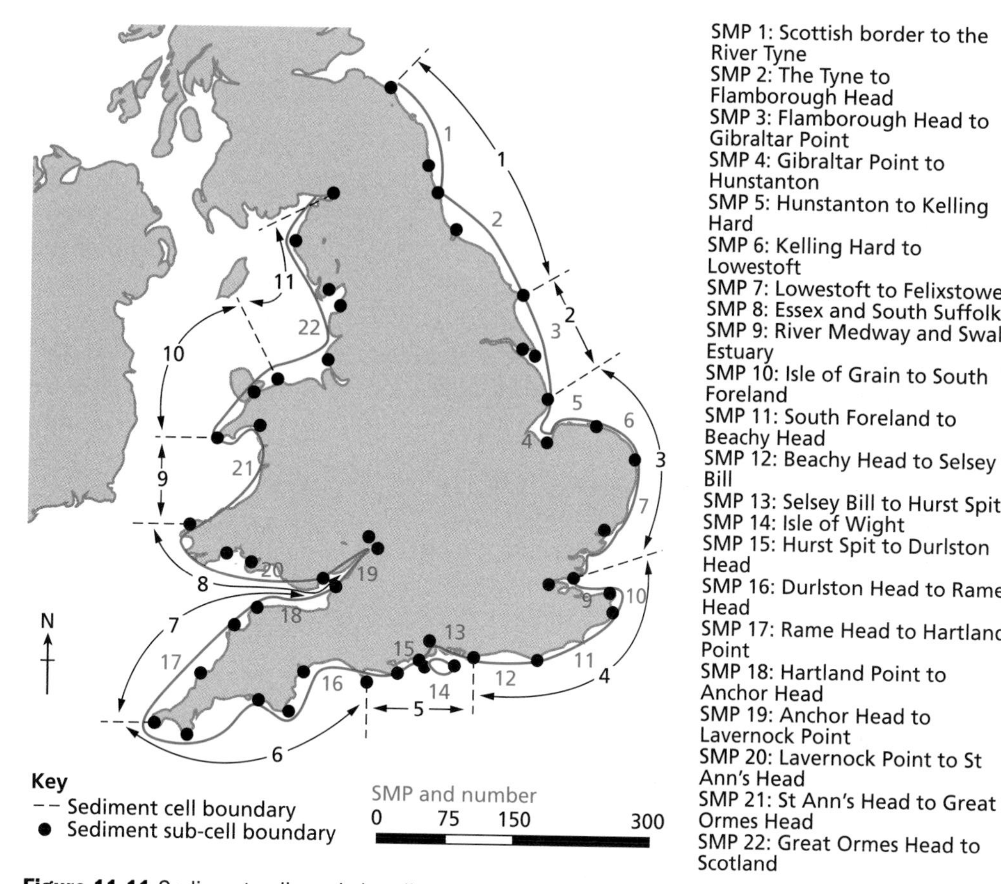

Figure 11.11 Sediment cells and shoreline management plans (SMPs) in England and Wales

2 It recognises the importance of the coastal zone to people's livelihoods, as globally very large numbers of people live and work on the coast – but their activities tend to degrade the coastal environment.

3 It recognises that management of the coast must be sustainable, meaning that economic development has to take place to improve the quality of life of people, but that this needs to be environmentally appropriate and equitable (to benefit everyone).

ICZM means adopting a joined-up approach across different stakeholders in coastal areas in order to harmonise policies and decision making. In practice this means that coastal management must:

- plan for the long term
- involve all stakeholders
- adopt 'adaptive management', changing plans and policies if threats change
- try to work with natural processes, not against them
- use 'participatory planning' where all stakeholders have a say in any policy decisions.

ICZM has been adopted worldwide. The Mediterranean is increasingly managed using ICZM. The ICZM Protocol is an agreement between fourteen Mediterranean countries to manage the sea and its coastline in an integrated way that was signed in 2008. The principles have also been widely adopted in New Zealand.

Littoral cells and shoreline management plans

From a physical geography perspective, ICZM works with the concept of littoral cells or sediment cells (see Chapter 9, page 126). The coastline can be divided up into littoral cells and each managed as an integrated unit. In England and Wales there are eleven sediment cells (Figure 11.11). Each cell is managed either as a whole unit or a sub-unit. In both cases a shoreline management plan (SMP) is used. The SMP area is

Key term

Littoral cells: All coastlines divide up into distinct littoral cells containing sediment sources, transport paths and sinks. Each littoral cell is isolated from adjacent cells and can be managed as a holistic unit.

further divided into sub-cells. SMPs in England and Wales broadly correspond to the physical geography of sediment cells. SMPs extend across council boundaries, so councils must work together on an agreed SMP to manage an extended stretch of coastline.

Policy decisions

Adopting the principles of ICZM and identifying large coastal units to manage in an integrated way (littoral cells) is the easy part of coastal management. The hard part is the decision-making process to decide on what actions to take.

Coastal management in the UK is overseen by The Department for Environment, Food and Rural Affairs (Defra). Since it introduced SMPs in 1995 there have been only four policies available for coastal management. These are very different in terms of their costs and consequences, as shown in Table 11.3.

Table 11.3 Coastal management policy options

No active intervention	No investment in defending against flooding or erosion, whether or not coastal defences have existed previously. The coast is allowed to erode landward and/or flood.
Hold the line	Build or maintain coastal defences so that the position of the shoreline remains the same over time.
Managed realignment	Allow the coastline to move naturally (in most cases to recede) but manage the process to direct it in certain areas. It is sometimes called 'strategic realignment' or 'strategic retreat'.
Advance the line	Build new coastal defences on the seaward side of the existing coastline. Usually this involves land reclamation.

Making decisions about which policy to apply to a particular location is complex. It depends on a number of different factors:

- The economic value of the assets that could be protected: on the rapidly eroding Holderness Coast assets such as the nationally important natural gas terminal at Easington will be protected, as will towns such as Hornsea, but not farmland and caravan parks.
- The technical feasibility of engineering solutions: it may not be possible to 'hold the line' for mobile depositional features such as spits, or very unstable cliffs.
- The cultural and ecological value of land: it may be desirable to protect historic sites and areas of unusual biodiversity, despite such locations having questionable economic value.
- Pressure from communities: vocal local campaigning can get results. A famous example is the small Holderness village of Mappleton, protected at a cost of £2 million in the 1990s despite the village being valued at £650,000.

In some cases protection from erosion and flooding may be forthcoming despite an economic case not being made, because of the political difficulty of the TV news showing pictures of people's homes falling into the sea.

SMPs plan for the future using three time periods called 'epochs'. These are up to 2025, 2025–55 and beyond 2055. A hold-the-line policy applied to an area up to 2025 may become a managed realignment policy after 2025. This is because, by 2025, rising sea levels are likely to have made hold the line a much more expensive policy to apply.

Hornsea and Mappleton

The impact of the policy options can clearly be seen on the Holderness Coast at Hornsea and Mappleton (Figure 11.12). The policy decision at Hornsea is hold the line. This is because:

- it is a regional economic centre with a population of about 8500 people
- there are important historic sites in the town and at Southorpe medieval village
- inland is Hornsea Mere, a very important lake habitat for birds that is designated as a Special Protection Area and Site of Special Scientific Interest (SSSI).

In the future Hornsea will be protected from erosion. So too will Mappleton, due to a decision to protect it in the early 1990s. Two rock groynes, rip-rap and cliff-regrading built in 1991 ensured Mappleton was saved from the sea. If that decision was revisited today, Mappleton may well not be defended. In the 2010 SMP for this area the economic case for defending Mappleton was described as 'marginal'. Figure 11.12 shows that in 'no active intervention' policy areas:

- up to 400 m of land could be lost to erosion by 2105
- in the North Cliff area of Hornsea, some coastal properties could be at risk by 2055

- by 2105 the main coastal road (the B1242) is likely to have been destroyed by erosion south of Rolston and south of Mappleton
- by 2055 around 200 hectares of farmland and 32 properties are likely to have been lost to erosion.

It is interesting to note the potential situation in Mappleton by 2105. The small village of 50 properties, a church and garage could be cut off apart from the minor road running west from the village. It would be a very high-risk location due to the possibility of erosion outflanking its defences. Properties would be very difficult to sell and their value would have been significantly reduced.

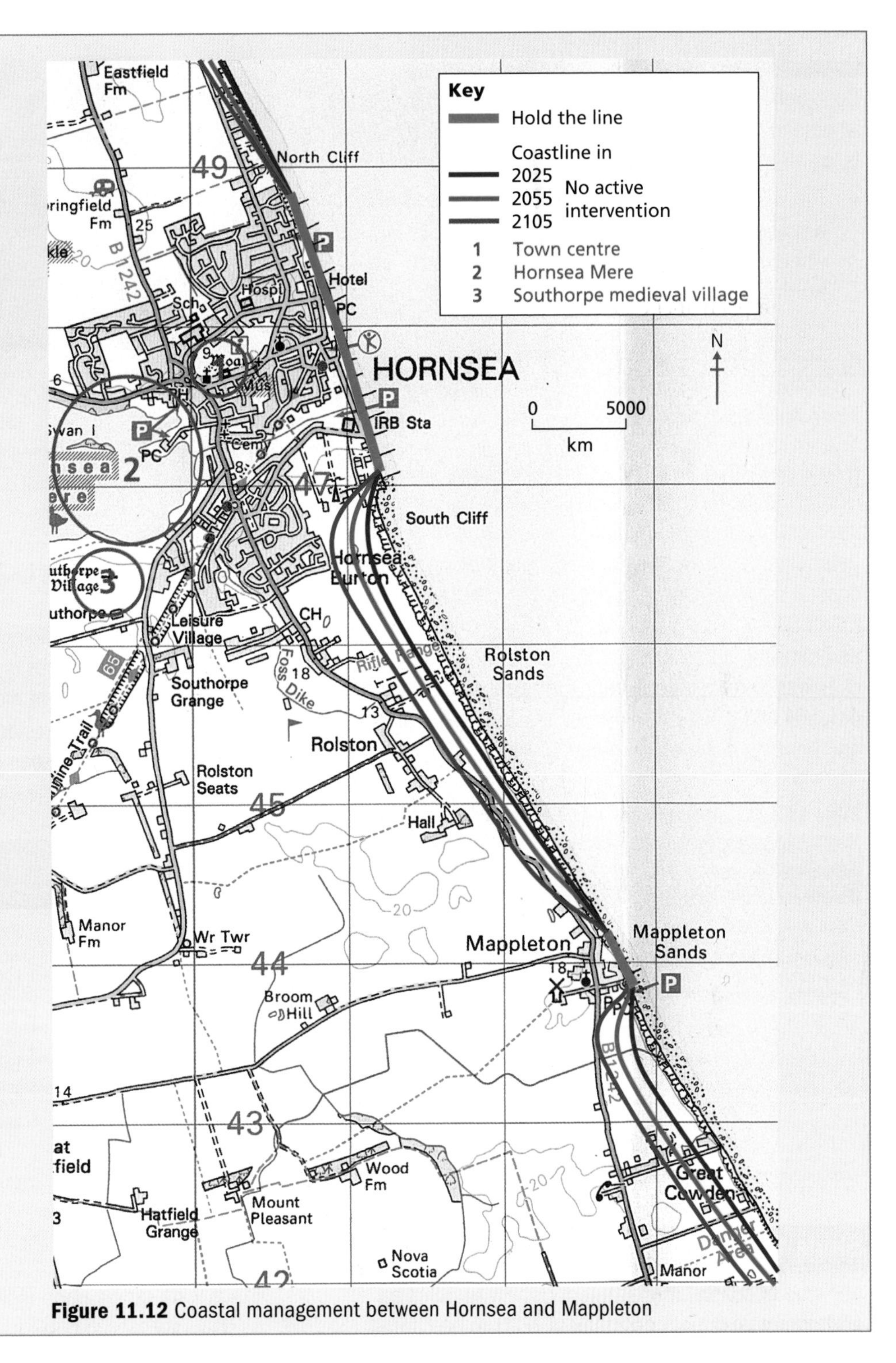

Figure 11.12 Coastal management between Hornsea and Mappleton

Key term

Outflanking: Occurs when erosion gets behind coastal defences at the point where they stop, leading to rapid erosion inland and undermining of defences.

Skills focus: Cost-benefit analysis

Cost-benefit analysis (CBA) is a tool used to help decide whether defending a coastline from erosion and/or flooding is 'worth it'. It often proves very controversial because:

- The value of property is dependent on how at risk it is: a house safe from erosion could be worth £300,000 whereas an identical unprotected home could be worth almost nothing if the threat of destruction was imminent.
- Some human costs (worry, stress) and environmental costs (value of biodiversity, scenic beauty) are very hard to quantify in financial terms.

A case in point is Happisburgh (pronounced 'haze-boro') in North Norfolk. The policy adopted in this area is 'no active intervention' in the immediate future. This is largely because defending the village would have an impact on the wider coastal management plan (Figure 11.13). Happisburgh would end up as a promontory, blocking longshore drift and causing further erosion downdrift. Longer term the plan is managed realignment, although this would still involve property being lost to the sea by erosion. Some of the costs and benefits are shown in Table 11.4.

Key
Cliff position in
2025 — 2055 — 2105

Figure 11.13 Satellite image of Happisburgh, North Norfolk, showing cliff positions to 2105 (Google and the Google logo are registered trademarks of Google Inc., used with permission. Imagery © 2016 Getmapping plc, DigitalGlobe, Infoterra Ltd and Bluesky, Map data 2016 Google)

The median cost of building coastal defences at Happisburgh is around £6 million. This is very close to the value of property that could be saved, and much higher than the compensation costs payable to local residents. Coastal managers argue that Happisburgh must be seen in the wider context of the whole SMP, further justifying the decision not to defend the village.

Table 11.4 Cost benefit analysis for Happisburgh

Hold the line costs for a 600 m stretch of coastline	Costs of erosion	Benefits of protection
Sea wall: £1.8-6 million Rip-rap: £0.8-3.6 million Groynes: £0.1-1.5 million	£160,000 could be available to the Manor Caravan Park to assist in relocating to a new site. Affected residents could get up to £2000 each (£40,000-70,000) in relocation expenses, plus the cost to the council of finding plots of land to build new houses. Grade 1 listed St Mary's church and Grade 2 listed Manor House would be lost. Social costs as the village is slowly degraded, including health effects and loss of jobs.	By 2105, between 20 and 35 properties would be 'saved' from erosion with a combined value of £4-7 million (average house price £200,000 in 2015). Around 45 hectares of farmland would be saved, with a value of £945,000. The Manor Caravan Park would be saved, which employs local people.

Skills focus: Environmental Impact Assessment

Any type of coastal management usually requires an environmental impact assessment (EIA) to be carried out. This is quite separate from any cost-benefit analysis, although it might inform the final CBA. EIA is a process that aims to identify:

- the short-term impacts on the coastal environment of construction
- the long-term impacts of building new sea defences or changing a policy from hold the line to no active intervention or managed realignment.

EIA is wide ranging and includes assessments of:

- impacts on water movement (hydrology) and sediment flow, which can affect marine ecosystems due to changes in sediment load
- impacts on water quality, which can affect sensitive marine species
- possible changes to flora and fauna including marine plants, fish, shellfish and marine mammals
- wider environmental impacts such as air quality and noise pollution, mainly during construction.

Conflict, winners and losers

Coastal management involves decisions that directly affect people's lives. These effects can be positive or negative. This means that any decision is likely to divide stakeholders into one of two groups:

- Winners: people who have gained from the decision, either economically (their property is safe), environmentally (habitats are conserved) or socially (communities can remain in place).
- Losers: people who are likely to lose property, or perhaps see the coastline 'concreted over' and view this as an environmental negative.

Feelings about decision making can run high. In 2006 when the SMP for Happisburgh was published, one resident said:

> *'Quite simply, we are not having it. If some idiot from the government comes along here and says we've got to sacrifice everything in the public interest, they can think again.'*

Reactions like this are understandable because:

- communities and homeowners have a strong attachment to place; they fear losing their homes and social networks, as well as being financially worse off
- businesses risk losing customers as erosion forces people to leave, and may lose their business permanently.

On the other side of the argument:

- coastal managers need to produce plans for entire SMP areas, which inevitably means protecting some areas but not others
- local councils and government (Defra) have limited funds to implement coastal protection schemes, so all places cannot be protected

Occasionally, difficult coastal management decisions can be made with minimal conflict, as was the case with the Blackwater Estuary in Essex.

Coastal management in the developing world

In developed countries such as the UK there are frameworks for coastal management and, in many cases, compensation packages for people whose property is threatened. This is not the case everywhere. In many areas of developing Asia coastal erosion is rapid:

- Rates of erosion in Jiangsu, China, exceed 80 m per year in some places, with 40 m per year reported in Hangzhou Bay and 16–56 m per year in Tianjin.
- Around 30 per cent of Malaysia's coast is eroding, and rates of 50 m per year have been reported in southern Vietnam.
- In West Bengal, India, rates of 6 m per year are common.

Although the causes are specific to particular locations, some common themes can be found:

- The construction of upstream dams on Asia's major rivers have reduced sediment supply to the coast and disrupted local sediments cells.
- Rapid coastal development, urbanisation and the development of tourist resorts has led to haphazard construction of defences with no overall plan.
- Widespread destruction of mangrove forests for fuelwood, and especially for shrimp ponds, has exposed soft delta and estuary sediments to rapid erosion.

In many cases the main 'losers' are the poorest people. Farmers and residents usually lack a formal land-title so cannot claim compensation (even if it were available). Coastlines become more vulnerable to rising sea levels, the impact of tropical cyclone storm surges and even tsunami. When these disasters strike it is the poorest that lose everything.

In the important tourist resort of Phuket in Thailand, erosion is a serious problem. Beach sand has been lost to such an extent that the very beaches tourists come to lie on may disappear. Hotel owners often resort to sand-bagging to try and minimise erosion. Local villagers use ad-hoc methods such as bamboo fencing to try to reduce the power of the eroding waves (Figure 11.14).

Figure 11.14 Bamboo fencing built by villagers to reduce erosion in Thailand

The Blackwater Estuary

The Blackwater Estuary in Essex is an area of tidal salt marsh and low-lying farmland. Prone to flooding and coastal erosion, the farmland was traditionally protected by flood embankments and revetments. Over the last 30 years it has become clear that responding to rising sea levels and greater erosion by building more and higher coastal defences in places such as Blackwater is not sustainable because of 'coastal squeeze' (Figure 11.15):

- As sea levels rise, estuary salt marshes naturally respond by migrating inland.
- Sea defences prevent this, so the salt marsh is 'squeezed' and would eventually disappear.
- This would remove the natural protection the salt-marsh provides against flooding and erosion, exposing the sea defences to direct wave attack.

Coastal squeeze

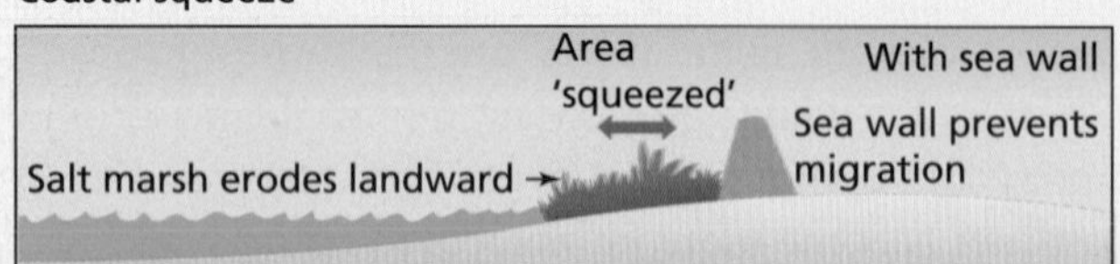

Natural salt marsh migration

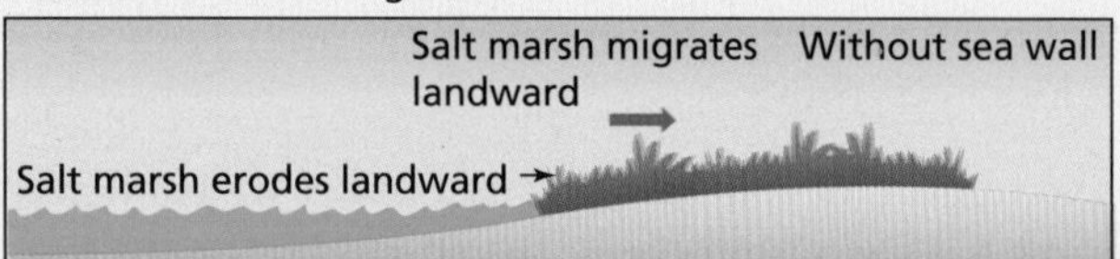

Figure 11.15 Coastal squeeze

Long term the salt-marsh squeeze is likely to make erosion and flooding even worse especially as 60 per cent of the Essex marshes have been eroded in the last ten years.

The solution adopted was radical. In 2000 Essex Wildlife Trust purchased Abbotts Hall Farm on the Blackwater Estuary, which was threatened by erosion and flooding. A 4000-hectare managed realignment scheme was implemented by creating five breaches in the sea wall in 2002. This allowed new salt marshes to form inland. The scheme has a number of benefits:

- The Abbots Hall Farm owners received the market price for their threatened farm.
- The very high costs of a hold-the-line policy were avoided, but flood risk was reduced.
- Water quality in the estuary improved due to the expansion of reed beds, which filter and clean the water.
- New paths and waterways were created for leisure activities.
- Additional income streams from ecotourism and wildlife watching were created.
- Important bird habitats (dunlin, redshank, geese) and fish nurseries (bass and herring) were enhanced.

The Blackwater Estuary shows that environmentalists, landowners, coastal managers, and local people and businesses can all be kept happy even when radical plans are adopted.

Review questions

1 Describe the variation in the value of different coastal land uses.

2 Explain why the costs of coastal defences are usually very high.

3 Comment on the compensation offered to people, such as those in the East Riding of Yorkshire, whose homes are threatened by erosion.

4 Explain the different aims of hard- and soft-engineering coastal defences.

5 Explain why coastlines, such as that of the UK, are increasingly managed as large units using shoreline management plans.

6 Comment on the similarities and differences between Mappleton and Happisburgh in terms of coastal decision making. Why was one village protected but not the other?

7 Explain why it is unlikely that coastal decision making can meet the needs of all stakeholders.

Further research

Investigate your local shoreline management plan through this government portal website: www.gov.uk/government/publications/shoreline-management-plans-smps/shoreline-management-plans-smps

Examine the role of mangrove replanting in the Maldives and other countries: www.mangrovesforthefuture.org/countries/members/maldives

Explore the views and concerns of the residents of Happisburgh in Norfolk: www.happisburgh.org.uk

Explore the work of Essex Wildlife Trust in the Blackwater Estuary at: www.essexwt.org.uk/reserves/abbotts-hall-farm

Exam-style questions

AS questions

1 Name one erosion process that occurs at a coast. [1]

2 Study Table 1 below. Calculate the average annual rate of erosion for the ten locations shown. [1]

3 Suggest one reason for the differences in average annual erosion rate in Table 1. [3]

4 Explain the importance of vegetation in the stability of coastal systems. [6]

5 Assess the importance of mass movement in influencing the rates of coastal recession and landform change. [12]

6 Study Figure 1, below. Explain one strength and one weakness of the method used. [4]

7 You have carried out field research investigating coastal landscapes and change. Assess the value of the primary data you collected in helping you form conclusions to your research question. [9]

Table 1 Average annual erosion rates for ten locations on the Holderness Coast, 2004–2014

Location	Annual erosion rate (metres)
Auburn Farm	0.69
North Barmston	2.03
Skipsea	2.97
Skirlington	0.72
Atwick	0.72
Hornsea South	2.91
Cowden	1.85
Grimston Park	3.42
North Withernsea	0.31
Holmpton	3.80

A level questions

1 Study Figure 2. Explain the likely causes of coastal recession at Aldborough. [6]

2 Explain why some coastlines are more vulnerable to coastal flooding than others. [6]

3 Explain how geological structure influences the development of coastal landforms. [8]

4 Evaluate the extent to which all coastlines can be protected using sustainable management approaches. [20]

Figure 2 Aldborough on the Holderness Coast

Fieldwork evaluation of sea defences								
Negative impacts	−3	−2	−1	0	+1	+2	+3	Positive impacts
Increases erosion					Dune stabilisation	Wooden groynes		Prevents erosion
Increases coastal flood risk				Wooden groynes		Dune stabilisation		Decreases coastal flood risk
Unattractive		Wooden groynes					Dune stabilisation	Blends in with coastal environment
Limits beach access		Wooden groynes	Dune stabilisation					Improves beach access
High maintenance costs			Wooden groynes			Dune stabilisation		Low maintenance costs
Construction disturbs people and ecosystems		Wooden groynes				Dune stabilisation		Minimal impacts during construction

Key
Wooden groynes
Dune stabilisation

Figure 1 A student's field assessment of two coastal management schemes

Topic 3
Globalisation

12 The acceleration of globalisation

What are the causes of globalisation and why has it accelerated in recent decades?

By the end of this chapter you should:

- understand the concept of globalisation and associated developments over time in transport, communications and the way businesses operate on a global scale
- understand how political and economic factors have contributed to globalisation's acceleration
- explain how and why globalisation has affected some places and organisations more than others.

12.1 The acceleration of globalisation

The umbrella term globalisation is used to describe a variety of ways in which places and people are now more connected with one another than they used to be.

Globalisation is viewed in a positive light by some groups of people. 'Hyper-globalisers' applaud the fact that millions of people have escaped dollar-a-day poverty since the 1970s. They celebrate the way cultures are mixing and, at the local scale, often becoming more diverse. However, there are downsides to globalisation

Key concept: Globalisation

Globalisation is the latest chapter in a long story of how people and states have become connected in economic, social, cultural and political ways (Figure 12.1). In the past, global connection was achieved through:

- trade – especially after 1492 when Columbus reached the Americas and the traditional world economy began to take shape
- colonialism – by the end of the nineteenth century, the British Empire directly controlled one-quarter of the world and its peoples
- co-operation – ever since the First World War ended in 1918, international organisations similar to today's United Nations have existed.

Economic globalisation
- *The growth of transnational corporations (TNCs)* accelerates cross-border exchanges of raw materials, components, finished manufactured goods, shares, portfolio investment and purchasing
- *Information and communications technology (ICT)* supports the growth of complex spatial divisions of labour for firms and a more international economy
- Online purchasing using Amazon on a smartphone

Social globalisation
- *International immigration* has created extensive family networks that cross national borders — world city-societies become multi-ethnic and pluralistic
- *Global improvements in education and health* can be seen over time, with rising world life expectancy and literacy levels, although the changes are by no means uniform or universal
- *Social interconnectivity* has grown over time thanks to the spread of 'universal' connections such as mobile phones, the internet and e-mail

Political globalisation
- *The growth of trading blocs (e.g. EU, NAFTA)* allows TNCs to merge and make acquisitions of firms in neighbouring countries, while reduced trade restrictions and tariffs help markets to grow
- Global concerns such as free trade, credit crunch and the global response to natural disasters (such as the 2011 Japanese tsunami)
- *The World Bank, the IMF and the WTO* work internationally to harmonise national economies

Cultural globalisation
- *'Successful' Western cultural traits come to dominate* in some territories, e.g. the 'Americanisation' or 'McDonaldisation' of tastes and fashion
- *Glocalisation and hybridisation* are a more complex outcome that takes place as old local cultures merge and meld with globalising influences
- *The circulation of ideas and information* has accelerated thanks to 24-hour reporting; people also keep in touch using virtual spaces such as Facebook and Twitter

Figure 12.1 Four strands of globalisation you will learn about in this chapter

Modern (post-1940s) globalisation differs from the global economy which preceded it due to the following reasons:

- **Lengthening** of connections between people and places, with products sourced from further away than ever before (in one extreme case, bottled water is now brought all the way to the UK from Fiji, 10,000 miles away).
- **Deepening** of connections, with the sense of being connected to other people and places now penetrating more deeply into almost every aspect of life. Think about the food you eat each day and the many places it is sourced from. It is more or less impossible not to be connected to other people and places through the products we consume!
- **Faster** speed of connections, with people able to talk to one another in real time, using technologies such as Skype, or travel quickly between continents using jet aircraft.

This sense of global connectivity is not shared by everyone in the world though. Some nations and regions (for example, parts of the Sahel) experience a much more 'shallow' form of integration (Figure 12.2). There can also be great disparity among a country's citizens in terms of how 'global' they feel. For instance, many citizens in Brazil's core cities of Rio de Janeiro and São Paulo are globally connected, either as producers of goods or as consumers of music and sports (the 2016 Olympic Games are in Brazil). However, some Amazon rainforest tribes, such as the Korubo people, have little or no knowledge of the outside world, and lack any connectivity with other places.

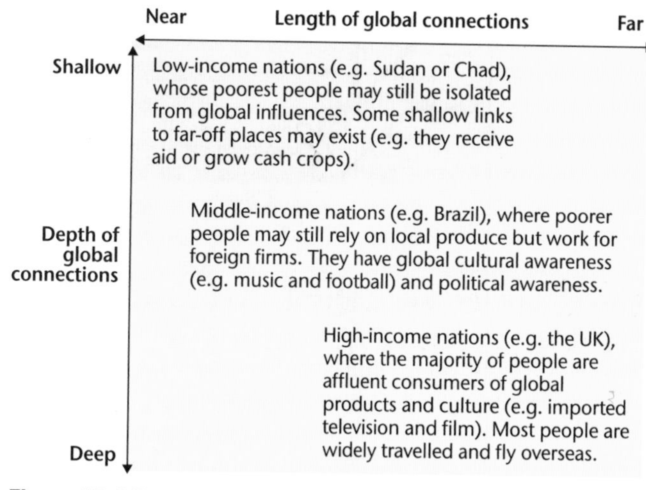

Figure 12.2 Types of global connection

What is globalisation?

'The increasing integration of economies around the world, particularly through the movement of goods, services, and capital across borders. There are also broader cultural, political, and environmental dimensions of globalization.'

International Monetary Fund

'It might mean sitting in your living room in Estonia while communicating with a friend in Zimbabwe. Or it might be symbolized by eating Ecuadorian bananas in the European Union.'

World Bank

too. A world in which people have greater freedom to migrate across political borders – as is the case for European Union nations – is not to everyone's taste, for instance. There are fears too that transnational corporations (TNCs) are responsible for a growing trend towards cultural homogeneity (uniformity) on a global scale. As a result, many people are sceptical about the merits of globalisation.

Global flows and global networks

In the study of globalisation, it is useful to conceptualise the world as consisting of networks of places and their populations. Imagine a global network of the most well-connected countries and cities drawn like the famous London Underground map (Figure 12.3).

The connections shown between places represent different kinds of network flow. These flows are movements of:

- **Capital:** At a global scale, major capital (money) flows are routed daily through the world's stock markets. A range of businesses, including investment banks and pension funds, buy and sell money in different currencies to make profits. In 2013, the volume of these foreign exchange transactions reached US$5 trillion per day.
- **Commodities:** Valuable raw materials such as fossil fuels, food and minerals have always been traded between nations. Flows of manufactured goods have multiplied in size in recent years, fuelled by low

Key term

Transnational corporations: Businesses whose operations are spread across the world, operating in many nations as both makers and sellers of goods and services. Many of the largest are instantly recognisable 'global brands' that bring cultural change to the places where products are consumed.

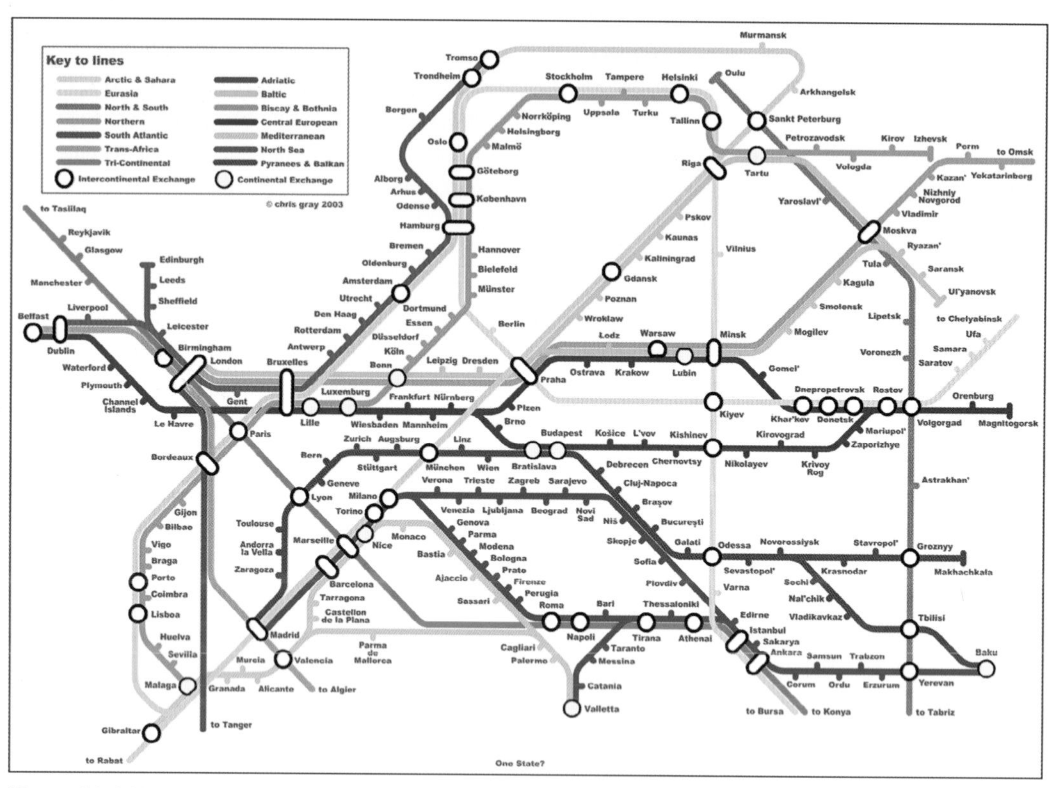

Figure 12.3 Europe redrawn as a network (based on the London Underground design)

production costs in China and even lower-waged economies, such as Bangladesh and Vietnam. In 2015, global gross domestic product (GDP) fell just short of US$80 trillion in value. Of this, around one-third was generated by trade flows in agricultural and industrial commodities.

- **Information:** The internet has brought real-time communication between distant places, allowing goods and services to be bought at the click of a button. Social networks have ballooned in size and influence, with Facebook gaining 1.5 billion users by 2015. On demand TV has increased data usage further. Information is stored in enormous 'server farms' such as the Microsoft Data Centre in Washington State and Facebook's data centre in Luleå, Sweden (where cold temperatures reduce the cost of cooling the hard drives).
- **Tourists:** Many of the world's air passengers are holiday makers. Budget airlines have brought a 'pleasure periphery' of distant places within easy reach for the moneyed tourists of high-income nations. Increasingly, people from emerging economies travel abroad too, using budget airlines such as AirAsia and East Africa's Fastjet. China is now the world biggest spender on international travel, with 120 million outbound trips made in 2014.
- **Migrants:** Of all global flows, the permanent movement of people still faces the greatest number of obstacles due to border controls and immigration laws. As a result, most governments have a 'pick and mix' attitude towards global flow: they embrace trade flows but attempt to resist migrant flows unless there is a special need (such as Qatar's encouragement of Indian construction workers). Despite restrictions, however, record flows of people are recorded every year. The combined number of economic migrants and refugees worldwide reached almost one-quarter

Key terms

Gross domestic product: A measure of the financial value of goods and services produced within a territory (including foreign firms located there). It is often divided by population size to produce a per capita figure for the purpose of making comparisons.

Emerging economies: Countries that have begun to experience high rates of economic growth, usually due to rapid factory expansion and industrialisation. There are numerous sub-groups of emerging economies, including Brazil, Russia, India, China and South Africa (the 'BRICS' group). They are sometimes called newly industrialised countries.

Skills focus: Proportional flow-lines

Figure 12.4 shows global flows of migrant remittances. The width of the arrow is directly proportional to the value of the flow (measured in US$). Draw a table to show the source region, host country and value for each remittance flow.

UK
USA
MEXICO
SAUDI ARABIA
UAE
CHINA
JAPAN
INDIA
HONG KONG
VIETNAM
PHILIPPINES
BANGLADESH
0 5000
km
Key
Remittances (% of GDP)
>10
5.1–10.00
1.1–5.0
0.6–1.0
0–0.5
No data
Main flows of remittances
Trade flow-lines (width shows value of the flow in $bn)
30
20
10
0

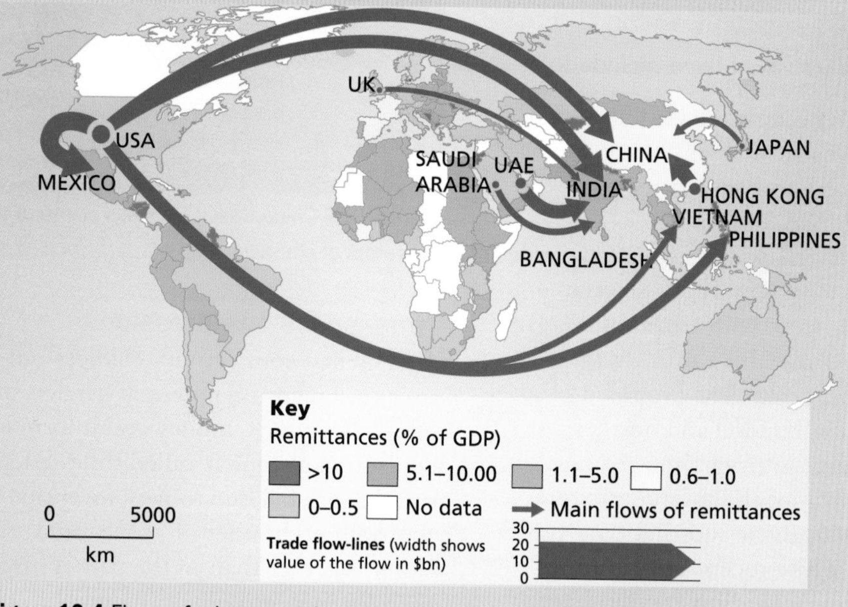

Figure 12.4 Flows of migrant remittances, 2011 (Sources: IMF, World Bank)

of a billion in 2013. The same year, around US$500 billion of remittances were sent home by migrants.

The combined effect of these global flows has been to make places interconnected. One result of this is the increased interdependency of places.

Key terms

Remittances: Money that migrants send home to their families via formal or informal channels.

Interdependency: If two places become over-reliant on financial and/or political connections with one another, then they have become interdependent. For example, if an economic recession adversely affects a host country for migrant workers, then the economy of the source country may shrink too, due to falling remittances.

Transport and trade in the nineteenth and twentieth centuries

Developments in transport and trade went hand-in-hand during the nineteenth and twentieth centuries. While transport improvements have allowed the value of trade to increase, it is also the case that major trading powers, such as the USA, seek to maintain their competitive edge through continued transport innovation. The result is a constant feedback loop (Figure 12.5).

As transport developments occur, TNCs prosper too. In the 1900s, large manufacturing companies, such as Ford and General Motors, were able to export products more widely. Over time, connectivity allowed multiple sites

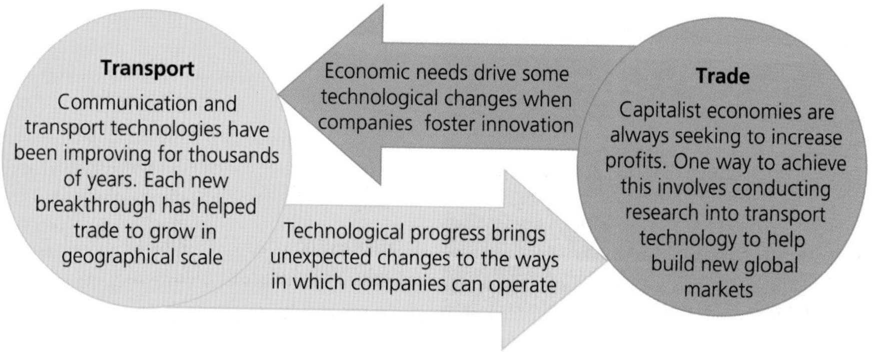

Transport
Communication and transport technologies have been improving for thousands of years. Each new breakthrough has helped trade to grow in geographical scale
Economic needs drive some technological changes when companies foster innovation
Technological progress brings unexpected changes to the ways in which companies can operate
Trade
Capitalist economies are always seeking to increase profits. One way to achieve this involves conducting research into transport technology to help build new global markets

Figure 12.5 The interrelationship between trade and transport growth

of production to be established, both to reduce transport costs and to take advantage of cheap sites of production where wages were lower. Transport has been essential in allowing TNCs to establish a spatial division of labour on a global scale.

Important innovations in transport have included:

- **Steam power:** Britain became the leading world power in the 1800s using steam technology. Steam ships (and trains) moved goods and armies quickly along trade routes into Asia and Africa.
- **Railways:** In the 1800s, railway networks expanded globally. By 1904, the 9000 km Trans-Siberian Railway connected Moscow with China and Japan. Today, railway building remains a priority for governments across the world. The proposed High Speed 2 railway (linking London and northern England) will halve some journey times.
- **Jet aircraft:** The arrival of the intercontinental Boeing 747 in the 1960s made international travel more commonplace, while recent expansion of the cheap flights sector, including easyJet, has brought it to the masses in richer nations.
- **Container shipping:** Around 200 million individual container movements take place each year (Figure 12.6). Some commentators describe shipping as the 'backbone' of the global economy since the 1950s. Everything from chicken drumsticks to patio heaters can be transported efficiently across the planet using intermodal containers. The Chinese vessel *Cosco* is 366 m long, 48 m wide and can carry 13,000 containers.

Figure 12.6 China Cosco Holdings is one of the world's largest operators of container ships

Time-space compression

Heightened connectivity changes our conception of time, distance and potential barriers to the migration of people, goods, money and information. This perceptual change is called time-space compression. As travel times fall due to new inventions, different places approach each other in 'space-time': they begin to *feel* closer together than in the past. This is also called the shrinking world effect (Figure 12.7).

ICT and mobile phone use in the twenty-first century

One important aspect of globalisation is the way that it can make us think of the world as a potentially borderless place. The famous photograph Earth-rise was taken by Apollo 8 astronauts in 1968 (Figure 12.9). It was the first time people living on Earth had ever seen

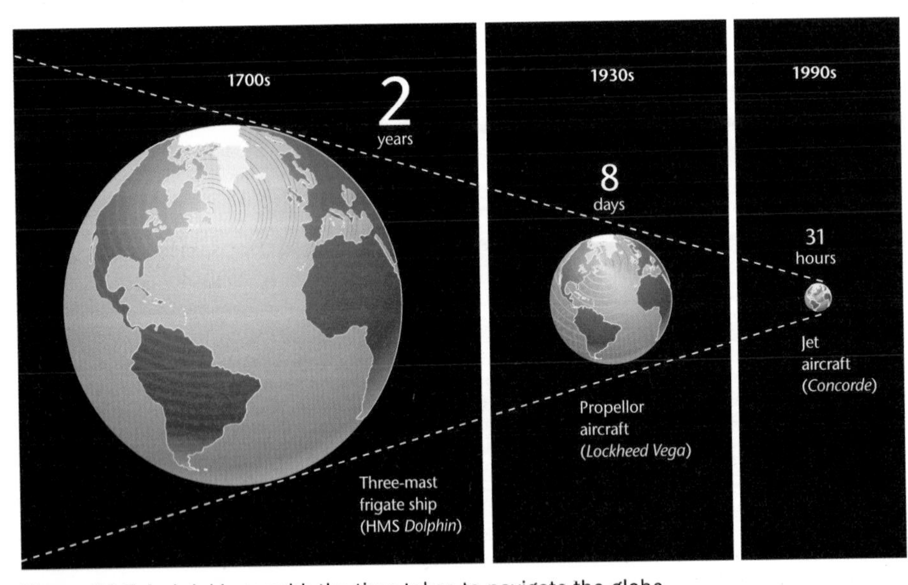

Figure 12.7 A shrinking world: the time taken to navigate the globe

Key terms

Spatial division of labour: The common practice among TNCs of moving low-skilled work abroad (or 'offshore') to places where labour costs are low. Important skilled management jobs are retained at the TNC's headquarters in its country of origin.

Intermodal containers: Large-capacity storage units which can be transported long distances using multiple types of transport, such as shipping and rail, without the freight being taken out of the container.

Shrinking world: Thanks to technology, distant places start to feel closer and take less time to reach.

easyJet

The easyJet airline was founded by Sir Stelios Haji-Ioannou in 1995 (Figure 12.8). It began as a small venture running flights solely within the UK. Most of Europe's major cities are now inter-connected via easyJet's cheap flight network.

Figure 12.8 easyJet began with just two aircraft

At the start, the airline had just two aircraft. Inaugural flights from London Luton to Edinburgh and Glasgow were supported by the advertising slogan 'Making flying as affordable as a pair of jeans – £29 one way'. In 1996, flights to Barcelona commenced and thereafter the company expanded at breakneck speed. It now has around 300 flight routes within the EU and several that extend beyond, to Egypt, Morocco, Turkey and Israel (Table 12.1).

Table 12.1 Growth in easyJet passenger numbers, 1995–2014

Year	Passengers numbers
1995	30,000
1996	420,000
1998	1,880,000
2000	5,970,000
2002	11,400,000
2006	32,950,000
2010	48,800,000
2014	65,000,000

By 2014, the company owned 200 aeroplanes (both Airbus and Boeing), carrying 65 million people to their destinations that year and bringing in revenues of nearly £4 billion. Places that easyJet adds to its flight network immediately become switched-on to a greater degree. For instance, Tallinn in Estonia is a city that is home to 400,000 people. In 2004, easyJet began to fly British tourists there for just £40 each. Suddenly, the city became an affordable destination for UK citizens, especially groups of young men and women seeking a cheap but interesting destination for 'stag' and 'hen' weekends.

Figure 12.9 Earth-rise from the Moon (1968)

the world as a single entity, helping to foster a sense of global citizenship. Since then, successive innovations in information and communications technology (ICT) have further transformed how citizens, businesses and states interact with one another. Some important elements of ICT history are detailed in Table 12.2 below.

Technology is used by different players in a vast array of ways which contribute to globalisation. Some of these include:

Table 12.2 Important elements of the growth of ICT over time

Telephone and the telegraph	The first telegraph cables across the Atlantic in the 1860s replaced a three-week boat journey with instantaneous communication. This revolutionised how business was conducted. The telephone, telegraph's successor, remains a core technology for communicating across distance. In parts of Africa, where telephone lines have never been laid in many places, people are technologically 'leap-frogging' straight to mobile phone use.
Broadband and fibre optics	With the advent of broadband internet in the 1980s and 1990s, large amounts of data could be moved quickly through cyberspace. Today, enormous flows of data are conveyed across the ocean floor, by fibre optic cables owned by national governments or TNCs such as Google (Figure 12.10). More than 1 million kilometres of flexible undersea cables, about the size of garden watering hoses, carry all the world's emails, searches and tweets.
GIS and GPS	The first global positioning system (GPS) satellite was launched in the 1970s. There are now 24 situated 10,000 km above the Earth. These satellites continuously broadcast position and time data to users throughout the world. Deliveries can be tracked by companies using vehicle-tracking systems, helping the growth of global production networks to be managed.
The internet, social networks and Skype	The internet began life as part of a scheme funded by the US Defence Department during the Cold War. The early computer network ARPANET was designed during the 1960s as a way of linking important research computers in just a handful of different locations. Since then, connectivity between people and places has grown exponentially. By 2014, 5 billion Facebook 'likes' were being registered globally every day.

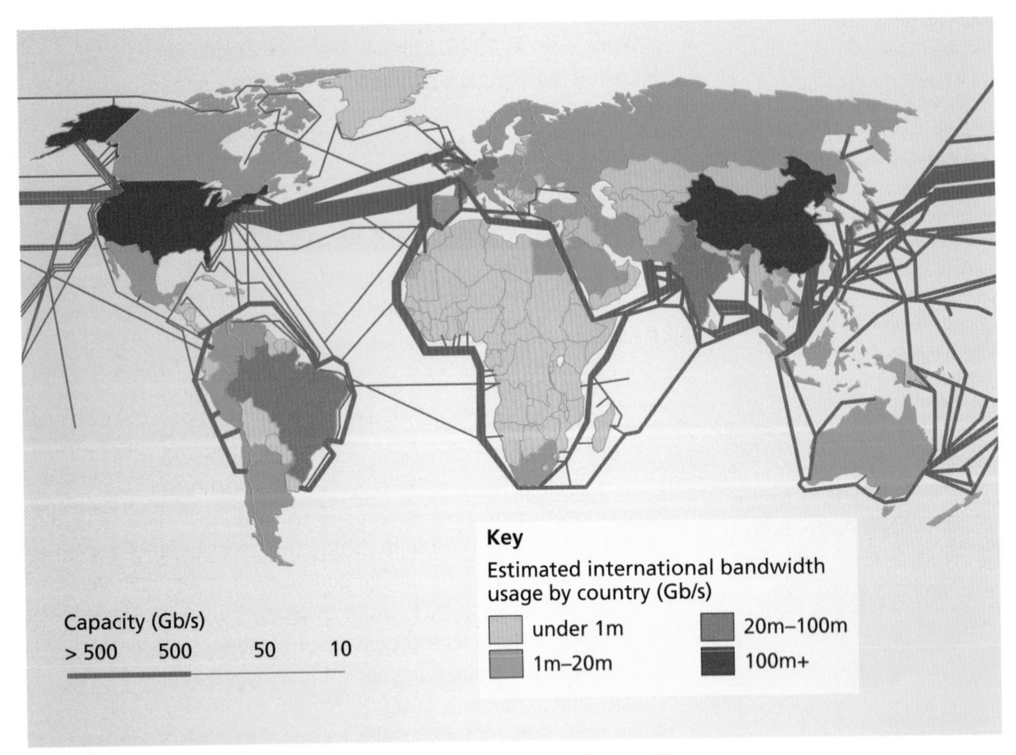

Figure 12.10 The uneven global distribution of undersea data cables

Figure 12.11 Psy, a global cultural phenomenon

- **Economic globalisation:** ICT allows managers of distant offices and plants to keep in touch more easily (for example, through video conferencing). This has helped TNCs to expand into new territories, either to make or sell their products. Each time the barcode of a Marks and Spencer food purchase is scanned in a UK store, an automatic adjustment is made to the size of the next order placed with suppliers in distant countries like Kenya.
- **Social globalisation:** The maintaining of long-distance social relationships through ICT use is a factor that supports migration. Since 2003, Skype has provided a cheap and powerful way for migrants to maintain a strong link with family they have left behind.
- **Cultural globalisation:** Cultural traits, such as language or music, are adopted, imitated and hybridised faster than ever before. During 2012, South Korean singer Psy clocked up over 1.8 billion online views of 'Gangnam Style', the most-watched music video of all time (Figure 12.11).
- **Political globalisation:** Social networks are used to raise awareness about political issues and to fight for change on a global scale. Environmental charities like Greenpeace spread their message online, while the militant group Daesh (or 'Isis') has used social media to spread its message of terror globally, and to gain new recruits.

The mobile phone revolution and electronic banking in developing countries

In countries where the lack of communications infrastructure has traditionally been a big obstacle to economic growth, mobile phones are now changing lives for the better by connecting people and places. The scale and pace of change is extraordinary. In 2005, six per cent of Africans owned a mobile phone. By 2015 this had risen more than ten-fold to 70 per cent due to falling prices and the growth of provider companies, such as Kenya's Safaricom. Rising uptake in Asia (in India, over 1 billion people are mobile subscribers) means there are now more mobile phones than people on the planet (Figure 12.12).

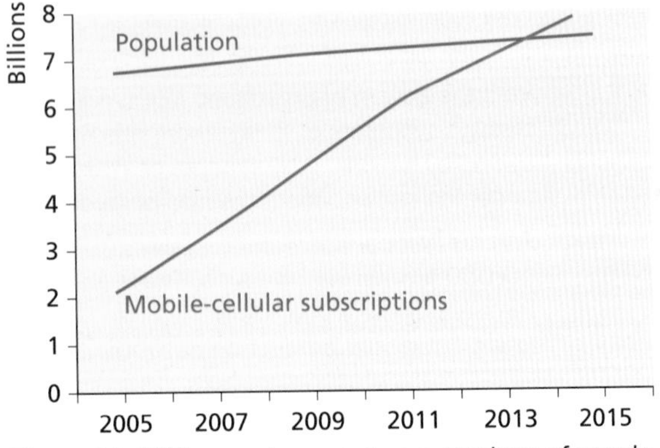

Figure 12.12 The great convergence: numbers of people and mobile phones

In 2007, Safaricom launched M-Pesa, a simple mobile phone service that allows credit to be directly transferred between phone users. This has revolutionised life for individuals and businesses in Kenya:

- The equivalent of one-third of the country's GDP is now sent through the M-Pesa system annually.
- People in towns and cities use mobiles to make payments for utility bills and school fees.
- In rural areas, fishermen and farmers use mobiles to check market prices before selling produce.
- Women in rural areas are able to secure microloans from development banks by using their M-Pesa bills as proof that they have a good credit record. This new ability to borrow is playing a vital role in lifting rural families out of poverty.

12.2 The politics and economics of globalisation

Globalisation is not an automatic outcome of 'shrinking world' technologies. Not everywhere has the population density or market potential to attract international investors in search of new markets. Moreover, demographic and economic disparities can result in some places becoming ICT 'have nots' as part of a global 'digital divide'.

Political factors also matter: sometimes, global flows cannot take place due to political barriers that national governments have created. On the face of it, this may seem surprising, given that global flows can stimulate economic growth. However, global flows may be viewed as threats too:

- Imports of raw materials and commodities can threaten a nation's own industries.
- Migrants can bring cultural change and religious diversity: not everyone welcomes this.
- Information can provide citizens with knowledge that their government finds threatening.

Many examples can be found of national governments (both authoritarian and democratic) attempting to isolate or protect themselves from global financial and trade flows, including foreign direct investment (FDI) from overseas TNCs. Conversely, international organisations such as the World Bank work hard to persuade them to take a different approach.

The work of international organisations

For many decades, three international organisations have acted as 'brokers' of globalisation through the promotion of free trade policies and FDI. Together, the International Money Fund (IMF), World Bank and World Trade Organization (WTO) have collectively striven to build a 'free trade consensus' (Table 12.3). The 'Bretton Woods institutions' were established after the Second World War (at a meeting in Bretton Woods, USA). The guiding principle was to restabilise the world economy and avoid a return to conditions that prevailed during the Great Depression of the 1930s, when a global economic downturn led to free trade becoming replaced by protectionism. Nations had blocked foreign imports with tariffs, damaging export markets for other countries, resulting in a vicious downward spiral of economic output for all major players. Over time, the remit of the Bretton Woods institutions has grown to include persuading developing countries to embrace free market economics and globalisation.

Will the Bretton Woods players maintain their influence in the future? When assessing the likelihood of this, there are several issues to explore.

- The global financial crisis (GFC) of 2008–09, which originated in US and EU money markets and undermined the entire world economy. As a result, governments in developing countries have become more sceptical of the financial advice that the IMF and World Bank offer.
- Geopolitical changes mean that new alternatives are emerging to the Bretton Woods institutions. Developing countries in search of assistance can instead approach the China Development Bank (CDB). China loaned more than US$110 billion to developing countries in 2010, a value that exceeded World Bank lending. In 2014, the BRICS group of nations announced the establishment

Key term

Foreign direct investment: A financial injection made by a TNC into a nation's economy, either to build new facilities (factories or shops) or to acquire, or merge with, an existing firm already based there.

BRICS group: The four large, fast-growing economies of Brazil, Russia, India and China, recently joined at their annual summit meeting by South Africa.

Table 12.3 Evaluating the role of international organisations in globalisation

	IMF	World Bank	WTO
Role in globalisation	Based in Washington, DC, the IMF channels loans from rich nations to countries that apply for help. In return, the recipients must agree to run free market economies that are open to outside investment. As a result, TNCs can enter these countries more easily. The USA exerts significant influence over IMF policy despite the fact that it has always had a European president.	The World Bank lends money on a global scale and is also headquartered in Washington, DC. In 2014, a US$470 million loan was granted to the Philippines for a poverty-reduction programme, for instance. The World Bank also gives direct grants to developing countries (in 2014, help was given to the Democratic Republic of the Congo to kick-start a stalled mega-dam project).	The WTO took over from the General Agreement on Trade and Tariffs in 1995. Based in Switzerland, the WTO advocates trade liberalisation, especially for manufactured goods, and asks countries to abandon protectionist attitudes in favour of untaxed trade (China was persuaded to lift export restrictions on 'rare earth' minerals in 2014).
Evaluation	IMF rules and regulations can be controversial, especially the strict financial conditions imposed on borrowing governments, who may be required to cut back on health care, education, sanitation and housing programmes.	In total, the World Bank distributed US$65 billion in loans and grants in 2014. However, like the IMF, the World Bank imposes strict conditions on its loans and grants. Controversially, all World Bank presidents have been American citizens.	The WTO has failed to stop the world's richest countries, such as the USA and UK, from subsidising their own food producers. This protectionism is harmful to farmers in developing countries who want to trade on a level playing field.

of the New Development Bank (NDB) as another alternative to the World Bank and IMF.

- The WTO's continuing lack of success in getting its 159 member countries to reach a global agreement on any aspect of trade, especially in relation to food, raises questions about its long-term role.

How foreign direct investment works

The Bretton Woods institutions have created a global legal and economic framework that is suited to free trade and foreign direct investment (FDI). TNCs have thrived in this environment, helped by changes in the rules that dictate how they can operate. In the 1980s, financial deregulation for many European countries and the USA led to banks and finance companies globalising rapidly, using strategies shown in Table 12.4. All involve the injection of capital into the economy of a foreign state in the absence of any political barriers.

Figure 12.13 Guitars being manufactured in Fender's factory in Mexico

Synoptic themes:

Players

- World Trade Organization (WTO)
- International Monetary Fund (IMF)
- World Bank

Table 12.4 Different types of foreign direct investment

Offshoring	Some TNCs build their own new production facilities in 'offshore' low-wage economies. For instance, US guitar-maker Fender opened its Mexican plant at Ensenada in 1987 (Figure 12.13).
Foreign mergers	Two firms in different countries join forces to create a single entity. Royal Dutch Shell has headquarters in both the UK and the Netherlands.
Foreign acquisitions	When a TNC launches a takeover of a company in another country. In 2010, the UK's Cadbury was subjected to a hostile takeover by US food giant Kraft. The UK has few restrictions on foreign takeovers. In contrast, the Committee on Foreign Investment in the USA closely scrutinises inbound foreign takeovers.
Transfer pricing	Some TNCs, such as Starbucks and Amazon, have sometimes channelled profits through a subsidiary company in a low-tax country such as Ireland. The Organisation for Economic Cooperation and Development (OECD) is now attempting to limit this practice.

The attitudes and actions of national governments

National governments become key players in globalisation when they adopt policies that allow TNCs to grow in size and influence using the strategies shown in Table 12.4. These government policies include:

- **Free-market liberalisation:** Also known as neoliberalism, this governance model is associated with the policies of US President Ronald Reagan and Margaret Thatcher's UK government during the 1980s. Essentially, they followed two simple beliefs. Firstly, government intervention in markets impedes economic development. Secondly, as overall wealth increases, trickle-down will take place from the richest members of society to the poorest. In practice, this meant restrictions being lifted on the way companies and banks operated. The deregulation of the City of London in 1986 removed large amounts of 'red tape' and paved the way for London to become the world's leading global hub for financial services (Figure 12.14) and the home of many super-wealthy 'non-dom' billionaires.
- **Privatisation:** Successive UK governments have led the way in allowing foreign investors to gain a stake in privatised national services and infrastructure. Until the 1980s, important assets, such as the railways and energy supplies, were owned by the state. However, running these services often proved costly: they were sold to private investors in order to reduce government spending and to raise money. Over time, ownership of many assets has passed overseas. For instance, the French company Keolis owns a large stake in southern England's railway network and the EDF energy company is owned by Électricité de France. Since the global financial crisis, the UK government has approached Chinese and Middle Eastern sovereign wealth funds (SWFs) to help fund new infrastructure projects (Figure 12.15).
- **Encouraging business start-ups:** Methods range from low business taxes to changes in the law allowing both local and foreign-owned businesses to make more profit. When Sunday trading was introduced in 1994, the UK became a more attractive market for foreign retailers, from Burger King to Disney Store. Italy has eased restrictions on Chinese investors wanting to start up textile companies inside the EU; as a result, the city of Prato now has the largest Chinese population in Europe.

Figure 12.14 London's Canary Wharf

The growth of free trade blocs

National governments have also promoted the growth of trade blocs (see Figure 12.16). To trade freely with neighbours or more distant allies, agreements have been drawn up allowing state boundaries to be crossed freely by flows of goods and money. Within a trade bloc, free trade is encouraged by the removal of internal tariffs. This brings numerous benefits for businesses:

- By removing barriers to intra-community trade, markets for firms grow. For instance, when ten new nations joined the EU in 2004, UK firm Tesco gained access to 75 million extra customers.

Key terms

Trickle-down: The positive impacts on peripheral regions (and poorer people) caused by the creation of wealth in core regions (and among richer people).

Sovereign wealth funds: Government-owned investment funds and banks, typically associated with China and countries that have large revenues from oil, such as Qatar.

Trade blocs: Voluntary international organisations that exist for trading purposes, bringing greater economic strength and security to the nations that join.

Tariffs: The taxes that are paid when importing or exporting goods and services between countries.

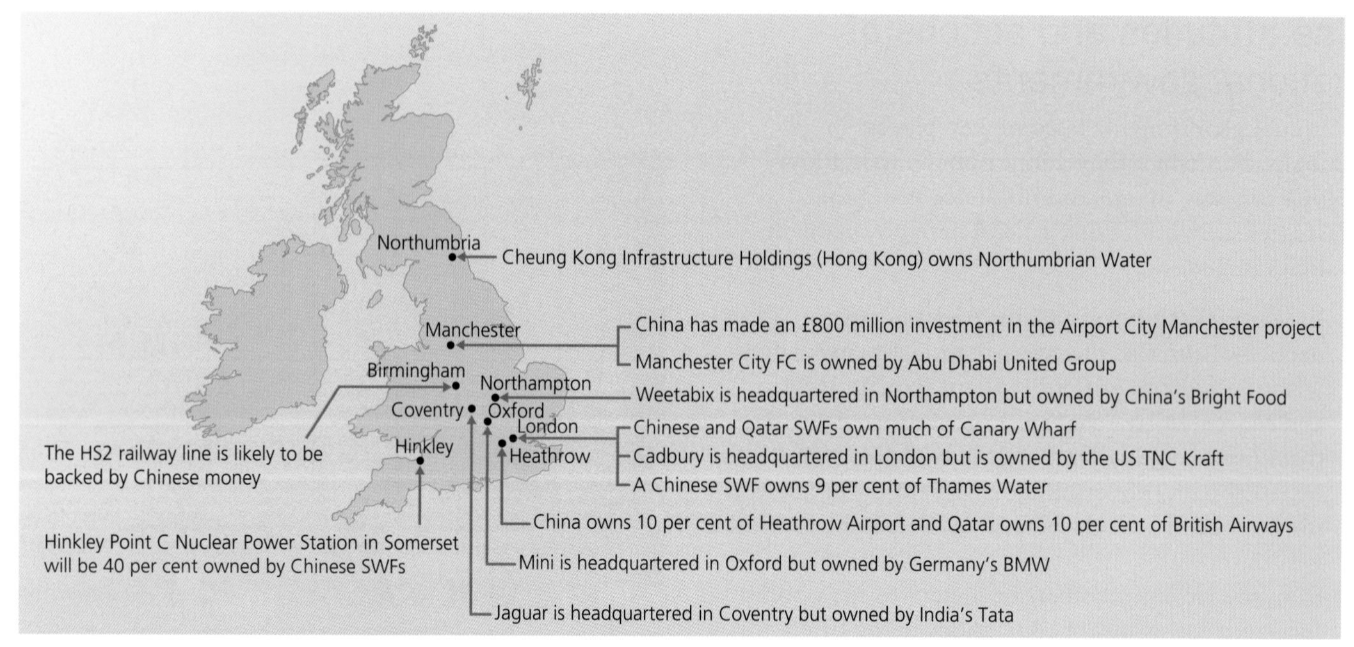

Figure 12.15 Foreign investment by TNCs and sovereign wealth funds in the UK

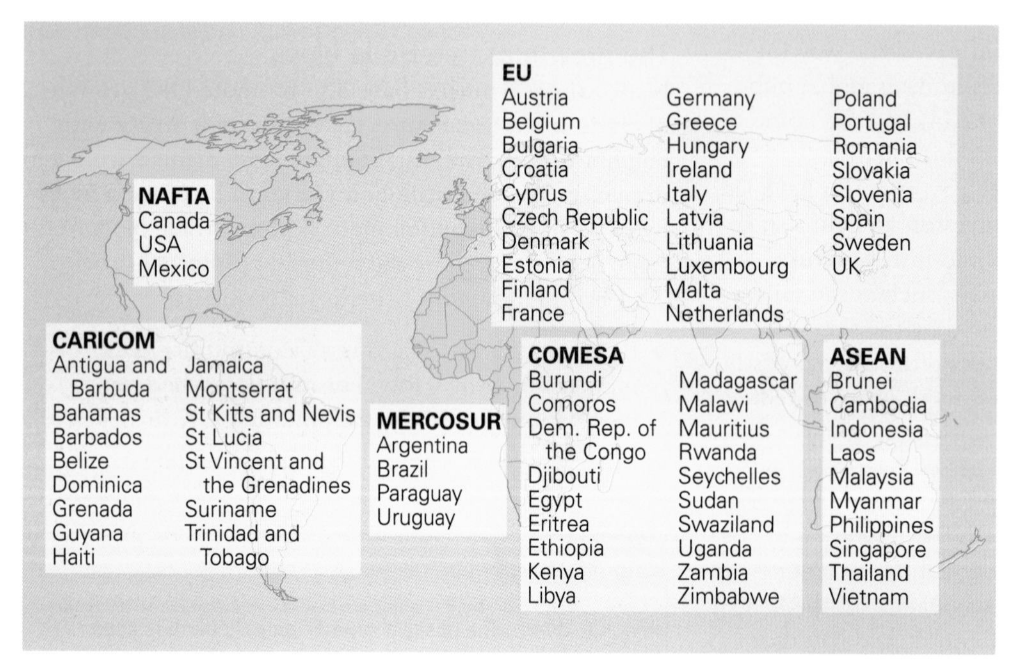

Figure 12.16 Selected trade blocs and regional groupings

- Firms that have a comparative advantage in the production of a particular product or service should prosper. Thus French wine-makers, thanks to their advantageous climate and soil, produce a superior product that is widely consumed throughout a tariff-free Europe.
- An enlarged market increases demand, raising the volume of production and thereby lowering manufacturing costs per unit. An improved economy of scale results, meaning products can be sold more cheaply and sales rise even further for the most successful firms.

There are other benefits too. Smaller national firms within a trade bloc can merge to form TNCs, making their operations more cost effective. Airtel Africa is a mobile phone company headquartered in Kenya whose expansion into seventeen African states has been helped by the existence of the EAC and COMESA trade blocs (Figure 12.17).

Trade bloc members may also agree a common external tariff and quotas for foreign imports. In 2006, the EU blocked imports of underwear from Chinese manufacturers on the basis that the annual quota had

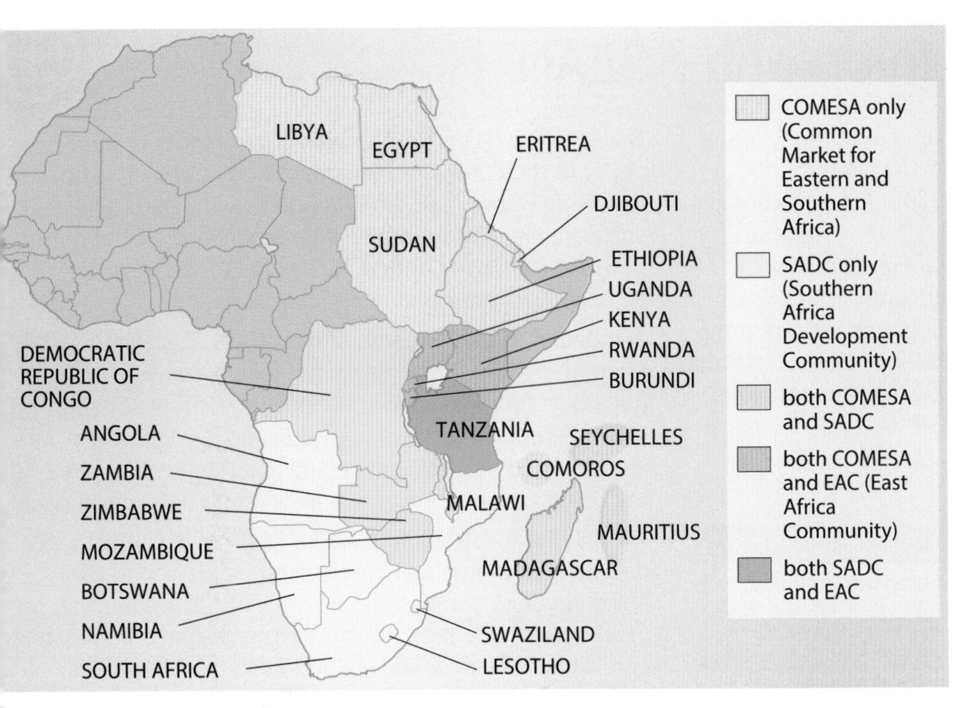

Figure 12.17 African trade blocs

Key term

Special Economic Zone: An industrial area, often near a coastline, where favourable conditions are created to attract foreign TNCs. These conditions include low tax rates and exemption from tariffs and export duties.

been exceeded, jeopardising sales of EU clothing makers (the media dubbed the incident 'bra wars').

Some trade blocs include nations at varying levels of economic development. For instance, Mexico and the USA are both part of NAFTA (North American Free Trade Agreement). This makes sense because Mexico is an emerging economy with a cheap labour force, while the USA has management and research expertise. This allows American TNCs, such as General Electric and Nike, to exploit the human resources of both nations, cheaply manufacturing goods in branch plants in Mexico (called *Maquiladoras*) that are designed and marketed by white-collar staff in the USA (shown in Figure 12.13).

The actions of governments in new global regions

One very important reason for the acceleration of globalisation in recent decades has been changing government attitudes in regions outside of Europe and North America. Asia's three most populated countries – China, India and Indonesia – have all embraced global markets as a means of meeting economic development goals. In all three cases, the establishment of Special Economic Zones (SEZs), government subsidies and changing attitudes to FDI have played important roles.

- Indonesia provides a striking example. In the late 1960s, President Suharto turned his back on communism and opened up Indonesia's markets. American and European TNCs met with Suharto's advisors and collectively built an attractive new legal and economic framework for foreign investors. Indonesia instantly became a popular offshoring location for TNCs like Gap and Levis. World Bank lending funded the speedy modernisation of its roads, power supplies and ports. However, human rights campaigners expressed concern that capital city Jakarta's export zone had become a low tax haven for sweat shop manufacturing.
- Globalisation began in 1991 for India, when sweeping financial reforms took place. Since then, Indian TNCs have grown in their size and influence. Tata and Bharti Airtel, India's mobile network operator, have both become major global players. Until 2013, however, foreign retailers could only gain a presence on India's own high streets by agreeing to form a partnership with a local Indian business. Thus, McDonald's restaurants in North India and East India are a joint venture between Vikram Bakshi and the McDonald's Corporation. India's high street rules have deterred many other foreign retailers, such as IKEA, however. As a result, 90 per cent of India's shops are still family owned.

The European Union and ASEAN

The European Union (EU) has evolved over time from being a simple trade bloc into a multi-governmental organisation with its own currency (the Euro) and some shared political legislation (Figure 12.18). Member states are eligible for EU Structural Funds to help develop their economies, while agricultural producers in the region all benefit from farm subsidies issued under the Common Agricultural Policy (CAP). The EU also helps cities gain a global reputation by awarding prestigious titles such as 'Capital of Culture' or 'European Capital of Innovation' (given to Barcelona in 2014).

Year	Events
1950	Schuman Plan proposes a European Coal and Steel Community (ECSC).
1952	ECSC is created.
1957	Treaty of Rome establishing the European Economic Community (EEC) is signed by Belgium, France, Germany, Italy, Luxembourg and the Netherlands.
1958	EEC comes into operation.
1962	Common Agricultural Policy (CAP) is agreed.
1968	Customs union completed.
1973	UK, Ireland and Denmark join the EEC.
1979	European Monetary System established; first direct elections to the European Parliament.
1981	Greece joins.
1993	European Union (EU) is established as the Maastricht Treaty comes into force.
1995	Austria, Finland and Sweden join the EU, which now has fifteen members.
2002	Euro notes and coins come into circulation in twelve of the fifteen EU member states.
2004	Ten new member states join the EU – Cyprus, Czech Republic, Estonia, Hungary, Latvia, Lithuania, Malta, Poland, Slovakia and Slovenia; EU Constitutional Treaty agreed.
2007	Romania and Bulgaria join the EU.
2008	Kosovo declares independence supported by the EU.
2009	Eurozone financial crisis begins – Greece, Spain and Ireland all face difficulty.
2011	Germany agrees to financial rescue package to help tackle crisis.
2013	Croatia joins the EU, bringing the total number of members to 28.
2016	UK government holds referendum on EU membership.

Figure 12.18 EU timeline

The decision taken by European governments to hand power to the European Parliament was not taken lightly. Two world wars prompted European countries to seek political unity and economic interdependency. What better way to avert further armed conflict in Europe?

The EU is the only group of nations that grants all citizens of member states freedom of movement. Elsewhere in the world, free flows of people do not take place as a result of trade bloc formation. Most national borders were removed within Europe in 1985 when the Schengen Agreement was implemented (the UK and Ireland had remained outside the Schengen Area so were provided with opt-outs).

ASEAN (the Association of South East Asian Nations) has ten member states and a combined population of 600 million people. Established in 1967, ASEAN's founding members include high-income Singapore and the emerging economies of Indonesia, Malaysia and the Philippines. Over time, they have worked to eliminate tariffs in favour of free trade. The enlarged ASEAN market has helped Indonesia's manufacturing industries to thrive, while the Philippines has gained a global reputation for its call centre services. ASEAN is now expected to develop further into a single market called the ASEAN Economic Community (AEC). This will operate along similar lines to the EU and may ultimately allow free movement of labour and capital. The ASEAN agreement also promotes peace and stability: its members have pledged to not have nuclear weapons (Figure 12.19).

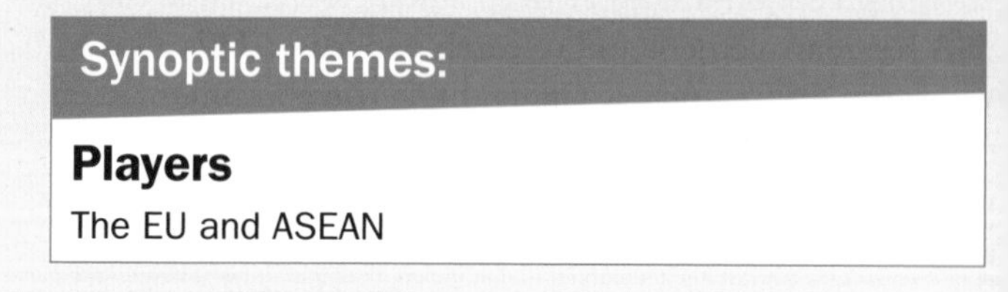

Synoptic themes:

Players

The EU and ASEAN

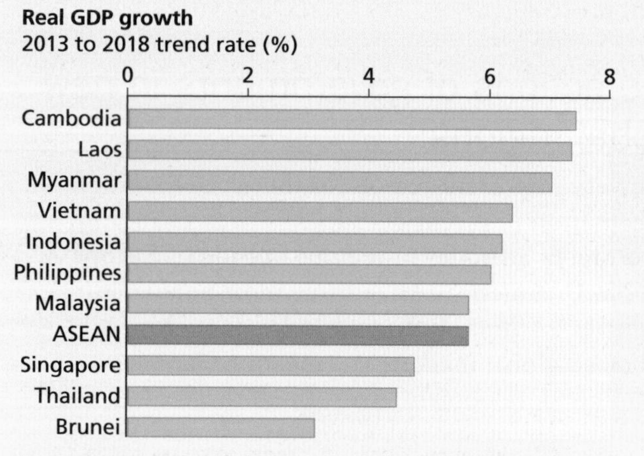

Figure 12.19 Predicted growth rates for ASEAN member states, 2013–18 (source: ASEAN)

China and its 1978 Open Door Policy

Prior to 1978, China was a poor and politically isolated country, 'switched-off' from the global economy. Under the communist leadership of Chairman Mao Zedong, millions had died from famine. Most people lived in poverty in rural areas. This changed in 1978 when Deng Xiaoping began the radical 'Open Door' reforms which allowed China to embrace globalisation while remaining under one-party authoritarian rule.

The earliest reforms occurred in rural areas. Agricultural communes were dismantled and farmers were allowed to make a small profit for the first time. Strict controls on the number of children were also introduced, to curb population growth.

China's transformation into an urban, industrialised nation gained rapid momentum. Over the next 30 years, the largest migration in human history took place. 300 million people left rural areas in search of a better life in cities. Only a strict registration system called *hukou* prevented rural villages from emptying altogether. Soon there will be 200 Chinese cities with 1 million inhabitants or more. Many are new, rapidly built 'instant cities'. An urban mega-region of 120 million people has grown around the Pearl River Delta. It includes the conjoined cities of Shenzhen, Dongguan and Guangzhou.

Initially, urbanisation fuelled the growth of the low-wage factories that gave China the nickname 'workshop of the world'. The world's largest TNCs were quick to establish branch plants, or trade relationships with Chinese-owned factories, in newly-established coastal special economic zones (SEZs), see Figure 12.20. By the 1990s, 50 per cent of China's GDP was being generated in SEZs. Since then, the Chinese economy has matured quickly. By 2015, many workers were earning US$40 a day or more making quality goods, such as iPhones, for employers like Foxconn in the Shenzhen SEZ.

Today, China is the world's largest economy. With 400 million people said to have escaped poverty since the reforms began, China's story lends support to the 'hyper-global' view that global-scale free trade can sometimes cure poverty. However, China is still not entirely open to global flows, as Table 12.5 shows.

Priority development areas
Special economic zones
Open cities
Autonomous regions
0 km 800
Three Gorges Zone
Shanghai
Xiamen
Shantou
Shenzen
Zhuhai
East China Sea

① North China Energy Industrial zone
② Huaihai Economic Zone
③ Yangtse Delta Region
④ Shanghai Economic Zone
⑤ Minnan Delta Economic Zone
⑥ Pearl River Delta Zone

Figure 12.20 Economic development areas in China

Synoptic themes:

Players

The world's nations, trade blocs and international organisations have re-written their own rules, or created new ones, in order to foster globalisation

Table 12.5 China's varied attitudes towards globalisation and global flows

'Open door' approach to global flows	'Closed door' approach to global flows
FDI from China and its TNCs is predicted to total US$1.25 trillion between 2015 and 2025 (of this, over US$100 billion is destined for the UK).	Google and Facebook have little or no access to China's market (instead, Chinese companies like Youku provide social network services).
China agreed to export more 'rare earths' minerals to other countries, in line with a WTO ruling.	China's Government sets a strict quota of only 34 foreign films to be screened in cinemas each year.
Foreign TNCs are now allowed to invest in some sectors of China's domestic markets, including its rail freight and chemicals industries.	There are strict controls on foreign TNCs in some sectors. China's Government blocked Coca-Cola's acquisition of Huiyan Juice in 2008.

Similarly, many other countries in Africa, the Middle East and South America have introduced legal and economic reforms in recent years in order to attract FDI and increase their level of global participation. Saudi Arabia is an interesting case, having recently abandoned Thursday and Friday as its official weekend in favour of Friday and Saturday. This brings it more in line with other countries for the purpose of doing business.

12.3 Uneven globalisation

Globalisation has affected some places more than others. Spatial variations in poverty, physical factors and the policies of national governments have inevitably introduced geographical bias into world markets. Studies show that the effects of globalisation are highly uneven.

Measuring globalisation

Uneven levels of globalisation can be measured using indicators and indices.

- The Swiss Institute for Business Cycle Research, also known as KOF, produces an annual Index of Globalisation. In 2014, Ireland and Belgium were the world's most globalised countries according to the KOF index (Figure 12.21). A complex methodology informs each report, using diverse data such as participation in UN peace-keeping missions and TV ownership (Table 12.6). While there is merit in KOF's multi-strand approach to measuring globalisation, the validity of these criteria might be debated.
- The A.T. Kearney World Cities Index ranks New York, London, Paris, Tokyo and Hong Kong as the top five 'Alpha' world cities for commerce. The ranking is established by analysing each city's 'business activity', 'cultural experience' and 'political engagement'. The data supporting this include a count of the number of TNC headquarters, museums and foreign embassies, respectively.

The KOF Index and Kearney Index combine many data sources that can be critiqued on the grounds of either reliability or validity. Some data suffers from crude averaging and statistical gaps. Other indicators are arguably poor proxies for globalisation (hours spent watching TV, for instance). However, they provide an interesting starting point for the statistical analysis of globalisation.

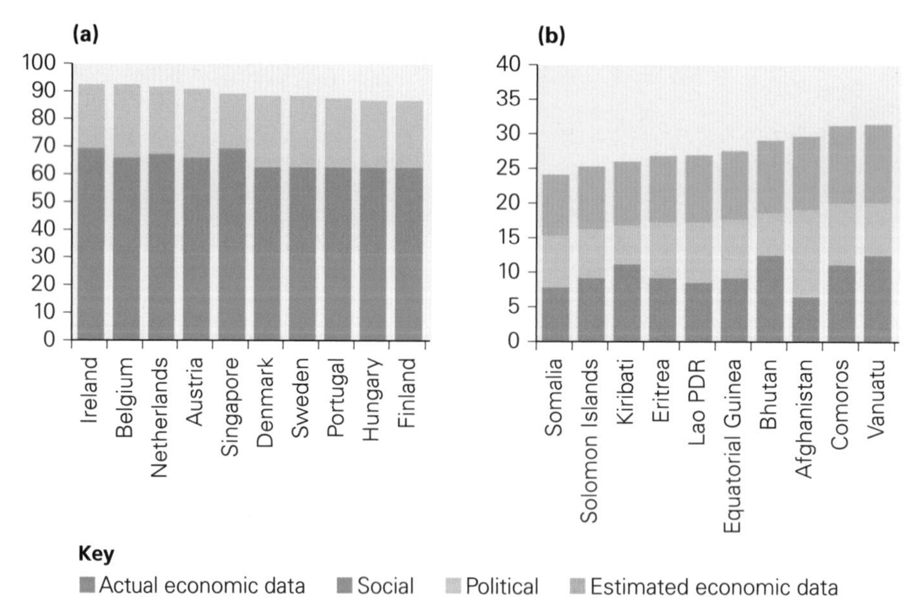

Figure 12.21 The KOF top ten and bottom ten globalisation rankings for 2014

Table 12.6 Calculating a country's globalisation score using the KOF index

1	Collect data relating to: • economic globalisation (trade and FDI figures from the World Bank; import tariff rates) • political globalisation (the number of foreign embassies in a country; the number of UN peace-keeping missions a country has participated in) • social and cultural globalisation (the volume of international ICT traffic, tourist flows and international mail; households with a TV set; imports and exports of books)
2	Analyse the new data by: • converting all 24 variables into an index value with a scale of one to 100 • substituting missing data with the most recent data available • averaging the individual scores to give a final score out of 100
3	Compare the new scores with previous scores dating back to 1970

The uneven geography of TNCs and their global production networks

TNCs are important agents of global change. Along with trade blocs, they can be described as 'architects' of globalisation, helping to 'build bridges' between nations. Table 12.7 shows examples referred to in this chapter and their countries of origin. Making connections, TNCs bolt together different economies and societies through their supply chains and marketing strategies. However, some parts of the world have benefited far more than others from FDI from TNCs because:

- not all places are suitable sites of production for goods, for a range of physical and human reasons (including accessibility, natural resources, government policies and levels of education)
- not all places have enough market potential to attract large retailers (due to low incomes, or culture).

Therefore TNCs do not, or cannot, always adopt the strategies outlined in Table 12.4 (page 168) in different national contexts. Internet companies have failed notably to gain access to China's market, for instance.

Other strategies come into play as TNCs attempt to build their global businesses. Rather than investing directly in the offshoring of branch plants, or acquiring foreign firms, TNCs can instead forge business partnerships with existing companies in other countries. Many of the world's biggest brands do not, in fact, make their own products. Instead, they use outsourcing as their strategy.

Table 12.7 Selected TNCs and their headquarters

TNC	Headquarters
Apple	USA
Kraft	USA
Tata	India
Samsung	South Korea
Airtel Africa	Kenya
BMW	Germany
Shanghai Electric	China
Lego	Denmark
Carrefour	France

Large corporations ranging from Dell to Tesco have established tens of thousands of outsourcing partnerships while building their global businesses. The resulting series of arrangements is called a global production network (GPN). A TNC manages its GPN in the same way the captain of a team manages other players (Figure 12.22). As globalisation has accelerated, so too has the size and density of global production networks, spanning food, manufacturing, retailing, technology and financial services. Food giant Kraft and electronics firm IBM both have 30,000 suppliers providing the ingredients they need. An amazing 2500 different suppliers provide parts to assemble BMW's Mini car, from the engine right down to the windscreen-wipers (Figure 12.23). Some parts are outsourced from suppliers within the EU (to avoid import tariffs). In contrast, the engine comes from an offshore factory in Brazil, owned by BMW.

GPN growth owes much to trade liberalisation and the changing attitudes of national governments, as outlined earlier in the chapter (see page 168). Developing countries have benefited from GPN growth because

Key terms

Offshoring: TNCs move parts of their own production process (factories or offices) to other countries to reduce labour or other costs.

Outsourcing: TNCs contract another company to produce the goods and services they need rather than do it themselves. This can result in the growth of complex supply chains.

Global production network: A chain of connected suppliers of parts and materials that contribute to the manufacturing or assembly of the consumer goods. The network serves the needs of a TNC, such as Apple or Tesco.

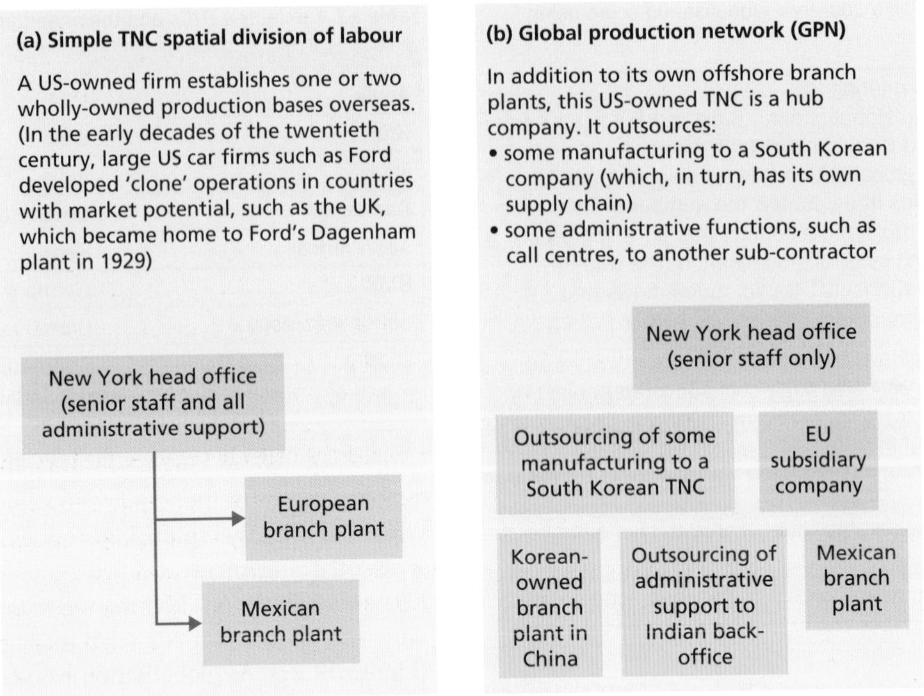

(a) Simple TNC spatial division of labour
A US-owned firm establishes one or two wholly-owned production bases overseas. (In the early decades of the twentieth century, large US car firms such as Ford developed 'clone' operations in countries with market potential, such as the UK, which became home to Ford's Dagenham plant in 1929)
New York head office (senior staff and all administrative support)
European branch plant
Mexican branch plant
(b) Global production network (GPN)
In addition to its own offshore branch plants, this US-owned TNC is a hub company. It outsources:
• some manufacturing to a South Korean company (which, in turn, has its own supply chain)
• some administrative functions, such as call centres, to another sub-contractor
New York head office (senior staff only)
Outsourcing of some manufacturing to a South Korean TNC
EU subsidiary company
Korean-owned branch plant in China
Outsourcing of administrative support to Indian back-office
Mexican branch plant

Figure 12.22 A simple TNC division of labour compared to a GPN

Windscreen
Made in: Belgium
Company headquarters: France
Wing mirrors
Made in: Germany
Company headquarters: Canada
Bonnet
Made in: Netherlands
Company headquarters: Austria
Radiator
Made in: Germany
Company headquarters: Germany
Exhaust system
Made in: UK
Company headquarters: USA
MINI
Wheels
Made in: Italy, Germany
Company headquarters: USA
Front and rear bumpers
Made in: UK
Company headquarters: Canada
Engine
Made in: Brazil
Company headquarters: Brazil

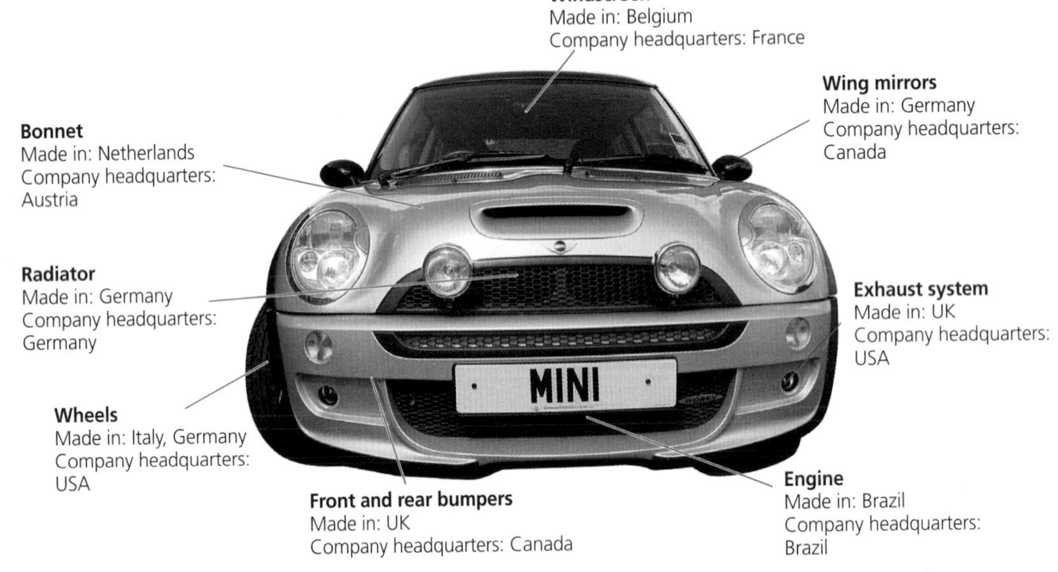

Figure 12.23 BMW's Mini and the global production network that contributes to its construction

outsourcing arrangements are economically beneficial. The local owners of factories in China's SEZs have profited from the work that foreign TNCs have outsourced to them.

However, some TNCs have discovered that outsourcing brings new risks. A poorly monitored GPN can damage corporate profits and image.

- Natural hazards, such as the 2011 Japanese tsunami, can disrupt global supply chains.
- UK supermarkets were stunned to find horsemeat had entered their supply chains in 2013.
- The collapse of the Rana Plaza textile factory in Bangladesh in 2013 killed 1100 people making clothes for Benetton and Wal-Mart, among others, as part of an outsourcing arrangement.

As a result of events like this, some TNCs are now 're-shoring' their manufacturing closer to home.

Developing new markets

The strengthening of Latin American, Asian and Middle Eastern economies has prompted an explosion of TNC interest in these emerging markets, where over 2 billion people have moved from dollar-a-day poverty into higher income brackets since 1990. Building a global production network helps TNCs gain access to these new markets. In order to maximise profit, many TNCs have, additionally, adapted their products to suit local tastes. This is called glocalisation.

Key concept: Glocalisation

This refers to changing the design of products to meet local tastes or laws. It is an increasingly common strategy used by TNCs in an attempt to conquer new markets. Glocalisation makes business sense because of geographical variations in:

- people's tastes (Cadbury makes its Chinese chocolate sweeter, as it is preferred that way)
- religion and culture (Domino's Pizza only offers vegetarian food in India's Hindu neighbourhoods; MTV avoids showing overtly sexual music videos on its Middle Eastern channel)
- laws (the driving seat should be positioned differently for cars sold in US and UK markets)
- local interest (reality TV shows, such as *Big Brother* and *Jersey Shore*, gain larger audiences if they are re-filmed using local people in different countries)
- lack of availability of raw materials (SABMiller, a major TNC, uses cassava to brew beer in Africa; this cuts the cost of importing barley but changes the taste too).

Far from rolling out an undifferentiated product across the world, many TNCs actively view localising strategies as integral to globalisation. However, when evaluating the importance of glocalisation for TNCs, remember that not all companies need to glocalise products. For some big-name TNCs, the 'authentic' uniformity of their global brand is what generates sales (see Table 12.8). For others, including oil companies, glocalisation has little or no relevance for their industrial sector.

Table 12.8 Different approaches used by TNCs

The Walt Disney Company	McDonald's	Lego
In 2009, Disney released its first Russian film, *Book of Masters*, based on a Russian fairy tale and produced using local talent. Disney acquired *Marvel* in 2009, gaining the rights to superhero characters that have sometimes been glocalised. 'Spiderman India' is an example. In a story made for Indian children, Mumbai teenager Pavitr Prabhakar is given superpowers by a mystic being. The story is different from the version UK and US children are familiar with.	By 2012, McDonald's had established 35,000 restaurants in 119 countries. In India, the challenge for McDonald's has been to cater for Hindus and Sikhs, who are traditionally vegetarian, and also Muslims who do not eat pork. Chicken burgers are served alongside the McVeggie and McSpicy Paneer (an Indian cheese patty). In 2012, McDonald's opened a vegetarian restaurant for Sikh pilgrims visiting Amritsar, home to the Golden Temple.	Unlike Disney and McDonald's, Lego has not glocalised its products. Since 1949, the Danish plastic brick-maker has been producing gradually more complex designs in four locations: Denmark, Hungary, Czech Republic and Mexico. However, Lego exports identical products to all global markets, including China. Like Apple and Samsung, Lego makes products with a genuine global appeal. The company does not take local tastes into account.

Switched-off places

A few of the very poorest nations of the world remain relatively switched off from global networks (aside from small elite groups of citizens). For physical, political, economic or environmental reasons, these countries still lack any strong flows of trade and investment with other places and economies.

Switched-off places

North Korea

For nearly 70 years, North Korea has been ruled as an autocracy by a single family (King Jong-un is its current leader). They have chosen deliberately to remain politically isolated from the rest of the world.

- Ordinary citizens do not have any access to the internet or social media.
- There are no undersea data cables connecting North Korea with anywhere else.
- A visiting journalist observed it that was the only country he had ever travelled to where nobody knew the song 'Yesterday' by The Beatles.

North Korea divided from South Korea in 1948. South Korea has since become a developed country which is home to Samsung and other global brands. A comparison of the two countries, and the policies of their governments, illustrates clearly how political decision making affects globalisation.

The Sahel region

Poverty affects the overwhelming majority of people in some of Africa's Sahel nations, such as Chad, Mali and Burkina Faso. These are some of the world's **least developed countries** (LDCs) and there are many reasons for the development challenges they face. The mismanagement of natural resources and human resources has played a role, dating back to colonial times. LDCs lacking a coastline, such as Chad, may struggle to attract FDI. Arid conditions and desertification give rise to further development challenges. In particular, extreme environmental conditions increase the cost of providing infrastructure, such as railways or ICT networks, in regions where poverty has meant there is limited market potential to begin with.

When people in these countries do become linked with the wider world, it tends to be a shallow form of integration. Subsistence farmers may become dependent on flows of food aid from charities in OECD nations. Some farmers grow cash crops for TNCs, for instance cotton producers in Mali. Wages are so low, however, that workers have negligible spending power. Thus, global brands do not yet view these places as viable markets, leaving them relatively switched off from consumer networks.

Change may soon come to the region though. Rapid economic growth is happening in neighbouring countries like Nigeria. Already, a minority of Sahelian people do interact with the rest of the world in surprising ways:

- Mali's folk musicians have a large global following on YouTube
- conflicts in the region involve groups linked with al-Qaeda's global terror network.

Key term

Least developed countries: The world's very poorest low-income nations, whose populations have little experience of globalisation. A number of these nations are described as 'failed states' by politicians, for example Somalia and South Sudan.

Synoptic themes:

Players

The links between different factors and players sometimes perpetuate poverty in places that are switched off. For instance, the Democratic Republic of the Congo is rich in resources. However, these resources have brought political corruption and conflict. Political factors thus rob physical factors of their potential to create wealth.

Review questions

1 Draw a table with two columns headed '1815' and '2015'. List the leading transport and communications technologies associated with each time period.

2 Compare the strategies used by Chinese and British governments to attract foreign investment.

3 Briefly outline the advantages that trade bloc membership brings to nations.

4 Using examples, describe ways in which the economies of different nations are linked together by:
   (i) the activities of TNCs
   (ii) the work of the Bretton Woods institutions.

5 Discuss the reasons why some places have become more globalised than others.

Further research

Explore easyJet's travel network: www.easyjet.com/EN/routemap

Find out more about FDI in the UK: www.bbc.co.uk/news/business-25299230

Study the EU timeline in greater depth: http://ec.europa.eu/ireland/events/photogallery/images/40-years-membership-mar-2013/time-line.jpg

Research urbanisation in China: http://ngm.nationalgeographic.com/2007/06/instant-cities/

Examine how glocalisation has affected Spiderman: www.geographyinthenews.rgs.org/news/article/?id=325

A study of North Korea's isolation: www.newstatesman.com/2014/02/voyage-town-where-no-one-knows-beatles

13 Geographical impacts of globalisation

What are the impacts of globalisation for countries, different groups of people and cultures?

By the end of this chapter you should:

- understand why global shifts in economic activity bring a range of environmental, economic and social impacts
- be able to explain how globalisation is linked with the increasing scale and pace of economic migration, and results in a range of impacts to places of varying scales
- be able to assess the global and local cultural changes associated with globalisation, and the reactions they bring.

13.1 Winners and losers of the global shift in economic activity

The term 'global shift' describes the international relocation of different types of industrial activity, especially manufacturing industries. Since the 1960s, many industries have all but vanished from Europe and North America. Instead, they thrive in Asia, South America and, increasingly, Africa. Global shift stems from a combination of off-shoring, outsourcing and new business start-ups in emerging economies (see Chapter 12).

Accompanying this movement is the 'exporting' of unethical economic practices that an early-industrialising nation like the UK abandoned long ago. These include dangerous working conditions, child labour and highly unequal pay for men and women. All can currently be found in Bangladesh, Vietnam and India. There are sometimes severe costs for the environment too. The global shift of polluting industries to lower-income countries has meant that TNCs are, in general, subject to fewer environmental rules and regulations.

Costs and benefits for emerging Asia

Average incomes have soared for successive waves of new Asian 'tiger' economies. Japan's success came first in the 1950s. South Korea followed soon after. Foreign investors began working with local firms called *Chaebols*. As national revenues soared, so too did South Korea's spending on education and health. Today, the country is an OECD member with the world's eleventh largest economy. Between 2000 and 2010, most large Asian economies sustained exceptionally strong annual growth rates, in part due to global shift (Table 13.1). More recently, growth

Table 13.1 Nominal income and economic growth trends in Asia

	Per capita income (US$) 1990	Per capita income (US$) 2015	Average growth rate 2000–10 (%)
Indonesia	600	3500	15
Malaysia	2400	10,800	8
Philippines	700	2800	8
South Korea	6100	28,100	6
Taiwan	7500	22,600	4
China	300	7500	17
India	400	1600	11

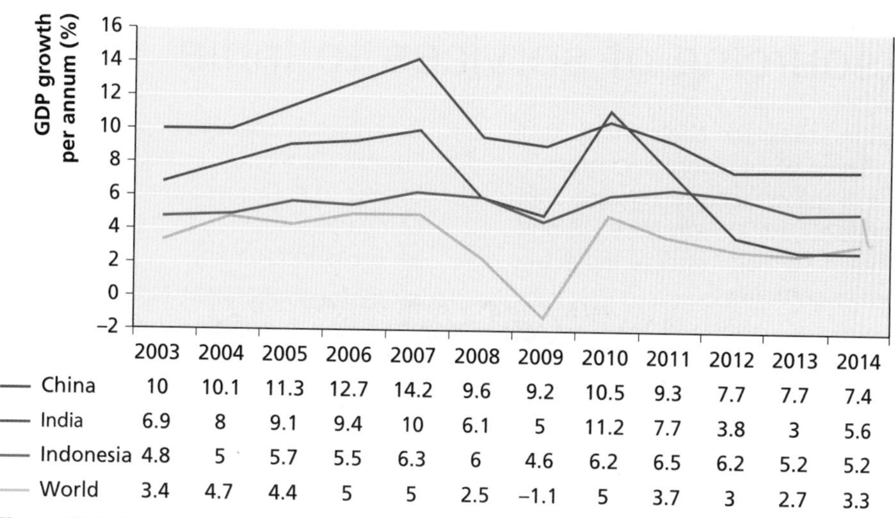

	2003	2004	2005	2006	2007	2008	2009	2010	2011	2012	2013	2014
China	10	10.1	11.3	12.7	14.2	9.6	9.2	10.5	9.3	7.7	7.7	7.4
India	6.9	8	9.1	9.4	10	6.1	5	11.2	7.7	3.8	3	5.6
Indonesia	4.8	5	5.7	5.5	6.3	6	4.6	6.2	6.5	6.2	5.2	5.2
World	3.4	4.7	4.4	5	5	2.5	−1.1	5	3.7	3	2.7	3.3

Figure 13.1 Annual economic growth for China, India and Indonesia compared with world average

has slowed but, in most cases, remains higher than in developed countries (Figure 13.1).

Across Asia, urban environments have been transformed by rapid industrialisation and the establishment of SEZs (see Chapter 12). Major economic, social and environmental changes are associated with globalisation:

- **Poverty reduction and waged work:** Worldwide, 1 billion people have escaped US$1.25-a-day poverty since 1990. The majority are Asian: over 500 million have escaped poverty in China alone. The term 'new global middle class' is used to describe the growing mass of urban, working people who have escaped rural poverty. Some work in the manufacturing sector in Bangladesh and China. Others belong to service industries in India and the Philippines. Many earn between US$10 and US$100 per day (far more than their parents did). By 2030, it is predicted that Asia will be home to 3 billion middle-class people.
- **Education and training:** High school achievement in Singapore and Hong Kong is envied by governments around the world, including the UK. Throughout Asia, education has improved in recent decades, albeit unevenly (illiteracy remains a problem in rural India and Bangladesh, for instance). Around 2500 universities in China, India and South Korea award millions of graduate degrees each year. China alone awarded 30,000 PhDs in 2012, and Asian countries now play a leading role in quaternary sector research in biotechnology and medical science.
- **Environment and resource pressure:** The flip-side of global economic growth is the acceleration of environmental decline. Forested land has been sacrificed to urbanisation, logging and cash cropping. Since 1990, Togo has lost 60 per cent of its forested area; Nigeria's forest has halved in size. Elsewhere, productive crop land has been ruined by over-exploitation, soil erosion or mining. From 1990 to 2008, globalisation helped drive a 'commodities supercycle'. Demand for raw materials – from soy beans to iron ore – rose steeply each year. However, global resource pressure has recently slackened, due to reduced demand in China (where economic growth has halved since 2008).
- **Infrastructure, the built environment and unplanned settlements:** Alongside economic take-off, infrastructure development has taken place, bringing modern motorways, high-speed railways and airports to major cities including Jakarta. There is a growing trend for extreme high-rise development in city centre 'hotspots' in many Asian cities, including Hong Kong, Singapore and Shanghai. Often these developments are accompanied by the loss of recreational spaces and older, unplanned neighbourhoods. Beijing's traditional *hutongs* (narrow lanes) are now all but lost. Mumbai's Dharavi slum is a cramped and chaotic place that is home to families who live on little more than £200 a month. It is also the location of a thriving recycling industry worth as much as £700 million a year and employing 250,000 people. However, city authorities are determined to replace the Dharavi slum with modern flats.

Global outsourcing of services to India

By 2040, India is expected to be the second-largest economy in the world. Some of its recent economic success is attributable to the call centre services that Indian workers provide (Table 13.2). Why have US and UK businesses outsourced so much work to India, and to the city of Bangalore in particular?

- Many Indian citizens are fluent English speakers. This is a legacy of British rule, which ended in 1947. It gives India a comparative advantage when marketing call centre services to the English-speaking world.

Table 13.2 Evaluating India's call centre success story

Costs	Benefits
Some call centre workers complain they are exploited. Their work can be highly repetitive. Business is often conducted at night – due to time zone differences between India and customer locations in the USA or UK – sometimes in ten-hour shifts, six days a week. Despite overall growth, the gap between rich and poor has widened sharply. India has more billionaires than the UK, yet it also has more people living in absolute poverty than all of Africa. In 2015, half a billion Indians lived in homes that lacked a toilet.	India's call centre workers earn good middle-class wages by Indian standards. Nightclubs and 24 shopping malls in Bangalore testify to the relatively high purchasing power of a new Indian 'techno-elite' typically earning 3500 rupees (£40) a week (Figure 13.2). Indian outsourcing companies have become extremely profitable. Founded in 1981, Infosys had revenues of US$9 billion in 2015. It is one of the top twenty global companies for innovation, according to the US business analyst Steve Forbes.

- Broadband capacity is unusually high in Bangalore. This city is a long-established technology hub, thanks to early investment in the 1980s by domestic companies such as Infosys and foreign TNCs such as Texas Instruments.

Today, large independent Indian operators conduct contract work for all kinds of firms, from travel companies to credit card providers. Dell, Intel and Yahoo have also built their own call centres here.

Figure 13.2 A call centre in India

Global outsourcing of manufacturing to China

The global shift of manufacturing has played an important role in extreme poverty in China falling from 60 per cent in 1990 to sixteen per cent by 2005 (Table 13.3).

- China first gained its reputation as the 'workshop of the world' in the 1990s. Cities like Shenzhen and Dongguan offered foreign investors a massive pool of low-cost migrant labour. At this time it was common to hear stories of Chinese workers suffering in factory conditions similar to those of Victorian England.
- Between 2000 and 2010, conditions improved markedly for many workers. The disposable income of urban citizens rose threefold following a series of protests. In 2010, workers walked off production lines for Honda, Toyota, Carlsberg and other global brands. Actions such as these led to wage increases of between 30 and 65 per cent (Honda employees now earn US$300 a month).
- Since 2010, strategic planning by China's government has helped some companies move further up the manufacturing value chain. The country's economy is maturing rapidly. 'Hi-tech' manufacturing is booming, bringing improved pay for skilled workers. Increasingly, high-value products such as iPhones are made in China, not just 'throwaway' cheap goods. Many less-desirable 'sweatshop' jobs have migrated to Bangladesh where labour costs remain much lower.

Table 13.3 Evaluating China's 'workshop of the world' status

Costs	Benefits
In the early years, many workers were exploited in sweatshops. Around 2500 metal-workers in Yongkang lost a limb or finger each year due to dangerous factory conditions. It gained a reputation as China's so-called 'dismemberment capital'. Since then, conditions have improved for many Chinese workers.	As conditions improve, people are enjoying large income gains. More people can now afford smartphones and fridges. Car ownership has grown from one-in-a-hundred families to one-in-five since 2000. Increasingly, China's economic growth is driven by this domestic consumption, and not just by the value of its exports.
The environment continues to suffer greatly. Dubbed 'airpocalypse' by the Western media, air pollution in cities reduces Chinese life expectancy by five years. The WHO is concerned with very high average levels of small particulate matter known as PM2.5. These deadly particles settle deep in the lungs, causing cancer and strokes.	A transfer of technology has taken place since the early days of manufacturing-led industrialisation. Local companies have adopted technologies and management techniques brought to China by TNCs. Increasingly, Chinese companies are developing their own products. A leading example is smartphone maker Xiaomi. Chinese banks are now some of the world's largest TNCs.

Environmental challenges for communities in developing countries

It is not only China that has experienced an 'airpocalypse'. Communities within many developing economies have experienced major environmental problems as a result of global shift. Adverse impacts on the health and well-being of people have resulted from pollution, over-exploitation of resources and the dumping of industrial waste.

Global shift has, in part, been driven by TNCs seeking low-cost locations for their manufacturing and refining operations. Weak environmental governance has sometimes been an attractive location factor. In high-income nations, bodies such as the UK Environment Agency have a well-funded remit to monitor industrial operations and fine polluters. Elsewhere, there is less red tape:

- **China:** In Dongguan, workers for Wintek – the firm that makes touchscreens for iPhones – were poisoned by chemicals used to treat the glass. In Hunan province, many people were poisoning by a lead-emitting manganese smelter (manganese is used to strengthen steel, one of China's major exports).
- **Ivory Coast:** Tens of thousands of Ivorians suffered ill health after toxic waste alleged to produce hydrogen sulphide was dumped by a ship in the employ of Trafigura, a European TNC. A £28 million cash settlement followed.
- **Indonesia:** Land degradation and biodiversity loss are widespread in Indonesia, where an area of rainforest as big as 100,000 football pitches is lost each year. Room is being created for oil-palm plantations and mining operations. The scale of forest burning has created transboundary smoke pollution affecting neighbouring states. More mammal species are threatened in Indonesia than in any other country. The government has been very slow to act and corruption remains widespread.

Social and environmental problems for deindustrialised regions

Global shift creates challenges for developed countries too. Economic restructuring has brought a wave of economic and social problems to inner-city areas. These are also explored in detail in Chapter 16.

During the 1970s, many European and American factory workers lost their jobs. Western factories closed in large numbers once Asia became the focus of global manufacturing. As inner-city unemployment soared in places like Sheffield (UK) and Baltimore (USA), local communities abruptly ceased to be significant

Key term

Deindustrialisation: The decline of regionally important manufacturing industries. The decline can be charted either in terms of workforce numbers or output and production measures.

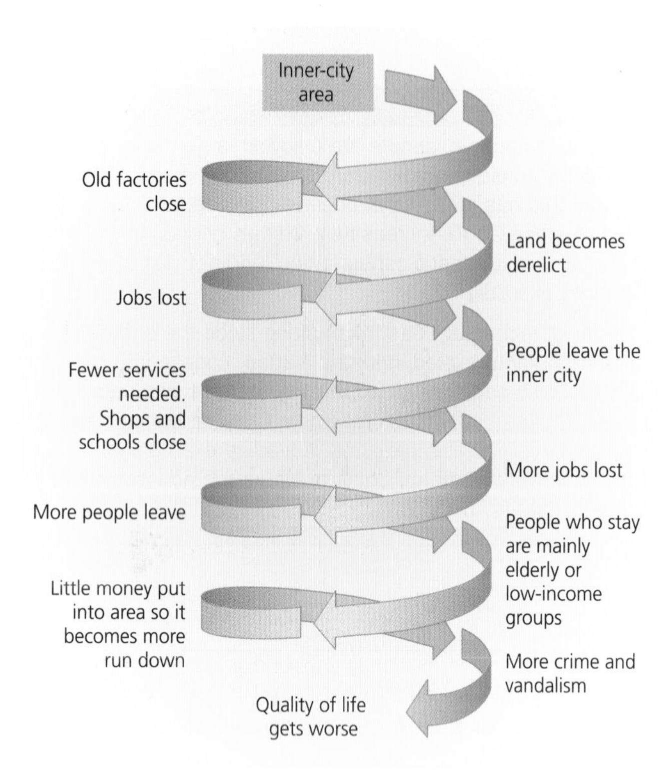

Figure 13.3 The inner-city spiral of deprivation

producers or consumers of wealth. The worst-affected neighbourhoods were now home to 'switched-off' communities who had become structurally irrelevant to the global economy.

Other cities remain caught in a spiral of decline (Figure 13.3). In the USA especially, the economic and social health of urban areas varies greatly (Figure 13.4). Particular challenges include:

- **High unemployment:** Detroit has yet to replace large numbers of jobs lost when global shift led to the disappearance of many of the city's automobile industries.
- **Crime:** Rising gun crime reminds us that 'losers' of globalisation can be found in all nations, not just poorer ones. In some low-income US urban districts, life expectancy is 30 years lower than in affluent districts. Drug-related crime is now the basis of an informal economy in some poor neighbourhoods of failing US cities. When areas are 'switched off' to legitimate global flows, they may instead become 'switched on' to illegal global flows of drugs and people trafficking.
- **Depopulation:** Middle-class Americans have migrated out of failing neighbourhoods in large numbers. Detroit has lost 1 million residents since 1950. One result of this depopulation has been a catastrophic collapse in housing prices. In Baltimore, which has lost one-third of its population, there are 20,000 abandoned properties. Homes in some districts have been sold for just one dollar. Those who stay become trapped in a state of negative equity (their home is worth much less than they paid for it). Increasingly, depopulation in US cities has become linked with race. Dubbed 'white flight' by the media, the process of out-migration has left some districts populated mainly by African-Americans. The economic problems triggered by global shift have, over time, reignited racial tensions in cities such as Baltimore and Jackson.
- **Dereliction:** The combination of manufacturing industry closures, falling house prices and rising crime results in widespread environmental dereliction. A 'broken windows' scenario develops (at first small acts of vandalism are tolerated; soon, more serious problems like arson become commonplace).

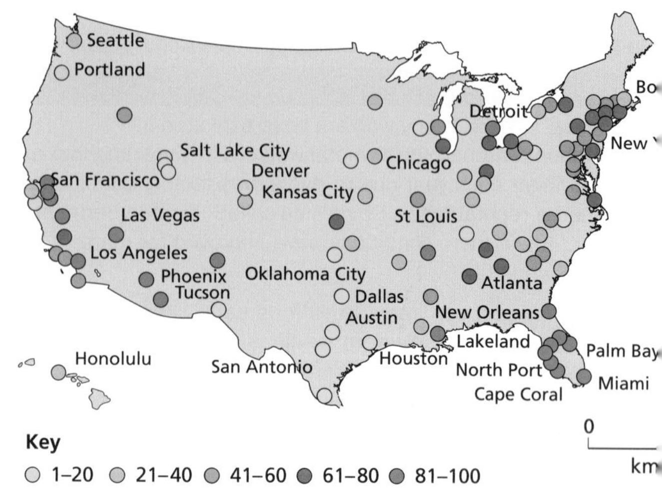

Figure 13.4 The uneven economic performance of US cities, 2008–15, showing percentage economic growth

13.2 The increasing scale and pace of economic migration in an interconnected world

In 2013, 750 million internal migrants were residing in cities across the world (around one-third were Chinese rural–urban migrants). Global urbanisation assed the threshold of 50 per cent in 2008, meaning ıat the majority of people now live in urban areas ee http://graphics.thomsonreuters.com/RNGS/2011/)CT/POP5_BH.jpg). Additionally, nearly 250 million ıternational migrants now live in countries they were orn outside. The overwhelming majority of movers, oth at international and internal scales, are economic ıigrants. However, 2014 also saw the largest isplacement of forced migrants since the Second Vorld War. Around 14 million new refugees were riven from their homes by natural disasters and onflict in countries such as Syria, bringing the global otal to 60 million displaced people.

Key terms

Internal migrant: Someone who moves from place to place inside the borders of a country. Globally, most internal migrants move from rural to urban areas (rural–urban migrants). In the developed world, however, people also move from urban to rural areas too (a process called counter-urbanisation).

Urbanisation: An increase in the proportion of people living in urban areas.

Economic migrant: A migrant whose primary motivation is to seek employment. Migrants who already had a job may have set off in search of better pay, more regular pay, promotion or a change of career.

Refugee: People who are forced to flee their homes due to persecution, whether on an individual basis or as part of a mass exodus due to political, religious or other problems.

Intervening obstacles: Barriers to a migrant such as a political border or physical feature (deserts, mountains and rivers).

Natural increase: The difference between a society's crude birth rate and crude death rate. A migrant population, such as that found in developing world megacities, usually has a high rate of natural increase due to the presence of a large proportion of fertile young adults and relatively few older people reaching the end of their lives.

Rural–urban migration and megacity growth

By 2050, three-quarters of us will be city-dwellers. Table 13.4 and Figure 13.5 explain why rapid urbanisation s still happening in many places, and examines how spects of globalisation link with this process.

Megacity growth

A megacity is home to 10 million people or more. In 970 there were just three; by 2020 there will be 30. They grow through a combination of rural–urban migration and natural increase.

Table 13.4 Causes of rural-urban migration

Urban pull factors	The main factor almost everywhere is employment. FDI by TNCs in urban parts of poorer countries provides a range of work opportunities with the companies and their supply chains. We can distinguish between formal sector employment (working as a salaried employee of Starbucks in São Paulo, for instance) and the informal sector (people scavenging material for recycling at landfill sites in Lagos). Urban areas offer the hope of promotion and advancement into professional roles that are non-existent in rural areas. Additionally, schooling and health care may be better in urban areas, making cities a good place for young migrants with aspirations for their children.
Rural push factors	The main factor is usually poverty, aggravated by population growth (not enough jobs for those who need them) and land reforms (unable to prove they own their land, subsistence farmers must often relocate to make room for TNCs and cash crops). Agricultural modernisation reduces the need for rural labour further (including the introduction of farm machinery by global agribusinesses such as Cargill). Resource scarcity in rural areas with population growth, such as the Darfur region of Sudan, may trigger conflict and migration (people are classed as refugees and not economic migrants, however).
'Shrinking world' technology	Rural dwellers are gaining knowledge of the outside world and its opportunities. The 'shrinking world' technologies we associate with globalisation all play important roles fostering rural–urban migration. Satellites, television and radio 'switch on' people in remote and impoverished rural areas. As poor individuals in Africa and Asia begin to use inexpensive mobile devices, knowledge is being shared. Successful migrants communicate useful information and advice to new potential migrants. Also, transport improvements, such as South America's famous Trans-Amazon Highway, have removed intervening obstacles to migration.

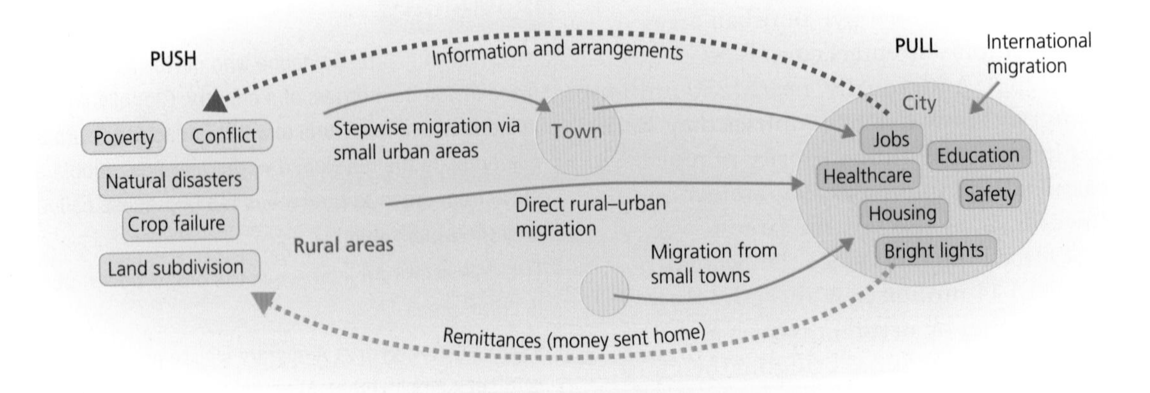

Figure 13.5 A model of rural–urban migration

Megacities in low-income (developing) and middle-income (emerging) countries have grown especially rapidly (Figure 13.6 and Table 13.5). São Paulo gains half a million new residents annually from migration. New growth takes place at the fringes of the city where informal (shanty) housing is built by the incomers. Centripetal migration brings people to municipal dumps (Lagos), floodplains (São Paulo), cemeteries (Cairo) and steep, dangerous hill slopes (Rio de Janeiro). Over time, informal housing areas may consolidate as expensive and desirable districts. Rio's now-electrified shanty town Rocinha boasts a McDonald's, hair salons and health clinics.

International migration continues to bring population growth, albeit far more slowly, to megacities in the developed world (for example, Poles moving to Greater London, or Mexicans to Los Angeles). There is residual internal migration too, for instance from the rural heartlands of the USA to New York. Environmental rules, such as UK's green belt policy, prevent further

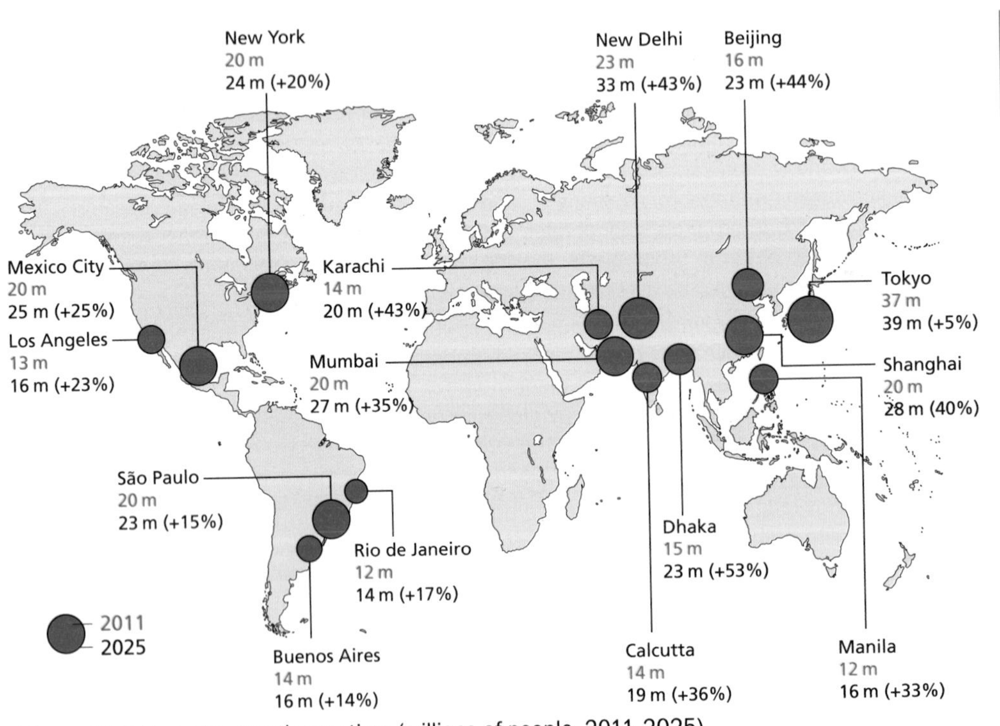

Figure 13.6 Megacity growth over time (millions of people, 2011–2025)

Key term

Centripetal migration: Movement of people directed towards the centre of urban areas.

Mumbai and Karachi

Table 13.5 Examples of rapid megacity growth – Mumbai and Karachi

Mumbai	Karachi
India's Mumbai urban area in 2015 was home to 22 million people, having more than doubled in size since 1970. People flock there from the impoverished rural states of Uttar Pradesh and Bihar. Urban employment covers a range of economic sectors and skill levels. Big global brands such as Hilton and Starbucks are present in Mumbai. In retail areas, like Colaba Causeway, large numbers of local people work selling goods to the country's rising middle class. Some very wealthy people live in Mumbai, including Bollywood stars and the senior management of large TNCs such as Tata and Reliance Industries, some of whom are billionaires. Their spending helps drive up housing prices in affluent areas such as Altamount Road. In contrast, Dharavi is a slum housing area. It has a buoyant economy: 5000 people are employed in Dharavi's plastics recycling industries. However, rising land prices across Mumbai mean there is great pressure to redevelop this and other slum areas.	Before Islamabad was founded in 1960, the port city of Karachi was the capital city of Pakistan. Approximately 24 million people lived in Karachi in 2015, making it the most highly populated city in Pakistan and the second most populous megacity in the world (after Tokyo). This colossal megacity is Pakistan's centre of finance, industry and trade (Figure 13.7). People flock to the city for work from rural areas all over Pakistan, including the Sindh and Punjab provinces. Once there, they find work in a range of industrial sectors including shipping, banking, retailing and manufacturing. Karachi's population increase over time is due mainly to internal migration, though international migrants from other South Asian countries play a role in its growth too. Karachi is a famous university city, producing skilled graduates who have helped it to become a hub for media and software companies. The TV channels Geo TV and CNBC Pakistan are based in the city.

Figure 13.7 A view of Karachi

suburban growth, however. High-rise redevelopment of brownfield sites therefore becomes the only way to meet new housing needs.

Social and environmental challenges of megacity growth

Continued urban growth is inevitable. As wealth grows in developing and emerging economies, more young rural folk will develop aspirations beyond agriculture. Can continued urban growth be made sustainable? Can the growth of these

Key term

Brownfield site: Abandoned or derelict urban land previously used by commercial or industrial companies.

Figure 13.8 Youth unemployment in North Africa

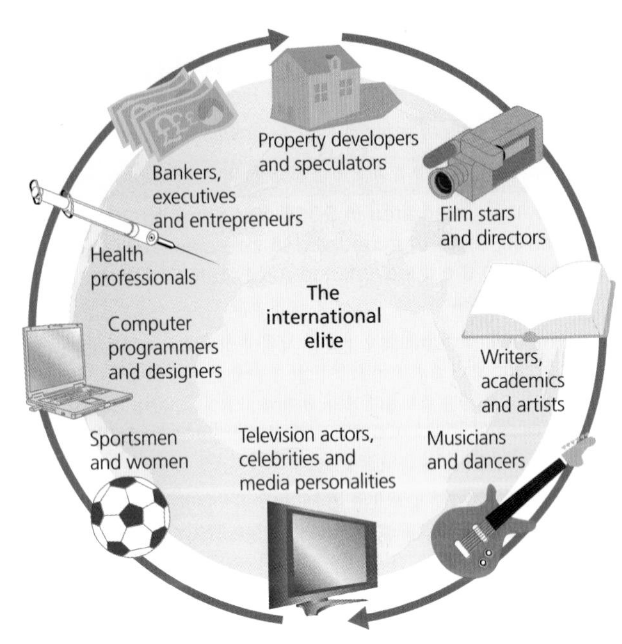

Figure 13.9 Elite migration

'instant cities' – as well as established megacities like Jakarta and Seoul – be made sustainable? Two important goals must be met:

- **Environmental sustainability:** Water pollution from untreated sewage, and air pollution from industry and exhausts, create challenges for city planners. The severity of these problems depends on their economic and physical context: Mediterranean cities such as Athens and Rome still suffer from smog, due to anticyclonic weather conditions; in contrast, cities in India and Pakistan's monsoon belt experience high-intensity rainfall and flooding due to sewer failures.
- **Social sustainability:** Provision of adequate urban housing, health care and education is a major challenge for planners in developing countries. Mass migration into Lagos (Nigeria) and Kinshasa (Democratic Republic of Congo) means these cities have doubled in size since 2000. In many European and North American capital cities, the challenge is to regulate the housing market to make affordable housing available for low-income groups. Finally, social sustainability is hard to achieve anywhere unless urban employment needs are met. In the North African cities of Tunis and Cairo, youth unemployment exceeds 25 per cent (Figure 13.8). This is the driver for many social problems, including the growth of extremist political movements.

International migration into global hubs

A global hub is a highly globally-connected city; or the home region of a large, globally-connected community.

Global hubs are found in countries at varying levels of development and are sometimes called 'World Cities'. Examples include New York, Mumbai, London, Beijing, Tokyo and São Paulo. Three types of population movement have led to the growth of global hubs:

- **Elite international migrants** are highly skilled and/or socially influential individuals (Figure 13.9). Their wealth derives from their profession or inherited assets. Some elite migrants live as 'global citizens' and have multiple homes in different countries. They encounter few obstacles when moving between countries. Most governments welcome highly skilled and extremely wealthy migrants (many skilled and affluent Americans and Russians oligarchs live or work in the UK, even though these countries are not part of the EU).
- **Low-waged international migrants** are drawn towards global hubs in large numbers. London, Los Angeles, Dubai and Riyadh are all home to large numbers of legal and illegal immigrants working for low pay in kitchens, construction sites or as domestic cleaners (Table 13.6).
- **Internal (rural–urban) migration** is the main driver of city growth in global hubs in developing and emerging economies but plays a lesser role in Europe and North America.

The costs and benefits of migration

In order to assess the costs and benefits of migration, a range of information must be structured and analysed.

- There are impacts for both the source (sending) and host (receiving) places to consider.
- These are further sub-divided into economic, social, political and environmental effects (which may all be interconnected).

Examples of low-wage international migration

Table 13.6 Low-wage international migration into global hub cities and regions

Indian workers moving to the UAE	Over 2 million Indian migrants live in the United Arab Emirates, making up 30 per cent of the total population. Many live in Abu Dhabi and Dubai. An estimated US$15 billion is returned to India annually as remittances. Most migrants work in transport, construction and manufacturing industries. Around one-fifth are professionals working in service industries.
Filipino workers moving to Saudi Arabia	Around 1.5 million migrants from the Philippines have arrived in Saudi Arabia since 1973 when rising oil prices first began to bring enormous wealth to the country. Some work in construction and transport industries, others as doctors and nurses in Riyadh. Around US$7 billion is returned to the Philippines annually as remittances. There are reports of ill-treatment of some migrants, however.

Key concept: Global hubs

A global hub is a settlement or region that has become a focal point for activities with a global influence, such as trade (Shanghai), business (London), international governance (The Hague) or education and research (Cambridge). Unlike a megacity, a global hub is recognised by its influence rather than its population size. Washington, DC is a relatively small city yet, as home to the White House, the Pentagon, the World Bank and the IMF, it is the world's premier global hub. Flows of money, goods and workers help link the world's global hubs together to form a network of important places.

Physical resources and human resources help explain the geographical location of global hubs, along with government policies (Figure 13.10). In developing and emerging economies, global hubs such as Jakarta and Cairo are places where the parent companies of major TNCs have established subsidiary firms or forged alliances with local companies. For instance, Disney, which is headquartered in California, has established an Indian subsidiary (Walt Disney Company India) in the global hub of Mumbai.

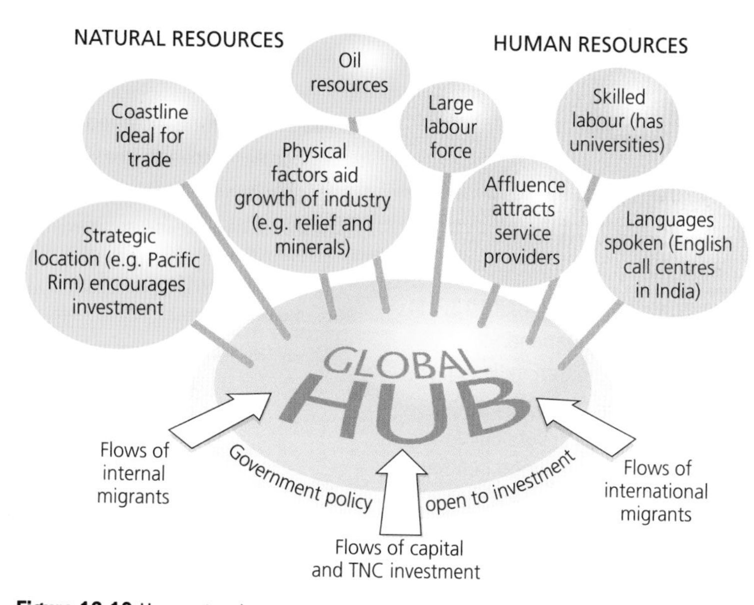

Figure 13.10 How natural resources and human resources help global hubs develop

- Another important outcome to consider is the way that migration leads two places to become interdependent on one another (this impact may deepen over time).

The sheer diversity of impacts stemming from migration makes it a divisive political issue: with so many different costs and benefits to account for, people rarely agree on whether the 'big picture' is negative or positive. Table 13.7 shows some important costs and benefits for international economic migration flows.

Table 13.7 Selected costs and benefits of migration for source and host countries

	Host region	Source region
Benefits	Fills particular skills shortages (e.g. Indian doctors arriving in the UK in the 1950s). Economic migrants willingly do labouring work that locals may be reluctant to (e.g. Polish workers on farms around Peterborough). Working migrants spend their wages on rent, benefiting landlords, and pay tax on legal earnings. Some migrants are ambitious entrepreneurs who establish new businesses employing others (in 2013, fourteen per cent of UK business start-ups were migrant-owned).	Migrant remittances can contribute to national earnings significantly (in 2014, remittances made up 25 per cent of Nepal's national earnings). Less public spending on housing and health (in 2004, prior to joining the EU, unemployment in Poland was twenty per cent; it has since halved). Migrants or their children may return, bringing new skills (young British Asians have relocated to India to start health clubs and restaurant chains). Some government spending costs (education, health) are transferred to the host region.
Costs	Social tensions arise if citizens of the host country believe migration has led to a lack of jobs or affordable housing (a view adopted by the UKIP party and some UK newspapers). Political parties change their policies to address public concerns (e.g. pledges to reduce migration). Local shortages of primary school places due to natural increase among a youthful migrant community (e.g. London boroughs that have become Eastern European migration 'hotspots'). New markets can develop for ethnic food (e.g. Korean food markets in Los Angeles) bringing visible changes to the urban built environment.	The economic loss of a generation of human resources, schooled at government expense, including key workers such as doctors, teachers and computer programmers. Reduced economic growth as consumption falls. Increase in the proportion of aged dependents and the long-term economic challenge it creates. Closure of some university courses due to a lack of students aged 18–21. The closure of urban services and entertainment with a young adult market, bringing decline and dereliction to urban built environments (many nightclubs closed in Warsaw, Poland, in 2004).

Key concept: Interdependence

Over time, international migration makes places interdependent. Each country depends on the economic health of the other for its own continued well-being.

- Firstly, economic interdependency may develop (Figure 13.11). Some sectors of the UK economy are highly dependent on Eastern European labour; Eastern Europe, in turn, relies on migrant remittances from the UK. In 2009, during the global financial crisis, many UK building projects were cancelled. The knock-on effect was that many migrants stopped sending money home; some even returned to their countries. Estonia's economy shrank by thirteen per cent.
- Secondly, social and political ties between two countries can be strengthened through migration. The arrival of a large Indian diaspora population in the UK has deepened the country's enduring friendship with India.

Writing in the 1990s, Thomas Friedman argued that economic and political interdependency are linked. In the 'golden arches theory of conflict prevention' he asserted that two countries with McDonald's restaurants would maintain good relations because their economies had become interlinked. While the recent conflict between Russia and Ukraine has weakened Friedman's argument (both countries have McDonald's restaurants), it remains an idea worth exploring.

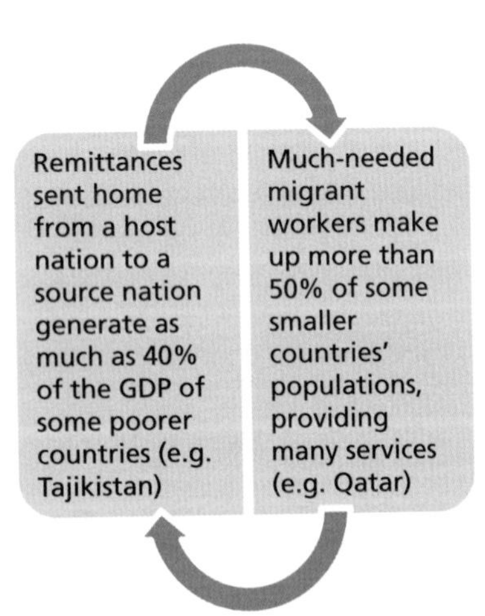

Figure 13.11 Economic interdependency between two nations

13.3 The emergence of a global culture

The word 'culture' describes what writer Raymond Williams called a society's 'structure of feeling'. Various shared cultural traits are held in common by different local or national societies (Figure 13.12). Cultures change and evolve over time naturally; globalisation has accelerated the rate of cultural change for many places, however. Perspectives differ on both the degree of change that is occurring, and its desirability.

Cultural diffusion and its causes

Powerful civilisations have brought cultural change to other places for thousands of years. This spread is called cultural diffusion. Sometimes it is achieved through coercion, using legal or even military tools. Forced assimilation of culture is also called cultural imperialism. Languages, religions and customs were spread around the world using force by the Roman and British empires, for instance. Today, countries like the USA and UK play a role in bringing cultural change to other places through their use of soft power. No force is involved. Instead, these powerful, wealthy states shape global culture through their disproportionately large influence over global media and entertainment.

The growth of a global culture

The specific cultural influence of the USA on other places is called 'Americanisation'. The joint role played by European and North American countries in bringing about cultural change on a global scale is called 'Westernisation'. Several factors help explain the emergence of a Western-influenced 'global culture' (Table 13.8).

Key terms

Cultural traits: Culture can be broken down into individual component parts, such as the clothing people wear or their language. Each component is called a 'cultural trait'.

Cultural imperialism: The practice of promoting the culture/language of one nation in another. It is usually the case that the former is a large, economically or militarily powerful nation and the latter is a smaller, less affluent one.

Soft power: The global influence a country derives from its culture, its political values and its diplomacy. Much of the USA's soft power has been produced by 'Hollywood, Harvard, Microsoft and Michael Jordan'.

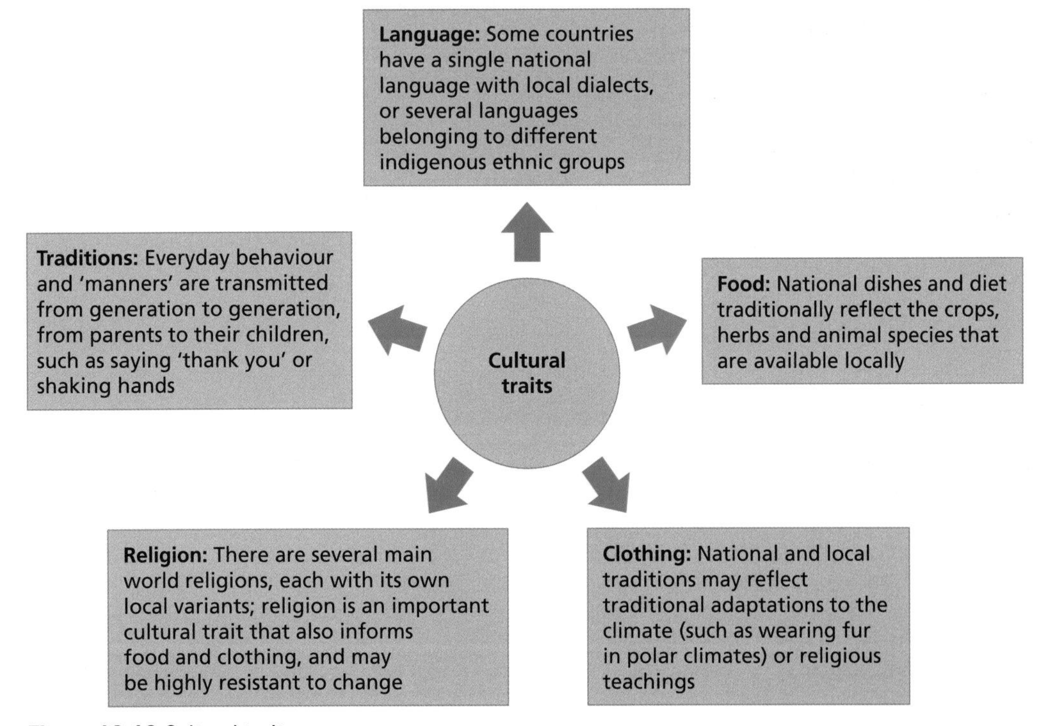

Figure 13.12 Cultural traits

Table 13.8 Evaluating the emergence of a global culture

Factor	TNCS	Global media	Migration and tourism
Influence	The global dispersal of food, clothes and other goods by TNCs has played a major role in shaping a common culture. Some corporations, such as Nike, Apple and Lego, have 'rolled out' uniform products globally, bringing cultural change to places.	Media giant Disney has exported its stories of superheroes and princesses everywhere. Western festivals of Halloween and Christmas feature prominently in its films. The BBC helps maintain the UK's cultural influence overseas (especially the World Service radio station).	Migration brings enormous cultural changes to places. Europeans travelled widely around the world during the age of empires, taking their languages and customs with them. Today, tourists introduce cultural change to the distant places they visit.
Evaluation	Chapter 12 introduced the concept of glocalisation. When TNCs engage with new markets and cultures, they often adapt their products and services to suit different places better. As a result, the products that are sold in different places increasingly reflect local cultures. You will be familiar with examples of this, such as McDonald's menus. In your view, is glocalisation merely a sophisticated form of cultural imperialism?	Other places gain a 'window' on American and British culture through shows such as period drama *Downton Abbey*. However, many reality and celebrity shows, such as *Strictly Come Dancing*, are entirely re-filmed for different national markets. Also, there are many non-Western influences on global culture, including the TV channels Russia Today and Qatar's Al Jazeera. Japanese children's TV has been highly influential, notably *Pokémon*.	Migrants can affect the culture of host regions, but the change may only be partial. British migrants took their language and love of cricket to many places but often had little effect on other cultural traits, notably religion. When carrying out an evaluation, it is important to ask if cultural changes for places are superficial, or more meaningful? We can also explore the effect of out-migration on the culture of source regions.

Synoptic themes:

Players

The role of TNCs

One manifestation of a 'global culture' is the way 4 billion people speak 'Globish' (a form of basic English consisting of around 1500 words). This language has a long history of adoption by the citizens of more than 60 ex-British colonies and countries under US influence, such as Singapore. Since the 1990s, however, Globish has diffused into countries which lack much shared history with either the UK or USA. They include Japan, China or Brazil. This is because English has:

- dominated internet communication from its outset
- become a global language of commerce, technology and education, in part due to the (English-speaking) USA's superpower status.

Globish is not an entirely uniform global language, however. Words, syntax and grammar vary from country to country because of the way English has blended with different native languages ('Singlish' is the Singaporean variant of Globish, for instance). Also, Globish is not *replacing* other local languages. Instead, people have adopted it in *addition* to their native tongues.

Changing diets in Asia: how cultural change affects people and the environment

Traditional Asian diets are often low in meat and high in vegetables. This healthy mix is giving way to more meat and fast food among the emerging middle classes, especially in China. During the 1990s, China's annual meat consumption per capita increased tenfold from 5 to 50 kg. By 2015, China had also become the world's biggest market for processed food.

The physical environment is affected by this at both the local and global scale. Livestock farming has become the new focus of Asian agriculture, bringing a steep rise in emissions of methane, a powerful greenhouse gas. Crops are imported from across the world to feed China's farm animals. Vast tracts of pristine Amazonian rainforest have been cleared during the last decade to make space for soya cultivation to feed Chinese cattle (Figure 13.13). China's food demands will only continue to grow as more people escape poverty. Mindful of this, the Chinese government has embarked on a programme of land acquisition in poorer countries, including Cuba and Kazakhstan.

Rising affluence also puts pressure on particular plant and animal species if their use or consumption is culturally linked with social prestige. Shark fin soup is an important but expensive dish traditionally consumed at Chinese weddings by those who could afford it. As incomes have risen, the number of sharks killed worldwide to meet growing demand has doubled.

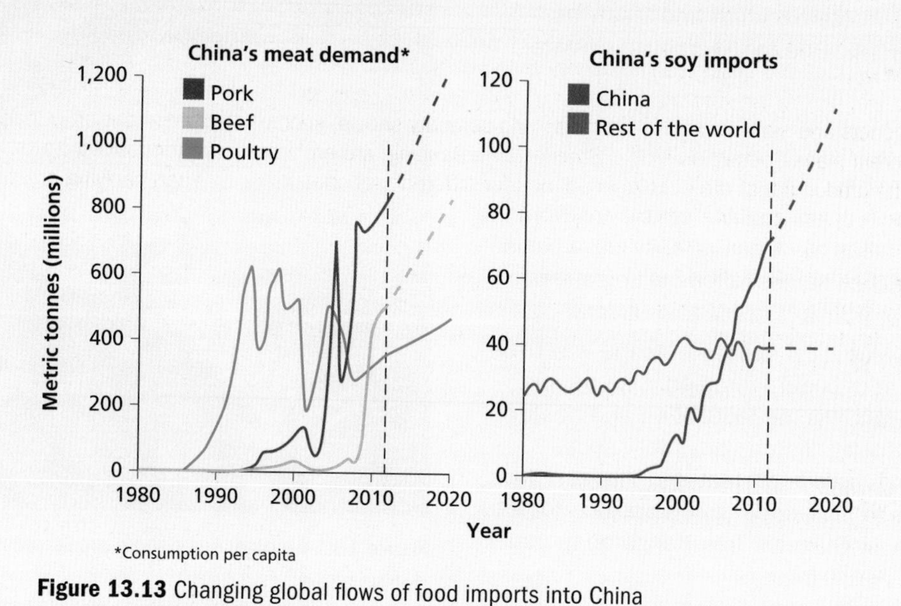

Figure 13.13 Changing global flows of food imports into China

Skills focus: Global brand analysis

The world's largest global brands have enjoyed a 'golden age' since 2000. Global shift, and the rise of the new global middle class, have helped the 100 biggest brands to grow in combined value to $US3.5 trillion. Table 13.9 shows the 2015 Global Brand Index produced by WPP, a market research company. This can be analysed for evidence of the continued dominance of Western (European and North American) brands. There are also signs that non-Western brands are beginning to influence global culture too. Can you spot them?

Table 13.9 Top 30 global brands in 2015

1	Apple	11	Tencent	21	Baidu
2	Google	12	Facebook	22	ICBC
3	Microsoft	13	Alibaba	23	Vodafone
4	IBM	14	Amazon	24	SAP
5	Visa	15	China Mobile	25	American Express
6	AT&T	16	Wells Fargo	26	Wal-Mart
7	Verizon	17	GE	27	Deutsche Telekom
8	Coca-Cola	18	UPS	28	Nike
9	McDonald's	19	Disney	29	Starbucks
10	Marlboro	20	Mastercard	30	Toyota

The costs of cultural erosion

The idea that a largely Westernised global culture is emerging as a result of cultural erosion in different places is called *hyperglobalisation*.

One viewpoint sees this as a negative development. Pessimistic hyperglobalisers are concerned that languages around the world are disappearing as use of English continues to spread. They also fear a global trend in the

Indigenous people of Amazonia and Papua New Guinea

Amazonia and Papua New Guinea's tropical rainforest tribes are among the world's last isolated groups of indigenous people. These ethnic groups have occupied the place where they live for thousands of years without interruption. More members of rainforest tribes are becoming aware of Western culture and lifestyles, however (Figure 13.14). Due to the tropical climate, indigenous people traditionally wore little in the way of clothing. Today, many Amazonians and New Guineans are wearing modern, Westernised clothing. The T-shirt has become ubiquitous.

Increasingly, many young Amazonians are moving from the rainforest to urban areas like Manaus. They leave behind their traditional thatched homes, often built on stilts.

One outside view of the changes is that indigenous people no longer value local ecosystems the way they used to, on account of cultural erosion. Like people everywhere, they want income, education and health improvements for their children. Inevitably, social goals are becoming more important and this can drive indigenous people to hunt endangered species for food or to sell. Papua New Guinea's Tree Kangaroo is under threat; so too are Peru's jaguars.

Figure 13.14 Indigenous people are increasingly exposed to global culture

Athletes at the Rio 2016 Summer Paralympic Games

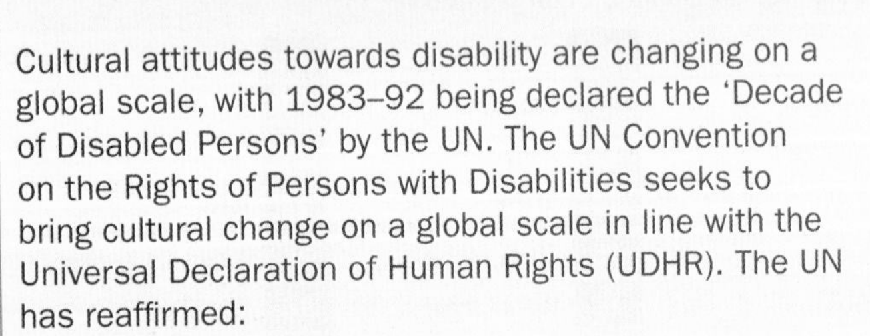

Cultural attitudes towards disability are changing on a global scale, with 1983–92 being declared the 'Decade of Disabled Persons' by the UN. The UN Convention on the Rights of Persons with Disabilities seeks to bring cultural change on a global scale in line with the Universal Declaration of Human Rights (UDHR). The UN has reaffirmed:

> *'the universality, indivisibility, interdependence and interrelatedness of all human rights and fundamental freedoms and the need for persons with disabilities to be guaranteed their full enjoyment without discrimination.'*

It was not always the case that disabled people enjoyed equal rights. In the USA, sterilisation programmes that sometimes targeted disabled people lasted until well into the twentieth century.

Since then, a seismic shift in cultural attitudes has taken places in the USA and elsewhere. Global media has helped turn the Paralympic Games into one of the world's biggest sporting events by celebrating the physical achievements of elite athletes with disabilities (Figure 13.15).

Sporting events specifically for those with disabilities first began in 1948 with Second World War veterans participating. The first official Paralympic Games were held in 1960 in Rome with participants from just 23 countries. Today, the event has grown significantly and athletes from 159 nations took part in the fifteenth Summer Paralympic Games in Rio, 2016 (with 107 medals, China was the winner).

Attitudes towards disability are changing in more and more places.

Figure 13.15 The Paralympic Games

evaluing of ecosystems, at varying scales. The capitalist hilosophy behind globalisation makes economic growth s primary goal; economies which do not grow are eemed failures. This global culture of consumerism fundamentally at odds with sustainable development oals: further devaluing of nature becomes inevitable.

Iowever, the opposing viewpoint is that globalisation nd cultural erosion can bring positive change on a vorldwide scale. Optimistic hyperglobalisers see merit n the emergence of a global culture that values equality, reedom of expression and reduced discrimination on he grounds of gender, sexuality or disability.

Resisting and reacting to cultural and environmental change

At a range of geographical scales, there is opposition o globalisation. Individuals, pressure groups and governments may all experience some degree of concern with the cultural impacts of globalisation, n addition to its social, economic or environmental consequences.

Cultural change is associated with an entire spectrum of reactions (Figure 13.16). In the case of North Korea, almost *all* change is resisted (see Chapter 12). In other cases, the reaction has been more selective (Table 13.10).

One positive way in which local culture can be preserved is through the UNESCO World Heritage Site (WHS) list. Since 1972, the UN has given special recognition to places that have unique cultural or physical significance. Over 1000 of these important places and cultural landscapes have now been recognised. Examples in the UK include the Liverpool waterfront, the city of Bath and the 'Jurassic Coast' which is valued for its physical geography. Due to their designation, policies have been established that protect these important places from too much change.

Table 13.10 Selected reactions against globalisation

France	France is fiercely protective of its culture and language, particularly in a world heavily influence by the internet and the English language. The French government is extremely supportive of French filmmakers and subsidises works filmed in the French language. Under local content law, 40 per cent of television output must consist of French productions. French language music is heavily promoted on radio stations.
China	The 'great firewall of China' prevents internet users from using BBC or Facebook services. China's government sets a strict quota of 34 foreign films a year. Western culture is still gaining a foothold in China though. Many Chinese people now celebrate Christmas as a good time for friends to get together.
Nigeria	Reports of serious degradation of Nigeria's Ogoniland due to oil spillages first began to emerge. Indigenous writer Ken Saro-Wiwa led the protests that gained media attention; he was executed by Nigeria's government in 1995, causing an international outcry. Since then, oil firms including Royal Dutch Shell and ExxonMobil have been accused by Amnesty International of bringing great environmental damage to Nigeria and other countries.

Key term

Cultural landscape: The landscape of a place that has been shaped over time in characteristic ways by the combined action of natural and human processes.

Progressive acceptance of new diaspora/ immigrant cultures	**'Melting pot' (or hybridism)**	Positive view of American culture as organic or hybrid — it adopts and absorbs new migrant values
	Pluralism	EU nations tolerate equal rights for all migrants to practise their religious and cultural beliefs
Cautious acceptance of diaspora/ immigrant culture with some controls	**'Citizenship' testing**	UK rules for migrants are becoming stricter in reaction to popular concerns over immigration
	Assimilation	A belief that minority traits should disappear as immigrants adopt host values
	Internet censorship	Preventing citizens from learning about other global viewpoints using online sources, e.g. China
Resistance to increased cultural diversity (right-wing view)	**Religious intolerance**	Notably lower levels of religious freedom for minority groups exist in some places, e.g. Iran
	Closed door to migration	Stopping any immigration altogether for fears of cultural dilution, e.g. Cambodia (the Pol Pot years)

Figure 13.16 The cultural continuum: differing responses to cultural diversity and change

Review questions

1 Using examples, compare the effects of global shift on (i) the economies and (ii) the environments of developed countries and developing countries.

2 Using examples, explain how globalisation has affected the scale and rate of economic migration flows (i) within countries and (ii) between countries.

3 Study Table 13.7. Using this information and your own knowledge, construct an alternative table with three columns labelled 'social impacts', 'economic impacts' and 'environmental impacts' of migration.

4 Explain what is meant by a global culture. To what extent is it inevitable that many of the world's local cultures will disappear from different places over time?

5 To what extent do you agree with the views of either the 'pessimistic' or 'optimistic' hyperglobalisers? Refer to a range of examples and evidence to help support your viewpoint.

Further research

Find out more about the rise of the global middle class: www.bbc.co.uk/news/business-22956470

Study changes taking place in India: www.bbc.co.uk/news/world-south-asia-12557384

Explore the culture of threatened indigenous people: www.theguardian.com/artanddesign/2015/may/24/photographing-the-omo-valley-people

Learn more about the world's vanishing languages and languages at risk: http://ngm.nationalgeographic.com/2012/07/vanishing-languages/rymer-text

The development and environmental challenges of globalisation

What are the consequences of globalisation for global development and the physical environment, and how should different players respond to its challenges?

By the end of this chapter you should:

- understand how globalisation has complex outcomes for development and the environment at different geographical scales
- be able to assess the tensions for individuals and societies resulting from the rapid changes that globalisation brings to places
- be able to explain the importance of the concepts of sustainability and localism.

14.1 Globalisation, development and the environment

Critics of globalisation assert that 'the rich get richer while the poor get poorer'. Is this true? Oxfam calculated that the richest one per cent have seen their share of global wealth increase from 44 per cent in 2009 to 99 per cent in 2016. The result is an 'explosion in inequality' at a time when 1 billion people still live on less than US$1.25 per day. This indicates that development gap *extremities* (the range of values between the world's very richest and poorest people and countries) have increased.

However, it does not necessarily follow that poorer people must actually *lose* money in order for the rich to make gains; globalisation is not a 'zero-sum' game:

- Overall, the global economy has grown enormously, far faster than its population (Figure 14.1).
- The already-rich do, however, take a disproportionately large share of each year's *new* economic growth. They are in the best position to invest their capital in new opportunities, such as rising property prices in global hubs like London or Beijing. These tripled in value between 2005 and 2015.
- In contrast, the wealth and incomes of the majority of people have grown more slowly over time.

Globalisation has *not* pushed large numbers of people into absolute poverty (measured in 2016 as earning less than US$1.25 per day). Instead, global poverty has been halved since the introduction of the Millennium Development Goals in 2000, with the greatest progress being made in Asia. It is true that the number of people living in relative poverty has risen in many societies, however. When the assets and earnings of the hyper-rich balloon in value, the *average* (per capita) level of wealth rises. As a result, some poorer people – whose earnings

Key terms

Absolute poverty: When a person's income is too low for basic human needs to be met, potentially resulting in hunger and homelessness.

Millennium Development Goals: Eight specific objectives for the global community created at the UN Millennium Summit in New York in 2000.

Relative poverty: When a person's income is too low to maintain the average standard of living in a particular society. Asset growth for very rich people can lead to more people being in relative poverty.

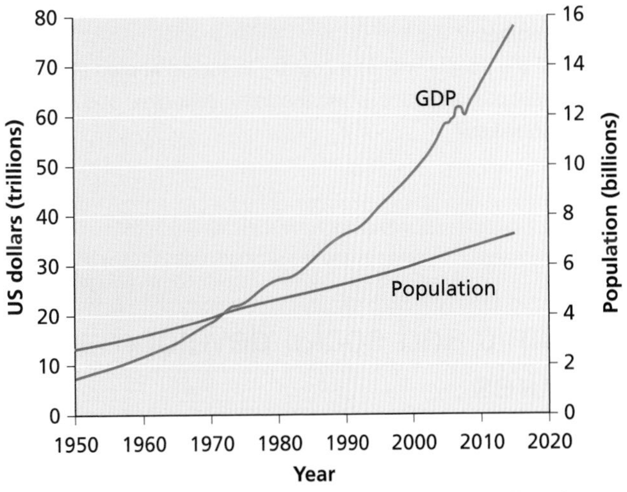

Figure 14.1 Comparing growth in world GDP (nominal) and world population, 1950-2015

are static or have risen modestly – are reclassified as having below-average incomes despite the fact they have experienced no material decline in wealth.

To summarise, 'the rich get richer while the very poorest do not' is perhaps a more accurate view.

Key concept: Development

Development generally means the ways in which a country seeks to progress economically and to improve the quality of life for its inhabitants. A country's level of development is shown firstly by economic indicators of average national wealth and/or income, but encompasses social and political criteria also. Figure 14.2 shows the 'development cable'. It presents the development process as a complex series of interlinked outcomes for people and places. In summary, it shows that in an economically developed society:

- citizens enjoy health, long life and an education that meets their capacity for learning
- citizenship and human rights are more likely to be established and protected.

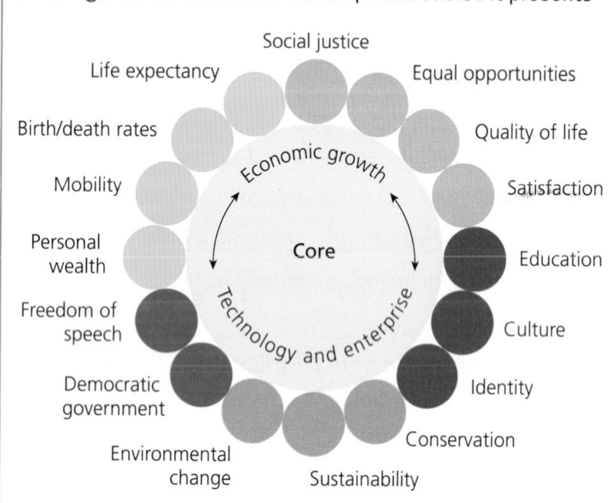

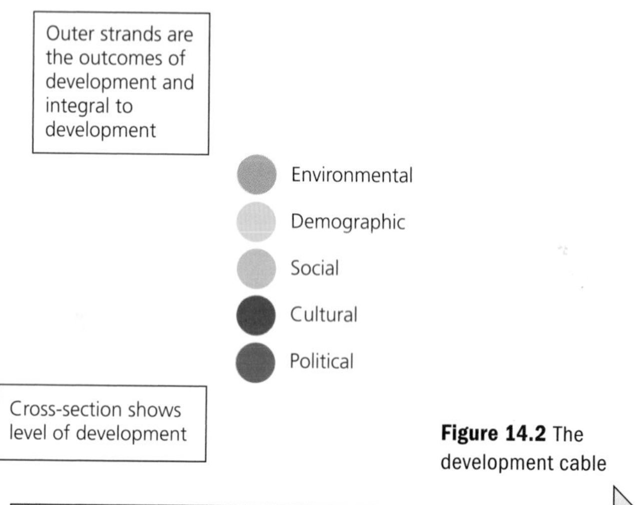

Figure 14.2 The development cable

Figure 14.3 shows the social changes that sometimes follow when the world's poorest farmers receive a boost in earnings. It highlights how different developmental changes are interlinked.

In development studies, the concept of geographical **scale** is important. Important questions to ask when investigating a country's level of development are:

- Do all local places *within* a country have the same level of development?
- Do all the people *within* a country, including women and different ethnic groups, share the same economic and social opportunities?

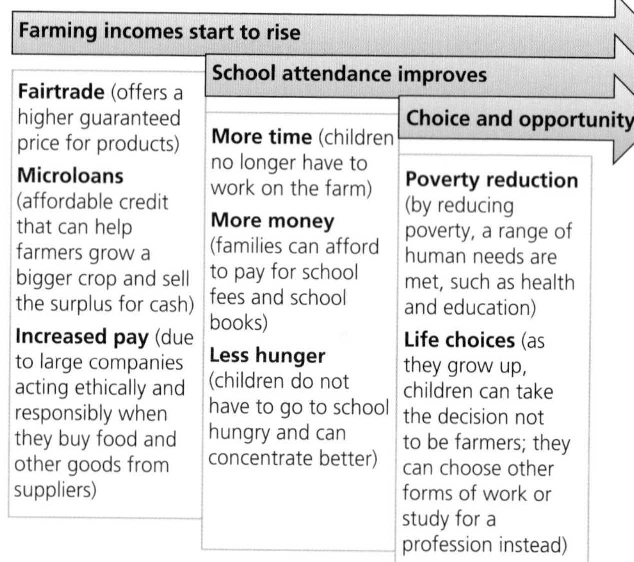

Figure 14.3 Economic and social development linkages

Economic and social development measures

Development is measured in many different ways using both single and composite (combined) measures. When assessing the value of different measures it is helpful to distinguish between issues of validity and reliability:

- For a measure to be *valid*, there should be broad agreement that it has relevance (do you agree that political corruption should be used as a measure of development, for instance?).
- To be *reliable*, a measure must use trustworthy data (do you think all countries' income and employment data are accurate?).

ıcome per capita and GDP

ıcome per capita is the mean average income of a group f people. It is calculated by taking an aggregate source of ıcome for a country, or smaller region, and dividing it by opulation size to give a crude average (which can give a ıisleadingly high 'typical' figure if large numbers of high-arners inflate the mean). GDP is a widely used aggregate ıeasure. It is the final value of the output of goods and ervices inside a nation's borders. Each country's annual alculation includes the value added by any foreign-owned usinesses that have located operations there. The World Bank recently estimated global nominal GDP in 2014 at bout US$78 trillion. Using this figure, can you make an stimate of global GDP per capita?

Estimating GDP is not easy because the earnings of every itizen and business need to be accounted for, including llegal or unregistered work in the informal sector. To ıake comparisons, each country's GDP is converted into JS dollars. However, some data may subsequently become ınreliable because of changes in currency exchange rates. Each country's GDP data is additionally manipulated to actor in the real cost of living, known as purchasing power arity, or PPP. Simply put, in a low-cost economy, where goods and services are relatively affordable, the size of its GDP should be increased and vice versa.

Economic sector balance

A country or region's economy can be crudely divided nto four economic sectors whose relative importance hanges as a country develops (Figure 14.4). Chapter 17 xamines industrial change over time in greater detail including the role of rebranding strategies). A country's conomic sector balance is also used as part of the ınnual GDP calculation. Every few years, each country levises a new formula that estimates the contribution hat different economic sectors, such as agriculture ınd manufacturing, make to total national income. For instance, Nigeria 're-based' its economic sector alculation in 2013 by reducing the share of agriculture rom 35 per cent to 22 per cent of its GDP. At the same ime, ICT services were increased from one per cent to ıine per cent of GDP, while Nollywood film earnings were included in the calculation for the first time. As a esult of these changes, Nigeria's nominal GDP value loubled overnight!

Key term

Informal sector: Unofficial forms of employment that are not easily made subject to government regulation or taxation.

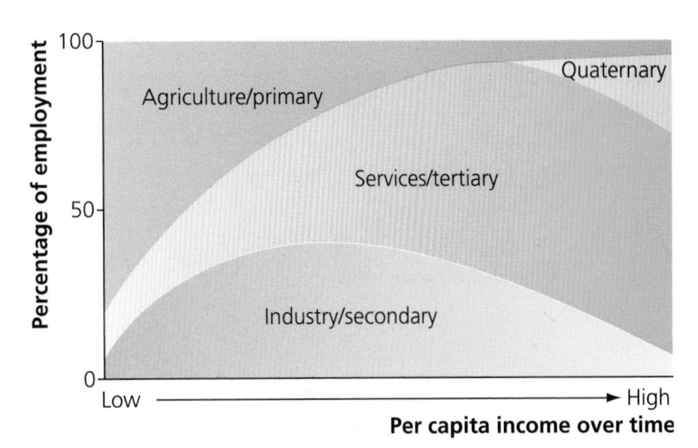

Figure 14.4 The Clark-Fisher model of economic sector change in a country over time

Human Development Index

The Human Development Index (HDI) is a composite measure that ranks countries according to economic criteria (GDP per capita, adjusted for purchasing power parity) and social criteria (life expectancy and literacy). It was devised by the United Nations Development Programme (UNDP) and has been used in its current form since 2010. The three 'ingredients' are processed to produce a number between 0 and 1. In 2014, Norway was ranked in first place (0.944) and Niger was ranked in last place (0.337).

Gender Inequality Index

The Gender Inequality Index (GII) is another composite index devised by the UNDP. It measures gender inequalities related to three aspects of social and economic development. These are:

- reproductive health (measured by maternal mortality ratio and adolescent birth rates)
- empowerment (measured by parliamentary seats occupied by females and the proportion of adult

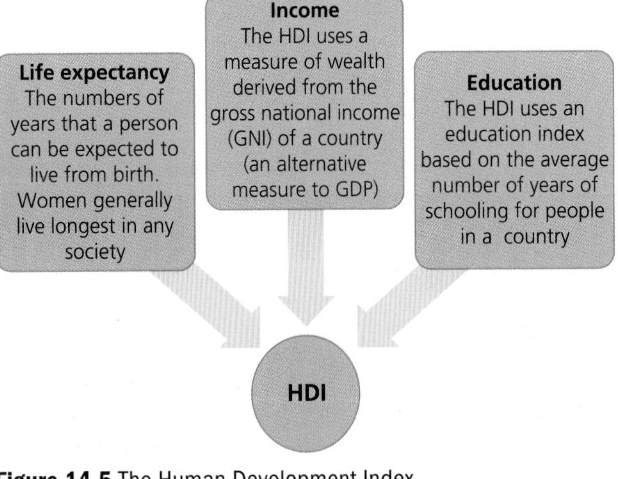

Figure 14.5 The Human Development Index

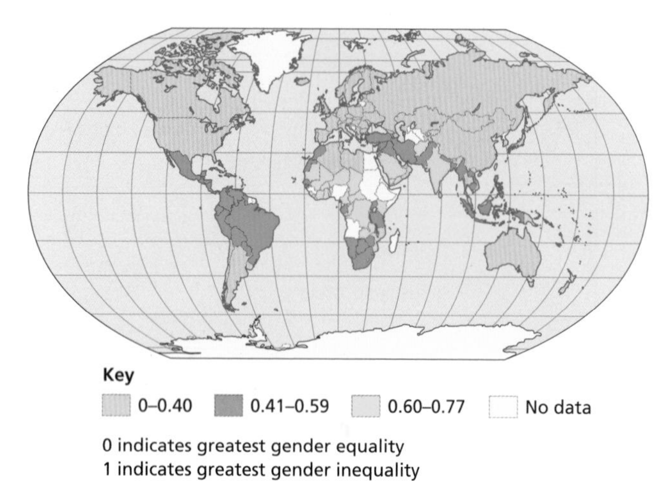

Figure 14.6 Global gender inequalities measured using GII, 2011

females and males aged 25 years and older with some secondary education)

- labour force participation rate of female and male populations aged fifteen years and older.

Internationally, views differ on the validity of gender inequality as a development measure. Why is this the case?

Environmental quality

Air pollution data show that environmental quality is often poor in developing and emerging economies. It usually improves as economic and social development occurs and places make the transition from industrial to post-industrial forms of economic activity. In 2014, the global Environmental Performance Index (EPI) used by Yale University ranked Ireland highly, with an air-quality index score of 98.3. Bangladesh was ranked in last place with a score of 9. The measurement takes into account the amount of pollution found outdoors and also the quality of air inside people's homes. Many lower-income countries score poorly on this index on account of the use of wood-burning stoves indoors. This is not always the case though: several Caribbean islands have very high air quality.

A multi-speed world at varying geographical scales

The steep rise in world money supply during the era of globalisation has been accompanied by a changing spatial pattern of global wealth:

- Average incomes have risen in all continents since 1950, but only very slowly in the poorest parts of Africa.
- The great gains made by European and North American nations over the same time period has resulted in a widening of the average income gap between people living in the world's wealthiest and poorest countries.
- Absolute poverty has fallen worldwide, but this statement hides the problem some countries still face.
- Many countries have advanced from low-income to middle-income status since the 1970s, resulting in a 'three-speed' world of developed, emerging and developing economies (Figure 14.7).

There is a growing wealth divide *within* nations. In China and Indonesia, the majority of people are better off than previous generations when their income is measured in real terms. Yet they are economically wors off than before *in relation to* the richest members of their society. The Gini coefficient is a useful analytical tool that can help us explore these patterns and trends further.

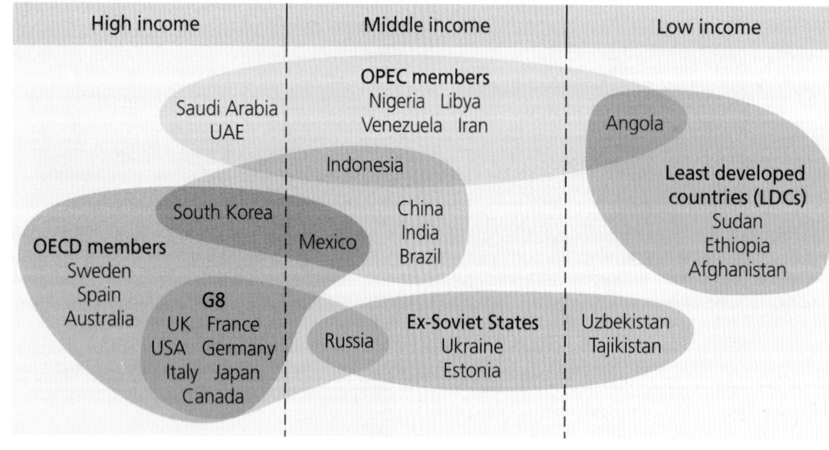

Figure 14.7 Economic groupings of nations, with examples of member countries

Environmental winners and losers

Major environmental issues are linked with globalisation, including climate change and biodiversity loss. Large-scale global flows of cheap food are good news for European and North American consumer nations. However, the transformation of 40 per cent of Earth's terrestrial surface into productive agricultural land has led to habitat loss and biodiversity decline on a continental scale. The negative impacts of large agribusiness operations penetrate deeply into many of the world's poorer regions, uch as east Africa and southern Asia (Figure 14.8). Intensive ash-cropping, cattle-ranching nd aquaculture bring damaging nvironmental effects ranging rom groundwater depletion to the emoval of mangrove forest (which ncreases flood risk in coastal areas).

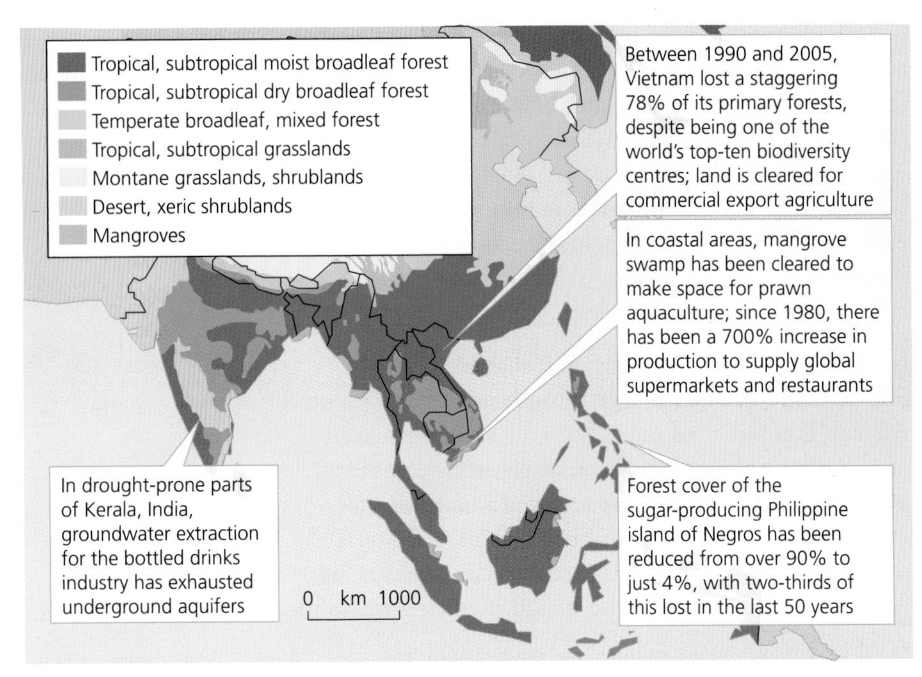

Figure 14.8 Impacts of agriculture, food and drink TNCs operating in parts of Asia

Development and environmental trends for global regions

Since 1970, the average income per capita in Asia has leapt over the absolute poverty threshold (driven in particular by the modernisation of Japan and South Korea). Asia's 2010 figure of roughly US$7000 per capita is equivalent to about US$20 per day, well in excess of the absolute poverty threshold of US$1.25 per day. Incomes in some African countries have remained closer to the poverty line (Figure 14.9). It is important to distinguish between different regions within Africa, however (Table 14.2).

Large income gains have been made in Tunisia, Algeria and other parts of the Maghreb region. North Africa in general

Skills focus: Gini coefficient

Table 14.1 shows how wealth varies between and within several nations. In addition, the Gini coefficient is provided for each. This is a number between 0 and 100. The higher the value, the greater the degree of income inequality. A value of zero suggests that everyone has the same income whereas a value of 100 would mean a single individual receives all of a country's income. When the Gini coefficient for each world continent is calculated, we find:

- Latin America is the most unequal region in the world (52), followed by Africa (44).
- Asia (37) and Europe (32) have the lowest Gini index scores.

Table 14.1 Wealth inequality between and within selected nations, 2014

HDI rank		GDP per capita (nominal, US$)	Share of national income going to *poorest fifth* of population (%)	Share of national income going to *richest tenth* of population (%)	Gini coefficient
High HDI					
12	Sweden	49,000	12	30	25
14	UK	46,000	6	29	38
Medium HDI					
91	China	7600	4	35	37
108	Indonesia	3500	8	29	33
135	India	1600	8	31	34
Low HDI					
180	Uganda	700	4	42	44
164	Burundi	300	7	33	33

is far more 'switched on' to globalisation (in part due to outsourcing by French TNCs). Elsewhere, some of Africa's coastal hubs, for example Lagos, Nairobi and Cape Town (Nigeria, Kenya and South Africa, respectively), are growth engines for a handful of important emerging economies. Economists predict a bright future for these places, provided political upheavals do not derail the economic development process. They expect strong African middle-class growth to rival Asia over coming decades.

In contrast, poverty remains entrenched in some Sub-Saharan countries, including Burundi and Central African Republic. In several places, a toxic cocktail of geographical isolation (lack of coastline), poverty and political extremism has resulted in falling life expectancy due to conflict. Human rights abuses are widespread and progress towards gender equality has been set back decades by the spread of extremist political organisations including the Lord's Resistance Army in Uganda and militant Islamists in Mali.

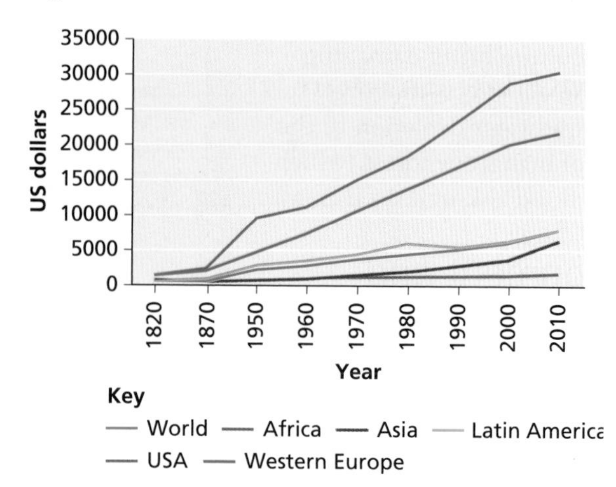

Figure 14.9 GDP per capita (PPP) growth for global regions since 1820

Table 14.2 Contrasting countries and regions of Africa

Country/region	Life expectancy	HDI rank	GDP per capita (nominal, US$)
Tunisia	76	90	4400
Egypt	71	110	3200
Libya	75	55	6600
Algeria	72	93	5400
North Africa (average)	**73**	–	**3400**
Chad	51	184	1200
Sudan	55	166	1900
Democratic Republic of Congo	52	186	430
Niger	59	187	460
Sub-Saharan Africa (average)	**62**	–	**1700**

Skills focus: Analysing environmental data using World Bank data sets

The World Bank website gives geography students an opportunity to study and learn independently:

- Follow the instructions at http://data.worldbank.org/about
- Experiment with the data sets to produce tables, line graphs and bar graphs showing environmental or development patterns and trends for selected regions and countries.
- To view more complex visualisations of World Bank and United Nations data, visit www.gapminder.org
- Figure 14.10 uses World Bank data to investigate patterns of biodiversity loss (plant species) at global and more local scales. Use the weblink to study the data online. You can change the time period or examine the data at varying scales: http://data.worldbank.org/indicator/EN.HPT.THRD.NO/countries/1W?display=map

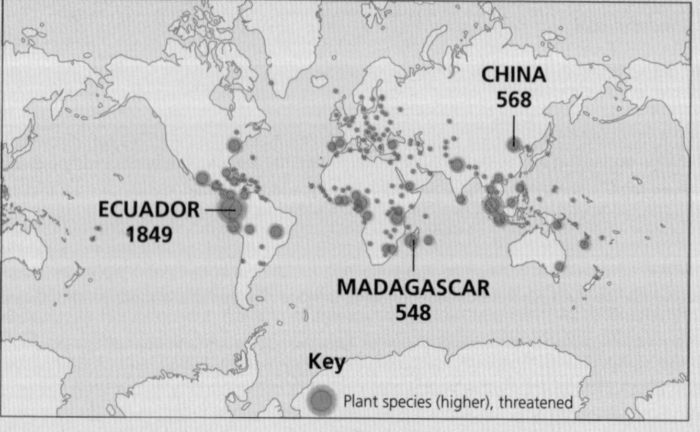

Figure 14.10 Proportional circles showing plant species under threat, 2015 (Source: World Bank)

.4.2 Social, environmental ınd political tensions :aused by globalisation

ſuman history is full of stories of places being changed y migration or new technology. For some societies, owever, the social changes accompanying globalisation :e unprecedented in their scale and rapidity. This has ›metimes created tensions.

/arying attitudes towards migration ınd cultural mixing

'he open borders of EU nations have brought rapid ultural change. In 2004, eight Eastern European nations ncluding Poland and Slovakia) joined the EU. An nprecedented volume and rate of post-accession in- ıigration followed for the UK and Ireland. The arrival f 1 million Eastern Europeans is an important reason hy the UK's population grew from 59.5 million to 64.5 ıillion between 2004 and 2015. Migrants have sometimes oncentrated in particular areas and enclaves. For instance, ew Polish migrants have joined a long-established iaspora community in Balham, London (first formed in he 1940s when the UK and Poland were Second World Var allies). Many non-Polish Balham residents have velcomed the new arrivals and their contribution to the ›cal economy and society. Small shopkeepers have visibly ourted Polish custom. However, some local people worry hat young migrants have increased the crude birth rate eyond the capacity of the area's primary schools.

n turn, a thriving British diaspora has seeded itself cross other EU countries (Figure 14.11). UK residents egan relocating to the Mediterranean coastline (France, taly and Spain) in 1993, when freedom of movement n the EU was first allowed. British enclaves can be dentified through local 'ethnoscape' features such as ars and cafes (Figure 14.12). However, poor behaviour f some younger 'Brits' has sometime strained cultural relations with indigenous communities (Calella, 48 km north of Barcelona, has suffered from this).

Key terms

Post-accession migration: The flow of economic migrants after a country has joined the EU.

Diaspora: The dispersion or spread of a group of people from their original homeland.

Crude birth rate: The number of live births per 1000 people per year.

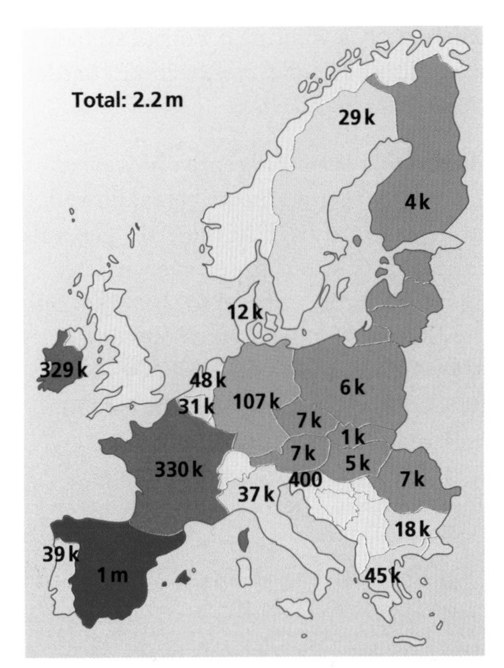

Figure 14.11 UK citizens living abroad in the EU, 2010

Figure 14.12 A British 'ethnoscape' in Spain

Tensions in London surrounding foreign investment and migration

The leaders of some of the UK's biggest TNCs have argued that migration restrictions threaten their own competitiveness and, more broadly, the UK's role as a global hub. Deregulation of the City of London in 1986 removed large amounts of 'red tape' for businesses (see Chapter 12). London's financial and legal firms began to regularly rotate staff between their different international offices in Asia, Europe and the Americas. Other companies have recruited large numbers of skilled people from overseas, such as Indian computer programmers. It is not just UK-headquartered TNCs that are alarmed by restrictions on migration. Indian,

Chinese and Brazilian TNCs wanting a European base may be less likely to choose London if it becomes harder to transfer staff to the UK.

Many other Londoners feel very differently, however. They believe too much in-migration has been allowed to take place. Some voted for UKIP in the 2015 general election. This political party wants to see even stricter controls on migration, including an end to the UK's full participation in the EU and the freedom of movement this has allowed. Approximately 30 per cent of London's 8 million residents were born in another country. On many London buses and trains you can now hear a variety of languages being spoken. Some Londoners judge the scale and rate of cultural change to have been too great.

Key terms

Nationalist: A political movement focused on national independence or the abandonment of policies that are viewed by some people as a threat to national sovereignty or national culture.

Post-colonial migrants: People who moved to European Countries from former colonies during the 1950s, 1960s and 1970s. The UK received economic migrants from the Caribbean (especially Jamaica), India, Pakistan, Bangladesh and Uganda.

Extremism in Europe

In some EU states, nationalist parties, such as France's Front National, command significant support. Nationalist parties often oppose immigration; some reject multiculturalism and openly embrace fascism. In the 1990s, the UK's British National Party voiced its opposition to the continuing presence of post-colonial migrants and their families.

Although race relations in the UK have improved over time, racially-aggravated assaults do, sadly, occasionally occur. The murder of Stephen Lawrence in Eltham in 1993 is one such example. Recently, tensions between some different communities have risen elsewhere in Europe. In France in 2015, staff of the satirical magazine *Charlie Hebdo* were killed by gunmen of Algerian descent. The murderers said that their Islamic faith had been mocked. Extreme events such as these are still rare but demonstrate tensions in multicultural Europe. Around 25 per cent of voters supported France's Front National party in the 2014 European Parliamentary elections.

Environmental tensions over water in south-east Asia

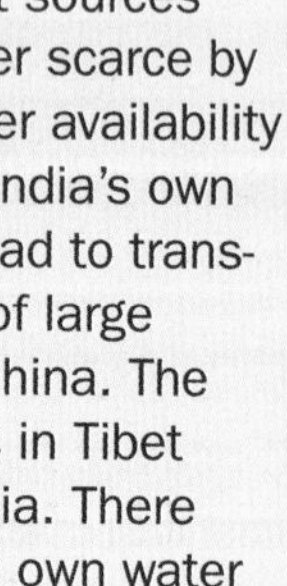

Trans-boundary water conflicts in south-east Asia can, in part, be linked with globalisation. In recent years, tension and conflict has grown between user groups both *within* and *between* countries.

- Globalisation has brought foreign investment to India while also has helping Indian-based TNCs such as Tata to thrive. These industries put pressure on water supplies. In the drought-prone Indian state of Kerala, an aquifer lies close to the village of Plachimada. In 2000, Coca-Cola's subsidiary firm Hindustan Coca-Cola Beverages established a bottling plant neraby. Six wells were dug, tapping into the precious groundwater store. Very soon afterwards, water shortages began to be reported.
- India's integration into global systems helps explain the income rise for hundreds of millions of Indians who now enjoy the use of flushing lavatories and showers. Improving the situation for the remaining 250 million Indians who still lack access to clean water will increase pressure on scarce water supplies in many places.

As a result of such changes, India's total demand for water is expected to soon exceed all current sources of supply; the country is set to become water scarce by the year 2025 (measured as per capita water availability of less than 1,000 cubic metres annually). India's own heightened water demands could, in turn, lead to trans-boundary tension and conflict over the use of large rivers shared with other countries, notably China. The transboundary Brahmaputra River originates in Tibet and flows through China before reaching India. There are real Indian concerns that China – whose own water needs have sky-rocketed recently – might build dams capable of diverting the Brahmaputra away from India.

Legislating against global flows

Governments may try to prevent or control global flows of people, goods and information, with varying success:

- Laws can be strengthened to limit numbers of economic migrants. However, illegal immigration is sometimes hard to tackle, as the USA has discovered. In 2015, large numbers of desperate refugees from Syria and poor African nations like Somalia arrived in Europe. Some had crossed the Mediterranean in overcrowded, leaky boats with great loss of life. Many more arrived at the borders of Hungary and Serbia, having walked there. European countries are obliged to take in genuine refugees, irrespective of economic migration rules. Since 1948, the Universal Declaration of Human Rights (UDHR) has guaranteed refugees the right to seek and enjoy asylum from persecution.
- Around 40 world governments limit their citizens' freedom to access online information. Violent or sexual imagery is censored in many countries. However, a 'dark web' also exists, which is harder to control.
- Trade protectionism is still common, despite the efforts of the Bretton Woods institutions (Chapter 12). Figure 14.13 shows extreme examples of trade protectionism that are in place around the world. Illegal smuggling of both legal and illegal commodities can be very hard to control, however. In 2014, global sales of illegal drugs are estimated to have exceeded US$300 billion.

Prohibited flows

Cuba → USA (until 2015)

The USA imposed a **complete trade embargo** on communist Cuba in 1962 as a result of Cold War antagonism between the two countries. The result? A commercial and financial blockade.

World → China

Not all **information flows** are allowed to enter China. For instance, internet users there are not allowed access to the BBC's Chinese-language website service.

Australia → New Zealand

For 50 years, imports of Australian **honey** were banned in New Zealand for fears of a 'bio-security threat' (Australia's bees suffer from a disease that New Zealand beekeepers have been keen to avoid).

China → Europe

In 2005, the EU briefly banned the further imports of **cheap Chinese textiles** — especially women's bras — in an attempt to protect its own manufacturers. This was dubbed 'bra wars' by the media.

Illegal and criminal flows

Afghanistan → UK

By some estimates, 60% of Afghanistan's GDP may come from illegal **opium trade**, feeding the demand for heroin among drug-users on the streets of European cities.

Colombia → USA

99% of **cocaine** reaching the USA is from Colombia, amounting to a billion-dollar trade. To fight this, the USA has in return given US$3 billion in mainly military aid to Colombia to fight the drugs trade.

Nepal → India

Girls as young as 10 years of age are kidnapped and taken to India where they are sold by **people traffickers** to brothels. There, the girls will work as prostitutes in a form of modern slavery.

Myanmar → Thailand

Each year, 100,000 **illegal migrants** escaping repression and poverty in Myanmar are intercepted by Thai border guards and promptly returned to Myanmar.

Figure 14.13 Prohibited flows and illegal flows

Internet censorship in China and North Korea

For nearly 70 years the People's Republic of China has been ruled by the Communist Party. China's rulers are intolerant of any criticism mounted against them by their own citizens. In 1988, Chinese students demonstrated against communism in Tiananmen Square: hundreds of people are thought to have died in the army crackdown that followed. Many Chinese people still do not know what happened that day, due to strict censorship of the press and internet. Google withdrew its services from China in 2010 when the Chinese government insisted that search engine results should be censored to hide information about Tiananmen Square.

Although Facebook, Twitter and YouTube remain unavailable there due to the 'great firewall of China', more than 400 million Chinese citizens interact with one another using local social media sites, such as Youku Tudou. In contrast, North Koreans have no access to the internet as a result of state controls. Restrictions on use therefore operate at two geographical scales: the national (China) and the personal (North Korea).

Migration controls in the UK

Since 2010, a five-tier point system has been in place in the UK designed to help control immigration by checking that economic migrants possess skills or resources that the UK economy needs. For example, tier 1 migrants must be prepared to invest more than £2 million in the UK or possess 'exceptional talent'. These rules do not apply to EU migrants, who are allowed free movement. The incoming UK government of 2010 pledged to cut net migration to 100,000 people a year. Figure 14.14 shows this target has not been met because:

- fewer British citizens have left the UK to live overseas since the 2008 global financial crisis (GFC). Also, the pound–euro exchange rate has weakened, meaning that the cost of living in the Eurozone has risen for UK citizens
- the government has no control over EU migrants wanting to work in the UK
- refugees are allowed to remain in the UK under human rights law.

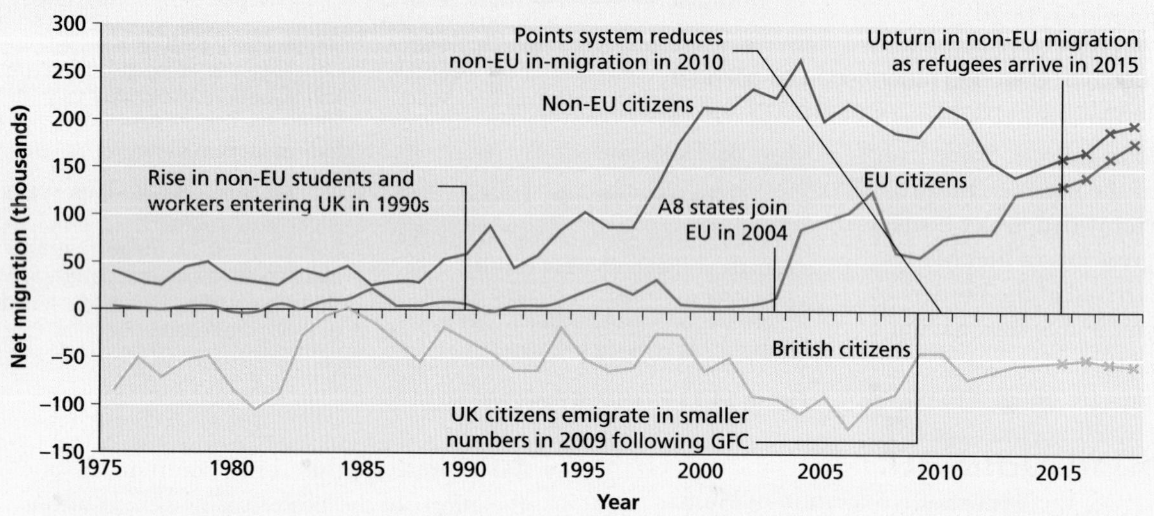

Figure 14.14 UK net migration, 1975–2015 (Source: Adapted from data from the Office for National Statistics Licensed under the Open Government Licence v.3.0)

Key term

Net migration: The overall balance between immigration and emigration.

Resource nationalism and protecting cultures

'Resource nationalism' describes a growing tendency for state governments to take measures ensuring that domestic industries and consumers have priority access to the national resources found within their borders. For instance:

- Hugo Chávez seized control of ExxonMobil and ConocoPhillips operations in Venezuela.
- In 2009, Canada-based First Quantum was forced to hand over 65 per cent ownership of a US$550 million copper mining project in the Democratic Republic of Congo to the country's government.
- Until recently, resource nationalism in China took the form of restrictions on rare earth exports. Japan, the USA and the EU all expressed concerns to the WTO. As a result, China finally relaxed restrictions in 2014.

Particular cultural groups within a nation may sometimes take a view on whether global forces should be allowed to exploit their resources. Opposition can be strong when an important landscape is threatened by the resource extraction process (see page 207). Examples include the Ogoni people's on-going struggle with oil companies in Nigeria and opposition to fracking (hydraulic fracturing) by Canada's First Nations people.

First Nations in Canada

Canada is home to six groups of indigenous people, known as the First Nations. Their occupation of the land long pre-dates the arrival of Europeans. Some First Nations people of the Mackenzie and Yukon River Basins oppose the attempts of global oil companies to 'switch on' their region (physically, an area of boreal forest and tundra). The Dene residents of the Sahtu Region have already experienced negative impacts of globalisation and petroleum development near the settlement of Norman Wells. Over 200 million barrels of conventional oil has been extracted there since 1920. Particular concerns include:

- the death of trout and other fish in oil-polluted lakes (a lifestyle based around subsistence fishing, hunting and trapping is fundamental to the Dene's cultural identity)
- the effects of alcohol and drugs (brought by oil workers) on the behaviour of young Dene people.

Oil TNCs, including Shell, ExxonMobil, Imperial Oil and ConocoPhillips Canada are now exploring the surrounding Canol shale and assessing its potential for shale oil (Figure 14.15). Shale 'fracking' (hydraulic fracturing) in other places has been linked with water pollution.

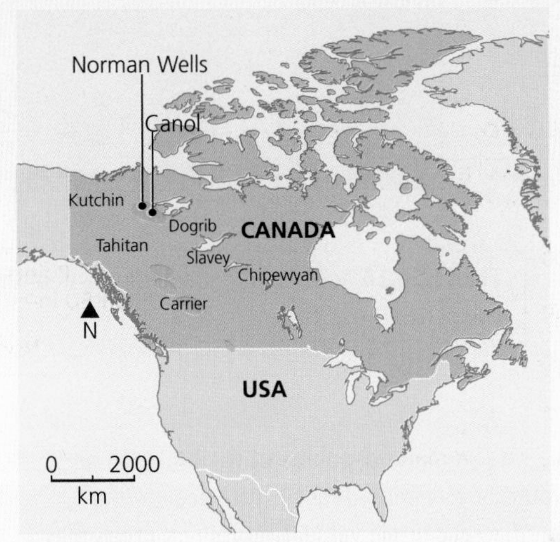

Figure 14.15 Unconventional fossil fuel resources in Canada

14.3 Globalisation, sustainability and localism

Globalisation is associated with a range of rising environmental stresses. These include growing food, water, energy and climate insecurity (Table 14.3). As a result of global trade and rising prosperity in emerging economies, almost 1 billion people in Africa, Latin America and Asia have attained 'new global middle class' status in the last 30 years; 2 billion more are on the cusp of it (Figure 14.16). Inevitably, this puts increased pressure on natural resources. Can the Earth cope with the growth of consumer societies?

The average US citizen has an ecological footprint twenty times larger than a subsistence farmer in Sub-Saharan Africa. In other words, the same area of land that supports ten US citizens with high-impact lifestyles supports 200 low-impact lifestyles. Increasingly, this 'US lifestyle' is an aspiration for people in developing and emerging economies. As

Table 14.3 Links between globalisation and rising environmental insecurity

Insecurity	Causes and symptoms
Food	By 2050, food demand is likely to double worldwide. Middle-class diets are characterised by their consumption of meat and dairy (Chapter 13) and have a larger ecological footprint.
Water	Food production also depletes water supplies. Animal husbandry and crop production can be water-intensive activities. Many increasingly popular global commodities, such as chocolate, coffee and wine, have a high water footprint. Additionally, as societies develop economically and urbanise, everyday household water use increases significantly.
Energy	A 50 per cent increase in global energy use is predicted by 2035. Unless significant innovation in renewable or nuclear energy is achieved, increased use of fossil fuels is inevitable, including 'dirty fuels' like oil shale. Their extraction scars local landscapes and threatens the transition to a clean energy economy.
Climate	The global diffusion and adoption of manufactured items, from energy-hungry televisions and fridges to throwaway plastic pens and bottles, has increased the average carbon footprint size of the Earth's population. In 2013, global concentration of carbon dioxide reached 400 parts per million. As a result, a harmful global temperature rise in excess of 2 °C now appears to be inevitable.

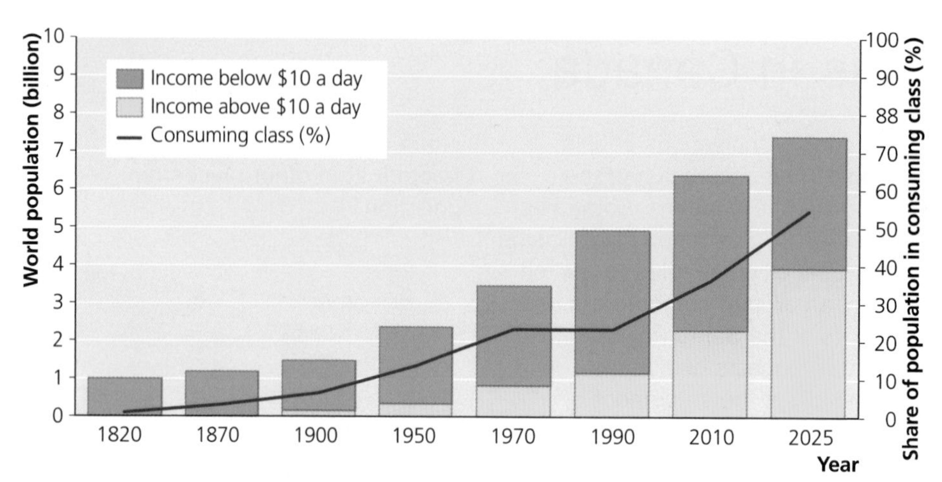

Figure 14.16 Actual and projected growth of the 'consuming class' or 'new global middle class', 1820–2025

Key terms

Natural resources: A material source of wealth, such as timber, fresh water, or a mineral deposit, that occurs in a natural state and has economic value. Natural resources may be renewable (sustainably managed forests, wind power and solar energy) or non-renewable (fossil fuels).

Consumer society: A society in which the buying and selling of goods and services is the most important social and economic activity.

Ecological footprint: A crude measurement of the area of land or water required to provide a person (or society) with the energy, food and resources needed to live, and to also absorb waste.

Water footprint: A measure of the amount of water used in the production and transport to market of food and commodities (also known as the amount of 'virtual water' which is 'embedded' in a product).

Carbon footprint: The amount of carbon dioxide produced by an individual or activity.

shown in Chapters 12 and 13, the combined power of TNCs, media corporations and new technology drives global consumerism. As a result, globalisation has become linked with a series of interconnected and thus far irresolvable 'twenty-first century challenges' (Figure 14.17).

The 'local sourcing' solution

The low pricing of containerised transport, allied with cheap labour and material costs in developing countries, has helped TNCs to develop extensive global production networks (Chapter 12). This strategy maximises profits, but it maximises carbon footprints too. Fiji water is a notorious example: this brand of bottled water is transported 20,000 km from Fiji to UK supermarkets. Critics of global capitalism say environmental costs should be accounted for in the pricing of products (a 'carbon tax' could be introduced, for instance). In the absence of this, some environmentally minded citizens adopt an

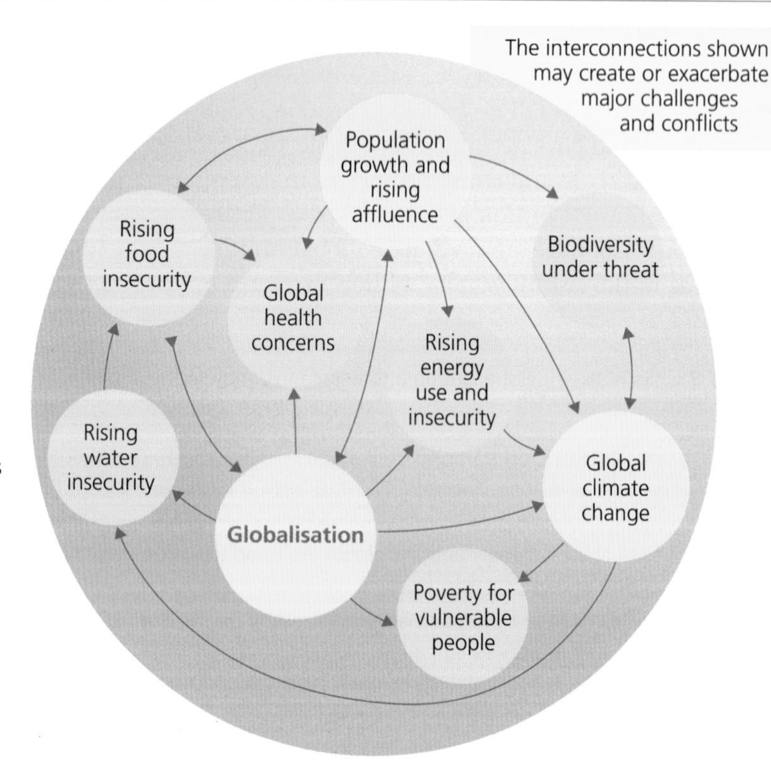

Figure 14.17 Globalisation is linked with many twenty-first century challenges for sustainability

Key concept: Sustainability

Widely adopted after the 1992 UN Conference on Environment and Development in Rio de Janeiro, the term sustainable development means:

> *'Meeting the needs of the present without compromising the ability of future generations to meet their own needs.'*

Three goals comprise sustainability, or sustainable development (Figure 14.18):

- **Economic sustainability:** individuals and communities should have access to a reliable income over time.
- **Social sustainability:** all individuals should enjoy a reasonable quality of life.
- **Environmental sustainability:** no lasting damage should be done to the environment; renewable resources must be managed in ways that guarantee continued use.

For the last of these goals to be met, there must be either a significant reduction in world economic output or new 'technological fixes', such as widespread carbon capture and storage. The former will not come voluntarily because a high economic growth rate is the goal of all free market economies.

As a result, government or corporate 'commitment' to sustainable development should, in general, be analysed with a critical eye. For instance, banning the use of throwaway plastic bags (as some governments have done) arguably does little to provide real environmental security for the future: far bolder moves are needed.

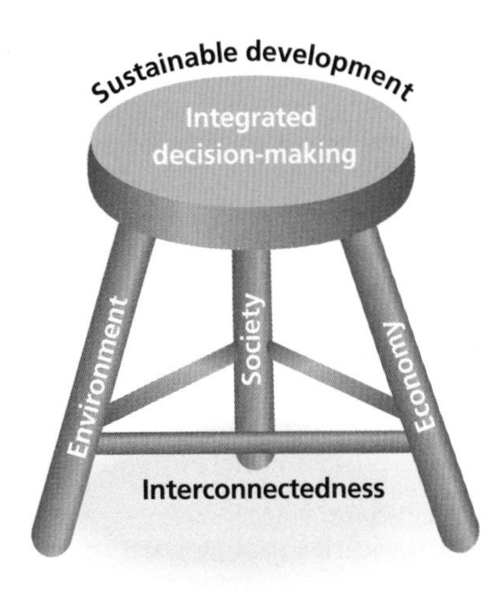

Figure 14.18 A model of sustainable development

thical consumption strategy by purchasing locally ourced food and commodities. They boycott upermarket products with high food miles. Local ressure groups play an important role in promoting ocal sourcing.

The Eden Project is a popular tourist site in Cornwall hat houses plants from all over the world in two normous biomes. For the 600,000 meals that are provided annually for visitors, 90 per cent of the produce s bought from local suppliers in Cornwall and Devon.

Locally-sourced produce is sometimes more expensive than globally-produced items. From an economic viewpoint, it is not 'rational' behaviour to purchase expensive items when cheaper ones are available. Pressure groups and NGOs such as Greenpeace counter this argument by saying that cheap imports of food and goods are *not* really cheap. This is because rising greenhouse gas emissions will generate long-term costs to society (an argument drawn from the 2006 Stern Review of climate change's long-term predicted impact on global GDP). For many people on low incomes, however, this is not necessarily

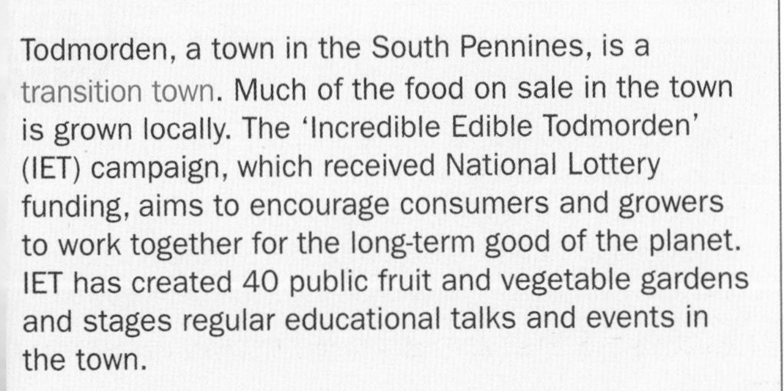

Todmorden: A transition town

Todmorden, a town in the South Pennines, is a transition town. Much of the food on sale in the town is grown locally. The 'Incredible Edible Todmorden' (IET) campaign, which received National Lottery funding, aims to encourage consumers and growers to work together for the long-term good of the planet. IET has created 40 public fruit and vegetable gardens and stages regular educational talks and events in the town.

Key terms

Food miles: The distance food travels from a farm to the consumer. The journey may be short and direct for some local produce, or may take longer, with food often crossing entire continents via a string of depots.

Transition town: A settlement where individuals and businesses have adopted 'bottom-up' initiatives with the aim of making their community more sustainable and less reliant on global trade.

Table 14.4 The costs and benefits of local sourcing

	Costs	Benefits
Consumers	Local sourcing of everyday meat and vegetables can be very expensive, especially for people on low incomes.	Many small producers in the UK have adopted organic farming methods. Crops are grown using fewer pesticides, which could have health benefits.
Producers	Less demand from UK consumers for food from producer countries means arrested economic development for places such as Ivory Coast.	UK farmers have moved up the value chain by manufacturing locally sourced items, including jams, fruit juices and wine.
Environment	Tomatoes in the UK are grown in heated greenhouses and polytunnels during winter, resulting in a larger carbon footprint than imported Spanish tomatoes.	The 1992 Rio Earth Summit introduced the slogan: 'Think global, act local'. Local sourcing sometimes helps people reduce their carbon footprint size.

a compelling argument. They may view locally sourced products as a 'middle-class luxury' (Table 14.4).

Ethical consumption and fair trade

While consumers benefit economically from global shift's cheap goods, many have ethical concerns about the social costs of worker exploitation. 'Opting out' of buying globally-sourced commodities is very hard to do in practice. However, ethical purchases are increasingly available thanks to the work of NGOs, charities and a growing number of businesses with a 'social responsibility' agenda (Table 14.5).

The Rana Plaza collapse

The Accord on Fire and Building Safety in Bangladesh is a significant recent development that shows Western retailers beginning to take more responsibility for working conditions in their supply chains. It was introduced following the collapse of the Rana Plaza building in Dhaka, Bangladesh, in 2013 (Figure 14.19). This led to the deaths of 1100 textile workers. On the day of the collapse, workers were sent back into the building to complete international orders in time for delivery, even though major cracks had appeared overnight in the building. Wal-Mart, Matalan and other major TNCs regularly outsourced clothing orders to Rana Plaza.

Since then, many British TNCs have signed the Accord, which is a legally binding agreement on worker safety.

Key term

Ethical purchase: A financial exchange where the consumer has considered the social and environmental costs of production for food, goods or services purchased.

Figure 14.19 The Rana Plaza factory collapse in Bangladesh

These companies now promise to ensure safety checks are carried out regularly in all Bangladeshi factories that supply them with clothes.

Recycling and resource consumption

At the end of their useful life, manufactured goods are often sent as waste to landfill. An alternative is to recycle them. This reduces the rate at which new natural resources are used. The recycling process does itself require the use of energy and water, however.

UK government actions

Local authorities in the UK run their own recycling schemes under Local Agenda 21 (established at the 1992 United Nations Conference on Environment and Development). In 2011, the Welsh Assembly banned shops in Wales from giving away free plastic bags. Instead, a 5 pence fee was introduced on all bags, both paper and plastic, provided by retailers. The charge was considered to be large enough to influence the behaviour of the shopper without harming trade for retailers. Consumers avoid paying for bags by simply reusing those they already have (though they must remember to take them). In 2013 Northern Ireland introduced a similar ban and charging scheme, followed by Scotland in 2014 and England in 2015.

able 14.5 Evaluating ethical consumption schemes

	Actions	Evaluation
Fairtrade	The Fairtrade Foundation's certification scheme offers a guaranteed higher income to farmers and some manufacturers, even if the market price changes. Examples of Fairtrade produce include coffee, chocolate, bananas, wine and even clothing items such as jeans. The Waitrose Foundation has also embraced fairer trading principles by improving pay for farmers in its own supply chains.	Fairtrade goods let shoppers know that what they spend will find its way into the pay packets of poor workers – but not all shoppers will pay more for it. However, as the number of schemes grows, it becomes harder to ensure that money has been correctly distributed. It is not possible for all the world's farmers to join a scheme offering a high fixed price for potentially unlimited crop yields.
Supply chain monitoring	Large businesses increasingly accept the need for corporate social responsibility. The largest TNCs have thousands of suppliers; this increases the risk of branded products being linked with worker exploitation. Apple investigated its iPhone touchscreen supplier, Lianjian technology, whose workers were poisoned by a chemical cleaning agent.	Firms such as Gap and Nike now prohibit worker exploitation in their own foreign factories, but it is hard to monitor the working conditions and pay for the workforce of every single supplier they buy from. It is especially hard to control what happens in the workplaces of their suppliers' suppliers (see Figure 12.22 p.176).
NGO action	Charity War on Want helped South African fruit pickers; it flew a woman called Gertruida to a Tesco shareholder meeting in London. Gertruida explained there was no toilet for female workers at the farm she worked at. Tesco told the farm it would use a different fruit supplier unless conditions improved.	NGOs have limited financial resources. This can limit the scale of what they can achieve, or result in slow progress. Although NGOs such as Amnesty International work hard to raise awareness of ethical issues, many people remain unaware of, or unconcerned with, worker exploitation.

The carbon footprint of recycled materials emissions is potentially high because energy is required:

- for the treatment of waste
- to transport waste to recycling sites (much of the UK's plastic, paper and glass waste is exported to China, for the same reasons that explain global shift, such as cheaper labour costs).

Beyond recycling

Recycling can be viewed as the first step towards the more ambitious goal of a circular economy (Figure 14.20). This approach to sustainable development calls for far more careful management of materials. The goal is to maintain or increase natural resource stocks by requiring manufacturers or retailers to do more to recycle, reuse or repair products that they ideally might lease, rather than sell, to consumers. Ultimately, this could result in a new global business model which will 'design out' waste altogether. Products would be designed in ways that allow them to be disassembled and reused or repurposed far more easily.

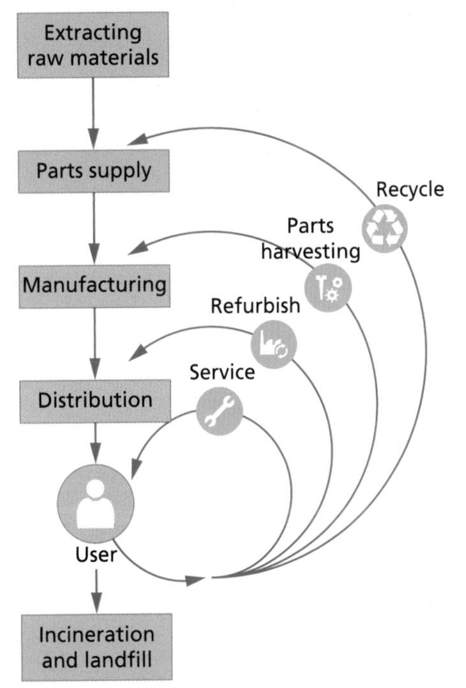

Figure 14.20 Recycling is just one element of a more ambitious 'circular economy'

The need to recycle could, in future, be reduced further through use of new substitute materials. Graphene is one newly invented 'super material' that is light and strong and may, in time, replace the need for older materials such as steel.

Assisting with progress towards a circular economy and new materials technology are countries like China and India. While the populations of the world's emerging economies have accelerated global consumption and waste production, they are also part of the solution. China especially is politically engaged with trying to tackle the sustainability challenges we all now face. Emerging economies are becoming leaders in resourcing efficient technology and renewable energy. China is the world's largest manufacturer of solar panels.

Review questions

1 a Explain how economic development can, in turn, help the social development process in different places.
  b In what ways can globalisation assist the development process?
  c Under what circumstances can globalisation hinder the development process?

2 Analyse the economic disparities that now exist between and within different African states.

3 Assess the strengths and weaknesses of local sourcing as a way of reducing some of the negative environmental impacts associated with globalisation.

4 'National governments have lost control of flows of people, information and goods across their borders.' To what extent do you agree with this statement?

5 Evaluate the strengths and weaknesses of different ethical consumption schemes. In your view, which scheme has the greatest potential for success?

Further research

Find out about globalisation and Canada's First Nations people: http://papers.ssrn.com/sol3/papers.cfm?abstract_id=2495900

Read about variations in development across Africa: www.undp.org/content/dam/rba/docs/Reports/MDG_Africa_Report_2014_ENG.pdf

Watch Hans Rosling explore global changes in social development: www.ted.com/talks/hans_rosling_shows_the_best_stats_you_ve_ever_seen?language=en

Exam-style questions

AS questions

1 Define 'shrinking world'. [1]

2 a Study Figure 1. Calculate the total volume of data flows linking the European Union (EU) with other regions. [1]

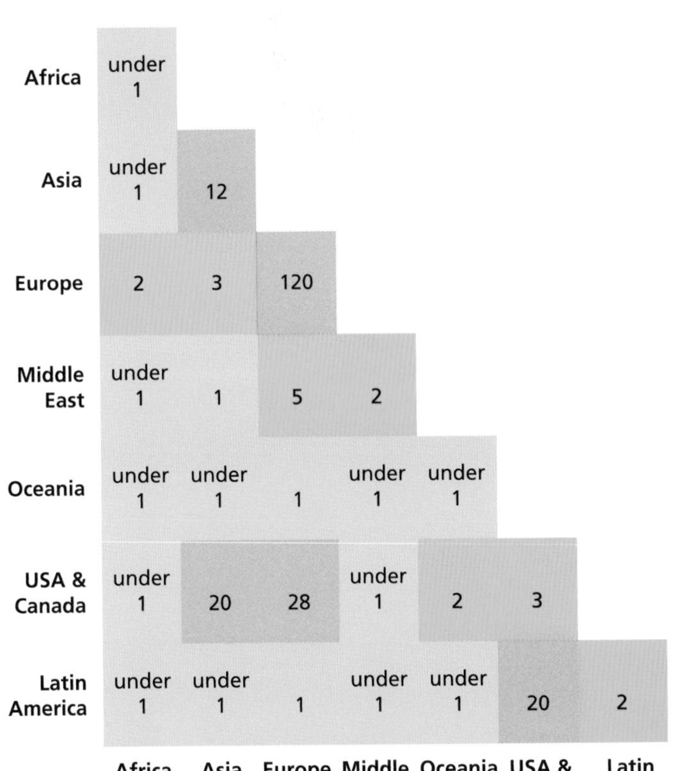

	Africa	Asia	Europe	Middle East	Oceania	USA & Canada	Latin America
Africa	under 1						
Asia	under 1	12					
Europe	2	3	120				
Middle East	under 1	1	5	2			
Oceania	under 1	under 1	1	under 1	under 1		
USA & Canada	under 1	20	28	under 1	2	3	
Latin America	under 1	under 1	1	under 1	under 1	20	2

Figure 1: Global data flows (thousand gigabits per second) between and within world regions, 2015 (Data source: McKinsey)

b Suggest **one** reason why the EU is a highly connected world region. [3]

3 Explain **two** ways in which individual states can benefit from trade bloc membership. [4]

4 Explain how the growth of a global culture may help improve opportunities for disadvantaged people in developing countries. [6]

5 Assess the extent to which globalisation is responsible for environmental degradation in developing and developed countries. [12]

A level questions

1 Explain why migration can lead to interdependence between countries. [4]

2 Assess the impact of international organisations on flows of free trade and foreign direct investment. [12]

3 Explain why some countries remain relatively 'switched off' from globalisation. [4]

4 Assess the extent to which ethical consumption schemes can minimise the environmental and social costs of globalisation. [12]

5 Assess the impact of globalisation on economic inequality both between and within countries. [12]

Topic 4
Option 4A: Regenerating Places

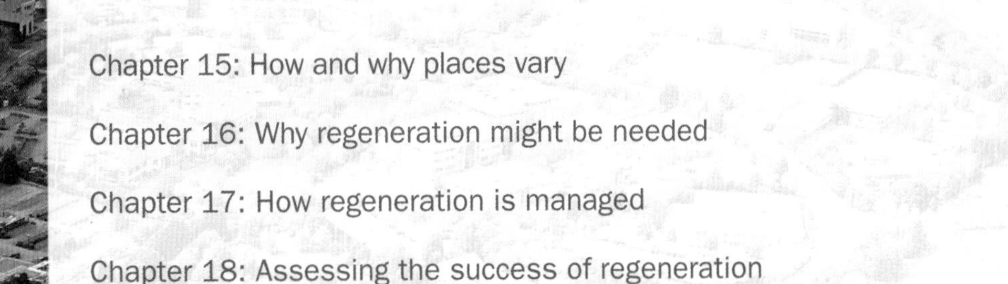

15 How and why places vary

An in-depth study of the local place in which you live or study, and one contrasting place

By the end of this chapter you should:

- be aware that economies vary
- understand how and why places have changed their functions and demographic characteristics
- be aware of the different ways these changes can be measured
- be able to compare how the identities of two specific places have been shaped by past and present connections at different scales.

Place and regeneration

The focus of this part of the specification is the economic and social changes affecting places, with an emphasis on the UK. You will also learn, in Chapters 16, 17 and 18, about the need for regeneration in some places and the effectiveness of policies designed to improve the quality of life for local people. A brief explanation of two key terms – place and regeneration – is needed.

Firstly, place may also be described as an area or location. They may be completely 'artificial' urban places or modified, as in rural landscapes, moulded by centuries of farming, forestry and mining. Places may vary in characteristics along a rural–urban continuum. Places are shaped by internal connections (between people, employment, services and housing) and external connections (such as government policies and globalisation). It is these linkages that drive much of the change that characterises a place. Place boundaries may be official, administrative ones such as an electoral ward or village boundary, or more functional, such as travel to work catchment areas.

A particularly important aspect of place is its *meaning*, to individuals and to defined groups of people. Meaning reflects how people perceive, engage with and form attachments to particular places. This means place boundaries may be perceptual as well as administrative or functional, distinctive in local people's minds and habits.

Places also vary in their *dynamism*, in other words, the rate at which they change. Smaller or more remote places may change socially and economically more slowly than larger cities, while villages close to cities will be affected by commuting. Places may lose or gain their attractiveness, value or attributes. This means there are differential needs for regeneration between places.

The changes may be driven by processes at three scales: local, national and global.

Key terms

Place: Geographical spaces shaped by individuals and communities over time.

Rural–urban continuum: The unbroken transition from sparsely populated or unpopulated, remote rural places to densely populated, intensively used urban places (town and city centres).

Figure 15.1 The rural–urban continuum of places, from countryside to megacity London

Key concept: Processes

These include the movements of people, capital, information and resources. These processes make some places socially and economically dynamic, while other places are left out or marginalised from wealth and opportunity. This can create, and exacerbate, economic and social inequalities both between and within local areas.

Synoptic themes:

Players

Regeneration involves a range of players (locals, planners, developers, pressure groups) who attempt to modify places to make them more productive and attractive places to live, work and use for leisure.

Key terms

Regeneration (or place making): Long-term upgrading of existing places or more drastic renewal schemes for urban residential, retail, industrial and commercial areas, as well as rural areas. This sometimes includes conservation to preserve a specific identity. It is connected with rebranding, which centres on place marketing, where places are given a new or enhanced identity to increase their attractiveness and socio-economic viability.

Quinary: The highest levels of decision making in an economy – the top business executives and officials in government, science, universities, non-profit organisations, healthcare, culture and the media. It is concentrated in STEM employment (science, technology, engineering and mathematics).

Regeneration is often designed to tackle inequalities in either urban or rural places, and make places economically productive and/or socially acceptable. Policies and programmes may have different impacts on people's lived experience of change and their perception and attachment to places.

15.1 Variations in places and economies

Classifying local economies

A key factor in the creation of 'place' is the structure of the local economy. This will affect directly and indirectly the income and lifestyle of individuals and communities and, hence, perception of place.

Employment sectors

You may already be familiar with the terms primary, secondary, tertiary and quaternary industries, and the reasons for the demise of employment in primary industries and manufacturing linked to deindustrialisation and global shift. Figure 15.2 summarises the key long-term changes in employment sectors in the UK. The quinary sector may be unfamiliar but is an important aspect of the increasing 'knowledge economy', creating prosperity in distinctive areas of the UK such as the Cambridge triangle, M4 corridor and London.

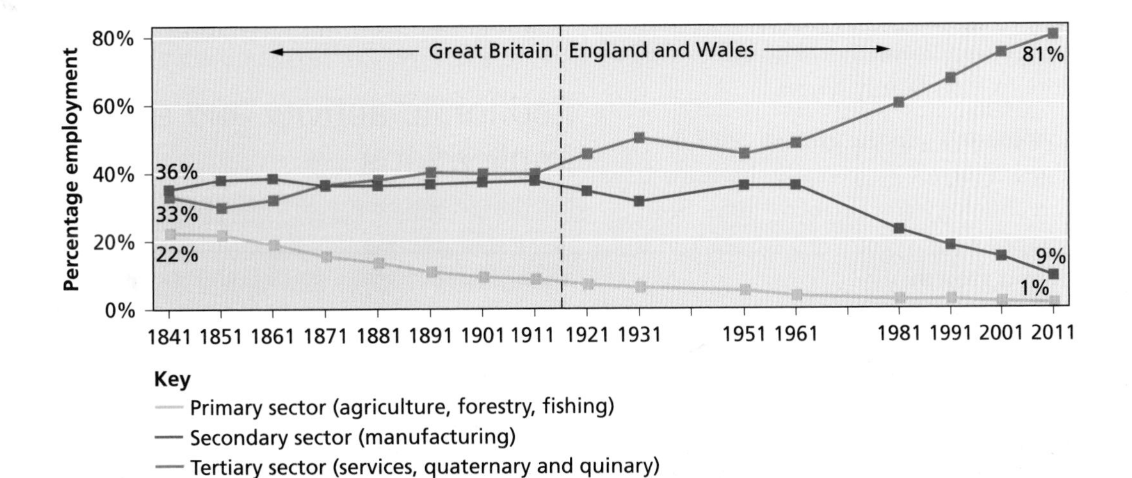

Figure 15.2 Sectoral change in the UK over 170 years (Source: Adapted from data from the Office for National Statistics licensed under the Open Government Licence v.3.0.)

Key concept: Economic activity

This may vary spatially both between places and within them, and can be classified in two main ways, by:

- **sector:** primary, secondary, tertiary and quaternary and, most recently, quinary
- **type of employment:** part/full time, temporary/permanent, employed/self-employed.

Skills focus: Categorisation

Line graphs like Figure 15.2 can help investigate temporal changes (over time). Categorisation of factors is an important skill, and geographers often use economic, social, political and physical categories.

Note that the data in Figure 15.2 excludes 'onshoring', the return of some previously deindustrialised functions (some manufacturing and call centres) back to the UK since 2011.

From 1948 the UK government has gathered data using a Standard Industrial Classification (SIC), with separate databases for England and Wales, Scotland and Northern Ireland. As tertiary industries, especially quaternary and quinary sectors, increase, social class is being replaced by levels of education and skills. Personal 'mobility' is now more dependent on access and opportunities for training than place of birth. This means that accessing higher levels of education at university and apprenticeships allows people from traditionally working class and unskilled families to access higher paid and skilled jobs.

Places have become less 'parochial', in other words, people are far less narrow minded and tied to their place of birth than a century ago. University graduates often settle in the place they were trained or where they find a job rather than close to their family home.

The process of globalisation and the growth of high tech and knowledge industries – especially the growth sectors of science, technology and finance – have been deliberately encouraged by governments, whose aim is economic stability and growth.

Places embracing growth employment sectors are able to become 'winners' in a competitive, interlinked world, for example Manchester, London and the 'M4 corridor'. Other places are relative 'losers', marginalised and even deprived in opportunity, facilities and standard of living, such as Cornwall. Some places may start as 'winners' and develop into more marginalised places, for example Teeside, and vice versa, for example the Lake District. Although larger cities occupy only 9 per cent of the UK's land area, they contain 54 per cent of the population, 60 per cent of jobs and 63 per cent of national economic output.

All these factors and trends will affect people's perception of the role they play in a place.

Employment type

In 2015 there were 32 million people in work in the UK, with 1.85 million unemployed (5.6 per cent). There are three main types of worker:

- employees with contracts (permanent or fixed); in 2015, 18.4 million people had full-time contracts and 9 million part-time contracts (a growing trend)
- workers (agency staff and volunteers)
- self-employed (freelancers, consultants and contractors).

There are several controversial aspects of work:

- The gender gap has narrowed but still exists; on average men are paid ten per cent more than women.
- Zero-hours contracts, designed for casual 'piece work' or 'on-call' work, mean no obligations by the employer or employee. The minimum wage, rebranded the 'living wage' in 2015, does now have to be paid, and in April 2016 the national living wage (NWL) was launched alongside the minimum wage policy in Britain. It is an increasingly popular form of work (used by companies such as Wetherspoon's, McDonald's and some councils).
- In 2015 the Government made illegal working a criminal offence in a crackdown on the black market. This relies on illegal migrant workers, often on very low pay and with poor conditions.
- Temporary and seasonal work usually has low pay, for example tourism and agriculture.

Economic activity and social implications

Economic activity in places has direct and indirect impacts on the key social factors affecting us all: health, life expectancy and levels of education. Economic activity may be measured by employment and output data (location quotients (LQ), gross domestic product and gross value added).

Social inequalities often result from concentrations. A large, high-LQ industry with a declining LQ over time may be detrimental to a local and national economy, for example the steel industry.

Key terms

Location quotient: A mapable ratio which helps show specialisation in any data distribution being studied. A figure equal to or close to 1.00 suggests national and local patterns are similar with no particular specialisation, such as retailing. LQs over 1 show a concentration of that type of employment locally.

Gross value added: Measures the contribution to the economy of each individual producer, industry or sector. It is used in calculating GDP.

Places specialising in modern high-tech industries, insurance and finance will generate 'new' money from their 'exports' and a positive spin off or multiplier effect on other services. There are distinctive patterns of certain economic sectors nationally, as shown in Figure 15.3. A North–South split may be identified in the location of manufacturing and financial services.

Such concentrations (for example in information and communication, insurance and high-tech industry) may cause congestion, overcrowding and increased house and land prices, as seen in the 'overheated' South East of England.

Deindustrialisation of the steel industry

Nationally, steel employs 30,000 people, often in areas with high unemployment rates. It supports many other manufacturers in the wider supply chain, including aerospace, defence and construction. However, in 2015 Thai-owned SSI at Redcar, Teesside, closed with 2000 redundancies. India's TNC Tata shut Scunthorpe's steel plant with 4500 redundancies. The branches of these TNCs were cut to reduce costs. Cheaper Chinese imports, high energy costs, green taxes and the strong pound were all factors. An estimated four other jobs will be lost for each steel worker redundancy as whole communities are affected (the negative multiplier effect). This demonstrates how original 'winners' may become 'losers' from external processes.

The overheated South

The lower relative importance of manufacturing for the economy of the South East means it has been less affected by deindustrialisation and recessions. During the economic boom from 1997 to 2007, the region generated 37 per cent of the UK's growth output. Since 2008, the region has increased to 48 per cent of growth output, while every other region, apart from Scotland, has experienced relative decline. This means that about a quarter of the population generates half of the UK's economic growth.

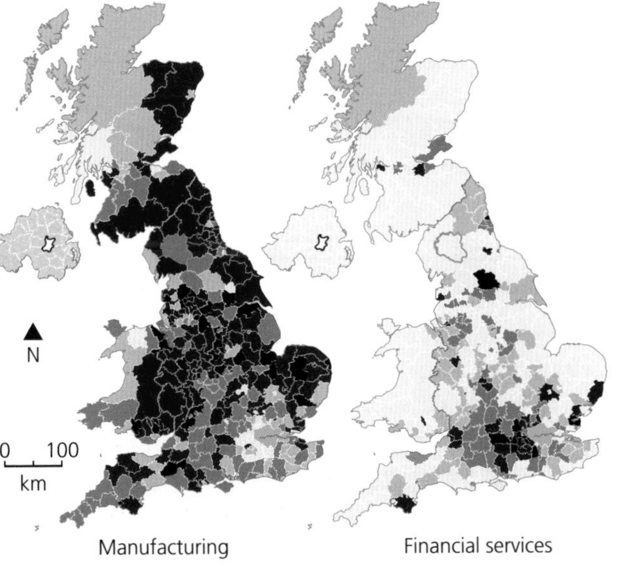

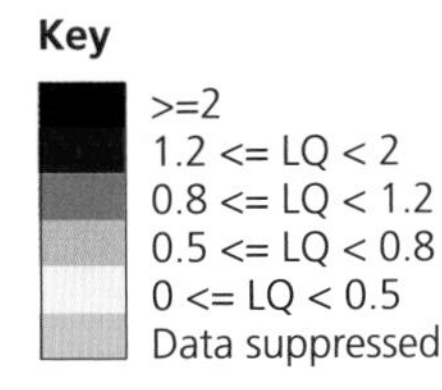

Figure 15.3 Choropleth 'heat maps' for manufacturing and financial services

Skills focus: Use of GIS

See Skill 7 in the Skills focus section (page 348).

Places needing regeneration may need to either increase economic specialisation or diversify their economic structure.

Differences in economic activities may be measured by variations in three social characteristics: health, life expectancy and education. GIS maps may help explore the spatial patterns of economic and social characteristics in your chosen place.

Health

Health may be measured by morbidity, the degree of ill health someone experiences, and longevity, how long a person's life expectancy is. There is a direct link between place, deprivation and associated lifestyles (Figure 15.4). There are many fewer 'blue collar' or manual jobs today in the UK, and far less pollution than the early twentieth century. However, those working long hours in manual jobs such as building and agriculture or exposed to harmful chemicals or pollutants will have a raised risk of poorer health and mortality.

Health is therefore linked to economic sectors and also the type of employment. Variations in income can affect the quality of people's housing and diets. Black and minority ethnic (BME) groups generally have worse health than the overall population, with one main driver being their often poorer socio-economic position.

A geographical factor is the spatial distribution of food. Some places, especially inner cities, may be 'food deserts' in terms of availability, with cheaper processed and take-away food dominating customer choice. Health may suffer as a result of access to food and lifestyle choices: obesity levels are soaring in the UK. In 2013 councils were given responsibility for encouraging people to stop smoking, eat better and drink less alcohol at a local scale, rather than just leaving it to national intervention polices.

Apart from environmental factors, population structure and lifestyle choices, there are also variations in healthcare nationally. The 2015 NHS Atlas of Variation highlighted the wide variations in healthcare: the so called postcode lottery.

Life expectancy

Longevity varies substantially between places, between regions, and both between and within settlements, especially larger cities. Average life expectancy in the UK is 77.2 years for men and 81.6 years for women. However, the 2011 census showed distinct North–South variations. In Harrow, northwest London, 65-year-old males can expect to live six years more than those in Glasgow – an example of the Glasgow effect. While much of the North East and North West have below average life expectancies of 75 years for men and 80 for women, Kensington and Chelsea in London, which has the highest rate of earnings (over £60,000 year), has rates of 80 and 85 years, respectively.

Gender (biological differences between the sexes), income, occupation and education are key factors, together with associated lifestyle choices, such as diet and smoking, although the longevity of males is improving generally (Figure.15.4).

The key factors that help explain these patterns are social (lifestyle choices and culture), economic (wealth of individuals) and locational (access to healthcare).

Education

Educational provision and outcome is also unequal in the UK. Outcome, measured by examination success, is strongly linked to income levels. Ofsted publishes data regularly to show the considerable regional variations in achievement nationally. Using data on free school meals, which are linked to low income, working-class white children in poverty have lower educational achievement and are more likely to continue to underachieve. By sixteen years old, only 31 per cent of this group achieved five or more GCSEs between A and C including English and Mathematics in 2013. Boys are more likely to have low results than girls, especially those of Bangladeshi, Pakistani and black African origin.

According to research by the Joseph Rowntree Foundation, only 14 per cent of variation in any individual's performance is due to the quality of the school attended. More disadvantaged children

Key terms

Postcode lottery: This refers to the uneven distribution of local personal health and health services nationally, especially in mental health, early diagnosis of cancer and emergency care for the elderly

Glasgow effect: The impacts of poor health linked to deprivation.

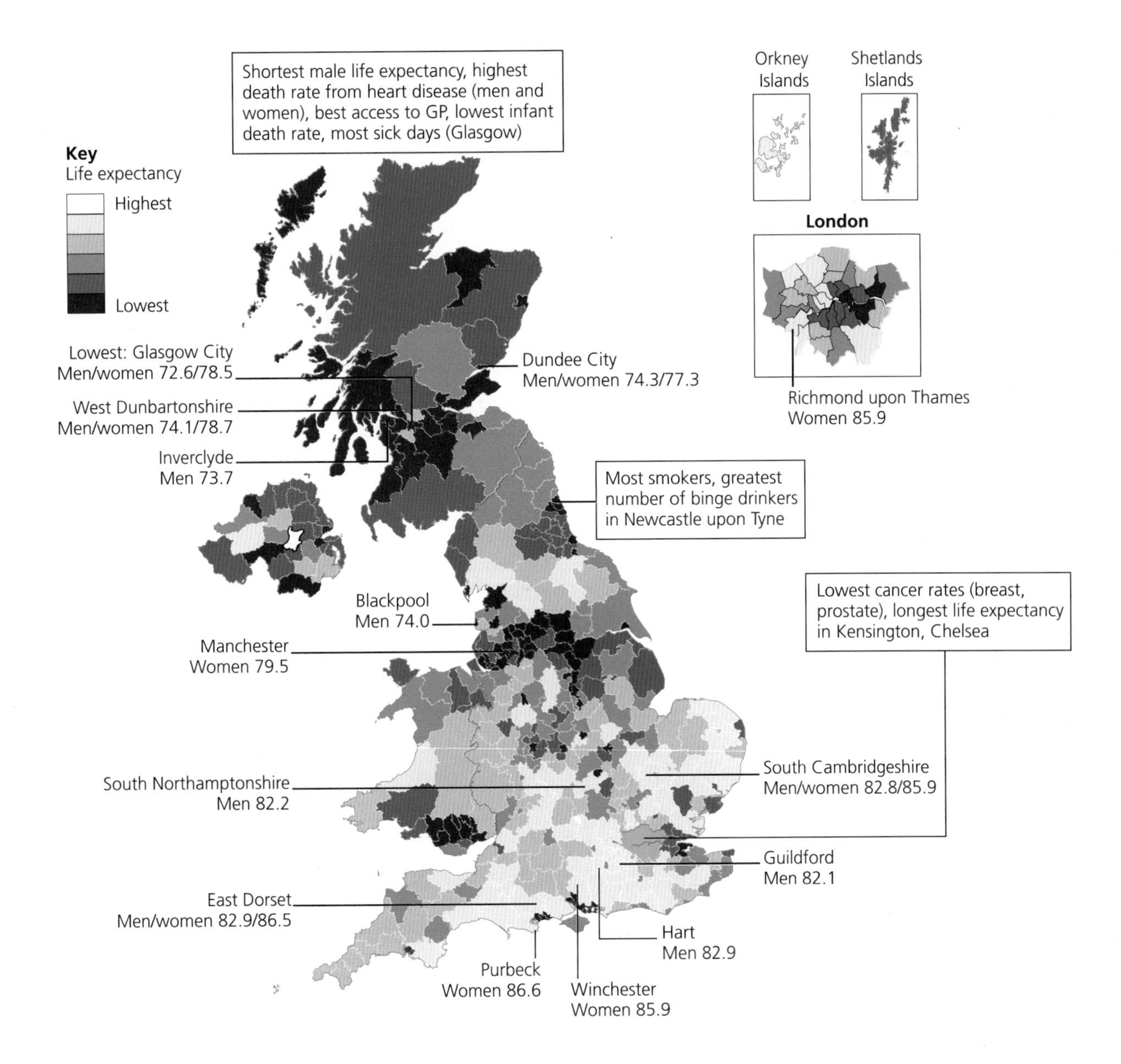

Figure 15.4 Health and life expectancy patterns

may feel a lack of control over their learning and may be reluctant to carry on to higher education academic studies.

Key concept: The intergenerational cycle

Educational underachievement and poor health may be intergenerational, meaning passed on from parents to their children. Breaking the cycle of poor educational achievement is a key goal of decision-makers.

Skills focus: Comparison statistical tables

Create simple comparison tables when profiling your local places. Use the headings of income, health, life expectancy and levels of education. The best starting point is the Office for National Statistics (ONS) and local authority websites. Public Health England and interpretations by media, for example the *Guardian* newspaper and BBC website, will help. Adding UK or England's averages may help gain a sense of perspective.

Inequalities in pay

Pay levels

Those working in the primary sector and low-level services receive lower pay than those in more skilled and professional sectors. Jobs may be seasonal and insecure compared with manufacturing and higher-level services. There is a huge disparity in incomes and cost of living nationally and locally. Prices for goods and services vary regionally. London and the South East are more expensive to live in than much of the rest of UK, hence the 'London allowance' in many jobs.

The richest one per cent of the population received thirteen per cent of all income and accumulated as much wealth as the poorest 55 per cent of the population put together in 2014. The top ten per cent of employees, mainly managers, directors and senior officials, earned over £53,248 annually. Their pay continued to rise through the 2008 recession and recovery. The UK now has the most billionaires per capita than any other country. Just five families control the same wealth as twenty per cent of the total population. This increasing minority have become super rich because of the opportunities provided by large, established transnational companies. Annual bonuses of over a million pounds are common for the elite executives of the FTSE 100 companies.

The bottom ten per cent of earners, with weekly wages of under £288, are concentrated in customer-service occupations such as carers. More than half a million people – over two per cent of the labour force – are on zero-hours or casual contracts. Many in this group lack savings and are forced into debt. The UK has record numbers resorting to food banks. The Trussell Trust recorded that the number of people receiving three days or more worth of emergency food increased from 26,000 in 2009 to over 900,000 in 2014.

Key concept: Inequality

High inequality in a place will almost inevitably reduce its potential for economic growth, and the benefits of growth do not necessarily trickle down across society. The Organisation for Economic Co-operation and Development (OECD) thinks that policies that help to limit or reverse inequality, targeting not just the bottom ten per cent but the bottom 40 per cent, may not only make societies less unfair but also wealthier. Regeneration programmes play a role in this.

Quality of life indices

The direct factors and processes contributing to quality of life and inequality, resulting from underlying factors, are shown in Table 15.1, together with some investigative profiling techniques.

Normally, but not always, poverty, inequality and quality of life rise and fall together. However, inequality can be high in a society without high levels of poverty, due to a large difference between the top and the middle of the income spectrum. This is very evident in London, and even small villages such as Easton outside Winchester with its influx of wealthy commuters, or Rock in Cornwall with its second home owners.

There are several respected international and national measurements or indices of quality of life; most well known is the United Nation's Human Development Index. Some focus on specific age groups, such as the Global Age Watch Index on the quality of life for the over 60s; however, for detail on places within the UK, the best statistical source is the ONS. Apart from its neighbourhood statistics and Index of Multiple Deprivation, it now produces an annual 'Measuring National Well-being: Life in the UK' index. It uses ten domains of well-being including 'health', 'where we live' and 'what we do'. The 2014 results showed the importance people put on economic security and job satisfaction, work-life balance, education and training, and local and natural environment. Figure 15.5 shows one result, on unemployment rates, regionally. You might be able to suggest some of the reasons for the differences shown after comparing it with the maps in Figure 15.4.

Key concept: Spatial inequality

This refers to differences across places at a neighbourhood or street level, as well as between cities or regions and countries.

Key term

Quality of life: The level of social and economic well-being experienced by individuals or communities measured by various indicators including health, happiness, educational achievement, income and leisure time. It is a wider concept than 'standard of living', which is centred on just income.

Table 15.1 Profiling the factors contributing to quality of life and inequality

Factors and processes	Fieldwork and secondary research opportunities
Economic inequality Employment opportunities, type of work and income	Employment/unemployment rates and types Average incomes Purchasing power (shopping basket surveys) Pound shop and loan shop surveys Land use surveys
Social inequality Segregation of people and marginalisation or exclusion of subgroups	Age, gender, health, longevity, disability and educational achievement data from ONS and health authorities www.police.uk and crime and design surveys ACORN, CAMEO, Zoopla's ZED Index Village/community centre activities, e.g. playgroups, elderly social groups, internet blogs and email lists Placecheck
Service inequality Health facilities, public transport, food may be unequally available and accessed	Functional surveys of services (high/medium/low order) Bus timetable survey Taxi and community shared transport surveys Supermarket and shop location survey
Environmental inequality Pollution levels, derelict land and access to open space have impacts on people's well-being	ONS: central heating provision Building quality surveys Environmental quality surveys on pollution, amount and access to parks and green space, graffiti, derelict land, litter

Skills focus: Questionnaires

Questionnaires using a mix of open and closed questions may be a useful source of data for all research into inequalities. Use a stratified sample and collect a large number. Questions could be framed using criteria in the sustainability survey called an 'Egan Wheel' (see page 268). This is a comparative technique needing changes over time, or you can compare your two chosen places.

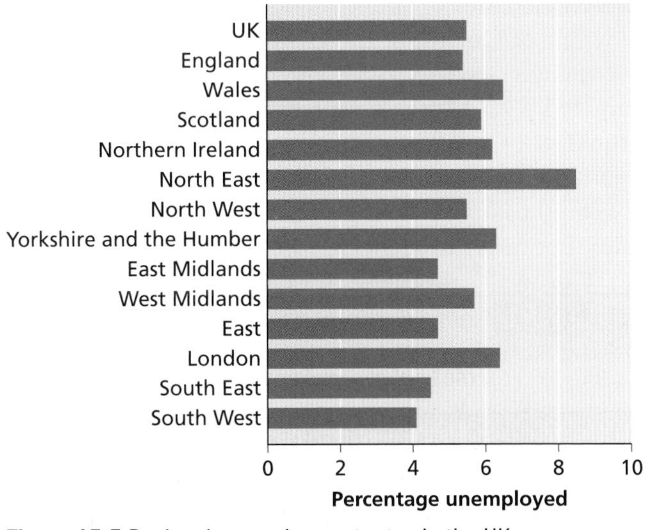

Figure 15.5 Regional unemployment rates in the UK

15.2 Changes in characteristics and functions

Places of all size and type, rural and urban, may be associated with one particular or dominant function, such as mining, steel, leisure and tourism, or a market. They will also have distinct demographic characteristics which are reflected in employment trends, land-use patterns and levels of inequality and deprivation. However, changes may occur in both functions and demography.

Key terms

Functions: The roles a place plays for its community and surroundings. Some, usually larger places, offer regional, national or even global functions. Functions may grow, disappear and change over time. There is a hierarchy according to size and number of functions.

Characteristics: The physical and human aspects that help distinguish one place from another: location, natural features, layout, land use, architecture and cultural traits.

Functional change

This section of the specification concentrates on four key functions: administrative, commercial, retail and industrial.

Historically, specialist functions such as banks, department stores, council offices and doctors surgeries are classed as high-order functions and located in larger settlements. More ubiquitous grocery stores, post boxes and pubs are classed lower-order functions and are found even in smaller villages.

However, the landscapes that these commercial functions have produced are rapidly changing because of internet and broadband services and changing customer habits. The retail landscape has transformed itself in the past decade with online shopping, click-and-collect and banking affecting city high streets. Regeneration may seek to counteract 'cloning' of land uses and encourage specific place identities to attract customers back. In rural settings, pubs may now double up as community centres, post offices and village shops. In both cities and in converted redundant farm buildings, small industrial units servicing light manufacturing, often high tech, have sprung up. There has been a large rise in small businesses nationally.

Functional areas are often different from administrative areas. This causes data comparison issues as well as obvious planning problems. Figure 15.6 shows how London and Leeds have much wider economic functional roles than their immediate administrative boundaries.

Skills focus: Statistics

Be careful when collecting and comparing statistics. Electoral boundaries and wards change as populations rise and fall to keep equal constituency numbers. Postcode areas do not always reflect the same boundaries. Functional areas, for example school catchments, may differ as well.

Demographic changes

In this section we focus on the characteristics of age structure, ethnic composition and gentrification in rural and urban places of the UK.

Urban and rural places have distinct demographic characteristics, which have also altered over time. Some headline characteristics and trends taken from the 2011 census compared with 2001.

- In England and Wales 86 per cent of people are classed as having white ethnicity (91.3 per cent in 2001). On average, the population grew by 7.1 per cent, but by 9.0 per cent in urban and 2.5 per cent in rural areas. These facts show the impacts of immigration and growth of a multicultural society.
- Approximately ten per cent of people were employed in trade and managerial or director occupations in urban areas, but fourteen per cent in rural areas. Such occupations generate higher earnings which, combined with improved technology and communications, has enabled more skilled people to live in rural areas.
- Rural areas have a higher elderly population (median age of 45 years, urban areas 37 years), more born in the UK (94.9 per cent) and lower unemployment rates than in urban areas.
- There has been a rise in youth unemployed in affluent areas such as Witney and Winchester because of less graduate employment.

Figure 15.7 shows two population pyramids: for one urban ward called Harpurhey, a deprived area of inner Manchester of about 16,000 people, and one for the whole of Cornwall. In Harpurhey, there are fewer people aged 15 to 34, and over 75 years old compared with Manchester, but a higher proportion in the 0 to 14 and 35 to 74 age groups. Compared to national figures, there are proportionately more between 0 to 4 and 15 to 39 than in England and Wales, but fewer residents aged 5 to 15 and 40 and over. Cornwall's profile shows the effects of an ageing population and outward migration of younger working adults.

Skills focus: Population pyramids

Population pyramids work best when using percentages for comparison purposes.

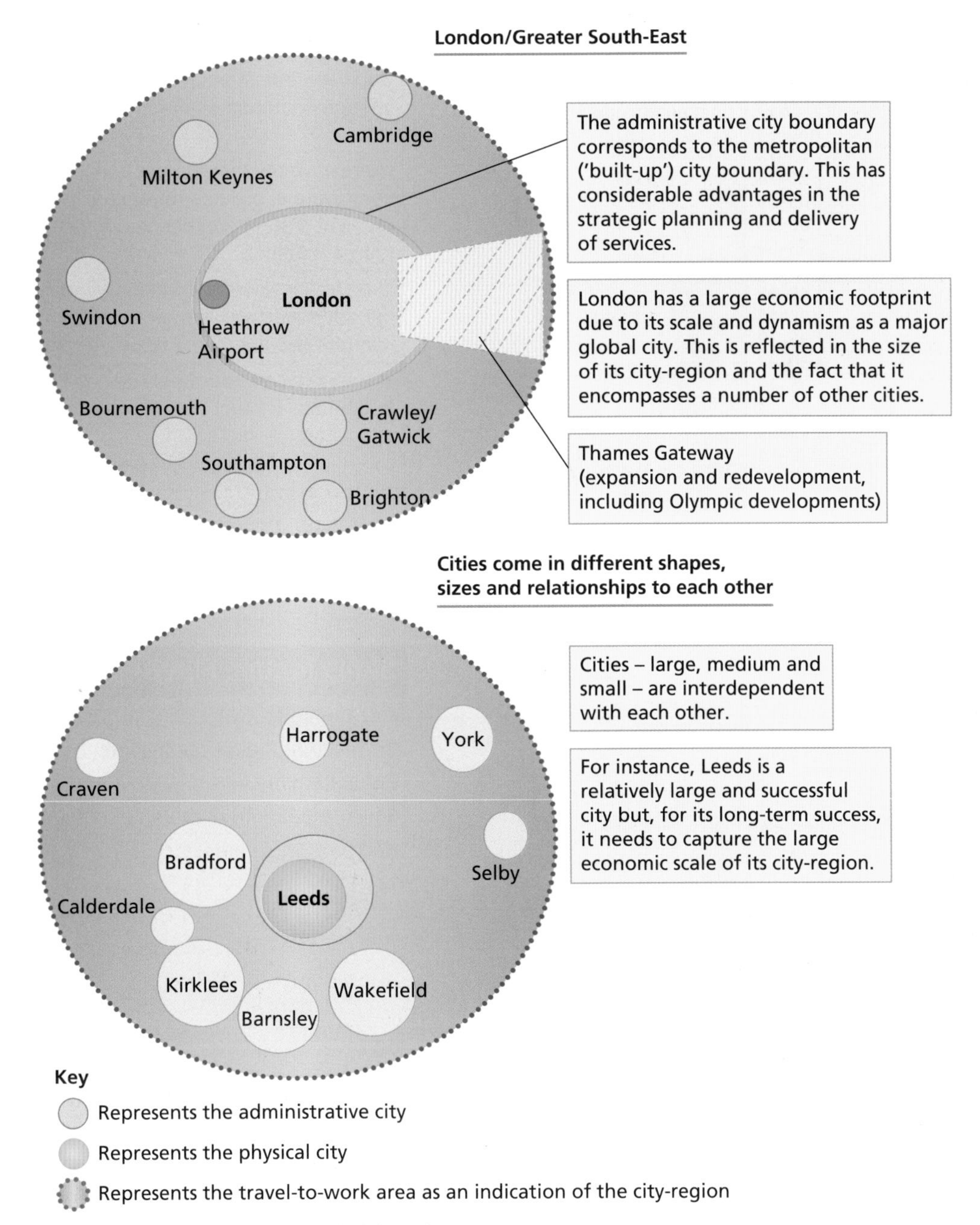

Figure 15.6 Functional versus administrative areas

Key concept: Demographic structure

The demographic structure of places may change by age, gender and socio-economic status. When places experience lower income groups moving in, such as lower-paid immigrants and students, it changes by the 'filtering down' process. The opposite is 'filtering up', or gentrification, where more affluent people take over an originally lower-income place.

Gentrification

Gentrification is a change in the social structure of a place when affluent people move into a location. Planners may allow developers to upgrade a place's characteristics, residential and retail, to deliberately attract people of a higher social status and income. A simple internet search on 'gentrification in the UK' will reveal a rich source of examples (Table 15.2).

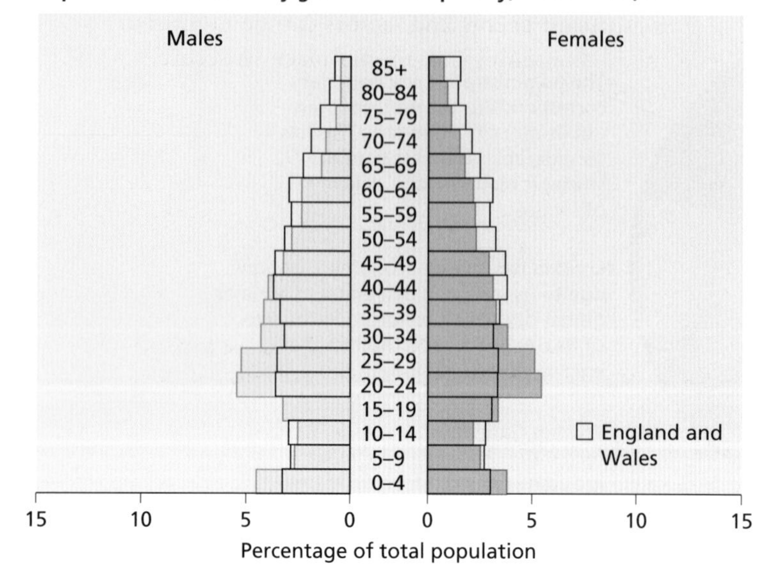

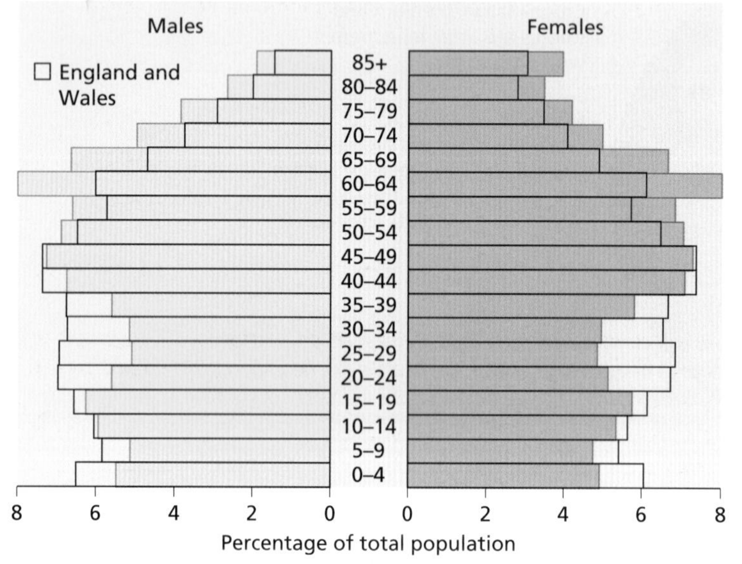

Figure 15.7 Contrasting local area population profiles

Table 15.2 Examples and views of gentrification

Super-gentrification in Portland Road, Notting Hill, London: Victorian slums are now sold for multi-million prices.
Aggressive regeneration of London's Soho: controversial policies to remove its reputation as a red light district and become a 'rich, middle-aged US tourist' hotspot.
The great inversion: London is getting turned inside out as inner-city areas become gentrified and outer suburbs become home to poorer people.
Is this civilised apartheid? Second home owners in the Lake District are forcing out lower-paid agricultural workers.
How sleepy rural Bruton in Somerset became 'hip'.
Bring on the hipsters! Professionals moving in know how to get things done and put pressure on schools, the police and their local place to improve.

Skills focus: 'Filtering' processes

You should find your chosen places will have evidence of these 'filtering' processes, as postcodes either become less desirable and offer cheaper accommodation, or more desirable and costly to buy and rent.

Examples of studentification

'Studentification' is a term now universally understood in places offering higher education provision. Students often cluster in certain areas of larger towns and cities. Their absence during most holidays and their sometimes antisocial behaviour when resident may cause conflicts. In some areas, students outnumber local residents, such as Queen's University, Belfast, where 50 per cent of immediate housing around the campus is student occupied. In Headlingly, Leeds, two-thirds of the 10,000 residents are students, concentrated in 73 streets of terraced houses. Nottingham, Southampton and Winchester have similar hotspots. The government wants to regenerate these areas by capping the number of houses in multiple occupation.

Reasons for functional and demographic changes

There are complex, often interlinked, reasons for changes in the functional and demographic characteristics of places. Significant events and factors can be identified in all economies that mould place structure (Table 15.3).

Table 15.3 Factors affecting the changing characteristics of places

Factors	Examples
Physical	**Location:** proximity to large cities and core economic zones. **Environment:** places vary in attractiveness. **Technology:** lifts allowing high rise, and motor vehicles, facilitating urban sprawl and counter-urbanisation. Fibre optic cables and broadband are shifting traditional 'landscapes' and relationships.
Accessibility, connectedness	**Access** to other places – by road (especially motorways), rail and air. **Connections** help competition for investment and visitors.
Historical development	**Post-production era:** once key factors in many places; primary production (agriculture, farming, fishing) and manufacturing has ended. **Competition** for the optimum site for functions: commercial, retail, residential, infrastructure. Land values and intensity of use historically increased towards the Central Business District (CBD) or the core of a village (church, marketplace) because access for most people pre-motor vehicle age was best here. **Changes in consumer trends:** • in retailing, from corner shop to supermarket to online shopping • in house types: increased demand for single homes due to demographic and cultural trends • role of big business and TNCs in shaping consumer demand and, hence, the character of places (cloned shopping malls). **Increased affluence** has increased leisure and tourism functions, so many houses and buildings converted, such as bars, B&Bs or second homes. **Historic buildings** (ex-warehouses, canals, old market squares) can be a physical asset for places seeking regeneration. Conversely, large areas of derelict buildings and the legacy of toxic waste from manufacturing may be a deterrent.
Role of planning by governments and other stakeholders	**National government** policies on restructuring the UK economy, trying to equalise the benefits and reduce the negative externalities of changes. The 1990's policy of increasing student numbers so that 50 per cent of children go on to higher education. **A 'plan led' system** with tight control over developments, zoning and segregating land uses began from 1948. Green belts introduced and new and expanded towns were developed to relieve population pressure from larger cities. The policy of state-funded council housing, industry and transport shifted from the 1980s towards privatisation and greater partnerships with private investors and speculators. **Conservation area** policies limit new developments and encourage conversions rather than renewal schemes. National interests may override local ones, for example HS2 and fracking. Larger schemes must have an Environmental Impact Assessment (EIA). **Central government intervention** in local places started to change in the late twentieth to early twenty-first century when the policy of 'localism' and individuality began. **Local planning** centres on elected parish and city councils, and on a few larger cities. There has been an increased input into local decisions through Local Area Plans and stakeholder meetings. The cumbersome planning process was streamlined in 2013, although criticised for fast-tracking decisions without full consultation. **Image** or the perception of a place may affect whether a place needs changing, or is able to change.

Skills focus: Oral accounts

See Skill 2 in the Skills focus section (page 346).

Measuring changes

Geographers measure changes within places using four key methods:

- land-use changes
- employment trends
- demographic changes
- levels of deprivation.

Measuring change can be carried out by quantitative indicators such as land-use maps and census data, but also through qualitative surveys and anecdotal oral histories, which breathe life into studies of places which, after all, are created by people.

YouTube and DVDs of TV programmes may give a flavour of an area, such as Channel 4's *Migration Street* based on Derby Road in Southampton and BBC's *People Like Us* based on Harpurhey in Manchester.

The following is an example of a transcript recorded to document personal memories of how connections have been involved in changes in the economic and social structure of inner-city Bristol. It would have been improved by a few structured questions rather than an unguided reminiscence.

'I'm 67 now. My family all lived and worked locally near the centre of Bristol, in or near Holmes Street, Barton Hill. Did you know Banksy did his first bits of graffiti there? Gran was one of 16 children, three died at birth; you wouldn't believe all crammed into a little two up two down with an outdoor privy [toilet]. We had a bit more room living two houses away cos Mum only had three kids, and so did I. It was normal for us then to leave school at 16. We had no qualifications but I got an apprenticeship in a local engineering firm. We only had a bike or walked everywhere then, no money like you have to buy and run cars. The area was badly bombed in the war [WW2] but lots of factories were still left; that was in the 1960s. But you wouldn't recognise the area now. We could of stayed in the terrace, cos our side of the road was saved, they were even going to have an indoor bathroom, but decided to move out to Bishopsworth [suburb], cos right opposite, half the street had been knocked down for some renewal scheme, tower blocks, flats. The community was gone by then [1970s]. I can hardly recognise Barton Hill now; it's got a lot of on the social [people on benefits], got a name for crime and violence AND it's full of Somalis [Somali refugees were resettled into a concentrated zone in the 1980s]. Bit of a no-go area now. But d'you know, the Settlement's still there where we had our family parties [a community centre established 1911, now funded by grants]. And there's a big government scheme trying to do up the shops and make better houses and stamp out the crime [New Deal and Community at Heart programmes]. They even knocked down some of those awful tower blocks and have rebuilt terraces, altho' they're bigger, got all mod cons!'

Skills focus: Index of Multiple Deprivation and GIS

See Skill 10 and Skill 7 in the Skills focus section (pages 350 and 348).
More quantitative and objective data may be found on a place by using the IMD. When investigating the IMD you will need the LSOA code or postcode to find your local place. The final figures can be mapped and used in GIS applications. Excel spreadsheets on the separate domains and overall IMD for your local place can be found for all English LSOAs.

Index of Multiple Deprivation

IMD is used by central government and especially by local authorities to target regeneration aid, to allocate resources to places and people (such as areas with low average GCSE scores) and target hotspots of crime. Places are ranked by their relative level of deprivation. This is a relative measure only. Not every person may be 'deprived' in a highly deprived area, and some deprived people may live in the least deprived areas.

In the IMD, 32,844 small areas or neighbourhoods (called lower-layer super output areas, LSOAs) are used, each having 1500 residents or 650 households. Thirty-seven indicators are grouped into seven 'domains' and ranked by importance: income and employment are higher than health and education. Crime and living environment are weighted the least.

The main findings from the September 2015 IMD, as identified by the Department for Communities and Local Government, were:

- There are pockets of deprivation within less deprived places in all English regions.
- Deprivation is still concentrated in large urban conurbations, areas that have historically had large heavy industry, manufacturing and/or mining sectors and coastal towns, for example large parts of East London, Middlesbrough, Knowsley, Kingston upon Hull, Liverpool and Manchester.

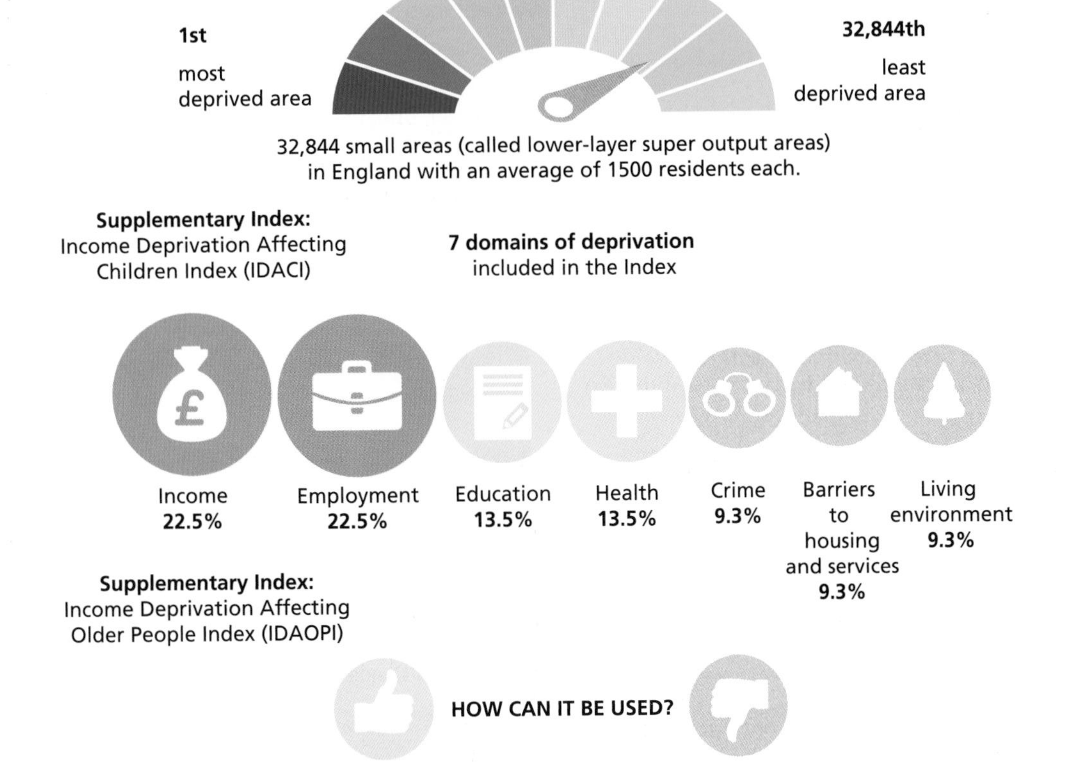

Figure 15.8 IMDs and the seven domains of deprivation

- The London Boroughs of Hackney, Tower Hamlets, Newham and Haringey have become relatively less deprived since 2010.
- For the first time the IMD has been measured in the 24 English Local Enterprise Partnerships (LEPs) to help target aid.

15.3 Local place studies (1)

The Specification requires you to study two places to examine the past and present connections that have shaped their economic and social characteristics.

You should have first-hand experience of one place and the other should be significantly different. The best starting point would be first to choose the place in which you live or go to school/college. That, in turn, will determine the choice of a contrasting place. There are many possibilities, such as comparing:

- two different parts of the same town or city
- two different types of rural settlement
- two different places (urban and rural).

The main watchpoint is that these places contrast enough in their characteristics. You also need to study the need for regeneration and/or the type of regeneration strategy used, so it would be sensible to choose one place where regeneration is involved.

Your choice may also be conditioned by practical matters. If you wish to study the contrasting place at first-hand, then obviously you will need to choose somewhere that is accessible from home. Or maybe your preference is to study a place that is truly remote and instead undertake a sort of 'virtual reality' investigation relying on secondary sources rather than field observations. No matter what your choice of contrasting place, you need to remember that the emphasis in both investigations is on regeneration. For this reason it is

Key term

Connections: Any type of physical, social or online linkages between places. Places may keep some of their characteristics or change them as a result.

Table 15.4 Some opening questions about your places

Question and lines of enquiry	Sources
What do you know already about your local/contrasting place?	Personal knowledge Concept map
Where is it on the rural–urban spectrum?	Maps: Google and OS 1:25,000 map
What is the boundary of your chosen place?	In urban areas, wards are of a manageable scale; parishes in rural areas. Census LSOAs may be relevant
Is there an obvious place identity?	Placecheck, descriptions by locals, your school/college, estate agents, tourist information
What economic activity and inequalities are there?	Census, neighbourhood statistics, ACORN
What are its functions and any changes?	Local authority website Land-use map

recommended that you define both of your places in terms of the spatial units used in the collation of census data, i.e. borough, local authority, parish or ward.

The following guidance is mainly in the form of questions to be asked of both your places. Useful sources are given where appropriate. The questioning is divided into four parts:

- opening questions
- regional and national influences
- international and global factors
- recent economic and social changes

Opening questions

The first questions to be asked of both places are set out in Table 15.4.

Skills focus: Concept mapping

An interesting exercise to establish your own sense of place, and your identity with it, is to draw your local area freehand and then annotate it with place names and features important to you. You could then compare this concept map with other people to see if age, gender or length of residence affects perception of the local place.

Figure 15.9 Conducting local studies

Figure 15.9 shows students investigating a local place to assess its value, asking other locals their views too.

Regional and national influences

Table 15.5 summarises the research you could carry out. (Note the references to the relevant skills in the Skills focus section which begins on page 345.)

Key concept: Place representations

Places can be represented in contrasting ways. Formal representations and official statistics of places (for example by government organisations) may differ from representations by media or locals.

Table 15.5 Questions about how regional and national connections shape economic and social characteristics of places

Scale	Questions or lines of enquiry	Sources
Regional	How well are your places connected in terms of transport?	OS maps and public transport timetables
	Are there major cities nearby and what is their impact on your places?	Think in terms of commuting and accessing services and leisure
	What are the regional enterprise zone, county council or metropolitan area's policies on growth and development?	Local authority and regional enterprise zone websites
National	What are the political inclinations of the national government?	Last local election results
	What are the political inclinations of the local/county councils?	Social media (Skills focus 1)
	How strong is the sense of community?	Local newspapers (Skills focus 3)
	What are the pressures for change?	

Table 15.6 Questions about how global and international connections shape economic and social characteristics of places

Scale	Questions or lines of enquiry	Sources
Global	Are any global brands present – retail outlets, branch factories and offices? Is there any involvement in global tourism? How well connected to the internet are your places?	Field observation Tourist information centres Broadband speed tests
International	Are your places directly affected by specific government/EU policies or designations? Is your place 'twinned' with a location abroad? Why was that place chosen? How do your places compare with the national averages of wealth and deprivation? How accessible are your places to London and other leading UK cities?	Look for signs in an area showing players involved Use the local authority websites Look at IMD results (Skills focus 10)

Table 15.7 Some questions about internal factors contributing to place identity

Questions or lines of enquiry	Source
Continuity: What architectural reminders are there of the past?	Field observation
Changes in functions and characteristics?	Old maps and field observation of age of buildings
What are the economic characteristics of your places, including sector, type of employment, unemployment?	ONS, 2011 census report
What are the population characteristics of your places, including gentrification, age structure and ethnic composition?	ONS, 2011 census report
What socio-economic inequalities exist: pay, health, life expectancy, levels of education?	ONS, 2011 census report, Public Health England
Do you think that each place has a distinctive identity? If so, sum up each of those identities in a few words.	Placecheck (Skills focus 12) Social media (Skills focus 1)

International and global factors

Table 15.6 summarises international and global questions to ask (again, note references to the Skills focus section in the following tables).

Place identity

Table 15.7 puts the spotlight on some aspects that can add character to a place and create a specific identity.

Recent economic and social changes

More questions follow in Table 15.8 (again, see more useful references to the Skills focus section on page 345). They focus on recent economic and social changes. These can have a powerful influence on the identity of today's places. A survey of 2500 people for the BBC's 'Who Do We Think We Are?' project (2014) found that, compared with ten years previously, 30 per cent felt a stronger identity with the local and global world and less with the national level. According to social anthropologist Kate Fox, humans are wired to live in small, close-knit tribal groups. New technology and social platforms such as Facebook, Twitter and Mumsnet, and local community email lists, allows a new sort of 'tribe' at a local scale, not just global. Local authorities are now actually tasked to use social media to help engagement in local politics and communication with local voters.

Key concept: Continuity and change

Your life, and those of others in a place, will be affected by continuity and change, both real and imagined. An example is varying viewpoints on migrants.

These last questions are challenging, but do your best to address them. You might pursue them by means of a limited questionnaire or by interviewing among, say, three different groups:

1 your peer group
2 adults with children living at home, and
3 retired people.

How differently do these groups perceive the changes?

The investigation of your chosen places continues in Section 16.3. It is important that the place investigations are undertaken by you, but with help from your teacher. Throughout them you should be mindful of the similarities and differences between your places. Furthermore, questions in the examination will ask for references to, and detail from, your place studies.

Table 15.8 Some questions relating to recent social and economic change

Questions or lines of enquiry	Sources
Have the employment sectors, types and pay levels of your two places changed over the last intercensal period? If so, how?	2001 and 2011 census reports
Does any inequality exist in your places? Has it increased or decreased?	IMD 2010, 2015 and census 2001–11 (Skills focus 10) Local authority GIS maps of IMD
What functional changes are evident?	Old maps, e.g. GOAD can be compared with your own land-use surveys
Any evidence of planning in these changes? For example refurbished town centres, new malls, changed housing, gated housing areas, new industrial or science park, new communications and infrastructure (broadband, park and ride etc.)	Local authority websites Regional enterprise areas Field surveys
Any environmental changes, for example parks, disabled access, crime and design developments?	Local authority websites Land-use surveys
In which direction do you think your place is moving: up- or down-market?	Look for evidence of filtering up: gentrification, improved quality of houses and shops, and 'café culture'; look for evidence of filtering down: deteriorating housing, multiple occupancy (Skills focus 4)
Are the economic and social changes having a noticeable impact on the basic identity of your places?	Oral accounts (Skills focus 2) Interviews (Skills focus 9)
In what ways have recent economic and social changes impacted on the identity of local residents?	Interviews (Skills focus 9) Social media (Skills focus 1)

Review questions

1 What factors affect the identity of places?
2 How and why have economic sectors changed in the UK over the last century?
3 What impacts have these changes had on place characteristics?
4 What are the factors creating differences in pay, health status and educational attainment in the UK?
5 How and why have functions shifted in their type and location in urban and rural areas over the past few decades?
6 Describe the different ways of portraying and statistically analysing changing places.
7 What are the connections which have shaped the economic and social characteristics of your two place studies?

Further research

Take a look at this rural–urban classification leaflet from the 2011 census: www.gov.uk/government/statistics/rural-urban-classification-leaflet

Explore these rich sources for statistics from censuses and more recent research by the UK Government Office of National Statistics: www.ons.gov.uk and www.nomisweb.co.uk

Have a look at the IMD 2010 deprivation mapper: http://apps.opendatacommunities.org/showcase/deprivation

Useful portraits of regions by the ONS: www.ons.gov.uk/ons/regional-statistics/index.html

A research organisation on cities in the UK: www.centreforcities.org

16 Why regeneration might be needed

Why might regeneration be needed?

By the end of this chapter you should:

- be aware how economic and social inequalities can change people's perceptions of an area
- understand the factors affecting people's lived experience and engagement with their home place
- be aware that there is a range of ways to evaluate the need for regeneration.

16.1 Inequalities and place perception

The first part of this chapter concentrates on how economic and social inequalities affect people's perceptions of places. Place has a huge impact on our quality of life, interactions with each other and the sustainability of communities. Place identity can be a source of strength but may also be parochial, where people feel trapped and cut off from opportunities. Often inequality and poverty rise and fall together. However, inequality can be high in places without high levels of poverty due to a large difference between incomes.

Key concept: Perception

Perception is a vital part of lived experience and affects how people engage with their place. It varies between individuals and groups of people and depends on factors including age, social class, ethnicity and overall quality of life. These factors may be real or imagined. People may have mainly positive or negative views about their place, which may alter over time.

Successful places

Regions perceived as successful tend to be self-sustaining as more people and investment are drawn to the opportunities created, both from inside the country and from other places. However, negative externalities may result: overheated property prices, congestion of roads and public transport, and skills shortages. The perception of residents in such places may differ according to the type and position of people and their needs and aspirations:

Key concept: Measuring success

Success is measured by high levels of employment, output levels (GVA, GDP – see Chapter 15 page 219 for their definitions), in-migration and quality of life, and low levels of deprivation.

- Younger people in high-earning jobs will enjoy the fast pace of life and plethora of opportunities offered by cities such as London or Manchester. Unskilled people, lower earners and the long-term unemployed will have more negative views about their quality of life in a successful place.
- Retirees may view places offering a slower pace of life with pleasant climate, sheltered accommodation and good access to healthcare as successful, for example Torquay in Devon or Christchurch in Dorset. Younger adults may wish to escape such places.
- Most will view the quality of the environment in rural places positively, since on the whole it is higher than in urban areas.

Urban places

Success is either due to market forces, as places compete in our globalised world, and/or from government-led regeneration policies. London and the South East is a good example of this, benefiting from its function as the capital but also enhanced by successive government polices to protect its competitive status, such as the Thames Gateway, the 2012 Olympic Games and Heathrow expansion plans. The fact that a place is popular shows it is viewed as largely attractive to people. Meanwhile, large cities such as Birmingham and Bristol have developed strong service and financial sector economies, following the lead of London.

People with lower incomes living in successful places will be especially disadvantaged from the higher cost of living and property prices, however. Another negative externality is that skills shortages may result

San Francisco

San Francisco has a reputation for economic energy, cultural vibrancy and tolerance. In the 1990s it became the focus of California's new 'gold rush', home to global dot-com businesses such as Dropbox and Twitter. There has been phenomenal job growth in STEM biotech, life sciences and digital media companies. The multiplier effect is fuelled by its technological and transportation infrastructure, high quality of life and highly skilled workforce.

However, not all have benefited: the so called 'Google effect' of gentrification of districts alongside Google buses transporting workers to its Mountain View campus, has created discontent from some established, less-affluent displaced locals.

Figure 16.1 'Successful' San Francisco: the gentrified historic area of inner-city Haight-Ashbury with the towering CBD skyline in the background, an indicator of success

Skills focus: Index of Multiple Deprivation

See Skill 10 in the Skills focus section (page 350).

from success, as seen across the UK in the sectors of IT, technology, creative, finance, engineering, plumbing, building and caring. This reflects a history of low take up educationally in these subjects, past government restrictions on skilled immigrants and, in London, inflated living costs. The Place Context on San Francisco and later content on rural commuter villages illustrates this paradoxical situation of success with negative spinoffs.

The official ONS Well-Being Index and IMD deprivation index quantifies 'success' while independent surveys give more subjective perceptions. The annual Halifax Rural Quality of Life Survey and Sunday Times Index reveal some interesting regional scores:

- Southern areas have higher ratings for weekly earnings, the weather, health and life expectancy.
- Northern areas rate well on education in terms of grades and smaller class sizes, lower house prices in relation to earnings, and lower traffic flows and population densities.

Rural places

The 2011 census showed that rural places generally were experiencing a reversal of a 250-year trend of urban areas dominating jobs, wages and productivity:

- Some small villages and towns such as Worcester have been growing faster than many larger urban areas, both in terms of population and economic output. Top of the Halifax survey list was Rutland in the East Midlands.
- Although affected by the global economic crisis, rural areas in general have lower rates of unemployment and insolvencies, with the exception of some ex-mining settlements. There has been much growth in smaller and micro businesses (under ten employees), and home working is more important than in urban areas. Higher-value food products are booming, as are leisure and tourism.
- Accessible and 'attractive' rural communities have seen in-migration of young families, commuters and retirees. This counter-urbanisation reverses the long-term trend of net out-migration from the countryside to urban areas. Transport and technology innovations, especially mobile networks, and government investment in high-speed broadband has allowed more highly skilled professionals to live in attractive rural locations.

Less successful urban and rural places

US president Barack Obama called economic inequality 'the defining challenge of our time'. It inevitably affects the perceptions of individuals and communities and, although difficult to quantify, their level of happiness and their identity with their local place. Very unequal societies, such as the UK and USA, have social consequences at the local, regional and national scale (Table 16.1).

able 16.1 Social consequences of inequality

Reduced	Increased
Trust in people with positions of power, especially police and planners	Segregation of different socio-economic groups, property damage and violent crime
Social and civic participation	Health issues: either because of lack of wealth, access to care or more deliberate lifestyle choices
Educational attainment and training	Higher infant mortality and shorter longevity
Social mobility	Status competition, which drives less-affluent people into debt to keep up with a peer group practising a higher level of consumerism
Attachment to place	

The results are intergenerational, passed down through families, unless something breaks the cycle, such as effective regeneration schemes, university education, random opportunity or sheer entrepreneurship.

Urban decline

In the UK, places like Hartlepool, a former shipbuilding and steel town in Teesside with an unemployment rate twice the national average (thirteen per cent), struggle with their Rust Belt legacy. Over a quarter of Hartlepool's high street shops are empty, showing the lack of spending power in the area.

Key concept: Rust Belt

The term was coined in the 1980s in the USA in reference to the once-powerful manufacturing region that stretched from the Great Lakes to the Midwest, famous for steel and car production. It fell into economic decline following automation, global shift and increased free trade.

Rust Belt refers to the concentration of problems associated with the loss of core employment and large scale deindustrialisation of manufacturing areas, characterised by derelict buildings and land.

Key concept: North–South and urban divide

The North–South and urban divide may widen over the next few years because of the reversal by governments of decades of subsidies and support for struggling areas. The safety blanket of many local government employees and high welfare payments is over in our 'austerity era'.

There is a marked North–South and urban divide in success. The Centre for Cities research group sees the take up of knowledge economy employment as pivotal. Figure 16.2 illustrates the four main groups of cities in the UK, plotting lower-knowledge industries in 1911, most affected by deindustrialisation, against private knowledge-intensive business services (KIBS) jobs in 2013. The size of the circles shows job growth changes.

- **Reinventor cities** have changed their economic base successfully by encouraging IT and digital media, have higher wages, graduate workers, new businesses and productivity.
- **Replicator cities** replaced cotton mills with call centres and dock yards with distribution centres and are less sustainable. They tend to have a higher share of workers with low qualifications and a working age population claiming benefits.

There is a distinct geographical pattern reflecting the difficulty in changing the legacy of a Rust Belt: 30 out of the 41 cities called 'replicators' in Figure 16.2 are in the North, Midlands or Wales, while eleven of the sixteen 'reinventors' are in the South.

Rural decline

Unlike urban places, rural areas do not have as many environmental issues, a lack of green space or conflicts centred on ethnicity. Decline centres instead on a faster ageing population than in urban places; remote rural communities continue to struggle with the out-migration of young people.

Rust Belts and urban divides

Some towns and cities have adapted since deindustrialisation and city authorities have managed to attract growth services and knowledge-based jobs, for example Pittsburgh in the USA and Manchester in the UK. However, many, such as Chicago, have struggled and are now characterised by increasing poverty, declining population and are even near bankruptcy, for example Detroit. In the UK, Bristol, with 39 per cent of its population having a degree, is more attractive to knowledge-based investors than Doncaster, with only 23 per cent being graduates. Teesside, with its declining steel industry, may well develop into a mini version of the American Rust Belt.

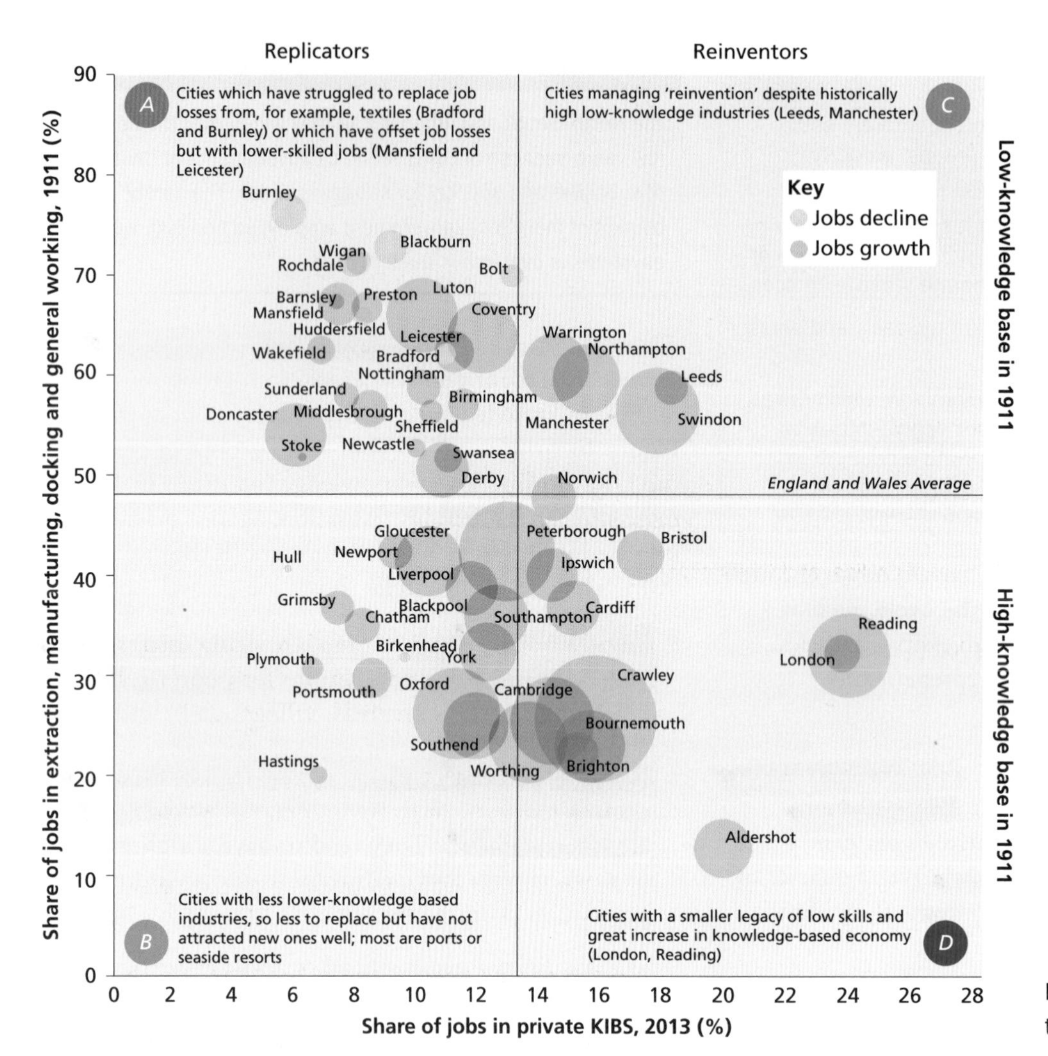

Figure 16.2 How cities respond to deindustrialisation

Whole regions may be classed as relatively unsuccessful, such as mainly rural Cornwall. 'Two countrysides' exist often side by side, and place perceptions by residents are likely to vary:

- Better-connected, well-off and growing places, such as the Itchen Valley in Hampshire, contrast markedly with less well-off remoter agricultural places such Llansilin on the Welsh border, or places once dominated by mining.
- Locally, many pockets of deprivation may be 'hidden' statistically, made up of a few houses, or streets or a small estate in even the most affluent of urban fringe villages or more remote settlements.

Struggling rural areas: Cornwall

Cornwall is famous for its tourist function, with hotspots such as Newquay and Padstow. As a whole, the county is not deprived; however, there are neighbourhoods like Redruth with such consistently high levels of deprivation that they receive European development funding (ERDF). In 2014 average wages were £14,300 annually compared with £23,300 in the UK (Eurostat). This puts it in the top ten most deprived places in Western Europe. Up to 40 per cent of households live on less than £10,000 a year. The relative wealth of the region is depressed further because of the high cost of living, so residents have less spending power than most of the UK and Europe. The loss of its mining industry and contraction in agriculture and fishing sectors are all important causes of the spiral of decline in Cornwall (Figure 16.3).

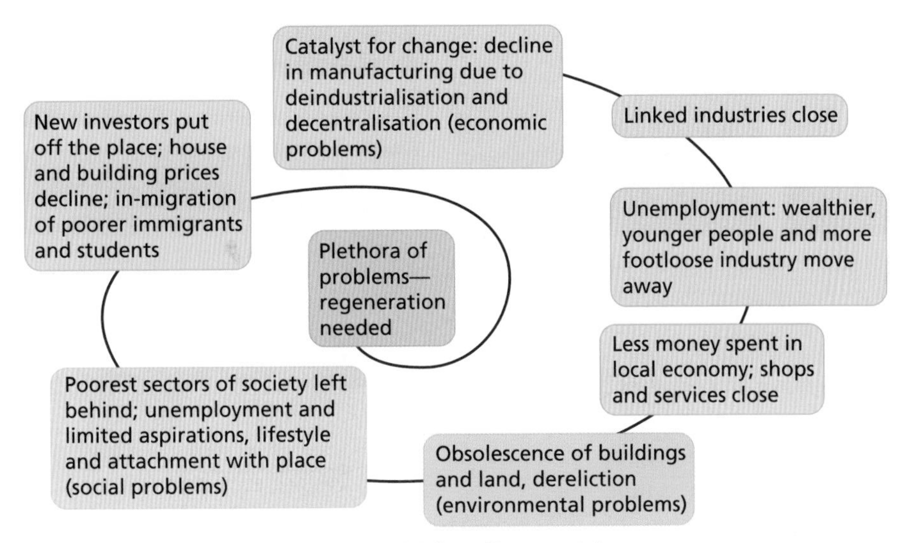

Figure 16.3 Spiral of decline, or de-multiplier effect, model

Priorities for regeneration

Economic and social inequalities create differing priorities for regeneration. The Specification focuses on four different places which either show regeneration or are in need of it. These include two extremes in social segregation and residential sorting:

- sink estates and declining rural settlements needing the greatest regeneration efforts
- gated communities and commuter villages with lower priorities.

The perception of those living in these two extremes will differ greatly.

The process tends to be self-reinforcing; as conditions decline in struggling neighbourhoods, people most able to move out are more likely to leave and less likely to return. The most extreme example is found in urban USA where white communities leave the inner cities, known as 'white flight'. Local perceptions on place will depend on ethnicity, family structure and age.

Sink estates

The term sink estate originates from 30 years of government policy that has resulted in segregating low-income groups needing social assistance from the rest of society. Ironically, many council housing estates were built to improve living conditions for poorer residents of inner cities, suburban areas, villages and rural towns. A report by the Fabian Society compared data on those born in 1946 and after 1970 in council homes. Children born after 1970 were twice as likely to have mental health problems, eleven times more likely to be unemployed and not be in training or education, and nine times more likely to live in a jobless household.

Examples of sink estates are the Barracks in Glasgow and Broadwater Farm in North London. Redruth was the first town in the UK to introduce a temporary curfew for youngsters centred on a sink estate of just six streets called Close Hill. Up to a third of families claim benefits in the ex-mining area of small rural towns in Cornwall's 'Camborne Corridor' (Figure 16.4a).

It must be remembered, however, that not all rundown estates are sink estates, and it is important not to stereotype these, or indeed any, places.

Key term

Sink estate: Housing estates characterised by high levels of economic and social deprivation and crime, especially domestic violence, drugs and gang warfare.

Key concept: Social segregation and residential sorting

These are a common feature of inequalities. Low-income households tend to seek out communities that provide lower-cost housing and have higher social welfare spending. Higher income groups similarly cluster together and, if they move into a previously lower income location, may gentrify it.

Figure 16.4 Extremes of social segregation: 16.4a Broadwater Farm, North London; 16.4b Honeycombe Chine, Boscombe

Gated communities

These have a long history, with whole cities being walled and gated in the past. The current trend stems from the 1980s in the USA where architects and planners created fortress-type architecture in regenerated inner-city locations. This was first replicated at a large scale in London's redeveloped Docklands. Nowadays they are found in many settlements, as either individual buildings or groups of houses, where wealthier residents have a secure building, perimeter wall or fence and controlled entrance for residents, visitors and cars. Where gentrification occurs, gated communities may be built to segregate the incomers from locals who are perceived to pose a threat (Figure 16.4b).

Commuter villages

According to the 2011 census, 19.8 million people live in rural areas in England. Of these 98 per cent live in accessible rural places, which are experiencing quite fast rates of population growth despite the recent economic downturn. Many property websites advise would-be commuters of the 'best' locations, and 'wealth corridors' have developed linked to high-speed railways and motorways. The Itchen Valley villages between Wincheste and Alresford are examples of successful commuter village Small towns such as Beaconsfield in Buckinghamshire, Sevenoaks in Kent and Alresford in Hampshire, are increasingly preferred over smaller villages.

Key terms

Gated communities: Found in urban and rural settlements as either individual buildings or groups of houses. They are landscapes of surveillance, with CCTV and often 24/7 security guards. They are designed to deter access by unknown people and reduce crime.

Commuter villages: Settlements that have a proportion of their population living in them but who commute out daily or weekly, usually to larger settlements either nearby or further afield.

Such places, and those with many second homes, tend t have affluent populations and low levels of deprivation measured by wealth and employment. However, they may need fewer services since commuters may not demand local shops, schools or bus services. This makes locals dependent on low-paid agriculture vulnerable. Any place with a more affluent population able to pay more for homes will pose a threat to lower-paid locals and force out younger adults. Local perception will depend on the wealth status and age of residents.

Regeneration in places dominated by commuters and second homes focuses on 'affordable housing' and boosting service provision.

The wealth corridor of the Itchen Valley, Winchester: A retirement and commuter community

This accessible rural place, with around 1900 residents in four adjacent villages, lies at the western end of the South Downs National Park. Its conserved scenic and ecological value, based on the meadows of the River Itchen, rolling chalk downland and picturesque settlements, are great attractions. Its location, a few miles from Winchester, the M3 and fast mainline rail services to London means that it has become a commuter hotspot over the past few decades. The influx of younger families has breathed new life into the local school and four pubs in the valley. However, it has also meant that house and land prices have rocketed, and the few remaining farmworkers have children unable to afford local housing. Zoopla's ZED Index (average house prices) in 2015 for England as a whole was £298,313. The Itchen Valley postcode's value was £588,882, more expensive than in Winchester (£505,039) and close to London prices (£636,172).

Figure 16.5 A typical commuter village: Easton in the Itchen Valley wealth corridor

Deprivation in remote areas: Llansilin in Powys

The 2011 census showed Powys, Wales, had 46.8 per cent of rural communities within the most 'access to service' deprived ten per cent in the country. This was measured by services such as banks and average travel time to get to a food shop, GP surgery and pharmacy, primary and secondary school, post office, public library, leisure centre and to a petrol station. Cars are an essential element of life.

The village of Llansilin, on the Welsh–Shropshire border near Oswestry, with a population of just under 700, typifies these characteristics. It has considerably less broadband and mobile phone coverage than the Itchen Valley in Hampshire. Zoopla's ZED Index for average house prices here was £230,076 in 2015. For every other indicator on the Welsh Index of Multiple Deprivation, this area scored in the 50 per cent least deprived in Wales (community safety, housing, health, employment and, especially, environment were good).

Declining rural settlements

Regeneration relates more to services and functions than population, at least in the most accessible places. An estimated 400 village shops and 700 rural pubs closed in Britain in 2010 alone.

16.2 Lived experience and engagement with places

Levels of engagement

Engagement may be measured in two main ways: by local and national election turnout, and the development and support for local community groups.

Election turnout

National elections

The Electoral Commission recorded that 7.5 million eligible voters were not registered in 2015. Poor, black and young people in urban areas are least likely to be on the electoral roll. Of eligible voters registered, 66.1 per cent did not cast a vote. The 2014 vote for Scottish independence, however, attracted 84.5 per cent of the electorate.

Traditionally, rural voters are more supportive of the Conservative and Liberal parties, and tend to have higher turnouts in elections than urban voters. The UK has one of the largest differences in voter turnout between the young and old in Europe with only 44 per cent of 18 to 24 year olds voting in 2015. These general points may be tested by comparing two very contrasting constituencies.

Contrasting electoral districts and their characteristics

- Westmorland and Lonsdale, Cumbria, is a sparsely populated, picturesque rural area including parts of the Lake District and Yorkshire Dales National Parks. It includes the market towns of Kendal, Kirkby Lonsdale and Windermere. The 2015 national elections produced a reduced majority to the Liberal Democrat Party.
- Densely populated inner-city 'Manchester central' has recently been expanding with more affluent communities moving in. Regeneration and gentrification have improved inner-city living. Areas such as Moss Side and Hulme, south and east of the CBD, remain some of the poorest however, comprising old terrace housing and many sink council estates. The 2015 turnout of registered voters was the lowest in the country, resulting in an increased majority to the Labour Party.

Table 16.2 Contrasting districts

Electoral data	Manchester Central, including Moss Side	Westmorland and Lonsdale, Cumbria
Turnout of registered voters in 2010 General Election (%)	46.7	76.9
Turnout of registered voters in 2015 General Election (%)	52.9	74.3
Median age in years	28	49
Deprivation (%)	70	50
Ethnicity (non-white %)	27	2
UK born (%)	70	95

Mayoral cities

Only four cities have directly elected mayors in England. The success of the local powers of the Greater London Authority (GLA) from its 1999 inception, with charismatic leaders like Boris Johnson, can be measured by improved strategic planning, including the 2012 Olympic Games. However, the 2012 referendum in nine large English cities invited to follow this local budgeting system only resulted in a 'yes' vote in three cities: Bristol, Doncaster and Liverpool. Whether a new system is wanted by locals largely depends on how effective local councils have been historically, but the very low voting turnout was also a key factor.

Skills focus: Electoral data

Contrast the characteristics of electoral areas and identify which factors lie behind the electoral engagement of locals. You can easily find data on electoral history from the local authority and electoral commission as well as local newspapers, and see if it follows national trends.

Local elections

In 2014, local election turnout was only about 36 per cent, and a mere fifteen per cent voted during the Police and Crime Commissioner elections in 2012. Such results have triggered calls for compulsory voting at local and national level. In a digital age, voting online will undoubtedly revolutionise voting habits.

Community groups

Support for local community groups varies across the country, depending on local willingness to participate and the main aim of the group. These range from committees running local allotments, open spaces and nature reserves to village shops and more powerful and vociferous 'NIMBY' (not in my back yard) groups protesting over planned developments such as new housing, fracking and wind farms. There are many groups focused on fundraising and helping more vulnerable people in the local community, such as meals on wheels, transport to hospitals and friendship groups. Residents in an estate may form a group, and can be effective in reducing antisocial behaviour.

Regeneration relies on community participation at all stages. There are some 9000 grant organisations – including the government, National Lottery, supermarket chains and charities – that may be able to help with basic administration and running costs. The Cabinet Office website has many examples of the partnership approach fostered by the government, and how crowdsourcing is encouraged by online investment. Charitable status is also an important funding mechanism. The internet has made information and support for such groups much more accessible.

Skills focus: Newspapers

Your local newspaper and other publications, as well as your local authority, will have adverts and lists of local community groups. Neighbourhood Watch signs and other adverts posted on local noticeboards and in shops, even big supermarket chains, are sources of data too. See also Skill 3 in the Skills focus section (page 346).

Synoptic themes:

Attitudes

The attitudes of people will affect the levels of engagement in a place by individuals and communities.

The voluntary sector has also been effective in youth mentoring schemes, addiction treatment centres and welfare-to-work organisations.

Factors affecting lived experience and levels of engagement

Two key factors affecting a person's sense of place, lived experience and resulting level of engagement with neighbours, community groups and elections are:

- Membership: a feeling of belonging, familiarity and being accepted.
- Influence: a sense of playing a part in a place, and hence caring about it.

However, there are other factors, as outlined in Figure 16.6.

Key term

Lived experience: The actual experience of living in a particular place or environment. Such experience can have a profound impact on a person's perceptions and values, as well as on their general development and their outlook on the world.

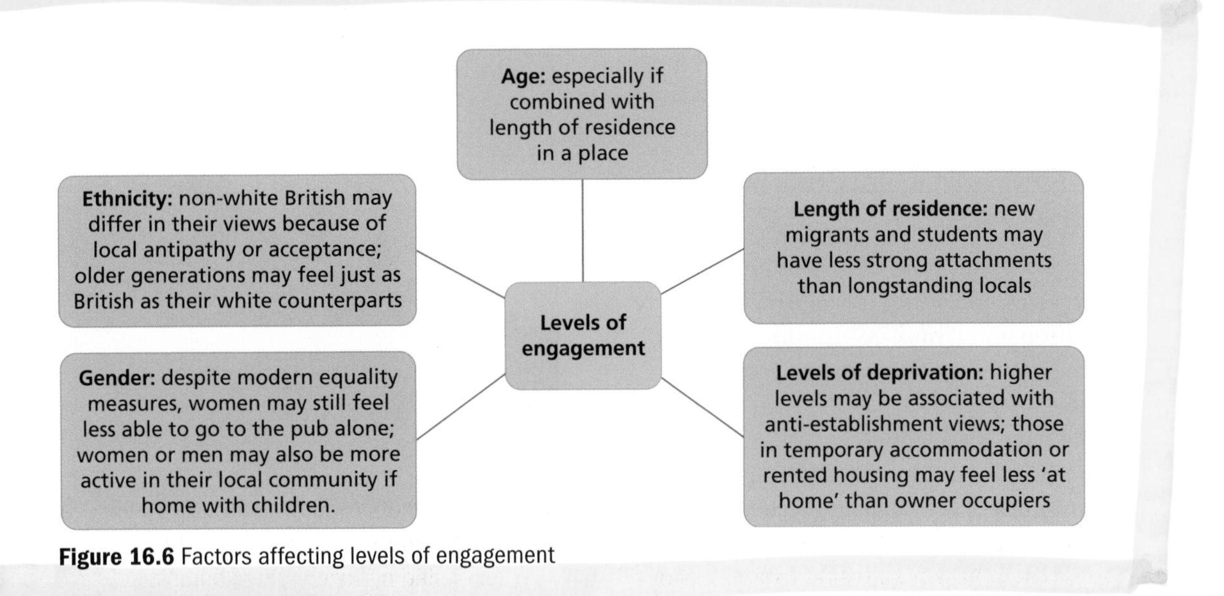

Figure 16.6 Factors affecting levels of engagement

Key concept: Attachment to place

The perception about, and attachment to, any place will depend on obvious factors such as age, length of residence, ethnicity and deprivation, but also the media's portrayal and whether government and private business policies, programmes and projects are successful for that particular person.

Place identity may be proudly flaunted by locals, or ridiculed by others, hence the terms 'Geordie', 'Del Boy', 'Scouse' and so on. Local language and dialect may foster a sense of place. Support for football clubs shows people's identity with both a local place and one which may be far removed. Team nicknames – Sheffield United are also known as the Blades because of the city's link with steel, Stoke City are known as the Potters because of ceramics – may indicate economic specialisms, even if they are now defunct.

Young people are probably the most directly affected by globalisation. It permeates everyday life affecting friendship groups, use of the internet (particularly for social networking) and wider cultural influences such as music, food and fashion. Paradoxically, young people are in one sense citizens of a global culture but at the same time may struggle for a sense of acceptance in the local societies in which they live.

Over the last decade there have been increasing debates about identity. This results from calls for political devolution, an escalation in economic and refugee migration, global terrorism and the impact of consumer culture, led by American companies. The UK government has specific policies to promote a sense of national place, which has become linked to citizenship.

Fieldwork opportunity

Neighbouring streets in any settlement may show huge wealth disparities, hence expressions such as the 'Two Islingtons' in London. Wealthy families in million-pound homes, often in gated houses and estates for security, may live literally next door to the most deprived households in the UK. The largest extremes at street scale, found by the data collection company CACI, which produces the ACORN consumer classification, were in St Albans in Hertfordshire, Birmingham, Edinburgh and Liverpool. Such micro differences could generate a rich source for fieldwork.

In 2014 it launched its 'promoting fundamental British values' scheme, deliverable by all state schools in England, to help counteract threats by fundamentalism and terrorism.

If you live in a different place to your school, there may be enough contrast in characteristics and need for regeneration to use these two places as your two chosen place examples.

Marginalisation, exclusion and social polarisation

Socially, people and groups may be marginalised or pushed out to the edges by the dominant, core culture they live in because of their language, religion or customs, and especially by wealth. Exclusion is the extreme form of marginalisation, when people's access to services and opportunities is restricted.

Social polarisation is the process of segregation within a society that emerges from income inequality and economic restructuring; it results in the clustering of high-income, elite professionals, or conversely of low-income social groups dominated by low-skilled services jobs.

Rural and urban areas help to shape each other's places and are interdependent:

- People in rural areas depend on towns and cities for many key services, including specialised healthcare, higher education and leisure. Commuter villages and towns may also depend on urban areas for employment.
- Urban people rely on the countryside for food and non-food products, and value the landscape and environment found in the rural areas for leisure and recreation. Urban dwellers may have more power than rural dwellers in a democracy like the UK, which can prove to be an issue, as with the long-running debate on fox hunting and, more recently, fracking. Urban dwellers may say they have a strong association and engagement with the countryside, through visits and holidays, but may not know it very well in reality.

Views and conflicts

Conflicts can occur among contrasting groups in communities that have different views about the priorities and strategies for regeneration. These have complex causes, including:

- a lack of political engagement and representation
- ethnic tensions
- inequality
- a lack of economic opportunity.

The following three contrasting examples show the types and level of conflicts that arise from many regeneration strategies.

Studentification

A process which affects many larger urban places, such as Leeds, Nottingham and Southampton, is that of studentification. Concentrations of transient, exuberant youthful groups, who may have little regard for their surrounding longer-term residents, have forced some local authorities to try and restrict the number of houses of multiple occupation.

Synoptic themes:

Players

Players vary in their attitudes and may have contrasting approaches.

Figure 16.7 Studentification: signs of a transient population include 'to let' signs, multiple door bells/entrances, temporary curtains, rubbish and uncared for gardens

Skills focus: Student conflicts

The impact of students and conflicts with locals would make an interesting study and require different skills. The location of student housing could be found through land-use maps, estate agents and the local university. Conflicts could be found by checking local news reports, police records, resident questionnaires and contacting the Student Union representative.

Barton Farm/King's Barton urban fringe regeneration

In 2014, the initial preparations for a new 93-hectare greenfield mixed development scheme began on Winchester's northern fringe. By 2025 there will be 2000 homes, 800 of which are designated 'social housing', a new academy primary school, a district shopping centre, light industrial units, a nursery, a district energy centre, nursing home, park and ride, and increased public open space.

The change of use from farmland to, effectively, a suburb of the city, was contested for fifteen years, especially by pressure group 'Save Barton Farm'. Winchester City Council's decision not to approve the development was overturned at appeal by the Secretary for State in 2012. Winchester was polarised during the debate. More affluent people tended to support the anti-development protests. Singletons and low-paid professionals, including teachers and nurses unable to afford inflated house prices, and those on council waiting lists, supported the project.

The Northern Powerhouse

The 'Northern Powerhouse' was a concept announced by Chancellor George Osborne in 2014 to empower cities in the North of England to work collectively

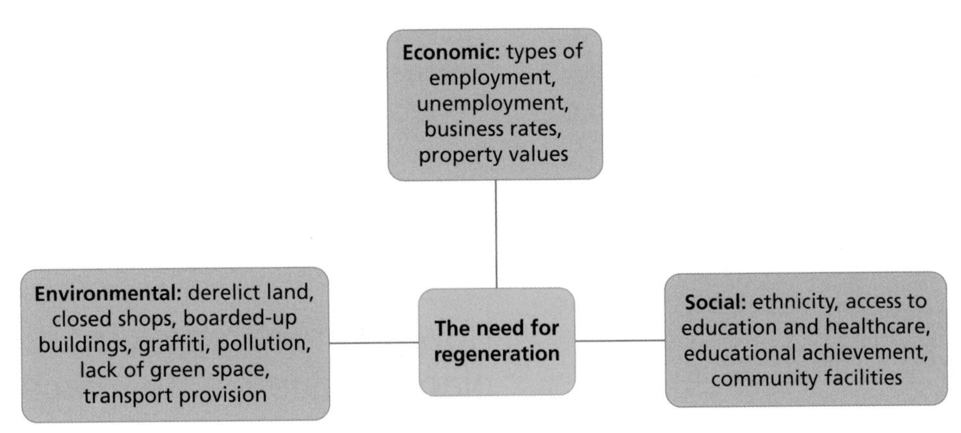

Figure 16.8 Criteria for the need for regeneration

to become a unified economic force to rival that of London and the South East.

The distinct local 'identities', built up over decades of industrial growth and decline, are likely to hinder plans, however. These separate identities are demonstrated by the fierce rivalries between football clubs and their supporters in Manchester and Liverpool. The choice of Greater Manchester, the biggest centre in the region, as the main hub is contested by Liverpool's City Council. One reason for these separate identities is poor inter-regional transport links. When built, High Speed 2 (HS2) will link the powerhouse to London. This high-speed rail link has also generated huge debate in all the locations it will run through.

In conclusion, inherent local characteristics and internal and external drivers of change result in economic and social inequalities, which are targeted by regeneration programmes. Inequalities in turn lead to differing perceptions of and attachments to places and levels of engagement. Some local areas will have places that are more in need of regeneration than others. The last section of this chapter will focus on the ways to evaluate the need for regeneration.

16.3 Local place studies (2): Evaluating the need for regeneration

There are some specific lines of investigation in this part of the Specification with respect to your two chosen places. These are not so much about place characteristics and identity, but rather about the evidence that can be used to evaluate the need for regeneration and any strategies that have been used (Figure 16.8). The following section will help guide you on collecting and using three specific types of evidence, shown as numbers 1 and 4, in Table 16.3 (note also the references to the relevant skills in the Skills focus section, which begins on page 345).

Use of statistical evidence to determine the need for regeneration

For information for any particular postcode and census output area, your main source will be the ONS census and neighbourhood statistics websites. Local authorities, both at county council and urban levels,

Table 16.3 Evaluating the need for regeneration in your local place studies

1	Is there a need to regenerate this place? (population, employment, housing, services, crime, environment or multiple deprivation; newspaper reports and other media sources – Skills focus 3 and 5)
2	What are the regional, national, international and global influences on the characteristics and lives of people in your place? (Oral histories – Skills focus 2)
3	How have economic and social changes influenced people's identity? (Social media – Skills focus 1)
4	Is there any statistical evidence showing the need for regeneration in your chosen local place (scattergraphs and Spearman's rank – Skills focus 8)
5	Are there any regeneration schemes in your locality?
6	Who are the stakeholders involved, and what is their role?
7	How effective have any strategies been? What do locals think?

Skills focus: GIS and testing the strength of relationships

See Skills 7 and 8 in the Skills focus section (page 348).

should also have a wealth of geo-located data derived from neighbourhood statistics and other sources, and may have interpreted them for you already. They may be more up to date than the 2011 census. They also often have simple GIS systems that you can manipulate yourself, for example Cornwall County Council and Manchester City Council.

Using the criteria in Figure 16.8, you should choose a range of economic, social and environmental data to see the need for regeneration. This will need comparison with the surrounding region. Ten to 30 data pairs should be used for Spearman's rank to be used successfully, such as twelve wards or twenty LSOAs. You may also be able to find comparable data to see if your chosen criteria have changed over time. Spearman's rank tests the strength of any statistical relationship, for example between employment levels and health, or life expectancy and education levels; your role as a geographer is to interpret the statistics and put them into context.

How different media can provide contrasting evidence, questioning the need for regeneration

Many sources may be inherently challenging of the need for either change or the type of change proposed in regeneration. The same data may be manipulated to give different 'spins' on any evidence.

Skills focus: Newspapers

See Skill 3 in the Skills focus section (page 346).

Skills focus: Evaluating different media

See Skill 5 in the Skills focus section (page 347).

How different representations of places influence the perceived need for regeneration

There is a large range in the types of representations that may influence people's perception about places and whether they need regeneration. Here are some possible sources: newspapers, as outlined before; news reports and documentaries on TV and the internet; YouTube video clips uploaded by individuals; local blog sites; estate agents; tourism and local enterprise offices; postcards depicting the locality; local authorities; and even graffiti.

Documentaries and video sources are very powerful image creators and may be particularly inflammatory if only a small number of people are interviewed or just a few images are shown. Local community internet forums may be biased towards the dominant writers; the 'silent majority' may not be represented.

As at the end of Chapter 16, you should remember that:

- these two place investigations should be undertaken by you
- you should try to draw out the similarities and differences between your places
- you will be expected to feed specific detail from these investigations into your examination answers.

Review questions

1. How does inequality affect people's perception of a place?
2. What factors affect people's varied lived experience with their home place?
3. Why does people's engagement with their home place vary?
4. Why do some local areas need more regeneration than others?
5. How can these inequalities be measured by secondary and primary research?

Further research

Take a look at the UK's gross disposable household income (GDHI) per head.
This graphic from the ONS is an interactive set of data from 1997–2013:
www.neighbourhood.statistics.gov.uk/HTMLDocs/dvc168/index.html

Read *Why Places Matter* by Living Streets, a national charity working to create safe, attractive and enjoyable streets:
www.livingstreets.org.uk/sites/default/files/content/library/Reports/why-places-matter-final-web.pdf

Look at the Office for National Statistics 2012 audit on quality of life in cities; find it by entering 'perceptions of city life' into the search box: www.ons.gov.uk/ons/index.html

The Joseph Rowntree Foundation is an NGO that studies social change and poverty in the UK; visit the website and look at reports on topics such as living through change in Bradford (enter 'living through change' into the search box): www.jrf.org.uk

Read the groundbreaking 2010 report on inequality by Professor Marmot:
www.instituteofhealthequity.org/projects/fair-society-healthy-lives-the-marmot-review

How regeneration is managed

How is regeneration managed?

By the end of this chapter you should:

- be aware of the key role national governments play in regeneration
- be aware of policies local authorities use in making places attractive for investment
- understand the role of rebranding as part of regeneration.

This section of the Specification focuses on the role of national and local governments in tackling inequalities and problems resulting from socio-economic changes through regeneration (see the key term in Chapter 15, page 217), rebranding and re-imaging. You need to know these key terms in order to understand the following concepts and examples, based on the UK.

Key terms

Rebranding: The 'marketing' aspect of regeneration designed to attract businesses, residents and visitors. It often includes re-imaging.

Re-imaging: Making a place more attractive and desirable to invest and live in or visit.

Infrastructure: The basic physical systems of a place:

- economic infrastructure includes highways, energy distribution, water and sewerage facilities, and telecommunication networks
- social infrastructure includes public housing, hospitals, schools and universities.

17.1 The role of national government in regeneration

Infrastructure

The government plays a key role in regeneration by managing the country's economic, social and physical environments through various political decisions. Investment in infrastructure and addressing issues of accessibility are seen as major factors in maintaining economic growth. Infrastructure projects have two main characteristics: high cost and longevity, hence needing government funding. Since the 1980s there has been increasing privatisation and partnerships between government and private financiers. The private sector is used to design, build, finance and/or maintain public sector assets in return for long-term payments or profit from the initial revenue generated.

The Infrastructure and Projects Authority, part of the Treasury, was formed in January 2016 by the merger of several departments. It oversees long-term infrastructure priorities and secures private sector investment, including Crossrail and broadband. There are many other departments involved in regeneration which, since devolution in 1999, differ slightly between Scotland, Wales, Northern Ireland and England. The list below shows an overview of the departments involved in both 'hard' regeneration (capital investment, physical buildings, infrastructure) and 'soft' regeneration (planning, skills and education).

- The Department for Communities and Local Government (DCLG) aims to create 'great places to live and work' and empower local people to shape their own places. It includes the Planning Inspectorate and Homes and Communities Agency, which oversees Environmental Impact Assessments (EIAs) and Local Enterprise Partnerships (LEPs).
- The Department for Culture, Media and Sport (DCMS) markets the UK's image abroad, and protects and promotes cultural and artistic heritage and innovation. It includes Sport England and the National Lottery.
- The Department for Environment, Food and Rural Affairs (Defra) overseas the Environment Agency, Natural England and the National Park Authorities. It advocates environmental stability as part of sustained economic growth.

- UK Trade and Investment supports UK businesses and encourages inward investment. It oversees the Regeneration and Investment Organisation (RIO), which is involved in large-scale **flagship regeneration projects**. By 2015, RIO had 40 large projects of over £100 million needing private investment, such as Liverpool Waters and Sutton Drug Discovery Complex.

When investigating the types of regeneration in your local area you will encounter some of these major government players as well as county and local councils, and other important players. These may be:

- mayors of metropolitan regions
- non-governmental organisations, such as pressure groups, environmental groups, charities and businesses
- local individuals.

There may be differences of opinions between people that need to be resolved by the government or the legal system, as exemplified in the long-running debate over airport and train infrastructure improvements to retain and generate economic growth.

Key term

Flagship regeneration projects: Large-scale, prestigious projects, often using bold 'signature architecture'. The hope is to generate a positive spin in a place.

Synoptic themes:

Players

There are many players involved in regeneration strategies, many with differing viewpoints, and some with more power than others.

Skills focus: Force field analysis

Force field analysis is a useful geographical tool: you can categorise the types of viewpoint (economic, social and environmental) then weigh up the strength of the viewpoints and plot them as a split bar chart. An example is given on page 270 (Figure 18.7).

Airport development

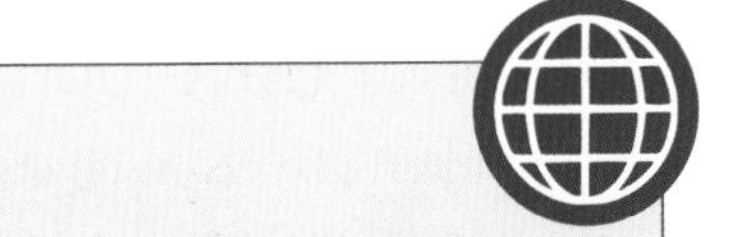

In 2015, after twelve years of debate, the Airports Commission gave a clear and unanimous recommendation for an expansion plan at Heathrow, including a third runway (Figure 17.1). The cost, an estimated £18.6 billion, will be privately funded but some of the support infrastructure will be publically funded. The recommendation has polarised views:

- Anti-expansion: London's mayor; many MPs; local and national protest groups such as Plane Stupid, Greenpeace and the Campaign to Protect Rural England (CPRE); and some high-profile celebrities such as Alistair McGowan.
- Pro-expansion: business leaders, British Chambers of Commerce and Richard Branson. Heathrow airport argues that the hub operates near full capacity and, since the South East is the main earner of GVA in the UK, expansion is essential to keep up with demand. It could generate £100 billion of benefits nationally, protect the current 114,000 local jobs and create over 70,000 new ones.

High Speed Two (HS2)

The Department for Transport's company HS2 Ltd is responsible for developing and promoting the UK's new high-speed rail network, High Speed Two, from London to Birmingham, Manchester and Leeds. It is key to the large-scale Northern Powerhouse regeneration scheme. The two phases, with end dates of 2026 and 2033, show the long period of time needed for such large infrastructure projects. There has been great controversy about the costs, exact route of the line and its effect on those living nearby (Figure 17.1).

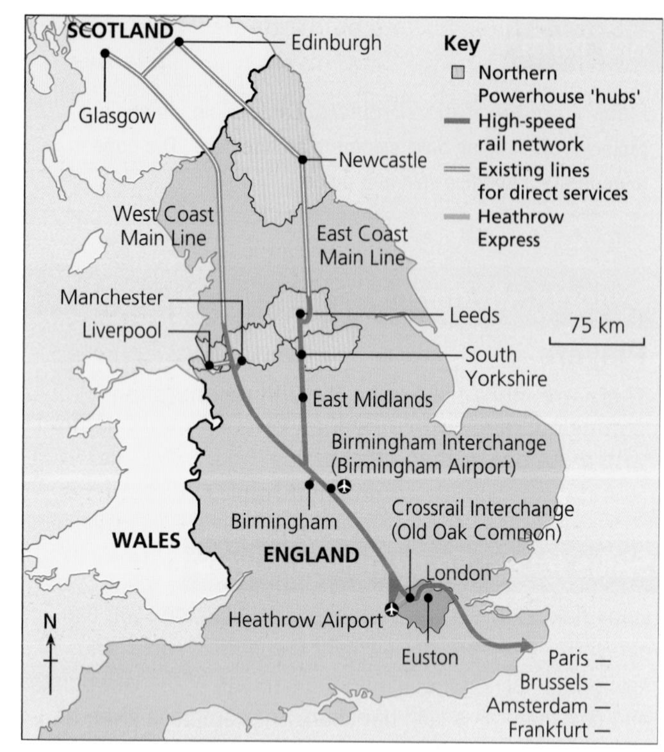

Figure 17.1 How infrastructural development is critical in economic growth: HS2, Heathrow and the Northern Powerhouse

Types of development

As the main players in regeneration, national or local governments make decisions that affect the rate and type of development which, in turn, affect the economic regeneration of places. These decisions may range from general planning laws and restrictions, house-building targets and house affordability, to permission for fracking.

There has been a long history of regeneration, replacing 'redevelopment' and 'renewal' schemes of the 1960s and 1970s. These were more restricted programmes of slum clearance and housing improvements at neighbourhood level, rather than large social or economic strategies. There has been a growing trend for less top-down, centrally dictated and funded schemes and more bottom-up, localised decision making. This has had varying effectiveness since local participation is rarely representative of all the people in any particular place. Party politics plays a great part in regeneration, and not all schemes are classed as successful, resulting in many changes in policy (see Chapter 18).

Skills focus: Past and present regeneration

When researching local regeneration, there will probably be some 'legacy' regeneration projects and strategies of the past to consider as well as current ones.

New Labour, in power from 1997 to 2010, was pro-public sector and social housing, and emphasised local democracy and 'active citizenship'. The Single Regeneration Budget and New Deal for Communities prioritised 'capacity building'. This means empowering people's participation in the urban regeneration process. They also introduced the controversial Pathfinder Programme, a scheme of demolition, refurbishment and new-building, which paradoxically caused more problems for some residents than improvements; it was axed in 2010.

Austerity regeneration policies have featured in the coalition and Conservative governments since 2010, with less funding from government and outside investment available. Any development funded has to prove it will generate economic gain. Four key policies began or continued in 2015:

1. In England, 39 Local Enterprise Partnerships (LEPs) were given responsibility for determining local priorities, including housing and infrastructure, and leading economic growth in the local area.
2. City Deals, which encourage local authorities to co-operate on a city-area basis on key economic development, regeneration and transport policies. These focus on promoting economic growth where conditions are favourable, but do not target specifically disadvantaged areas. Cities bid for funding to create more 'sustainable, smarter cities'.
3. The Coalfields Regeneration Trust and the Coastal Communities Fund are special cases and still receive funding.
4. Directly elected Mayors, such as in London and Liverpool (see Chapter 16, page 240).

Factors affecting regeneration policies

There are several factors affecting the type of regeneration policy in any place (Figure 17.2).

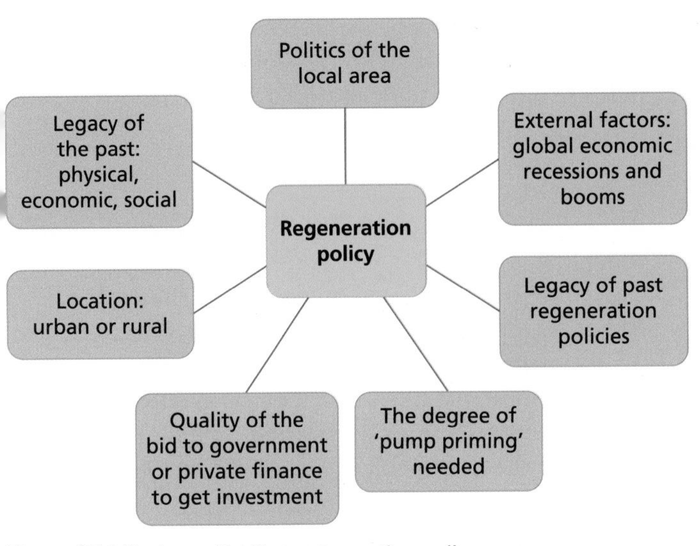

Figure 17.2 Factors affecting regeneration policy

Key concept: Pump priming

Pump priming means that the government allocates funds for regeneration expecting outside investment to help, especially needed if toxic waste needs removal, for example, or if the area is very large. Private and charitable investment is then expected at a higher ratio.

Three aspects of the role of government planning

Planning laws

Planning is about deciding how land is used. It helps create places that people want to live, work, relax in and invest into. It shapes places and includes 'place marketing' to either improve an existing place or completely change its image. Central and local governments have tightly controlled all aspects of development since 1948 through a plan-led system. Since the players involved may differ in views on its use, planning also includes 'mediation' to try to reconcile these differences.

National interests may override local interests in planning decisions. Since 2010 the government's National Planning Policy framework has focused planning on stimulating economic growth.

Planners may regulate markets by using a tool called 'planning gain' whereby they allow development if there is a benefit to the local community, for example, some social housing built as part of a new housing scheme or a new road system or community centre paid for by the developer.

Slow decision-making on the future of a place may result in planning blight: investors are unwilling to commit until a decision is made, house prices may fall and trap residents into not being able to move, landlords may not maintain properties and a downward spiral results.

When a developer wants to implement a scheme, they submit a proposal to the local authority. This decides whether it fits with the current local plan, which itself fits into national guidelines. There are appeal procedures if it is rejected, and sometimes a public inquiry is held, which can be very costly and even last decades. The EU also has a say in planning, and has made Environmental Impact Assessments compulsory for some developments. From 2013, where local authorities are considered to have a 'poor' planning record of decisions, developers were able to apply directly to the DCLG's Planning Inspectorate because the government sees delays as negative to economic growth.

Planning for fracking

The process of obtaining gas from shale rock by hydraulic fracturing is known as fracking. The government sees it as a national priority to increase secure energy supplies and economic prosperity. Exploration licences for oil and gas companies are given by the Department of Energy and Climate Change (DECC). EIAs are required, but it has caused a great uproar in many constituencies. There are many local and national anti-fracking pressure groups, for example Frack Off. From 2015 a new fast-track system was introduced to deal with licence applications. One of the first tests of the system was by Cuadrilla Resources, which was blocked from drilling by local authorities in Lancashire.

To look into this further, put your postcode into the Friends of the Earth website to see the state of play in your own 'back yard'.

Planning for housing needs

The government's involvement in housing supply started from the late nineteenth/early twentieth century. There have been many political twists and turns and changes in priorities since then.

- Labour-led governments have fostered social housing, and vast state social housing schemes were built from 1918 to 1940. Old-style Regional Planning Authorities were encouraged to set building

Figure 17.3 The range of planning for housing in an inner-city centre

targets for local authorities in the 1990s, often controversially, because they did not deliver enough new houses.

- Conservative governments have favoured a market-led approach, with the contentious Right to Buy policy resulting in the sale of more than 2 million council homes from 1980 to 1995. The expectation that charitable or private housing associations would replace local authorities in building lower-cost homes has not happened. There is now a shortfall in supply, with long waiting lists. The 2011 Localism Act abolished Regional Planning Authorities. Post-2011 policies centre around local decisions on housing supply. This has resulted in:
  - Underinvestment (partly due to the economic downturn since 2008).
  - Shortfall of private and rented accommodation.
  - Large number of empty properties.
  - Hotspots of inflated housing areas, especially in London and the South East.
  - Planning restrictions hindering developers.
  - Social changes, especially increases in the numbers of elderly people and single households, adding pressure on the housing market.
  - Paradoxically, those on lower incomes sometimes paying proportionately more for their rented housing than wealthier home owners, so affordability is a key issue nationally.

Government policies on international migration and deregulation of capital markets

Government policies have significant impacts on the potential for growth and both direct and indirect investment. Two examples are now considered: the direct degree of involvement in capital markets comprised of banking and the stock exchange, and more indirect migration policies that influence labour supplies and skills and, hence, GDP.

Government policy on the deregulation of capital markets

Capital markets are based on dealing in shares, bonds and other long-term investments, so any government involvement may have significant impacts on the potential for national and local growth. After decades of tight restriction on banking and capital markets following the depression in the 1920s, in 1986 the Conservatives began the process of deregulating financial markets by introducing a policy known as the 'Big Bang'. This was to encourage more investment as London was becoming uncompetitive and losing business to other financial centres; it also coincided with the rise of electronic trading.

Ending the Stock Exchange's monopoly and removing entry barriers encouraged European and US banks to open in London, resulting in banking, finance and business services creating almost 30 per cent of the UK's GDP by 2008, double that of 1986. The skyscrapers of London's Canary Wharf are the visible image of this new investment and prosperity.

The policy of deregulation was added to by subsequent Labour governments, but these 'light touch' regulations on banks are partly blamed for the 2008 financial crisis, subsequent low economic growth and austerity measures. In the UK, the government has increased monitoring and regulation through the Financial Services Act of 2012. This strengthened the role of the Bank of England, critical in maintaining the stability of financial institutions and markets through its use of the lender-of-last-resort function and regulatory committees. It also set up the Financial Conduct Authority.

Government policy on international migration

The economic argument for immigration is increased national GDP. Apart from extra taxes and production generally, well-qualified as well as lower-skilled people can fill skills shortages. Immigrants tend to be mainly younger adults and their families, making places with an ageing population structure more sustainable. Historically there have been many waves of immigration into Britain, but the pace has accelerated markedly due to the ten countries that joined the EU in 2004.

ven relatively small numbers of illegal immigrants, specially when clustered into certain places, can cause nedia hysteria and influence party politics and policies. ince the mid-twentieth century there have been gnificant changes in government immigration policies, which are simplified below:

- 1950–77: restrictive policy to limit the new and unexpected rise of immigration from New Commonwealth countries.
- 1997–2010: pro-immigration policies by the Labour administrations.
- 2010 onwards: restrictive policy by the coalition then Conservative administrations, described by Prime Minister David Cameron as 'good immigration, not mass immigration', meaning only the most 'beneficial' are allowed to stay in the UK. The aim of reducing overall net migration levels, from hundreds of thousands to tens of thousands by 2015, has proved difficult to achieve.

olicies centre on minimising opportunities for abuse nd being more selective about the criteria for entry. verseas students have been targeted by scrapping ost-study work visas. Incoming extended families nd 'benefit tourism' have also been reduced. From 015 there were proposals for restricting work visas to pecific skills shortages and specialisms, and a higher alary threshold before people are allowed residence.

The 2016 referendum on the UK's position in the EU may influence such trends.

By 2015 there were 117,161 refugees, pending asylum cases and stateless people in the UK – notably from Eritrea, Pakistan, Iran and Syria – just 0.24 per cent of the total population. The majority of asylum seekers do not have the right to work in the UK so must rely on state support with housing and a weekly cash payment of approximately £36 per person. They cannot choose their location; hard-to-let properties, which council tenants reject, are often used. These are often substandard homes in 'white' estates found in Liverpool, Middlesbrough and Glasgow, with resulting social issues.

17.2 Local government policies

Local plans

There is competition between local authorities to create attractive business environments for investors and workers who are highly skilled and paid and who can choose where to work more easily. They develop local plans which designate specific areas for development; Science parks are a good example, since knowledge-based industries underpin the UK's current economic growth.

Cambridge Science Park

The purpose of science parks is to represent areas as being attractive for inward investment. These private or public areas provide attractive environments, purpose-built buildings and infrastructure, and advice and networking groups. The first was built at Stanford University in the USA in the 1950s. In the UK, there were more than 100, employing around 42,000 people in 2015.

One of the first and largest is the Cambridge Science Park, closely linked to the university. Built in the 1970s on a redundant defence site, it grew rapidly in the 1990s when life sciences began to flourish globally. Expansion in the early 2000s has attracted many foreign TNCs, such as AstraZeneca. Life science is now the third largest UK growth sector economically. In the future, the Regeneration and Investment Organisation aims to consolidate the 'Golden Triangle of Life Sciences' between Oxford, Cambridge and central London by 2018. The existing campus of the Institute

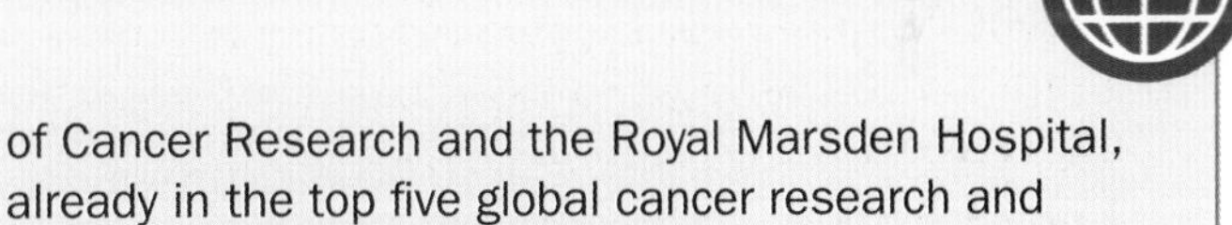

of Cancer Research and the Royal Marsden Hospital, already in the top five global cancer research and treatment facilities, will be the focus of the Sutton Drug Discovery Complex.

Figure 17.4 Cambridge Science Park

Tensions created by the 2012 Olympic Games

Clays Lane Estate was a housing co-operative development built in 1977, creating a community for vulnerable single people in Newham, London. Unfortunately the site was designated for the Olympic athletes' village and the 430 residents were forced to move. There was huge public opposition and even a public inquiry. Several small businesses were also evicted from the Olympic site, such as Forman's salmon smokery.

The role of local interest groups

Regeneration planning and management involves a range of players, sometimes with differing interests and aims. Local interest groups play varying roles in regeneration policies as we have seen already in the Heathrow expansion, HS2 and fracking debates. There are often tensions between groups that wish to preserve places and those that seek change. They may be categorised by their viewpoint or stance:

- socio-economic, for example city and town Chambers of Commerce, addiction treatment centres, youth and retirement groups, and trade unions
- environmental, for example local conservation or preservation societies in rural and urban places.

Key concept: Cold spots

Ironically some most needy areas lack active voluntary sector involvement, called 'cold spots'. An example is the ex-mining area of Camborne, Cornwall.

Areas with affluent retirees tend to have more vociferous and mobilised local interest groups: in Winchester the local council was taken on by a quickly formed pressure group called Winchester Deserves Better, delaying the Silver Hill mixed development scheme in the city centre (see Chapter 18).

The range of regeneration strategies

A variety of strategies are used; we will focus on retail-led plans, tourism, and leisure and sport, as shown in Figure 17.5.

Retail-led plans

National and local governments are heavily involved in retail planning (Figure 17.6). Local authorities decide on changes of use to buildings and can influence shop types and locations of malls, pedestrianised areas and alcohol-free zones.

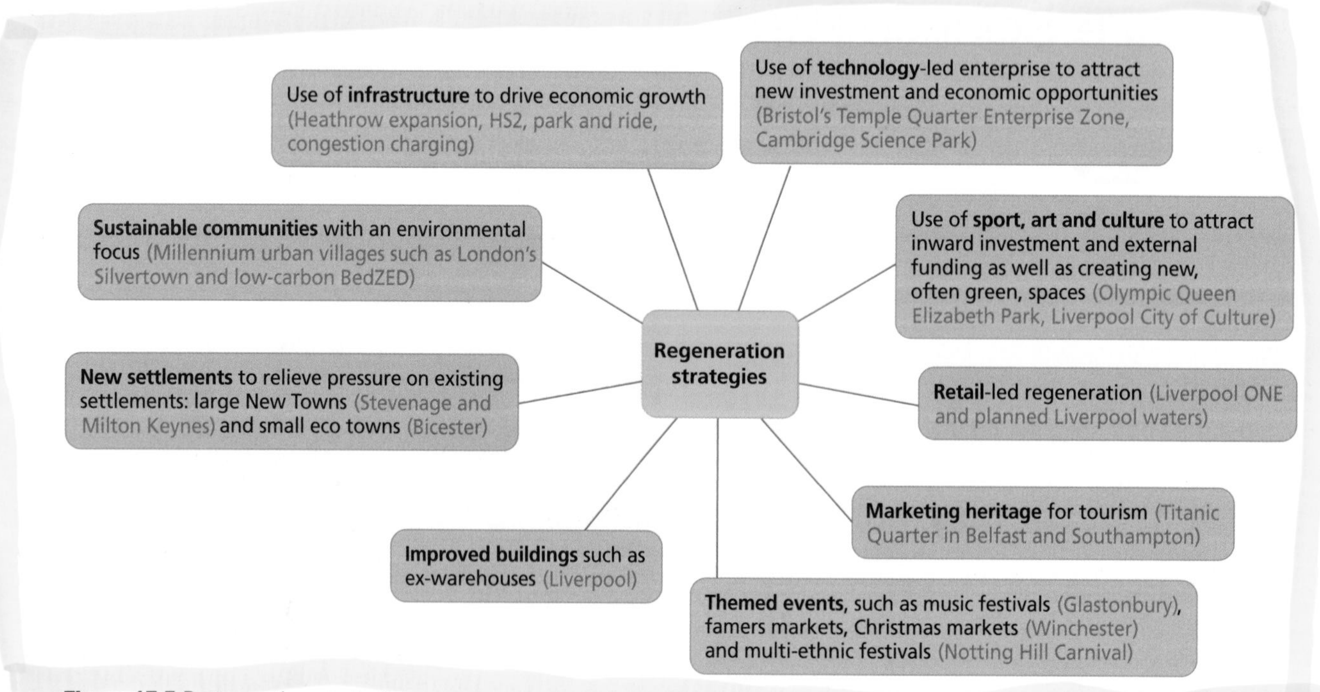

Figure 17.5 Regeneration strategies

The 2014 Portas Review highlighted the two main challenges to the high street as:

- competition from out-of-town centres
- the rapid growth of internet shopping.

It led to government support of £1 billion to ensure growth in high street jobs.

The University of Southampton's Retail Research Group shows that 'convenience' shopping has fundamentally changed from the late twentieth century one-stop shop, to a greater 'topping up' of goods in local stores. An increased interest in specialist retailers and an increased demand for leisure means that high streets offering a mixture of bars, restaurants and cafes, beauty services and gyms are more likely to prosper. Government actions in 2015 included allowing more click-and-collect locations, pop-up shops and gyms, encouraging street markets, changes to business rates to help smaller ones compete with chains, competitions such as Britain's Best High Street and the Future High Streets Forum.

Figure 17.6 The Great British high street

Skills focus: Surveys

Land-use surveys are easy to carry out on high streets, although you may need permission inside shopping malls. A clone town survey may show changes over time.

Tourism and leisure-led regeneration

There are few rural or urban areas in the UK which do not use this growth sector to help or lead regeneration. There is great diversity in types, ranging from informal individual households offering B&B, custom-built private centres such as Center Parcs, purpose-built leisure complexes in towns and cities, to

Declining coastal communities

These receive special attention from the government because often they have higher than average deprivation levels. There are various reasons for their economic decline: many are seaside resorts with a tourism legacy, but which largely fell into a spiral of deprivation by the 1970s when more holidays abroad and jet travel became commonplace. Large numbers of retirees, unemployment, transient students, immigrants and poor quality health and housing are all contributory factors.

Some resorts, such as Bournemouth in Dorset, have managed to reinvent themselves by diversifying into a business and conference hub while holding on to its family holiday image. It has also developed into a stag/hen and clubbing hotspot. Adjacent resort Boscombe was innovative in creating an artificial surf reef to try to re-image the town, funded by the sale of a car park to Barratt Homes, which built exclusive flats called Honeycombe Chine (Figure 17.7). In the 1990s central government schemes such as the Single Regeneration Budget, used in Boscombe's town centre, were important in tackling deprivation, but the problems have remained and even increased.

Current policies are overseen by the government's Big Lottery Fund with a focus on fostering economic growth. By 2015 it had spent £119 million and attracted £200 million in inward investment, creating an estimated 12,000 jobs. Projects include Europe's first National Coastal Academy in Bournemouth. In 2015, following the localism policy, Coastal Community Teams were set up whereby partnerships can apply for funding to develop plans and bid for capital funding for local projects. The Coastal Revival Fund started with an initial £3 million to help coastal heritage or community assets to have new economic uses.

Figure 17.7 The start of the Boscombe Spa regeneration project in 2008

whole settlements devoted to tourism such as seaside resorts. It is a volatile industry, however, dependent on the weather, its image and fast changes in preference which may reflect developments in technology, social forums and websites such as TripAdvisor and terrorist attacks.

Sport-led regeneration

Many areas have used sport-led regeneration, not only for the spinoff from the construction and running stages of a major sporting event and associated jobs, infrastructure and buildings, but as a catalyst for longer-term regeneration. This may be by one-off or regular events, such as World Cups, the Olympic Games and Commonwealth Games, or the building of long-running facilities like stadiums.

One of the reasons London won the bid for the Olympics in 2012 was because it had a 'legacy' plan in place, even if it has not all materialised. It is a model that Rio de Janeiro adapted for its Games in 2016. London also hopes to win the 2026 Commonwealth Games. The media coverage before and during such events helps put the place on an international stage, and inward investment is a critical spinoff. Chapter 18 weighs up the impacts of such regeneration.

Strategies to make the original Olympic park attractive for investment were centred on flagship developments:

- The International Quarter is a 37,000 sq m new office area, a new business 'frontier', next to Europe's largest shopping centre, Westfield Stratford City. This extends the axis from the City of London and the older regeneration projects of Canary Wharf and the Docklands.
- 'Here East' is a new digital and creative industry hub on the former Olympic media site. By 2019 a new cultural and education centre, Olympicopolis, will help expand the V&A Museum, Sadler's Wells, and possibly the first UK branch of the Smithsonian Institution.
- The main stadium with be home to West Ham United by the 2016–17 season.
- The Queen Elizabeth Olympic Park has 560 acres of attractions and has had 4 million visits since it opened. Permanent sports venues still in use include the Aquatics Centre, Velodrome and multipurpose Copper Box Arena.
- 10,000 new homes, two primary schools, a secondary school, nine nurseries, three health centres, plus multipurpose community, leisure and cultural spaces.

Culture-led regeneration

Culture is the background for many different strategies, from City of Culture and some music festivals at an international and national scale, flagship arenas and art galleries, to small conservation areas and 'cultural quarters' in cities. Since Victorian times, there has been demand from people wanting to visit places either associated with the life or the works of famous authors, musicians and painters, which has been exploited by local councils and private tourism providers. Some have marketed place associations, such as London with its Charles Dickens connection, or Bristol with the graffiti artist Banksy, which offers a self-guided tour through a smartphone app.

Rural areas, such as Thomas Hardy Country in Dorset, and villages, such as Grasmere in the Lake District where Wordsworth lived, are classic examples. One such large-scale area is Brontë Country in West Yorkshire. More recently, books turned into films, such as *Harry Potter*, or TV series such *Downton Abbey*, set in Hampshire's Highclere Castle, use places which then attract great numbers of visitors and help boost local rural economies. Other examples are very innovative, such as Cornwall's Eden Project.

Public and private rural diversification

Approximately half of all farms in the UK use some form of diversified activity in their faming businesses to boost income, dependent on location, land type and the entrepreneurial aspirations of individual farmers or landowners. This is supported by governments and the main trade union: the National Farmers Union.

Grants are available from Defra's Rural Development Programme (RDP) and also from commercial banks and charities. The EU has a policy of helping diversity as well through the reformed Common Agricultural Policy, administered through the Single Payment Scheme.

The National Trust is an influential charity in rural areas, the largest farm owner nationally and the second largest landowner in the UK. Indeed, 36,000 individual people – less than 1 per cent of the total population – own 50 per cent of rural land, so the decisions of landowners are critical in rural regeneration as well as official policies.

There are several types of diversification:

- **Agriculture-based**: producing and selling speciality cheeses; farming unusual animals (deer,

Powys Regeneration Partnership and the LEADER programme

This is an example of a co-ordinated and integrated approach to economic and community regeneration in a rural area. It is funded by the Welsh government and EU using the LEADER programme (Links between actions for the development of the rural economy). It is a key source of funding for deprived rural areas, using local knowledge of the value of a place to promote grass roots, community-led rural development.

Between 2011 and 2013, grants of over £4 million helped 310 business and community projects across Powys, creating 36 full-time jobs and safeguarding 80 more. The next phase runs from 2014 to 2020. Powys County Council helps to deliver this support through projects called Sustainable Tourism, Farm Diversification and Resilient Powys. Grants are given for new glamping sites, welding workshops, equine enterprises, wildlife tourism and projects showing a 'sense of place'.

llama); growing non-food crops (speciality flowers), pharmaceutical crops (opium poppies) or energy crops (rape); developing farm shops; craft-making facilities; shoots; training in rural crafts such as dry stone walling. Climate change is also allowing new opportunities in the UK, such as a vineyards, biofuels, olives and almonds.

- **Non-agricultural**: redundant farm buildings converted to offices, light industry or tea shops; campsites; horse livery. At a larger scale paintballing, clay pigeon shooting, golf, motocross, car boot sale sites and country parks may be developed, plus annual music and art festivals such as Glastonbury. Wind and solar farms may feature.
- **Environmental schemes**: funded by RDP, such as Natural England's Environmental Stewardship Scheme, and planting woodland (administered by the Forestry Commission).

The following place context on Powys shows a mix of public and private sources used in rural regeneration.

17.3 Rebranding

Re-imaging and rebranding

Rebranding attempts to represent areas as being more attractive for potential investors and visitors, and also locals, by changing public perception of them, called re-imaging. For the UK's deindustrialised cities, rebranding can stress the attraction of places, creating a specific place identity that builds on their industrial heritage. This can attract national and international tourists and visitors.

Skills focus: Evaluating different media portrayals

A simple internet and YouTube search should produce some useful data on media portrayals. See also Skill 5 in the Skills focus section (page 347).

Few schemes are solely devoted to one land use only, although the original catalyst may be focused on one theme, such as retail or sport. Waterside developments and town centres are good examples of mixed-use developments at varying scales.

Table 17.1 shows a range of media portrayals of the planned Liverpool Waters scheme, viewed as so important to the long-term regeneration of Liverpool that it was approved by the Secretary of State for Communities and Local Government without a public enquiry in 2015.

Rebranding deindustrialised places

Rebranding often focuses on the attractiveness of places. Specific place identities building on their historical heritage can attract national and international visitors. We will focus on one famous city here: Glasgow. There has been a long history of rebranding by planners and business leaders, using both long-running attractions and one-off events. Many of these have had positive legacies for local areas, the city and country as a whole. A mix of sport, leisure and cultural catalysts have been used, with infrastructure and improved physical environment being key to success.

Liverpool Waters

The Atlantic Gateway project is nearly a 65-km long, £75 billion growth corridor from the Port of Liverpool to Manchester. Liverpool Waters is part of this: it covers 2 km of waterfront with plans for 9000 flats, shops, office space, a new cruise terminal and cultural buildings designed to attract Chinese businesses, reflecting the twinning of the city with Shanghai. The flagship 55-storey Shanghai Tower will be the tallest skyscraper outside London. The mayor views the scheme as 'unprecedented in its ambition, scope and potential to regenerate a city'. Other players have contrasting views on the development (Table 17.1).

Figure 17.8 Liverpool Waters: A 'Waterfront for the World'

Table 17.1 Extracts from discursive media's views on regeneration: Liverpool Waters

City Council rebranding company, Liverpool Vision,	Aims to transform perceptions of Liverpool by building Liverpool Waters
Property developers, Peel	Regenerating a 60-hectare historic dockland site to create a world-class, high-quality, mixed-use waterfront quarter: a 'Waterfront for the World'
***Architects' Journal*, July 2015**	UNESCO wants a moratorium on the scheme because of the potential damage it could cause the existing World Heritage Site
Local press: *Liverpool Echo*, 2014	Property giant Peel will start the development of the £5.5 billion Liverpool Waters scheme and seek other developers in China for the 30-year project
Local blog: YoLiverpool!	Who cares, and who needs their [UNESCO's] approval of our city, anyway? I'm all for it, if it leads to jobs, etc.

Rural rebranding strategies

The twenty-first century has seen a 'new rural economy' develop, with rural areas much more like urban economies in their activities, with most employment in services rather than primary production. There has been a huge swing towards quality food products, leisure and tourism and environmental management. More accessible areas have benefited more than remote areas on the whole, hence the growing differences between rural areas, partly because of their response to these changes. Making money from the appeal of landscapes, rural environments and local cultural heritage is increasingly important.

Table 17.2 A timeline of Glasgow's rebranding and regeneration

1983	Glasgow's 'Miles Better' campaign helped it become European City of Culture
2004–13	The 'Scotland With Style' rebrand aimed, with considerable success, to attract trade to the city, with new hotel chains, conference centres and flight routes by easyJet. Its achievements include the UEFA Cup Final in 2007, World International Gymnastics and the European Cheerleading Championships.
2014	Glasgow's Commonwealth Games was the largest event so far: 1.26 billion people saw or read about Glasgow. The eight-year project, from winning the bid to hosting the event, had a very positive legacy with over £740 million gross injected into Scotland's economy, including £390 million in Glasgow. It supported on average 2100 jobs annually, with about half in the city, and helped youth unemployment. The East End of Glasgow, Rutherglen and South Lanarkshire, benefited most. Land remediation, transport infrastructure and sports facilities such as the Sir Chris Hoy Velodrome, Emirates Arena and Tollcross International Swimming Centre were implemented, all of which are now being used by the public and will be used for future events.
2015	Glasgow hosted the Turner Prize and was shortlisted for the 2018 Youth Olympic Games. The latest schemes encourage bio-medical science, financial and business services, low-carbon industries, higher engineering, manufacturing and design, and higher and further education, with public consultation on new brand images.

Synoptic themes:

Players

A key player is Defra, which works with a range of partners including local government networks, civil society organisations such as Pub Is The Hub and the Plunkett Foundation, local action groups, business groups, charities such as the National Trust and organisations like the Rural Coalition which includes the NFU and CPRE. Seventeen Rural and Farming Networks (RFNs) have been created in England to help organise rural policies.

Like urban regeneration, the traditional top-down model of rural development is now viewed as less effective than grass roots, diverse, networked partnership approaches.

Key concept: Rural proofing

Governments are increasingly seen as a facilitator and enabler rather than provider or manager. 'Rural proofing' is carried out, meaning checks on the design, development and review stages of national, devolved and local policies.

There is a range of rural rebranding strategies in the post-production countryside, in both accessible and remote areas, as shown below. These strategies are intended to increase the attractiveness of places to national and international tourists and visitors, improve services for locals as well as providing diversity in income other than the traditional employment sectors of agriculture and forestry. Some have conservation aims too. Two contrasting rural examples are outlined in the following section.

Kielder Water and Forest Park: An 'outdoor nature playground'

Kielder in Northumberland is one of England's most remote villages, dramatically altered by the creation of Europe's largest coniferous plantation in the 1930s and an 11 km-long reservoir in 1975. Kielder Water and Forest Park attracts 345,000 visitors annually. It differs from a National Park because it has no major national funding, and generates revenue from car parking and the facilities on site. Conservation is also integral to its plans, since Kielder has rare red squirrels. The Observatory for Dark Skies is attracting 'astrotourism'. This special, human-made place demonstrates how large numbers of players – in this case the Kielder Water and Forest Park Development Trust and ten others – can work together successfully.

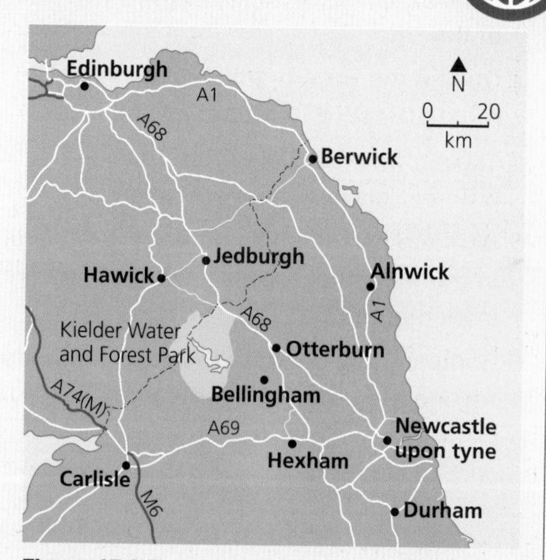

Figure 17.9 The location and scale of Kielder

Rebranding by literary associations: Brontë Country

The bleak, wild moorland of the Pennine Hills in West Yorkshire and East Lancashire was the inspiration for the classic literature of the Brontë sisters, including the works *Wuthering Heights* and *Jane Eyre*. It is marketed to attract visitors and revenue. The village of Haworth, the Brontës' birthplace, has become a hotspot for tourists. Its quaint cobbled high street with small stone houses was once used by sweatshop-type textile weavers and has been preserved from the early nineteenth century.

- Heritage and literary associations, e.g. Brontë Country.
- Farm diversification.
- Specialised, valorised products such as cheese or buffalo meat production.
- Outdoor pursuits such as equestrianism, paintballing, country parks.
- Adventure leisure and tourism such as climbing, white-water rafting, 'tough mudders'.
- Technology and infrastructure development, e.g. broadband access opening up mail order in rural areas.
- Themed events, e.g. famers markets and specialist food fairs, like Alresford Watercress Festival.
- Ecovillages and sustainable communities, e.g. Poundbury, Dorchester.

Skills focus: Kielder research and oral accounts

Investigate the positive and negative effects of the changes in this area by visiting Kielder's official website (www.visitkielder.com). More anecdotal sources would include online oral histories. See Skill 2 in the Skills focus section (page 346).

Review questions

1. Differentiate between regeneration, rebranding and re-imaging.
2. Create a summary table of the range of policies used by national and local government to manage the economic, social and physical environments of the UK.
3. Outline the ways in which places may be made more attractive to attract inward investment.
4. Explain why there are often tensions between different players in regeneration.
5. Why is a partnership approach now encouraged in the ways that regeneration policies are implemented?
6. Compare and contrast the main methods used to regenerate urban and rural areas.

Further research

Visits this page of the UK government website giving updates on regeneration schemes such as City Deals: www.gov.uk/government/policies/city-deals-and-growth-deals

Take a look at the DECC page on the UK government website: www.gov.uk/government/organisations/department-of-energy-climate-change

Visit this fracking page on the Friends of the Earth website; enter your postcode to see if your neighbourhood is at risk from fracking: www.foe.co.uk/campaigns/climate/issues/uk_fracking_map_41274

Explore Visit Britain, the official government-funded tourism promoter: www.visitbritain.com

Read *Clone Town Britain*, an investigation published by the New Economic Foundation: www.neweconomics.org/publications/entry/clone-town-britain

Visit the official website for Kielder Water and Forest Park: www.visitkielder.com

Visit this Kielder Oral History Report page: www.environmentalhistories.net/?page_id=599

18 Assessing the success of regeneration

How successful is regeneration?

By the end of this chapter you should:

- be aware of the range of measures used to evaluate regeneration projects
- understand that different urban stakeholders have their own evaluation criteria
- understand that different rural stakeholders have their own evaluation criteria.

18.1 Measures of regeneration success

The Specification focuses on three main measures: economic change, demographic stability combined with social progress and improvements to the living environment.

Economic measures

The term regeneration indicates a long-running process rather than a quick fix to economic, social and environmental problems, despite political and economic pressures for speed. Events designed as a catalyst, such as the Olympic Games, may be successful in attracting visitors and investment; creating a legacy of success, which tackles more systemic and longstanding issues of inequality and a poor environment, may be more problematic.

Key terms

Catalyst: The method used or event that starts a regeneration scheme, such as the building of a new shopping mall, leisure facility, creation of a country park or holding an event.

Area Based Initiatives: ABIs aim to improve selected people or places within a specific location and include educational attainment, enhancing crime prevention and reducing unemployment.

Poverty: Poverty is relative to the place and time people live in. The poverty threshold used in the UK is households with an income of less than 60 per cent of the national median, after housing costs are included.

Key concept: Legacy

Legacy refers to the longer-term effects of a regeneration scheme; it can be positive or negative. It is judged on the reuse of any landmark buildings built for an event, the amount of government support needed, the level of private investment and whether the local people benefit long term.

Evidence of success is difficult to quantify. There are many variables in regeneration and its outcomes. Success is measured by comparisons with other areas or with past conditions. Time for the regeneration scheme to have an effect also varies. Research on the success of Area Based Initiatives (ABIs) has been inconclusive. With an austerity focus, the Government has been less involved in large-scale regeneration programmes. One exception is the 2012 Olympic site in London.

The Royal Town Planning Institution (RTPI) supports the place making aspect of ABIs: living in safer, cleaner and more attractive places is likely to enable individuals to become more economically active and live more fulfilling lives in the long term.

The three economic measures of success stipulated in the Specification are: employment, income and poverty. These may be both absolute and/or relative changes, both within an area and by comparison to other more 'successful' areas.

Key concept: The aim of regeneration

The aim of regeneration is to increase income and employment, and decrease poverty. Whether this aim is achieved directly or indirectly is dependent on the type of regeneration scheme.

Schemes involving an immediate job focus, other than the construction phase, such as a shopping mall or new science park, will generate a greater initial rise in income compared with a refurbished or new housing scheme or new/upgraded park. Regeneration increases opportunities, but outsiders may take new jobs rather than locals.

If people's incomes have risen following a regeneration scheme it points to its success. However, if only certain groups have benefited then this may be relative. Regeneration programmes have rarely been created to tackle poverty directly. Getting out of the poverty trap depends in the short term on household income, but longer term on educational attainment.

Since these three aspects – employment, income and poverty – are so linked, the following examples will explore their link with regeneration.

Key concept: Indicators of success

You must look at a variety of indicators to measure success: inward migration may not be a useful indicator if poverty is perpetuated just with differing sets of people.

The success of regeneration must therefore be measured by a range of criteria over a short and longer time scale, both within areas, and by comparison to other more successful areas. Success often involves a strong brand and identity. A sense of place is important, with architecture that may help people identify with a location. The Eden Project, Olympic stadium and shopping malls are all examples. WestQuay shopping centre in Southampton has been a huge boost to the city's economy, with a building reminiscent of a city wall and ocean liner. It was opened in 2001, while the city's Watermark

The Gorbals in South Glasgow: Closing the economic gap?

By the 1950s the original tenements built for Victorian industrial workers had a reputation for poverty, overcrowding and poor public amenities, with associated gang culture and violence. A major redevelopment programme replaced the tenements with high-rise, concrete tower blocks, such as the Hutcheson Estate. Unfortunately these were poorly constructed and the design also fostered crime which, combined with poor management, led to alternative strategies such as the Crown Street regeneration project of the 1990s. The attractive varied designs, spacious flats and areas with employment spaces resulted in not just 'closing the gap' but exceeding Glasgow's average economic growth. Unemployment fell by 31 per cent between 2004 and 2012, while the percentage of 'income deprived' people, including those on welfare benefits, fell by 35 per cent. Meanwhile Glasgow 'only' had an average drop in unemployment of 16 per cent and a 21 per cent drop in income deprived citizens.

A hotspot of in-migration and poverty: Newham, London

This is an example of a continuing cycle of poverty. Once people better themselves, they move out, but are replaced by another set of poor people. In 2014, 36 per cent of residents in Newham had no recognised qualifications, double the city average; 50 per cent earned less than the London Living Wage and 20 per cent were illegal workers on less than the minimum wage.

Meanwhile, parts of the borough near the 2012 Olympic site, for example Westfield, saw the biggest price rises in the country in 2015. Newham Borough Council promotes its place abroad as an 'Arc of Opportunity'. In 2013 it attracted a £1 billion, 35-acre business park investment by the Chinese company Advanced Business Park to another of its derelict sites: the Royal Docks.

igure 18.1 Signature architecture in Southampton city centre: VestQuay mall and the brownfield site Watermark

cultural quarter, a brownfield site, lacked developer's nvestment until 2015 (Figure 18.1).

Internal and external measures of social progress

Social progress can be measured by:

- reductions in inequalities both between areas and within them
- improvements in social measures of deprivation
- demographic changes: improvements in life expectancy and reductions in health deprivation.

Previous chapters have explained the role of Index of Multiple Deprivation (IMD) statistics in measuring differences between places in the UK, and how they can be used to target regeneration. Looking at changes in indicators before and after a regeneration scheme will help measure its success. The IMD's health deprivation and disability domain measures the risk of premature death and the impairment of quality of life through poor physical or mental health.

Key concept: Social progress

Social progress relates to how an individual and community improve their relative status in society over time.

Skills focus: Index of Multiple Deprivation

Your place studies could focus on looking for changes in deprivation and life expectancy. Be careful when using IMD data: you cannot compare domain scores because they have different minimum and maximum values and ranges. You can compare ranks, however. See also Skill 10 in the Skills focus section (page 350).

Examples of smaller regeneration schemes are those tackling 'food deserts'. Stores may also help to break inequality and deprivation cycles. Tesco spearheaded this through the use of its stores in the Seacroft estate in Leeds in 2000, with marked improvements in local diet and health as a result.

Quality of the environment

A large 2012 survey by the ONS, Life in the UK, found that 73 per cent of respondents mentioned the local and global environment as an important factor in well-being:

- The local environment included having access to open, green space within walking distance of home and the quality of the local area.
- Global environment factors included air quality and climate change.

Regeneration that tackles the built environment, such as better transport links, provision and upgrading of retail space, creation of green space, parks and public areas and improvements in housing, will have positive impacts on health and also act as a draw to people to live there.

General improvements in aesthetics, security and safety via neighbourhood redesign (pedestrian zones, lighting, street furniture, public art) and tackling environmental stressors (graffiti, litter and noise) are also common components of regeneration programmes.

The Glasgow effect

The 2010 Marmot Review highlighted stark health inequities in Glasgow. Boys in the deprived area of Calton had an average life expectancy of 54 years compared with 84 years in affluent Lenzie, 12 km away. This pattern emerged in the 1990s. Efforts to combat this have centred on the psychology of health promotion rather than prohibition. The Scottish government launched four main initiatives: Equally Well, Achieving Our Potential, a Child Poverty Strategy and the Early Years Framework, designed to tackle poverty, income inequality, health inequities and to ensure that all children are given the best possible start in life. Other initiatives include supporting youth centres.

Key concept: Specific improvements

It must be remembered that separating out the importance of specific improvements in regeneration is very difficult, even at university level.

Key term

Baseline data: The information used to compare present-day characteristics with, for example, past land-use maps, photographs and statistics.

The Specification asks you to concentrate on the effectiveness of these improvements, in particular:

- reductions in pollution levels
- reductions in abandoned and derelict land (called 'drosscape' in the USA).

The IMD has a separate domain called Living Environment Deprivation, which measures the quality of the local environment. There are two subdivisions:

- **Indoors:** the quality of housing including the structure (walls, roofs, windows), facilities (modern kitchens and bathrooms), insulation and central heating provision.
- **Outdoors:** air quality (concentration of four pollutants: nitrogen dioxide, benzene, sulphur dioxide and particulates) and number of road traffic accidents (death or personal injury to a pedestrian or cyclist).

Individuals pay taxes to fund national environmental watchdogs and planners in order to control levels of pollution and overall environmental quality. The Environment Agency, Natural England and English Heritage come under the umbrella of Defra (Department for the Environment, Food and Rural Affairs). Austerity cuts to Defra and local authorities from 2015 will inevitably reduce these organisations' roles, meaning that businesses, community groups and individuals will need to play a larger role.

Traditionally dereliction is associated with ex-manufacturing areas and redundant infrastructure, such as power plants. However, unused buildings, houses, shops and discarded infrastructure are found in most places. In the countryside, redundant dairies and barns feature. The Campaign to Protect Rural England (CPRE) is a pressure group that campaigns for the greater use of such sites for new housing, rather than building on greenfield sites. In 2014 they estimated that 1 million new homes could be built on brownfield sites in England alone.

These factors and characteristics have an impact on the health of local residents, and on the attractiveness of a place to visitors, residents and investors. The surrounding environment is an important factor to the perception of, and attachment to, a place.

Skills focus: Before and after cross-sections

Research into how regeneration affects the local environment will include the interpretation of photographic and map evidence showing 'before' and 'after' cross-sections of places. Baseline data is required to assess the extent of change and, hence, success or failure. If a whole settlement or parts of a larger one is being investigated, then a transect method showing before and after regeneration characteristics may be used. See also Skill 4 in the Skills focus section (page 346).

Models may be used to assess why regeneration is needed in specific places (see Chapter 20, Figure 20.11, page 300), which may be tested against reality.

The spiral of rural decline is a useful model in rural places. It can be annotated to show whether it has been broken or reduced by regeneration programmes.

Skills focus: Quality of life lines

You could draw two quality of life lines: before and after a regeneration project. A composite cross-section may be created with different rows of information above the transect line for primary fieldwork and secondary data. An example is shown on page 300, Fig 20.11.

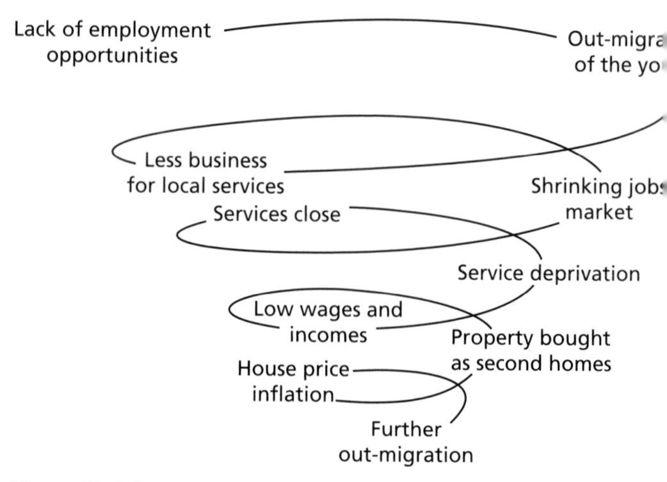

Figure 18.2 Regeneration attempts to break the spiral of rural decline

able 18.1 Regeneration success in inner Manchester: Harpurhey ward

Indicators	Changes
IMD	Within the poorest 1% of neighbourhoods in England in 2010; improved by one rank by 2015
Unemployment	Reduction in young people not in employment, education or training (NEET) from 23.1% in October 2006 to 8% in September 2011
Housing	400 new homes in progress by Redrow, giving existing residents the chance to own their own home and attracting new people to the area; 20% of the workforce local; Northwards Housing (a housing association) has invested £27 million in bringing all their social housing up to Decent Homes Standard
Health	Increasing numbers in Zest Healthy Living Projects
Education	Improved performance at Key Stage 4 (GCSE) and improved attendance levels every year since 2008
Crime	Steady decrease in reported crime and antisocial behaviour since 2007; Project Cove of 2011 was a partnership approach (police and locals) to deal with crime and disorder, problem families, tenants and businesses; gating of most alleyways has helped and the spin off has been many residents using their alleyways as extended gardens and entering 'Manchester In Bloom'
Open space	Boggart Hole Clough has had Green Flag status since 2002; new park and football pitch, Moston Vale
Pollution	Harpurhey Reservoirs, polluted old mill ponds, were restored in 2000 and are now a haven for wildlife, although not safe to swim in because of toxicity
Community spirit	Many tenants and activity groups, e.g. North City Residents Forum and the Factory Youth Zone, have increased activity and memberships

Environmental regeneration projects deserve recognition separately from general ABIs because, while improving the appearance and form of the built environment and public spaces can be goals in themselves, such improvements often have significant and diverse wider social and economic multiplier benefits. Regeneration based on physical upgrades in buildings and space have two basic effects:

- can force out locals because of unintended regeneration, as is happening now around the Olympic site in London
- good planning and place making has a direct impact on individuals' lives, rather than just delivering 'gentrification' effects, as in Glasgow and Broadwater, London.

Table 18.1 shows a summary of economic, social and environmental success for one place: Harpurhey in Manchester.

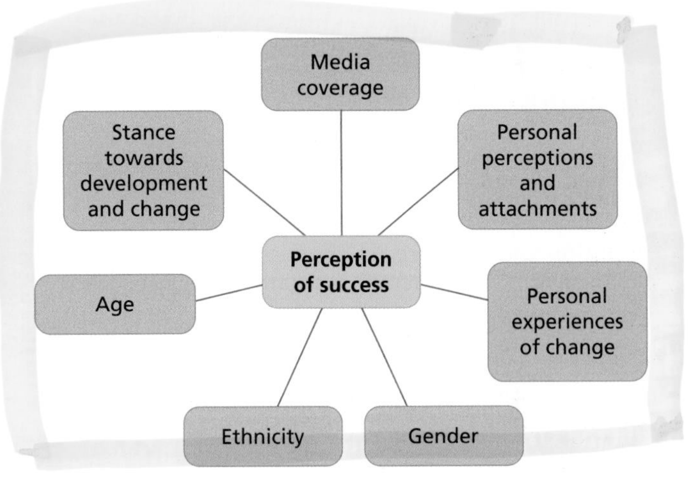

Figure 18.3 Factors influencing perception of success

18.2 Urban stakeholders' views on regeneration

The relative success of regeneration is often viewed differently by individuals and groups of stakeholders. Their views will depend on a range of factors, particularly their own perceptions, attachments and motives, as shown in Figure 18.3.

In a globalising world with an astonishing pace of technological change, leaders and planners have to create places to attract an ever-more mobile and educated population and customer base. This is often at a global level, not just regional, national or local.

Party politics may affect decisions and the longevity of any scheme. According to the Conservative government in 2015, successful regeneration involved 'achieving additional economic, social and environmental outcomes that would not otherwise have occurred'. It should represent 'good

Key concept: Age

Planners may also have to cater for an ageing population, often with distinct perceptions about the legacy of the past and what they see as successful, which may differ from younger generations.

Table 18.2 The viewpoints and roles of different urban players

	Viewpoints	Roles
National government and planners	Reconciling different interests, longer-term national goals take priority	Planning permission Pump priming to start large, nationally important developments
Local councils	Have a duty to tackle inequality in their communities Make local planning decisions They are supposed to balance out the economic, social and environmental needs of a locality	Small or local regeneration schemes 'Soft management' helping regeneration, e.g. 'alcohol free zones' Permissive arrangements, e.g. roller skating, street performers and artists
Developers	Economic standpoint: profit is needed	Funding of schemes
Local businesses	Views may be polarised: those expecting an increased customer base by the spin offs from regeneration will differ from those threatened by it The local Chamber of Commerce may give majority viewpoints of business leaders	Lobby councils Invest in schemes
Local communities	The silent majority may be represented by a few willing and able to give up their time to either be involved in the local council or a pressure group; Broadwater Farm is an example of many players working together	Lobby councils Vote for local and national political parties Form pressure groups

Key term

Benefit–cost ratios: The balance between investment and outcomes; a positive ratio is desirable.

value for money'. Intervention was only seen as needed if market forces failed with resulting inequity.

Benefit–cost ratios were based on cost per job and per hectare of open space improved. In England, unlike the rest of the UK, delivering regeneration became a local matter after 2010.

The national government has just a strategic and supporting role, and has stopped monitoring spatial inequalities or setting targets. The previous government's neighbourhood renewal programmes were cancelled or replaced by small-scale schemes to support coastal and coalfield communities. Spending on these schemes has been £32 million per year, on average, whereas the Labour government's Neighbourhood Renewal Fund alone cost £500 million annually.

Local economic growth is viewed as critical, overseen by Local Enterprise Partnerships (LEPs) able to fund housing and infrastructure developments. City Deal status gave 28 urban areas powers to attract private investment.

It is important to differentiate between small- and large-scale schemes.

How to measure success

Skills focus: Social media and blogs

The internet has allowed cyber activism with online polls and forums, and allowed a greater section of society to participate than in the past. See also Skill 1 in the Skills focus section (page 345).

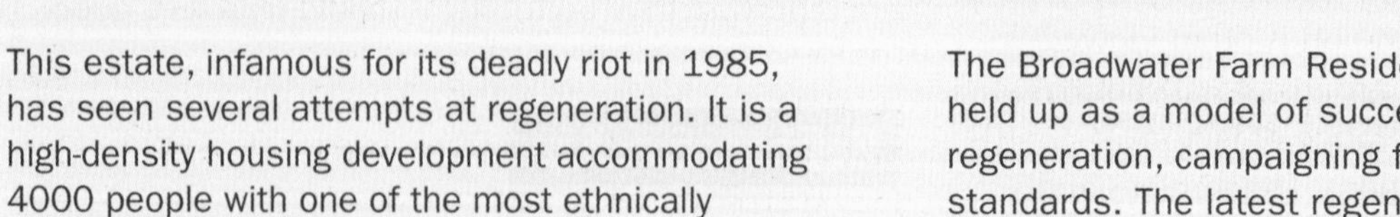

Broadwater Farm Estate in Tottenham, London

This estate, infamous for its deadly riot in 1985, has seen several attempts at regeneration. It is a high-density housing development accommodating 4000 people with one of the most ethnically diverse communities in Europe (300 languages are represented). It remains one of the poorest areas of London.

The Broadwater Farm Residents' Association is held up as a model of successful community-led regeneration, campaigning for better facilities and standards. The latest regeneration scheme involves some of the advisors and planners involved in the Olympic Park. Re-imagining the built environment is seen as essential.

Large- and small-scale schemes

Salford Quays is a successful, innovative cultural flagship project. In the nineteenth century Salford Docks, together with the Manchester Ship Canal, were integral to Manchester and, indeed, the North West's success. Deindustrialisation led to polluted waterways and derelict wasteland, transformed from the 1980s into a centre for commerce, retail, the arts and sports. The choice to relocate there by the BBC and the development of the UK's first 'media city', costing £550 million, has led to global prestige again. Salford Quays has become a desirable residential location with a growing population. The city authority was the key stakeholder.

The research group Rethinking Cities suggests success depends on place details, such as the availability of community activities, perception of safety and safe play areas. The role of greenery, sympathetic landscaping and even colour can be important. However, traditionally the level of funding for such projects has been far less, so charities and community volunteers have led their development. Akzo Nobel's (Dulux paint) Let's Colour campaign donated paint to the Humphry Davy School on the Treneere Estate, Penzance, Cornwall. The school is in one of the most deprived areas in the country. It is now highly visible from the main road, making a statement about its social transformation.

Key term

Sustainability: The definition of sustainable regeneration varies but in this context it may be thought of as regeneration that creates long-lasting economic, social and environmental benefits for a place.

Cities, especially in older developed economies such as the UK, have to adapt or lose out. Proximity to knowledge rather than to resources is now the primary driver of growth. Image is also critical in a competitive world. Most schemes have economic, social and environmental elements; larger schemes, in particular, may have sustainability built in to be most effective.

Figure 7.15, page 102, simplifies the elements needed. The 'legs' of the stool, i.e. components, may be divided into local and national in this context.

We have seen that the economic component has become the dominant driver of regeneration, with national benefits sometimes overriding local views. This was exemplified in the national fracking debate, the more regional HS2 project, and local Heathrow expansion (Chapter 17).

Synoptic themes:

Future

The future of a place depends on the mix of economic, social and environmental factors at play.

Skills focus: Town centre studies

You may choose a local town centre to study during fieldwork. Look out for business-led partnerships (BIDs), community-led schemes such as 'shop local' loyalty cards, pop-up shops and consumer changes, such as gyms and click-and-collect points.

The society component should cater for locals as well as incomers to an area, and not exclude or push them out of their locality, otherwise the regeneration really does not tackle local needs.

Without a good or enhanced environment, few with any choice will want to live, work in or visit the place, unless trapped by circumstance.

Flagship, high-profile regeneration schemes need large-scale planning and investment. Table 18.3 outlines the key indicators of London's Olympics. Was it worth £9.3 billion invested, including a council tax levy for all Londoners? Using the Games as a catalyst for the regeneration of the whole of East London was always going to be ambitious, and continued investment over a long period will still be required.

Viewpoints of urban stakeholders

Different stakeholders will assess success using contrasting criteria. Their views will depend on the meaning and lived experiences of the urban place and the impact of change on both the reality and the image of that place. You may have a current controversy over a regeneration scheme in your local place, such as the next Place Context based on Silver Hill in Winchester.

Table 18.3 Olympic success?

Economic	£13 billion injected into the national economy, including £130 million worth of new contracts for UK companies. The 2012 London Olympics were so successful that the UK was chosen to hold other world events, such as the 2017 World Athletics Championships. The one underperforming feature is the giant sculpture 'Orbit'; there are plans to reduce entrance fees and add a giant slide to attract more visitors.
Social	10,000 new homes, two primary schools, a secondary school, nine nurseries, three health centres and multipurpose community, leisure and cultural spaces. Broadened demographic base; more affluent incomers, who may demand better standards of education and services. The UK has not become 'healthier' as measured by increased take up of sport, especially by more disadvantaged people in Newham. Austerity cuts to local authority and school budgets has limited facilities and training. Gentrification has mixed benefits. The original athlete flats have been converted into East Village: 2800 new housing units. However, entry-level prices are £250,000 for a three-room apartment, so not affordable for most locals. The six Olympic Host Boroughs agreed a joint Strategic Regeneration Framework, aiming to achieve convergence in living standards with the rest of London by 2030. The Games legacy is seen as key to changing their local reputation and for re-imaging.
Environmental	The London Legacy Development Corporation (LLDC) plan – Sustainable Vision for 2030 – is based on people, places and performance. 2.5 sq km remediated brownfield land and wetland restoration along the River Lea; 200 buildings demolished and 100 hectares of open green space created. New housing is zero carbon and more water efficient in design. Easy walk and cycle design in the neighbourhoods with good public transport.

Silver Hill, Winchester

The 2008 mixed land-use proposal supported by Winchester City Council was contested by pressure groups, including Winchester Deserves Better, which attracted 1000 people on a Facebook petition. Regeneration is supported, but not this scheme, since it lacked affordable housing and had futuristic, insensitive architecture for the centre of the historic city. The High Court found the council to have acted unlawfully and the scheme delayed even further, creating planning blight (Figure 18.5).

Figure 18.5 Planning blight from lack of decision making

Figure 18.4 An aerial view of Silver Hill, Winchester

Skills focus: Conflict matrix

In a conflict matrix the players are categorised, then their views are compared. Comments are written into the boxes to explain the different viewpoints; scores are then allocated from zero, meaning no conflict, to five, meaning a high conflict.

Table 18.4 A conflict matrix based on a proposed redevelopment scheme

	Local residents	Conservation	Local businesses	Local councillors	M3 and BID
Main stance	NIMBY if very local Social and economic	Keep heritage	Economic	Split pro/against duty to reconcile conflicts	Economic
Local residents in Winchester and surroundings					
NGOs, pressure groups (Winchester Deserves Better)	Represent officially only a small proportion of the 116,000 population Some NIMBY elements 1				
Chamber of Commerce representing local businesses Pro scheme	New retail will generate businesses, which not all residents may see as necessary 1	Strongest conflicts here 5			
Local councillors are split between the scheme The majority council viewpoint is pro	The need to redevelop the area has become embroiled in politics Many conflicts here 4	Conservation of heritage conflicts with futuristic designs 4	Local businesses will generate employment 1		
M3, the regional Local Enterprise Partnership, and Winchester's Business Improvement District are pro-business generation	Locals will want prosperity but not if it is detrimental M3 represents regional aspirations, not necessarily local ones 2	Regeneration wanted but not at the level proposed by developer, which fits into regional aims 4	Businesses and economic growth are interlinked 0	Local councillors 'nest' inside a hierarchy of pro-business government organisations They also have to represent voters 3	
Totals	**8**	**13** **Inevitably the local pressure group conflicts with most other groups**	**1**	**3** **Some agreement between groups**	

Comments explain the different viewpoints; scores are allocated, from zero (no conflict) to five (high conflict).

Urban areas are a rich source for any geographical investigation. Two areas other than retail areas that might prove interesting and accessible are crime and design surveys, including gated developments and controversial garden infilling or densification.

18.3 Rural regeneration stakeholders

Contested rural regeneration strategies

Some decisions on regeneration strategies generate more conflicts within local communities than others. You will probably be aware of the internationally famous Eden Project in Cornwall. When first proposed some locals feared that the problems of increased traffic would outweigh employment benefits of this futuristic scheme. An effective management partnership between private and public investors and local planners ensured that this scheme developed into a global model of regeneration.

Conversely, there are more conflicts to be seen in restructuring other rural places, such as the North Antrim coast.

Lastly we will examine the success and failure of some other projects, including a Millennium regeneration project called The Earth Centre. This was, and still is, the focus of debate.

Judging the success of rural regeneration strategies

Figure 18.3 showed some criteria to evaluate success. The Egan Wheel is another useful technique which ma[y] be used in urban or rural settings.

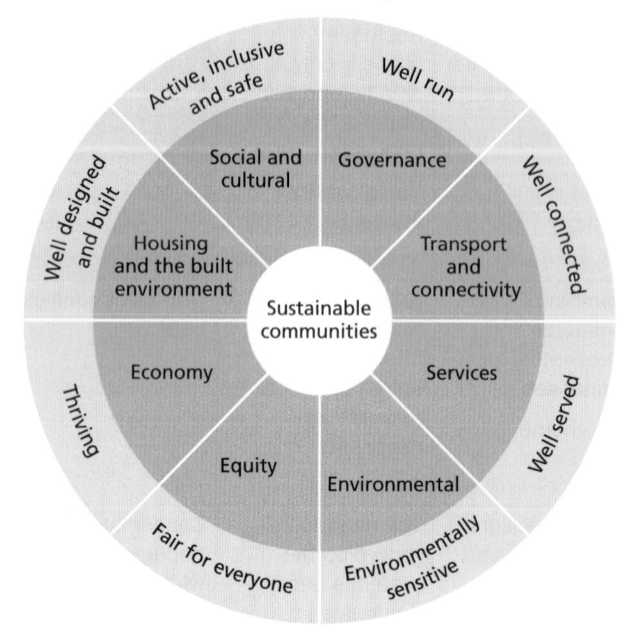

Figure 18.6 The Egan Wheel

North Antrim coast

The Giants Causeway area was designated a World Heritage Site in 1986 because of its unique geology and striking landscape. As such, there are a huge number of stakeholders involved in any development decisions.

Plans to develop a £100 million golf resort called Bushmills Dunes close to the conserved area were disputed in court for more than a decade before being approved in 2013. The opponents, the National Trust together with UNESCO, thought the landscape change so close to the protected coastline was inappropriate, despite the potential job creation from its proposed facilities. Although approved, the proposal was eventually shelved as it was unable to raise the necessary finances.

Mixed success projects

The Millennium Commission helped create 225 projects with £1.3 billion of Lottery money including the successful Eden Project, O2 arena, Tate Modern, The Lowry gallery in Salford and Cardiff's Millennium Stadium. High-risk, imaginative and futuristic ideas were used. Only three of the projects were really unsuccessful, and actually closed.

The Earth Centre in Doncaster was a rural ex-colliery, redeveloped at a cost of £55 million as an environmental tourist centre. However, it failed to attract enough target audience and shut in 2004. Since then it has been turned into a children's outdoor adventure centre and its car park may be redeveloped into a large housing scheme.

able 18.5 Applying the Egan Wheel to a rural community

Name of the community/rural location ______________________________

Is it a sustainable community? 1 = very good 6 = very poor

Ring a score for each of the following:

(1) Well run?
People are:

Included in decision making	1	2	3	4	5	6	Not included
Feel responsible	1	2	3	4	5	6	Don't care
Proud of local community	1	2	3	4	5	6	Not proud

(2) Well connected?
Getting in/out and around your community:

Excellent bus service	1	2	3	4	5	6	Non-existent bus service
Easy access to rail service	1	2	3	4	5	6	No access to rail service
Safe local walking routes	1	2	3	4	5	6	Lack of safe pathways
Safe local cycle-ways	1	2	3	4	5	6	Lack of safe local cycle-ways
Roads clear	1	2	3	4	5	6	Roads congested
Off-road parking	1	2	3	4	5	6	Parking on roads

Skills focus: The Egan Wheel

The Egan Wheel was designed by a government think tank looking into the future of communities. It creates an evaluative scoring system which could be used as part of a questionnaire, extended interview or focus group. In Table 18.5 just the first two criteria are expanded to give an idea of the range of statements that might be generated from each segment of the wheel. It could be used to compare your two chosen places.

Benefit–cost ratios and environmental impact assessments are also techniques that help assess success, as discussed in the previous urban section.

Rural stakeholder viewpoints

We have already met the key groups of stakeholders involved in an urban context and seen why their views may vary. In rural areas, specific players are landowners, often foreign, farmers and the government's Defra. Successful regeneration may mean shared or different aspects: better leisure, better retail, more jobs, more visitors, better housing or higher biodiversity. A hotly contested location is often the urban–rural fringe and greenbelts, as shown in Figure 18.7. National Parks are also usually a rich source for differing viewpoints.

Two examples of community-led rural regeneration schemes

Llanmadoc, Gower Peninsula, Wales, is a fairly remote rural village dominated by second homes. In 2007, 150 local residents paid £5000 for local shares to open a community shop, post office and cafe in an old barn run by 30 volunteer staff. The scheme proved so successful, benefiting tourists as well, that it moved to a larger purpose-built shop in the village. Grants were given by the Welsh Government and Swansea City Council.

The Butchers Arms in Crosby, Ravensworth, Penrith, is one of 600 registered UK community pubs. In 2011, 300 local people bought shares at £250 each to buy it after it had been shut for a year. Grants from the 'Pub is the Hub' policy of the Department for Communities and Local Government helped.

Very pro	Pro	Player	Anti	Very anti
	← Released greenbelt land for 500 homes and 35 hectares for industry	Kirklees Council		
← Profit for Church of England		Landowners: Church Commissioners own Chidswell site		
← Proposal for 400 homes; letters sent to locals; affordable component but still big profit		Construction firms Barratt and David Wilson Homes want to develop Soothill opposite Chidwell		
	← Development is needed to help successful firms expand and generate empl-oyment	Conservative Councillor Jim Dodds, Denby Dale, in 2012		
		Labour Councillor Hanif Mayet, Batley East	Concerns: traffic congestion and infrastructure strain (doctors surgeries, schools) →	
		Local NIMBY groups: Chidswell Action Group and Soothill Forum	Happy for some but not a huge number of houses →	
		Council to Protect Rural England	Use brownfield not greenfield sites →	

Figure 18.7 Force field analysis showing people's viewpoints in Chidswell, West Yorkshire, on the use of the greenbelt for housing

Skills focus: Measuring environmental success

Measuring environmental success is probably the easiest component to see through fieldwork, especially if baseline data is available to see what changes have occurred.

Rural areas have a tradition of local people identifying and acting on their own needs: the Big Society in action. Government policies channelled through Defra support this by strategies such as the Community Right to Bid, which enables locals to run their own community buildings and facilities, such as the village shop, pub, community centre, children's centre, allotment or library.

Synoptic themes:

Players

There are many players involved in rural regeneration.

Table 18.6 Research techniques for measuring regeneration success; those in italics are quantitative, the others are qualitative

Secondary research	Primary research
Economic – more prosperity and disposable income?	
Census, neighbourhood statistics and ACORN profiles	*Footfall surveys*
Estate agent websites, e.g. Zoopla	*Level of current land-use occupancy/derelict land/closed businesses*
Valuation Office Agency and local authorities (council tax and business tax bands)	*Retail clone town survey*
Historical land uses from Google maps and photos, GOAD maps, past school fieldwork	Bipolar attraction potential surveys
Past photos	Retail quality surveys
Past/extrapolated retrospective attraction surveys	Estate agent adverts, information and interpretation signs
Video clips, podcasts from local TV, radio stations, press	Questionnaires and interviews with local authorities, planners, police, focus groups, pressure groups
Local authority/private developer publicity	
Demographic – stability? Net in-migration? Improved safety, health, life expectancy? Reduced dependency ratio?	
Postcode census and neighbourhood geo-demographic statistics	*Service surveys, e.g. of community facilities, clubs, volunteer groups*
	Questionnaires of local residents and visitors, migration patterns
Social – better safety? Accessibility? Image?	
Crime statistics	Placecheck survey
Bus timetables	Mind maps
Broadband access and speed	Oral histories
Local blogs, Facebook and Twitter	Crime and design surveys
Old postcards	Disabled/family access survey
Publicity texts and information leaflets, films	Egan Wheel survey
	Interviews with focus groups
	Photos of neighbourhood identity signs, sculptures and other art
	Ethnographic surveys (studies of people and their actions), e.g. visitor activities
Environment – indoor and outdoor improvements? Less overcrowding? More central heating? Better access to open space and facilities like parks?	
Census data	*Land-use survey*
Purple, Green and Blue Flag designations (night life, parks and beach quality)	Environmental quality surveys
Historical maps and past land-use maps	*Pedestrianisation surveys*
Pollution data from the Environment Agency, Friends of the Earth, local council	Photo and videos of architecture and design: any signature buildings?
Past photos and videos	Street furniture surveys
	Annotated field sketches
	Desire lines
	Broken glass in windows, graffiti and litter surveys
	Air, noise and water quality surveys and, if appropriate, biodiversity and beach quality surveys

Fieldwork and further research

It will now be apparent that regeneration strategies – past, current or planned – are diverse. Most places will demonstrate either the need for, or the outcomes of, a regeneration strategy. In both urban and rural contexts, a number of simple data collection methods may be made to help assess success: primary fieldwork on the quality of the environment is relatively easy to carry out compared with social aspects, which will probably need questionnaire collections. This is where working in a group to gather data may be an advantage. Table 18.6 shows some techniques.

Baseline data is essential to compare current characteristics of any place. These include both quantitative and qualitative techniques. Careful sampling methods are needed to ensure that anecdotal data is not collected. There may well be a difference in attitudes of people depending on the distance they live from any scheme, so distance decay analysis may be useful.

Key concept: Distance decay analysis

Distance decay simply means the further away you are from something, the less influence there is likely to be. Hence the impacts of regeneration are probably highest in and immediately around the place targeted.

Review questions

1 Evaluate the range of measures that can be used to assess the success of economic regeneration.

2 What do you consider to be the best way of measuring social progress? Give your reasons.

3 Explain why improvements in the living environment are so important for both starting and continuing regeneration.

4 To what extent are strategies used to regenerate urban areas contested by differing stakeholders?

5 To what extent are strategies used to regenerate rural areas contested by differing stakeholders?

Further research

Read this BBC article about Glasgow's health issues – the 'Glasgow effect': www.bbc.co.uk/news/magazine-27309446

Read information about Covent Garden's regeneration in the 1970s: www.coventgardentrust.org.uk/resources/article

Take a look at this information and video on the Salford Quays regeneration from the local council: www.salford.gov.uk/regeneration.htm

Visit the official website of the Eden Project in Cornwall: www.edenproject.com

Investigate CPRE crowdsourcing data on the brownfield/greenfield debate: www.cpre.org.uk/how-you-can-help/take-action/waste-of-space

Exam-style questions

AS questions

1 In which employment sector is financial services such as banking? Tick the correct answer. [1]
   a Primary
   b Secondary
   c Tertiary
   d Quaternary

2 What is the technique most commonly used to map data showing a concentration or specialisation, such as concentration of manufacturing? [1]

3 Study the choropleth 'heatmap' for financial services shown in Figure 15.3b on page 219. Suggest one reason for the growth of financial services in some regions. [3]

4 Suggest two methods of assessing whether a place can be classed as 'successful' [4]

5 Explain two reasons why there are variations in people's perceptions of their local place, based on reality or imagination. [4]

6 Explain the consequences of a loss of manufacturing in urban areas. [6]

7 Study the resources on Gower in Wales shown in Figure PI.2 page 357 and Tables PI.3, PI.4 and PI.5 on pages 338 and 339. Assess the extent to which the information shows that this rural area experiences multiple deprivation. [12]

8 You collected primary data during your fieldwork relating to Regenerating Places. Assess the value of the primary methods you used when investigating your research question. [9]

A level questions

1 Study Figure 15.4 on page 221, which shows health and life expectancy patterns in the UK.
   a Suggest one reason why lifestyle choices may affect health. [3]
   b Suggest reasons for the concentration of lowest life expectancy in particular parts of London. [6]

2 Suggest how GIS maps may help in profiling places. [3]

3 Explain why there are different perceptions about lived experience in rural places. [6]

4 Explain why different groups would have contrasting views about regenerating greenfield sites in rural areas. [6]

5 Evaluate the importance of rebranding to the success of urban regeneration. [20]

Topic 4
Option 4B: Diverse Places

19 Population, time and place

How do population structures vary?

By the end of this chapter you should:

- be aware that population change, density and structure vary from place to place and over time
- understand the dynamics of population change
- be aware that population characteristics vary within and between places as well as over time
- be aware of the factors shaping the demographic and cultural characteristics of particular places.

Figure 19.1 The diversity of people

A brief explanation of the two terms in the title of this topic – Diverse places – is necessary to make clear the focus and content of this option topic. *Place* is an old term that has recently been revived in geography. Perhaps *area* or *location* are more familiar alternative terms! Places, like areas and locations, are parts of geographical space, but they do not necessarily have definite boundaries. Places are dynamic and ever changing; they are constantly being transformed. Places are shaped by internal connections (such as between people, employment, services and housing) and external connections (such as government policies and globalisation). It is these linkages that drive much of the change that characterises 'place' (see Figure 19.16 on p. 291).

Key concept: Rural–urban continuum

The unbroken transition from sparsely populated or unpopulated, remote rural places to densely populated, intensively used urban places (town and city centres).

A particularly important aspect of place is its meaning, both to individuals and to defined groups of people. Basically, meaning relates to how people perceive, engage with and form attachments to particular places.

The place diversity to be explored in this chapter is rooted in the rural-urban continuum. Places at different points along that continuum are certainly visually different, but those same places also show significant differences in the characteristics of their populations.

Throughout this chapter, and the next three, the spatial focus is exclusively on the UK.

19.1 Population growth and structure vary over time and from place to place

Populations vary over time and from place to place. They do so in terms of both their total numbers and basic characteristics. Numbers rise and fall over time. Over any period in time, those numbers will be increasing in some places and decreasing or remaining static in others.

Differential population growth

Since the first census of 1801, the UK's population has increased sixfold to reach a total of nearly 65 million today. Over the last 50 years, the population has grown by 10 million people.

Population growth is very rarely evenly distributed. The same is true for the spatial distribution of people at any one time. Related to both growth and distribution is population density. It varies:

- over time with changes in population numbers
- from place to place because density is the most widely used measure in plotting and analysing the distribution of population.

Present distribution

Table 19.1 takes a broad view of the distribution of the UK's population, namely in terms of the four constituent countries. England accommodates over three-quarters of the population on just over half of the total area. This results in a high average density of 406.5 persons per km^2. Scotland's share of the total population is a mere 8.4 per cent, but this is spread over nearly one-third of the UK's total area. As a consequence, population density is a mere 67.3 persons per km^2. The populations of Wales and Northern Ireland are even smaller, but so too are their shares of total area. Their population densities are a modest 149 and 130 persons per km^2, respectively. In short, Table 19.1 clearly makes two important demographic points:

- the countries of the UK do not have populations that are strictly proportional to their shares of the total land area
- as a consequence, population density varies considerably from country to country.

Key terms

Population density: The number of people per unit area (usually per km^2); i.e, the total population of a given area (country, region or city) divided by its area.

Demographic: Of or relating to some aspect of a population, for example its size, rate of change, density and composition.

Table 19.1 The distribution of population in the UK, by country, 2011

Country	Population (millions)	% of UK population	Area (km^2)	% of UK area	Population density (persons per km^2)
England	53.0	83.9	130,395	53.5	406.5
Wales	3.1	4.9	20,779	8.5	149.2
Scotland	5.3	8.4	78,772	32.3	67.3
N. Ireland	1.8	2.8	13,843	5.7	130.0
UK	**63.2**		**243,789**		**259.2**

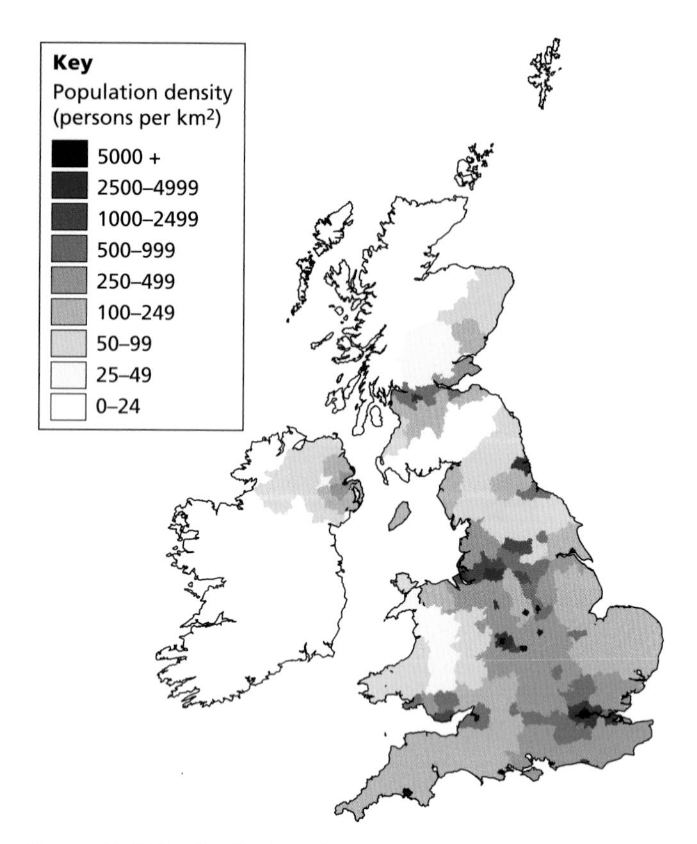

Figure 19.2 The distribution of population in the UK, 2011

Figure 19.2 zooms in on the distribution of population. In doing this, it is important to bear in mind the UK's curren average density figure of just over 250 persons per km^2 .

- In England, there are parts of the country where densities are below that figure, namely in the South West, over much of East Anglia and in the North. These areas of relatively low density help to define a broad belt of above-average density stretching from the South East to the North West and which includes the leading cities of London, Birmingham, Manchester and Liverpool.
- In Wales, the highest population densities are to be found in the South. Much of the rest of the country is sparsely populated except for an area in North Wales.
- Much of Scotland's population is concentrated in the Central Lowlands, particularly in the cities of Edinburgh and Glasgow and, to a lesser degree, along part of the east coast. Large areas of the country – the Highlands, Western Isles and the Southern Uplands – have a population density of less than 25 persons per km^2 (one-tenth of the UK average).
- In Northern Ireland, there appears to be a rather more even spread of population, but with higher densities around the city of Belfast.

Skills focus: Choropleth maps

Figures 19.2 and 19.3 are examples of choropleth maps. They represent data by means of a scheme of tonal shadings showing different degrees of density.

This technique is widely used to show spatial distributions. In this topic, you will see it has been used to show not only the distributions of total population and population change, but also of ethnic groups.

Recent population change in the UK

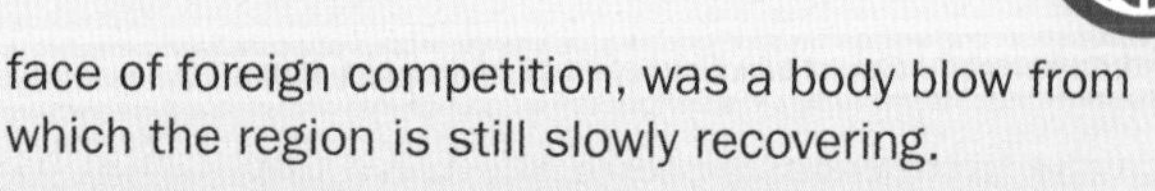

During the second half of the twentieth century population growth was very much concentrated in England. This was especially the case in the South of England. By comparison, much of the North of England experienced relatively little growth. Indeed, there was sustained population decline in some parts. Broadly speaking, these changes were produced by two powerful forces:

- The rising economic prosperity of London and the South East as a global centre of finance and business, as well as a hub of modern service industries.
- The decline of manufacturing industry in its former strongholds in the Midlands and the North of England. The collapse of traditional heavy industries in the North East, such as coal mining, iron and steel making, shipbuilding and chemicals, in the face of foreign competition, was a body blow from which the region is still slowly recovering.

The redistribution of population resulting from this spatial pattern of growth and decline is often referred to as the North–South drift (see Chapter 21).

One notable feature north of the border was the change in demographic fortunes of the north of Scotland during the 1970s and 1980s. This owed its origins to the expansion of the North Sea oil and gas industry. At the same time, however, the rural south of Scotland suffered population loss throughout most of the second half of the twentieth century.

Wales as a whole underwent something of a demographic turnaround. The old situation was one with growth concentrated in North and South Wales and widespread population decline in between. The new situation is now one where population growth has revived in mid-Wales, while parts of South Wales have gone into population decline.

Between the 2001 and 2011 censuses, the UK's population increased by just over 4 million (6.9 per cent). Figure 19.3 shows that most parts of the UK showed some population growth. The only noticeable area of population decline was in West Scotland, but the map also shows very small patches of decline scattered across England and Wales. These coincide with the inner areas of some cities. Perhaps it is surprising to see the very north of Scotland showing a high percentage increase. However, remember that a high percentage change figure can be produced by a very modest increase in a small existing population.

What happened to the distribution of population growth during the most recent intercensal period (between 2001 and 2011) might suggest that the UK has entered a new and rather different phase in its demographic history. This might be described as a light touch with the population rolling pin – a slight evening out of both population growth and population densities. It looks as if the North–South drift is over.

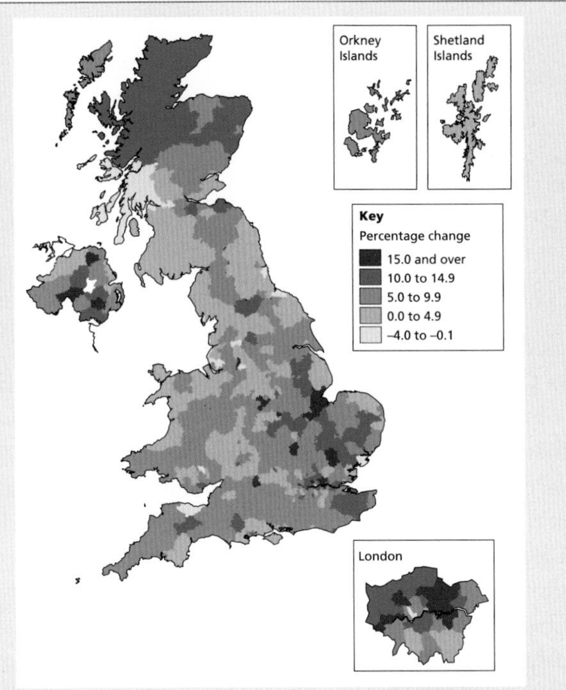

Figure 19.3 Change in the distribution of population in the UK, 2001-2011

Key concept: Population structure

Population structure is the composition of the population of a particular country, region or area. Significant aspects of this composition are how the population is made up in terms of different age groups and the balances between those groups and between the sexes within them. Other components of population structure include life expectancy, family size and marital status.

Variations in population density and structure

In this next section, we will begin to look at how the character of population varies along the rural-urban continuum. Population density is perhaps the most important single characteristic that changes, but there are others of varying significance. Most of these other characteristics are collectively referred to as population structure. It is important to have a sound grasp of what these characteristics are as population structure provides the key to understanding how and why population change varies from place to place and over time.

Population density

The fact that urban places are more densely populated than rural places comes as little surprise. However, it is perhaps less widely understood that there are considerable variations in density within urban areas. We might think that population densities generally decline from town and city centres to the outer edges of the built-up area. That is true to some extent. Within London, for example, centrally-located Westminster has a density of over 11,000 people per square kilometre, while Havering on the eastern outskirts has a density of only just over 2000 people per square kilometre. So, yes, there does seem to be a gradient of declining population densities with increasing distance from the town or city centre.

Unfortunately, this idea of an outward and downward gradient might be a dangerous oversimplification. It is important to recognise the following exceptions to that generalisation (Figure 19.4):

- a low-density crater coinciding with the central business district (CBD) where the permanent residential population has been squeezed out by commercial activities and the concentration of many public buildings

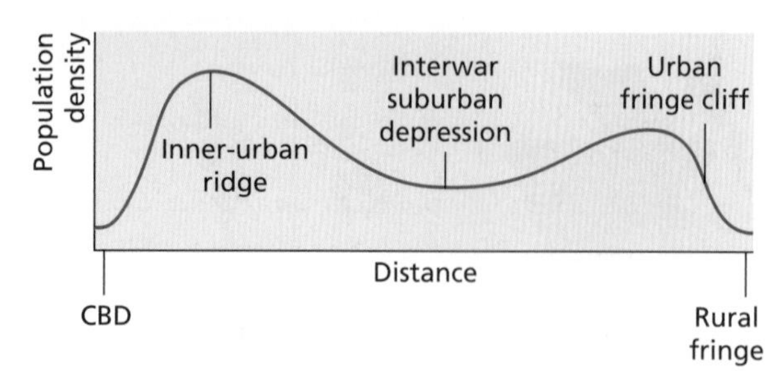

Figure 19.4 An idealised population density profile of an urban area

- beyond the CBD population densities peak in what were some of the earliest areas of housing. Originally, this may have been densely-packed terraced housing. Many of these areas may subsequently have been cleared and replaced, often by high-rise blocks of flats. In some areas, even these high-rise blocks have since been demolished and replaced by more conventional modern housing. Both these redevelopments are unlikely to have eroded this high-density rim around the CBD, except where non-residential uses have taken over some of the space
- a depression reflecting the low-density house building that prevailed throughout the interwar period (1918–1939)
- a kick-up in the density gradient towards the edges of the built-up area ending in a 'cliff' as densities fall away abruptly. This is the result of three main factors:
  - dwelling and residential densities have been raised since the end of the Second World War (1945) by planning controls – densities have certainly been much higher than those that prevailed in the interwar period (1918–1939)
  - the location of large estates of social housing on the urban fringe
  - the rural fringe of urban areas has been firmed up and controlled by the creation of green belts, designed to stop urban growth spilling over into the countryside.

Population densities generally decline beyond the urban fringe, reaching their lowest in remote rural areas. However, the downward density slope will be punctuated by small peaks that coincide with commuter settlements and then market towns and villages.

It needs to be remembered that Figure 19.4 is a generalisation. The population density profile is likely to vary in different directions from the urban centre, when other factors come into play:

- **Physical environment:** Fairly flat areas lend themselves to residential development; building houses on steep slopes and flood plains will be more expensive. They are therefore likely to be avoided until the need for residential space becomes urgent. Historically, the spread of urban areas has tended to finger outwards along lowland corridors.
- **Socio-economic status:** In general, the more wealthy members of urban society live in the most expensive housing. Typically, they will live in areas of low housing and population densities. They will also wish to be as far away as possible from various forms of pollution (atmospheric, visual, smell and noise). It is for this reason that in many of the UK's industrial towns and cities, the wealthy have always favoured the western (windward) side. Any atmospheric pollution generated within the urban area will inevitably collect and drift over the leeward side. Not surprisingly, this is where the poorer housing has tended to be located. Here too the costs of housing can be minimised by living at high densities, i.e. in small dwelling units.
- **Dwelling type and household size:** It is not only housing density that affects population density, but also the type of dwelling unit. A high incidence of flats (likely when a planning consent sets a high density figure), will generate higher population densities than estates of detached houses. Household size (the number of people living in each dwelling unit) will also have an impact.
- **Functions:** The distribution of non-residential activities has both direct and indirect impacts on population density. Population density will be directly lowered where housing is intermixed with non-residential activities. As noted earlier, the location of activities, such as manufacturing, that have a negative impact on the quality of the living environment of adjacent areas, can be significant indirectly. Such areas are likely to be avoided by low-density, expensive housing and occupied instead by higher-density, cheaper housing.
- **Planning:** Reference has already been made to the impact of green belts. But planners also control the density of all new residential development. Planning consent will stipulate how many dwelling units should be built per hectare. Thus planners and government policies on housing do have a significant impact on both housing and population densities, as well as on where particular activities can be located.

Population structure

Age and gender

So far, attention has focused on two ways in which populations vary, namely by changing:

- their numbers over time (increasing or decreasing)
- their spatial distributions and densities over time and space.

The various characteristics of a population that come under the heading of population structure also change and vary in the same two dimensions of time and space. Indeed, as will be seen elsewhere in the coverage of this topic, there is a relationship between changes in the size and changes in the characteristics of a population.

Of all the different components of population structure, age and gender are particularly significant. The population pyramid unravels the structure of a population in terms of these two criteria.

Skills focus: The population pyramid

The population pyramid is in effect a histogram, constructed in one-, five- or ten-year age groups, with males on one side, females on the other. The base of the pyramid represents the youngest age group and the apex the oldest. The detailed shape of a population pyramid can tell us much about what has happened to the population over the last 75 or more years. It can also provide some pointers as to how the population is likely to change in the foreseeable future.

Population pyramids

Figures 19.5 and 19.6 show how the population pyramid of the UK has changed over the last 200 years as it has moved through the demographic transition. Around 1800, with the Industrial Revolution well under way, the pyramid is broad-based but tapers fairly abruptly upwards (Figure 19.5). Its shape indicates a youthful population with a significant proportion of the population falling within the reproductive age-range (15 to 49 years) and, as a consequence, there are high rates of fertility and natural increase. In contrast, the elderly account for a small percentage of the population. This tells us that life expectancy was rather low, around 40 years.

The population pyramid for 2001 and 2011 (Figure 19.6) is drawn using a different horizontal scale of absolute numbers rather than percentages. But this does not alter the basic point that the UK's population structure has changed immensely over the last 200 years. The pyramid shows a bulge in the 40 to 50 age range. These people are the result of the so-called 'baby boom' of the 1950s. Below them, the shape of the pyramid is undercut, the outcome of low birth rates. Above them, the pyramid shape is somewhat flat-topped rather than tapered. Overall, the pyramid is showing a 'greying' population, with significant numbers of people aged over 65. Life expectancy is now 81.5 years, double what it was at the beginning of the nineteenth century.

Superimposed on the 2011 pyramid is the outline of the 2001 pyramid (Figure 19.6). The following changes are to be noted. The conspicuous bulge in the profile appears lower in the 2001 pyramid. The greater length of the age bars above the bulge in 2011 attests to the 'greying' of the population, while the very base of the pyramid clearly indicates a rise in the birth rate. This might be related to the small bulge in the population in the early reproductive age range of 15 to 25. Equally, it might reflect that the average age of mothers has passed the 30-year mark for the first time (Figure 19.7). Many women are deciding to start their families later in life.

Key terms

Demographic transition: A model representing changing rates of fertility and mortality over time, their changing balances and their net effect on rates of population growth.

Life expectancy: The average number of years from birth that a person born in a particular year can expect to live. In developed countries, women enjoy greater life expectancy than men by a margin of a few years.

Family size

This is the number of children and their parents or guardians living together in one household. This aspect of population structure also links with the population pyramid. In a youthful population, average family size is likely to be significantly larger than in an ageing population. Attitudes about the ideal family size vary enormously and are conditioned by cultural and socio-economic factors. Family size in the UK currently stands at 1.7 children. In 1900 it was 4.6 children, and by 1950 it had halved to 2.2.

Shrinking family size is partly the consequence of the changing status of women. The twentieth century saw more women enter the workforce and enjoy more choices about their lives. For example, society now largely accepts the choice of some women not to marry and have children. Divorce, unusual in 1900, has been made legally easier and divorce rates increased a thousandfold over the century. Contraception has been another factor. It means that those who do not wish to have children can choose not to have them. In addition,

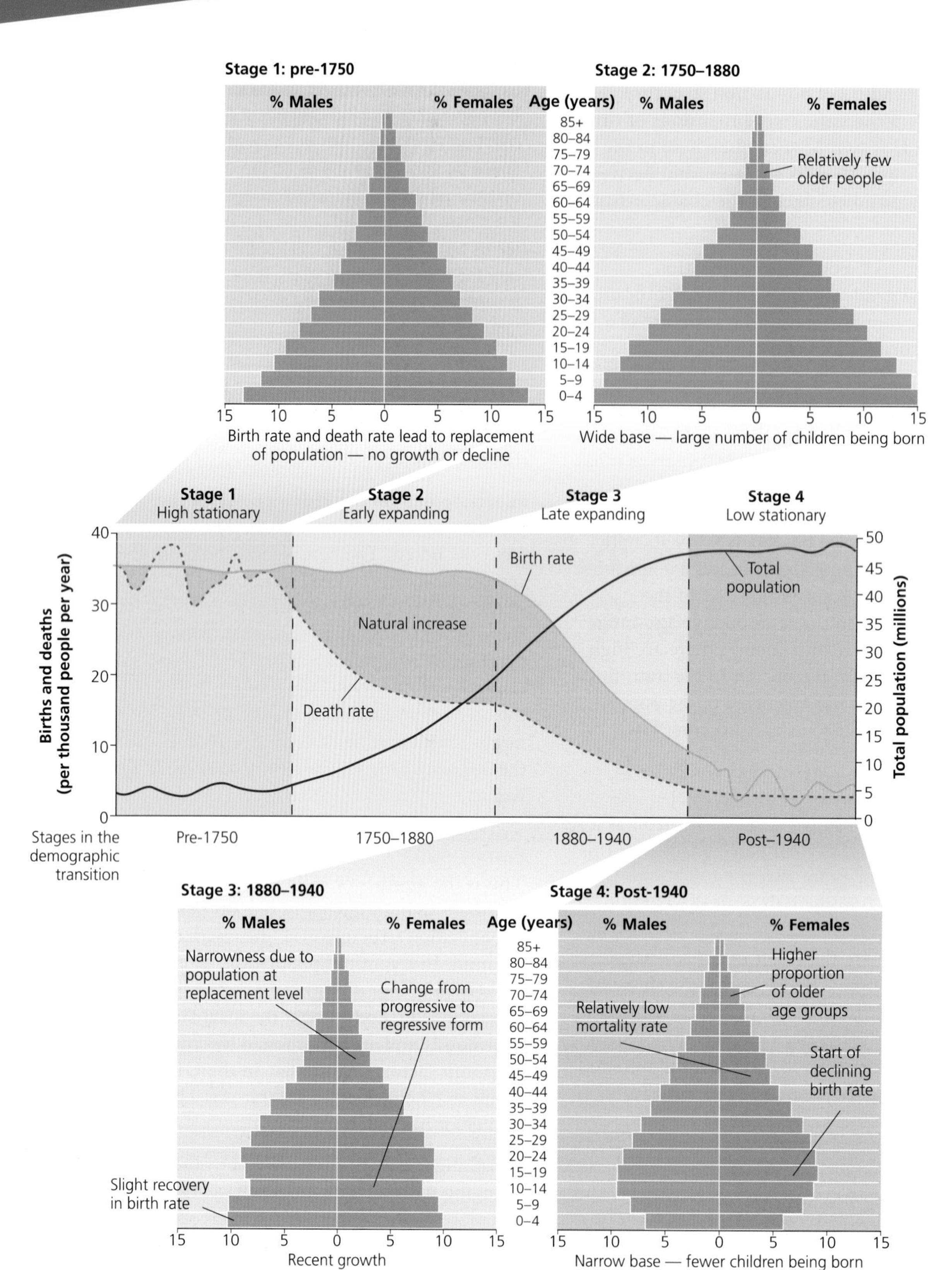

Figure 19.5 The UK's changing population structure during the demographic transition

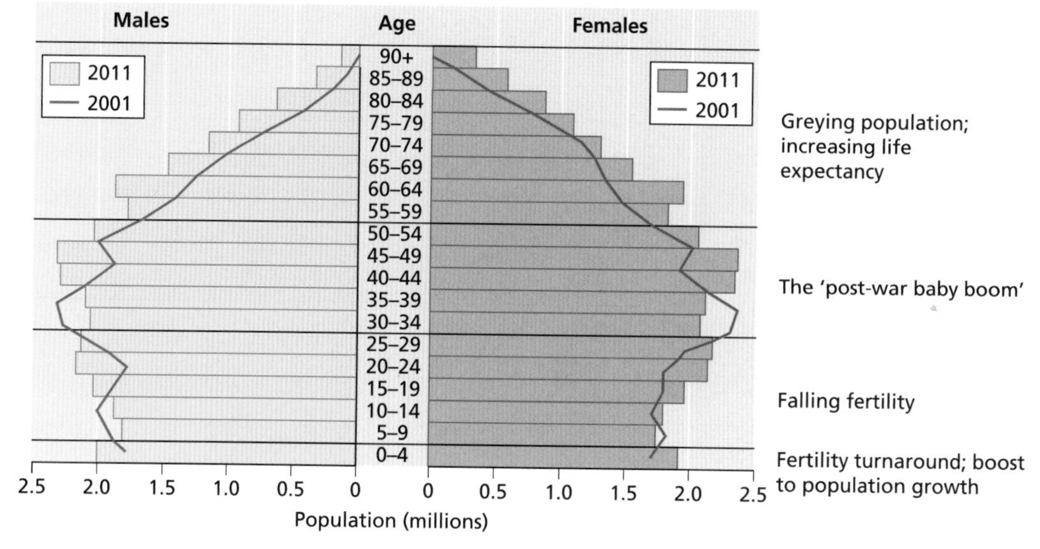

Figure 19.6 The UK's population pyramids, 2001 and 2011

Key term

Internal migration: The movement of population within a country, as distinct from the movement of people between countries (international migration).

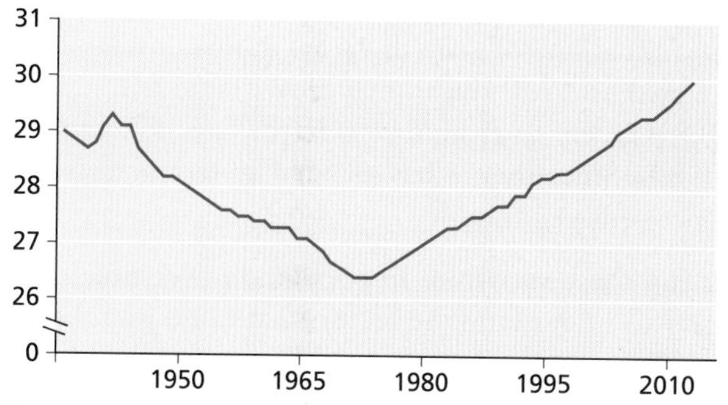

Figure 19.7 Average ages of women when they gave birth, 1945–2010

women are delaying starting families, often after pursuing their careers for fifteen or twenty years, but in sufficient time to have one or two children before they are biologically no longer able to do so.

At this point, it is necessary to remind you of the theme running through all of this chapter (as well as through chapters 20, 21 and 22), namely the demographic differences that exist along the rural-urban continuum. For the moment, let us simply identify some of the main differences between the populations of urban and rural places.

Skills focus: Line graphs

These are used to plot continuous data over time, such as population change (Figure 19.8) and changes in the average age of women when they gave birth (Figure 19.7). The variable is plotted on the *y*-axis (vertical) and time plotted on the *x*-axis (horizontal). It is possible to plot more than one set of a particular data type on one graph, such as population in two or more countries over a given period.

- Urban places show high population densities over large areas. In the UK, urban places have younger and more ethnically diverse populations, with a high proportion of young adults (20 to 40 years of age). Because of this age structure, urban populations tend to show high fertility rates and, therefore, higher rates of population growth. Mortality rates are conditioned by two opposing forces, namely the better availability of healthcare on the one hand and the stress and pace of urban living on the other.
- Rural places are rather more difficult to generalise about as population density and structure vary considerably. They depend particularly on the nature of the physical environment, accessibility and historical factors largely to do with function. Population densities are significantly lower in rural areas, but raised densities do occur in compact traditional villages and in those rural settlements that have become commuter dormitories. Rural places in the UK tend to have older populations, with relatively low numbers of young adults and a high incidence of older adults (over 50 years of age). This characteristic age structure means that rural populations show lower fertility rates and higher mortality rates than urban populations. Population change in rural places is often more the outcome of internal migration. Rural places are relatively untouched by international immigration. For this reason, the ethnic component of rural populations is minute compared with that in urban populations.

> **Key concept: Net migration**
>
> At a national level, this is the nature of the balance between arrivals (immigrants) and departures (emigrants) in a population. However, where areas within a country are concerned the term takes into account both international and internal migration. The balances can be either positive (arrivals exceed departures = net in-migration) or negative (departures exceed arrivals = net out-migration). It is sometimes referred to as the migration balance.

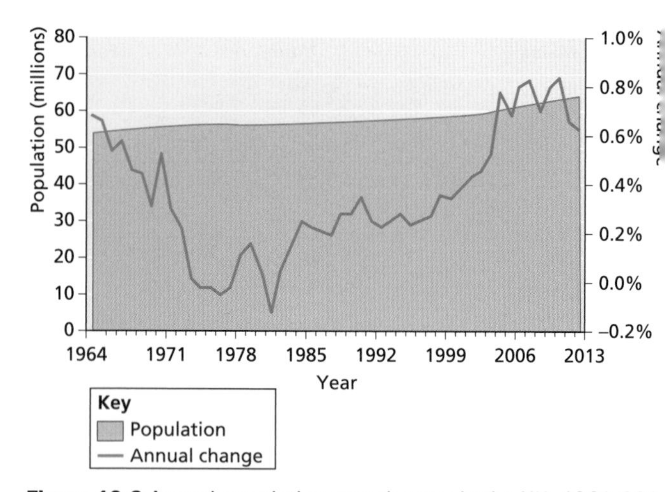

Figure 19.8 Annual population growth rates in the UK, 1961–201

Remember that the above are very broad generalisations. Later we will see that there are, in fact, considerable demographic differences within urban areas, for example between people living in the inner parts and those in the suburbs. Similarly, there are considerable demographic differences between the inhabitants of rural settlements that are accessible to towns and cities and those living in more remote rural places.

The dynamics of population change

Our understanding and explanation of these demographic variations in numbers and character lies in three key factors:

- Population structure, particularly the relative importance of different age groups.
- The difference between the fertility and mortality rates of a population.
- Migration is, in many instances, an influential factor. The influx or exodus of migrants can have significant impacts on a population.

Figure 19.8 looks at population change in the UK during the period between the 1961 and 2011 censuses. In that time, the total population grew by 18.8 per cent but the annual rate of growth varied quite considerably. From a peak in 1961, the annual growth rate fell from 0.8 per cent to around zero in the late 1970s and early 1980s. Since then it has speeded up to reach a rate of 0.6 per cent per annum.

To explain these variations in population over time, it is necessary to understand the factors that cause population change. A systems view of population at a national level sees change as the outcome of two processes: natural change and net migration. The inputs are births and inward international migration (immigration), while the outputs are deaths and outward international migration (emigration), see Figure 19.9.

Population change may be likened to changing levels in a water tank. The nature of the balance between natura

> **Key terms**
>
> Natural change: The outcome of the balance between births (birth rates, fertility) and deaths (death rates, mortality) in a population. Natural increase occurs when births exceed deaths; natural decrease occurs when deaths exceed births.
>
> International migration (immigration and emigration): The movement of people between countries. Immigration refers to the arrival of people from other countries; emigration refers to the departure of people to other countries.

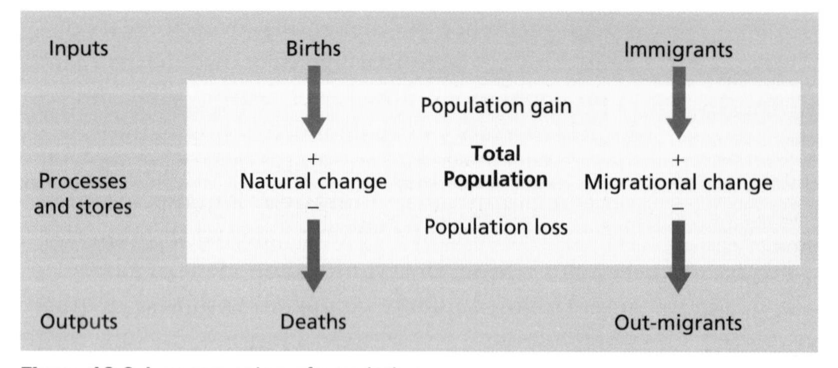

Figure 19.9 A systems view of population

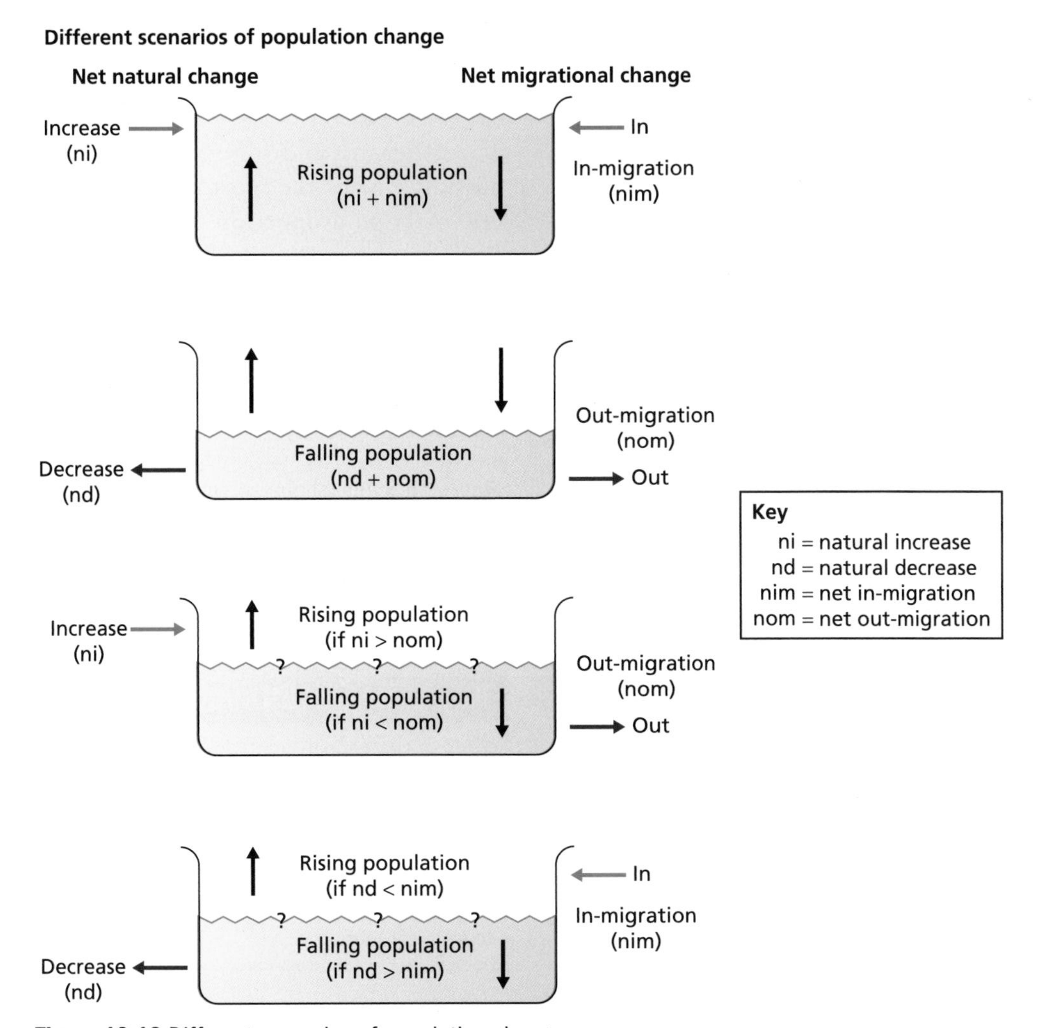

Figure 19.10 Different scenarios of population change

hange and net migration determines how the water evel changes, that is both its direction (up or down, plus r minus) and its speed or rate of change (Figure 19.10). Broadly speaking, these two processes can work either ogether or against each other. If they work together, he rate of population change (plus or minus) can be considerable. If they are in opposition, one process will tend to neutralise the impact of the other. Therefore, the result will be a reduction in the rate of population change, be it plus or minus.

Figure 19.11 analyses the drivers of recent population change in the UK. It clearly shows that the contribution of

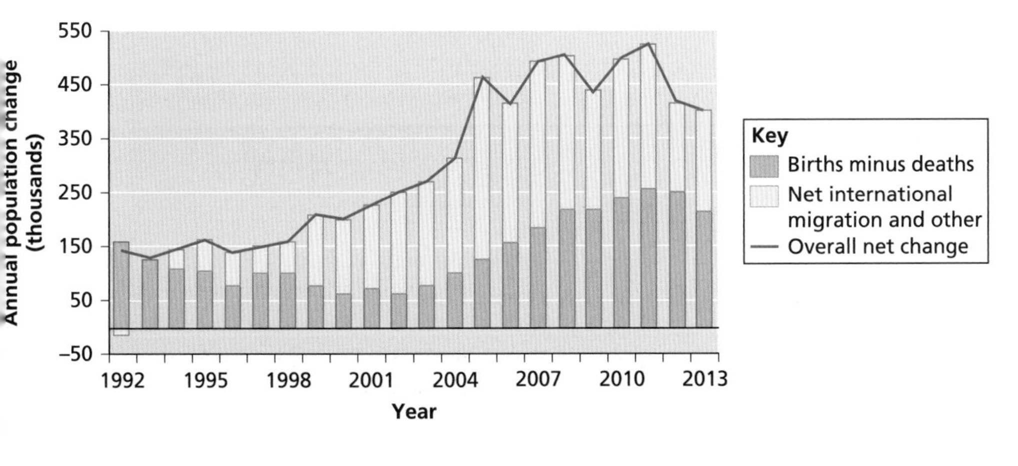

Figure 19.11 The drivers of population change in the UK, 1992–2013 (Source: Adapted from data from the Office for National Statistics licensed under the Open Government Licence v.3.0.)

net migration from overseas (immigration) has so increased that it now matches the contribution of natural increase. This is an important dimension of the changing population structure of the UK. Because many immigrants are young adults, they are contributing to the present rise in both birth rates and the rate of population growth.

19.2 Variations in population characteristics from place to place

Variations between and within settlements

Tables 19.2 and 19.3, each based on a place comparison, illustrate two important points:

- Settlements differ in the detail of their population characteristics.
- There are significant variations in those same characteristics within larger settlements, namely towns and cities.

Cultural diversity

The huge difference in population densities is one striking feature of Table 19.2, but it should be said that high population densities (but of less extent) are also encountered in many villages, no matter where they are located along the rural–urban continuum.

The other major difference is ethnicity. Brent is possibly one of the most multicultural parts of the UK. A large proportion of the population has their family roots in the Indian subcontinent, Africa and the Caribbean. As will be explained in later chapters ethnic immigrant groups have become concentrated in Britain's major cities. Social clustering has led them to become further concentrated in particular places within those cities, the borough of Brent being a prime example. The reasons for this clustering are to do with a variety of factors such as the availability of cheap housing and the momentum created by the established presence of relatives and friends.

Key concept: Ethnicity

This is the cultural heritage shared by a group of people that sets them apart from others. The most common characteristics of ethnicity are racial ancestry, a sense of history, language, religion and forms of dress. With the exception of racial characteristics, ethnic differences are learnt, not inherited.

Greater London: the boroughs of Brent and Bromley

Table 19.2 samples some of the data in the 2011 census and ONS update that can be used to investigate variations in demographic and cultural characteristics that occur within settlements.

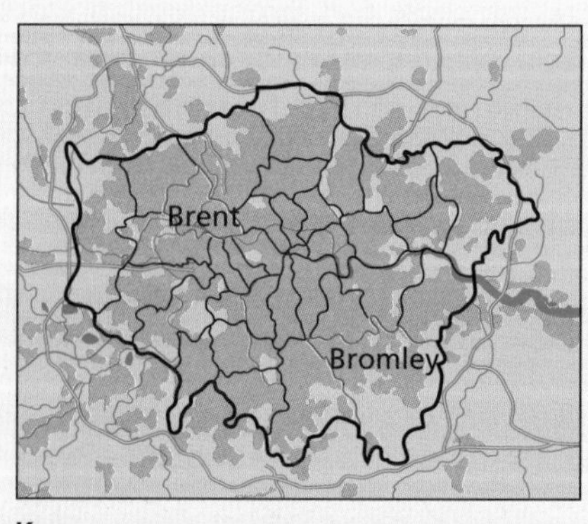

Figure 19.12 The locations of the Greater London boroughs of Brent and Bromley

Table 19.2 Demographic contrasts within the Great London conurbation

	Borough of Brent	Borough of Bromley
Population total (2015)	325,200	324,600
Population density (2015)	7520 persons per km^2	2160 persons per km^2
Population change (2001–2011)	18%	5%
General fertility rate (2014)	72 births per 1000 women	60 births per 1000 women
Average age (2015)	35 years	40 years
Retired population (2015)	11%	18%
Life expectancy	Men 80 years Women 84 years	Men 81 years Women 85 years
Deprivation (2014)	35% of children in poverty	18% of children in poverty
Gender balance	50 M:50 F	48 M:52 F
White British	36%	78%
English speaking	63%	94%
Most common second language (2014)	Gujarati	Polish
Christian	42%	61%
Married or cohabiting	45%	49%
Net internal migration (2014)	–6912	+1169
Net international migration (2014)	+6717	+728
Net natural change (2014)	+3694	+1486
Education (adults with no qualifications) (2014)	8%	4%
Top three occupations	Professional 18% Elementary 14% Associate professional and technical 13%	Professional 21% Admin and secretarial 16% Associate professional and technical 16%

The borough of Brent lies midway in a NNW direction between the centre and the fringe of Greater London. The borough of Bromley straddles the southeastern margins of Greater London (Figure 19.12). Housing in Brent mainly dates from the late nineteenth and the early twentieth century. Terrace and semidetached dwellings prevail, many being subdivided into flats. In contrast, most of the housing in Bromley dates from the inter- and post-war years and is either detached or semidetached.

The most striking differences between the two boroughs are:

- **Population density:** Brent is over three times more densely populated than Bromley. Such a remarkably high density is achieved by small dwelling units and by the subdivision of these units into flats. The flats are occupied by a mix of single people, cohabiting couples and families with children. Living at high densities is one way of bringing housing costs in this part of the capital more into line with earnings.
- **Rate of recent population change:** The rate of change in Brent was over three times that of Bromley. The higher rate is partly explained by the slightly younger population and the influx of immigrant families.
- **Migration:** Brent shows an interesting contrast of a substantial inward international migration and a strong outward internal migration. In Bromley there were relatively small inward internal and international flows.
- **Ethnicity:** Brent has a remarkably high level of ethnicity; immigrant families are prepared to live at high densities and their fertility rates are relatively high. High levels of ethnicity also help to explain the cultural differences in the borough (religion and language).

Brief comments might also be made on other aspects of the comparison:

- **Gender:** Little difference, but higher female life expectancy is part of the explanation in Bromley.
- **Age structure:** The indicators (average age and percentage elderly) are that the Brent population is a younger one.
- **Marital status:** Very little difference between the boroughs.
- **Education:** Single measures suggests a slightly higher level of educational attainment in Bromley.

A comparison of London's commuter villages and remote Welsh villages

In Table 19.3 we move out of London and into the UK's rural space, making use of a classification of rural space that is set out in more detail in Chapter 20 (page 300). The table examines the main generic differences between the two types of village that lie at either end of the rural accessibility scale. In terms of accessing relevant statistical data, this particular comparison is rather more challenging. Data for both settlements needs a finer spatial framework than that provided by district councils. Unfortunately, census data at a parish level is more difficult to access and less comprehensive.

Table 19.3 perhaps exaggerates some of the differences between the two places. However, there are some clear demographic differences with respect to:

- rates of natural increase
- age structure
- migration.

Those differences reflect, above all, the impact of decreasing accessibility on economy and occupation. Cultural differences as such are not to do with ethnicity (see later) but rather to do with the fact that commuter villages are part of the 'modern, dynamic' scene, while 'traditional' cultural customs linger on in remote villages.

Table 19.3 A village comparison in the UK

	London's commuter villages	Remote Welsh villages
Population trend	Growing	Stagnant or declining
Natural increase rate	Relatively high	Relatively low
Gender balance	Even	More women than men
Age structure	High incidence of economically active adults and young families	Top heavy; high proportion of elderly; low proportion of children
Migration	Strong inward, internal flows	Strong outward, internal flows, but some retirement moves in the opposite direction
Ethnicity	Largely white-British	Overwhelmingly white-British
Deprivation	Relatively low	Relatively high
Settlement features	Modern housing estates of varying densities wrapping around old village nucleus	Compact; little evidence of recent building; derelict properties
Occupations	Commuting; many jobs in service sector	Farming, tourism; many retired
Sense of community and identity	Diluted by steady influx of newcomers	Strong, but weakening as older residents die or leave
Planning policy	Controlled expansion	Supportive of local initiatives to sustain survival
Ten examples	Great Shelford (Cambridgeshire) Great Waltham (Essex) Stock (Essex) Ashwell (Hertfordshire) Barley (Hertfordshire) Penshurst (Kent) Chobham (Surrey) Windlesham (Surrey) Hurstpierpoint (West Sussex) Ticehurst (East Sussex)	Y Gyffylliog (Denbighshire) Aberarth (Ceredigion) Abergorlech (Carmarthenshire) Llanrhystud (Ceredigion) Llangybi (Monmouthshire) Parc (Gwynedd) Ysbyty Ifan (Conwy) Llandefalle (Powys) Glascym (Powys) Hirnant (Powys)

Figure 19.13 Multi-ethnic London

Outside the cities, the ethnic mix is greatly diluted. Indeed, there are very low levels of ethnicity in the UK's truly rural areas. This is due to the fact that those immigrants from Eastern Europe working in the 'picking, plucking and packing' activities of agricultural areas prefer to live in nearby towns. From there they are bused out on a daily basis to work in the fields.

The presence of these groups in our increasingly multi-ethnic country is reflected in distinctive cultural traits that are largely to do with language, religion, dress and codes of conduct. Their presence is also reflected in specific types of retailing along the high streets of those places where ethnic groups have become concentrated (see Chapter 21).

Pathways of cultural change

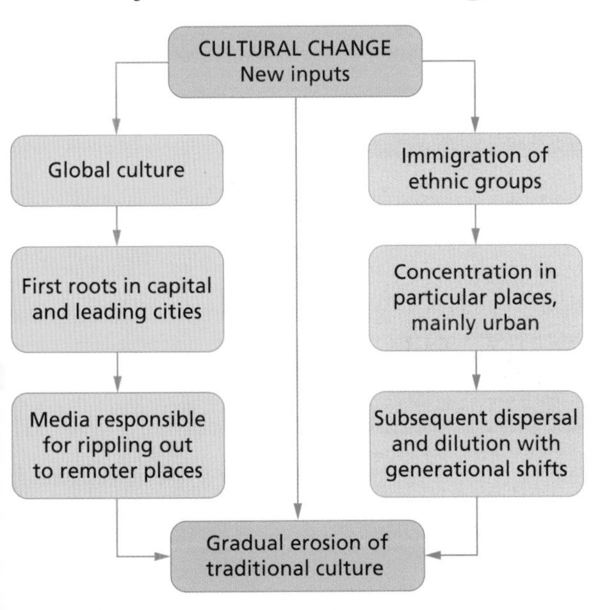

Figure 19.14 Pathways of cultural change

Cultural change (Figure 19.14) is driven along two different pathways by:

- **A changing mix of ethnic groups:** Initially, this results from the arrival of immigrants of an ethnicity different from that of the host country. The cultural change is consolidated if the immigrants become concentrated in particular places and if their rates of natural increase happen to be higher than the native ones. Subsequently, as offspring of the original immigrants move out to new places, so they are likely to take their culture with them, perhaps in a more diluted form.
- **The gradual dissemination of a constantly updated 'new' culture by the mass media:** Some of this updating might be seen as part of the globalisation of culture. Modern communications are promoting and spreading an international culture or lifestyle. The media are now the global exporters of a 'Western' culture rooted in Europe and North America. The outcome is referred to as 'Westernisation', 'Americanisation' or 'MacDonaldisation' (Figure 19.15). This growing global culture is distinguished by its emphasis on consumerism and consumption, democracy and technology. Within the UK, the tendency is for changes in global culture to first impact on London and other leading cities and then to ripple out from there.

Key term

Cultural change: The modification of a society through innovation, invention, discovery or contact with other societies.

Figure 19.15 Evidence of an Americanised global culture

Both pathways involve coming up against the 'core' culture, that is, the cultural values held by the majority of people in the UK. However, it is the former pathway – the immigration of ethnic minorities – in which we are more interested in this topic. It presents particular challenges and opportunities. When people from other cultural or ethnic groups first arrive in the UK they are, of course, suddenly exposed to the national 'core' culture. Initially, they are likely to be bemused by that culture but, over time, they will come to understand some aspects of it and slowly adapt to it. They will be slowly assimilated and gradually, over several generations, assume some, if not all, of the core cultural values. But cultural assimilation works both ways. The core culture will inevitably take in some aspects of immigrant culture. The British love of Indian and Chinese food is evidence of this cross-cultural exchange.

19.3 Local place studies (1)

The Specification requires you to study two places. You should have first-hand experience of one and the other should be significantly different. The best starting point would be first to choose the place in which you live. That, in turn, will determine the choice of a contrasting place. There are many possibilities, such as comparing:

- two different parts of the same town or city (see Table 19.2 on page 287)
- two different types of rural settlement (see Tables 19.3 and 20.2 on pages 288 and 303, respectively)
- two different places (urban and rural) within the same local authority area (see Breckland, Chapter 22, page 331).

Your choice may also be conditioned by practical matters. If you wish to study the contrasting place at first-hand, then obviously you will need to choose somewhere that is accessible from home. Or maybe your preference is to study a place that is truly remote and instead undertake a sort of 'virtual reality' investigation relying on secondary sources rather than field observations. No matter what your choice of contrasting place, you need to remember that the emphasis in both investigations is on people. For this reason it is recommended that you define both your places in terms of the spatial units used in the collation of census data, i.e. borough, local authority, parish or ward.

The essence of 'place'

Cultural geography studies why certain places have a particular meaning to individuals or groups of people. Places said to have a strong 'sense of place' have a strong identity and character that is felt deeply by local inhabitants and possibly by visitors. It is claimed that a sense of place exists independently of any individual's perceptions or lived experiences. The sense of place derives from the natural environment as well as from a mix of natural and cultural features in the landscape, and perhaps most importantly includes the people who occupy the place.

Meaning is another concept in place studies. Meaning derives from how a person or a group of people perceive a particular place, that is, what that place 'means' to them. It is a highly subjective aspect of place. For example, the meaning of a place to some might be security; to others neighbourliness or contentment.

As far as this chapter is concerned, the key place questions are:

- What distinguishes one place from another?
- What makes a place and its inhabitants distinctive?
- What influences people's perceptions of, and attachments to, a place?
- What causes places to change?

Some guidance on how to undertake the investigations required by the Specification is offered here. Figure 19.16 is a working framework or reminder of what you need to take into account when establishing:

- the essential identity or character of your chosen places – their sense of place; this is the outcome

Key concept: Sense of place

This grows from a person identifying themselves in relation to a particular location, most obviously the place where they live. It involves experience of a particular landscape and a knowledge of its people and their culture.

Key concept: Meaning of place

The meaning of place has two components: 1) the perceptions, associations and experiences of individuals, and 2) the common understanding of a place shared by social or culture groups.

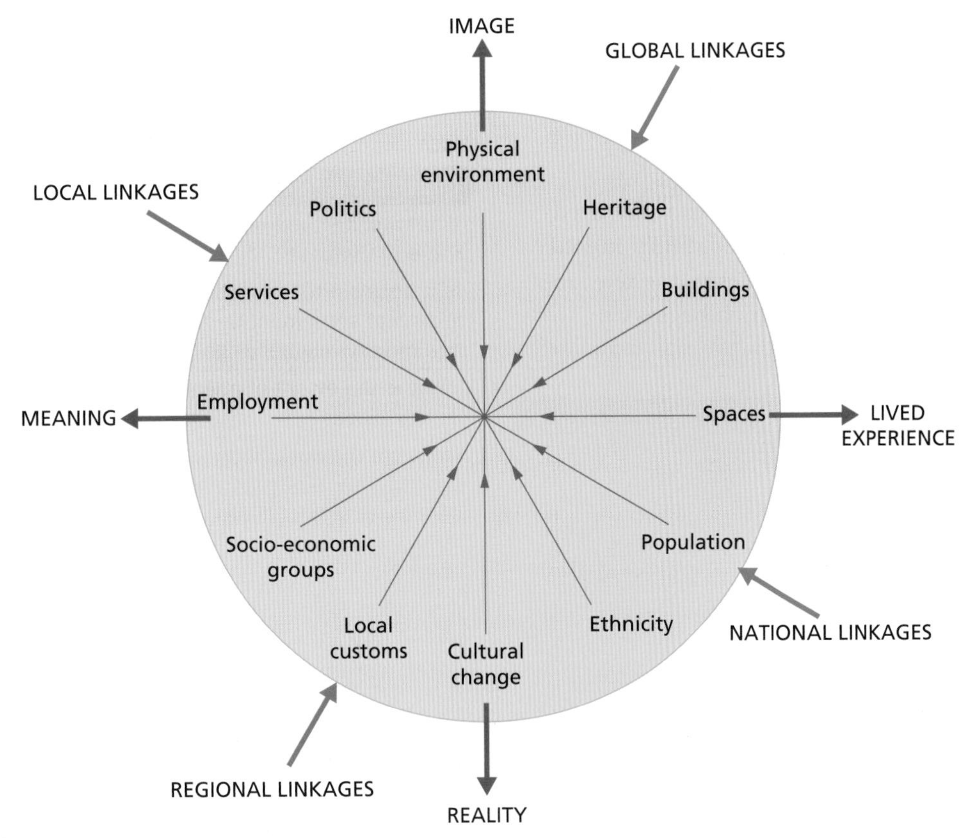

Figure 19.16 Components of place

of many connections involving interacting internal factors plus a range of external influences – global, national, regional and local

- the divergence, if any, between the popular image and the reality of each of your chosen places
- the impact of recent changes, particularly of a demographic and cultural nature, on the identity of a place and its inhabitants
- the lived experiences of inhabitants, their feelings towards the places they occupy and what the places mean to them (this is investigated in Chapter 20).

This guidance is mainly in the form of questions to be asked of both your places. Useful sources are given where appropriate. The questioning is divided into four parts:

- opening questions
- external influences
- internal factors
- recent demographic and cultural changes.

Key term

Sense of place: An overarching impression encompassing the general ways in which people feel about places.

Opening questions

Table 19.4 sets out the first questions to be asked of both places.

Table 19.4 Some opening questions about your places

Questions or lines of enquiry	Sources
What do you know already about your chosen places?	Walk about in both places
In what administrative units are you going to define your places? Borough, ward, parish or postcode?	In urban areas wards are of a manageable scale; parishes in rural areas
What are the physical landscape characteristics of both places?	OS 1:25,000 map plus field observation
Where are your places located on the rural–urban continuum?	
What has broadly happened to your places over the last 100 years or so?	Local libraries usually have collections about local history
What are their main functions or economic activities today?	Identify main employers and find out about commuting
What services are provided within your places?	Locate shops, schools, medical centres

External linkages or influences

The impacts of external linkages shown in Table 19.5 are generally less tangible than the introductory aspects just considered. For this reason, you may find it rather difficult to assess the degree to which the identities of your chosen places have been influenced by them. The following questions should help identify relevant linkages. (Note the references to the relevant skills in the Skills focus section, which begins on page 345.)

Synoptic themes:

Players

Places in the UK, as elsewhere, are experiencing the various impacts of external players, such as TNCs and IGOs – from employment and branded goods to rules and regulations.

Internal factors

Figure 19.16 shows that there are potentially many internal factors contributing to place identity. A few have already been dealt with in Table 19.5. Clearly, the relative importance of these factors in terms of their impact on place identity will vary from place to place. Indeed, there may be others that you might need to consider. Table 19.6 puts the spotlight on some aspects that can add character to a place (again, note references to the Skills focus section).

Table 19.6 Some questions about internal factors contributing to place identity

Questions or lines of enquiry	Source
What architectural reminders are there of the past?	Field observation
Is it possible to reconstruct the growth of settlement through the age of surviving buildings?	Old maps and field observation of age of buildings
What is the extent of open space in your places?	OS 1:25,000 maps
What are the population characteristics of your places – density, structure, life expectancy, fertility and mortality rates, and so on?	ONS, 2011 census report
What is the ethnic mix in your places and are minority groups segregated?	ONS, 2011 census report
How would you describe the prevailing socio-economic 'tone' of your places?	Field observation
Is there a residential segregation of different socio-economic groups?	Field observation
Do you think that each place has a distinctive identity? If so, sum up each of those identities in a few words.	Placecheck (Skills focus 12) and social media (Skills focus 1)

Table 19.5 Some questions about external influences contributing to place identity

Scale	Questions or lines of enquiry	Sources
Global	Are any global brands present – retail outlets, branch factories and offices? Is there any involvement in global tourism? How well connected to the internet are your places?	Field observation Tourist information centres Broadband speed tests
National	Are your places directly affected by specific government policies or designations? How do your places compare with the national averages of wealth and deprivation? How accessible are your places to London and other leading UK cities?	Look at Index of Multiple Deprivation results (Skills focus 10)
Regional	How well are your places connected in terms of transport? Are there major cities nearby and what is their impact on your places? What are the county council's policies on growth and development?	OS maps and public transport timetables Think in terms of commuting, accessing services, leisure Contact local planning offices
Local	What are the political inclinations of the local councils? How strong is the sense of community? What are the pressures for change?	Last local election results and relative representation of different political parties Social media (Skills focus 1) Local newspapers (Skills focus 3)

Recent demographic and cultural changes

Places are dynamic in that they fuse change with continuity. Questions focusing on recent demographic and cultural changes follow in Table 19.7. These changes can have a powerful influence on the identity of today's places (see more useful references to the Skills focus section).

These last questions are challenging and rather tricky, but do your best to address them. You might pursue them by means of a limited questionnaire or by interviewing among, say, three different groups (possibly of both white British and other ethnic origin), for example:

1 secondary school students
2 adults with children living at home
3 retired persons.

To what extent are these groups affected in the same way by the mix of continuity and change that prevails in almost all places? Do those groups perceive the continuity and change in the same way? Might it be that some aspects of that continuity and change are imagined rather than real?

Table 19.7 Some questions relating to recent demographic and cultural change

Questions or lines of enquiry	Sources
Have the populations of your places changed over the last intercensal period; if so, how?	2001 and 2011 census reports
To what extent has population change been the result of migration?	2011 census report
What cultural changes are evident?	Look for recently established services catering for specific minority groups
Has the mix of socio-economic groups changed?	ONS, 2001 and 2011 census reports
Are the recent demographic and cultural changes significantly changing the type of people resident in your places?	ONS, 2001 and 2011 census reports
In which direction do you think your places are moving: up- or downmarket?	Look for evidence of gentrification or areas of deteriorating housing; improved quality of shops (Skills focus 4)
Are the demographic and cultural changes having a noticeable impact on the basic identity of your places?	Oral accounts (Skills focus 2) or interviews (Skills focus 9)
In what ways have recent demographic and cultural changes impacted on the identity of local residents?	Interviews (Skills focus 9), social media (Skills focus 1)

The investigation of your chosen places continues in Chapter 20. It is important that the place investigations are undertaken by you, but with help from your teacher. Throughout them you should be mindful of the similarities and differences between your places. Furthermore, questions in the examination will ask for references to, and detail from, your place studies. It is also important to remember that your chosen places can be represented in different ways that will inevitably convey different images. For example, your places may be portrayed by descriptive text, by maps and diagrams, by photographs (old and new), by statistics, as well as through the observations of others, be they artists, historians or newspaper reporters (see Local place studies (2): evaluating perceptions of living spaces, p. 305).

Review questions

1 Identify three significant changes in the distribution of the UK's population between 2001 and 2011.
2 Explain and illustrate what is meant by the term 'population structure'.
3 Describe and explain how urban population pyramids in the UK differ from rural population pyramids.
4 Suggest reasons why the urban population of the UK grew at a faster rate than the rural population between 2001 and 2011.
5 Examine the factors causing population densities to vary spatially within cities.
6 In what ways are the functions of rural places in the UK changing?
7 Suggest reasons why the populations of rural places in the UK show less ethnicity than those of urban places.

Further research

Find out more about the rate of population growth in the UK between 2001 and 2011 compared with that in other EU countries: www.ons.gov.uk/ons/dcp171778_270487.pdf

Choose two examples from Table 19.3 and use the internet to research the demographic differences between accessible and remote rural places.

Use the internet to research the appeal and problems of a chosen 'honeypot' village. A good example is Lyndhurst in the New Forest National Park.

20 Living spaces

How do different people view diverse living spaces?

By the end of this chapter you should:

- be aware that different groups of people have their own perceptions and lived experiences of both urban and rural places
- understand that urban and rural living have their downsides
- be aware that there is a range of ways to evaluate how people view their living spaces.

20.1 Contrasting lived experiences and perceptions of urban places

The enquiry question that heads this part of the topic asks: how do different people view diverse living spaces? The word 'view' as used here is an alternative to a more commonly used verb in geographical literature, namely 'perceive'. People's lived experience and perceptions of specific places and the world around them are an important aspect of modern human geography.

It is important to remember that different individuals and different groups may well view a given place, situation or development in very different ways. Some will perceive urban living as being attractive and as offering advantages, while others will view it negatively as being unattractive and as presenting a range of disadvantages. In between those two extremes, there will be others who perceive urban living as involving both costs and benefits. Each person gradually accumulates a vast series of perceived images or mental 'pictures' of reality formed by memory, imagination and experience. They are a vital part of the lived experience and of the meaning of place.

Key terms

Living space: In a narrow sense, the term living space refers to land given over to housing. In its broader sense, the term embraces all that space given over to the day-to-day needs of a population, from work, shopping and leisure to education, healthcare and entertainment. Housing will certainly be in focus for much of this chapter, but the broader interpretation of living space will prevail.

Perception: An individual's or group's 'picture' of reality resulting from their assessment of information received.

Changing perceptions of urban places

The Industrial Revolution in the nineteenth century gave birth to many of the UK's larger cities. Their growth was fuelled by massive rural-to-urban population flows. This migration was based on the perception that the good life was to be found in urban rather than rural places. But that perception conveniently overlooked the considerable downside to urban living. Nowhere was this downside more apparent than in Victorian London.

Figure 20.1 Urban space meets rural space

Victorian London

Victorian London was a city of startling contrasts. New building and affluent development went hand in hand with horribly overcrowded slums where people lived in appalling conditions. The city's population exploded during the nineteenth century, rising from 1 to over 6 million. This growth far exceeded the capital's ability to look after the basic needs of its citizens (Figure 20.2).

While the Industrial Revolution had brought economic growth and technological advances, little or nothing of this was directed towards helping the poor. Many households struggled to survive. Children as young as five were often set to work begging or sweeping chimneys. Working conditions for manual workers were both dangerous and unhealthy (Figure 20.3)

A combination of coal-fired stoves and poor sanitation made the air heavy and foul-smelling. Immense amounts of raw sewage were dumped straight into the River Thames. It was hardly surprising that the general health of the capital's population suffered. There were frequent outbreaks of contagious diseases such as cholera, smallpox and typhoid, as well as regular influenza epidemics. Mid-century life expectancy in London was a mere 37 years. The poor were unable to pay for a doctor to attend them at home. Instead they had to go to a charitable hospital or workhouse infirmary.

The prevalence of poverty in parts of London encouraged a high incidence of crime. Most offenders were young males, and most offences were petty thefts. The most common offences committed by women were linked to prostitution and were, essentially, 'victimless' crimes – soliciting, drunk and disorderly, vagrancy. In short, there were good reasons why London was perceived to be a dangerous and threatening place by visitors.

Some important steps were taken to improve the general quality of life, however. These included:

- the construction of a proper sewage system of tunnels and pipes to divert sewage outside the city
- the founding of the Metropolitan Police in 1829, which led to improvements in law and order
- the building of new homes for the working class.

While the poor lived in the slums of the East End of London, the urban landscape of the West End was being improved by the completion of:

- Regent Street and Piccadilly Circus
- the new Houses of Parliament
- Buckingham Palace
- Trafalgar Square and the National Gallery
- opulent housing for the wealthy in Belgravia.

It is structures such as these that help to create our images of Victorian London. Thankfully the dreadful slums have long since been cleared.

Figure 20.2 Working-class family life in Victorian London

Figure 20.4 Slum housing in Victorian Kensington

Figure 20.3 The interior of a factory in Victorian London

Figure 20.5 The newly completed Regent Street

What about London today? Have perceptions of the city changed? It is viewed by many as an irresistible magnet drawing in migrants from overseas and from other parts of the UK. The attraction is made up of several elements:

- the wide range of employment opportunities
- the range and quality of commercial and social services
- the variety of entertainment and other leisure activities.

London today may be a safer, healthier and overall a more affluent place than it was in Victorian times. However, it still has a downside, but of a rather different complexion. This is explored in more general terms in the next section.

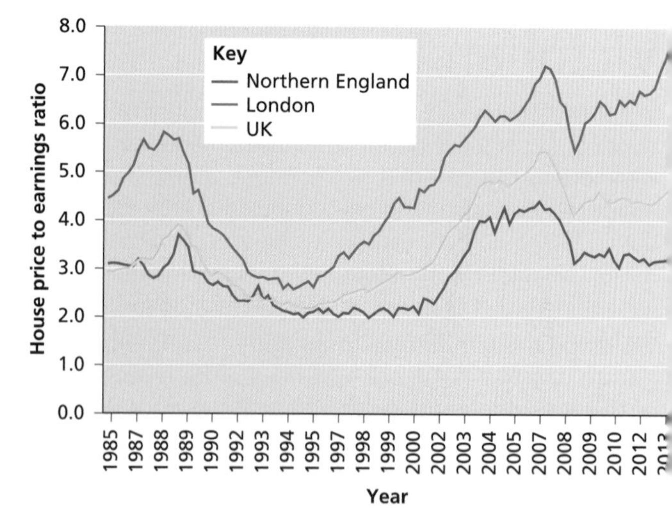

Figure 20.6 House price to earnings ratios, 1985-2013

The urban reality

Reference is often made to the 'bright lights of the city'. The term implies that towns and cities are places where you will find the good life. But what are the realities of actually living in urban places? What is the lived experience?

High living costs

Work in urban places is generally paid more than most jobs in rural places. But it would be wrong to claim that urban dwellers are better off than their rural equivalents. It is critical to offset this against the fact that urban living costs are significantly higher. These living costs have three components:

- The high cost of housing, which immediately neutralises the strong attraction of reliably high wage and salary levels. Critical here is the ratio between house prices and average earnings. Figure 20.6 shows just how disadvantageous this is in the case of London. Clearly, the North of England offers a strong attraction here!
- The financial and physical costs of commuting.
- The generally higher costs of food.

It is these high living costs that persuade so many urban people to make urban-to-rural residential moves.

Low environmental quality

The brutal reality of urban living for many people is having to live in a poor-quality environment involving:

- unsatisfactory or substandard housing – a fate suffered particularly by those towards the poverty end of the spectrum
- atmospheric pollution, particularly at street level, and despite various Clean Air Acts
- noise and light pollution
- antisocial behaviour.

Perhaps it is the case that urban living involves developing a sort of immunity to this and other negative aspects.

Crime

Figure 20.7 confirms what is widely believed: that crime rates are significantly higher in urban areas compared with rural areas. One noteworthy causal factor must be the higher incidence of poverty in our cities. It is poverty that so often drives people into crime.

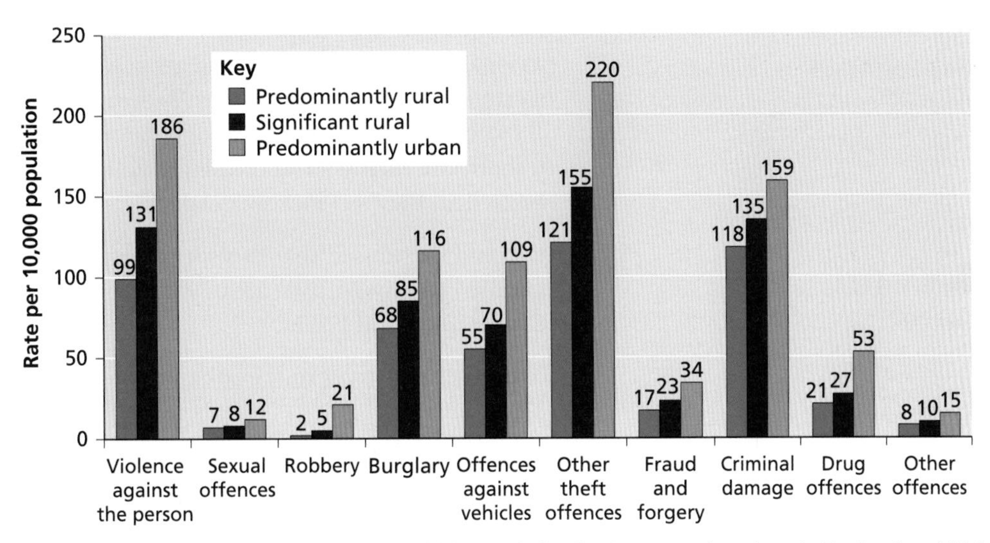

Figure 20.7 Recorded crime rates per 10,000 population in three type locations in England and Wales, 2009–10

While there are striking differences in the crime rates of urban and rural areas, Figure 20.8 illustrates another important point, namely that crime rates vary within urban areas. The crime map of Greater London clearly shows that the highest rates occur in central London and the areas immediately surrounding it. There is one notable exception in west London, however. The hub of the crime recorded there is Heathrow Airport where a host of different crimes, such as theft of luggage, the arrival of illegal immigrants and the carrying of illegal drugs, help boost the overall crime rate.

Skills focus: Use of GIS

See Skill 7 in the Skills focus section (page 348).

Ethnic diversity

The views of ethnic minorities on urban spaces will be discussed in Chapter 21. Minority groups continue to be strongly segregated within the UK towns and cities, either by choice or for reasons of mutual support. Despite the UK being acknowledged as a multi-ethnic society, there is still discrimination on the grounds of ethnicity. Not everyone is happy to see their local area being 'infiltrated' by members of a particular ethnic group. But there are some positive signs; subsequent generations of immigrants are now moving out of their original enclaves and in the direction of the suburbs. The likely reasons for this include:

- increasing self-confidence
- secure employment

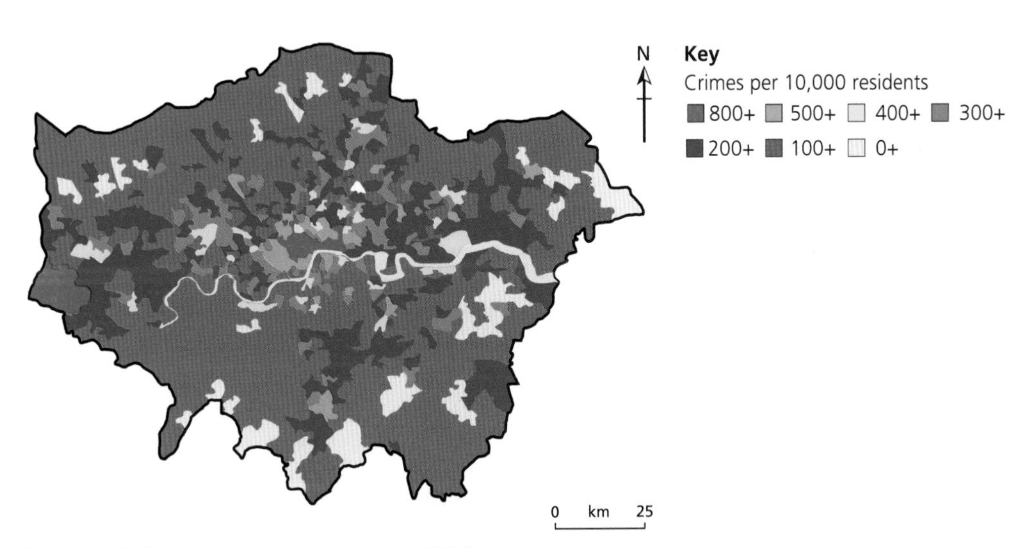

Figure 20.8 The London crime heat map, 2013

Figure 20.9 A portrait of loneliness

- increasing affluence
- a wish to put some space between themselves and their ethnic culture
- a wish to become more integrated into UK society
- the perception that the 'grass is greener' elsewhere.

Social isolation

Many newly-arrived immigrants initially have feelings of exclusion and social isolation. These feelings can be diluted over time by:

- a willingness to become assimilated into UK society
- living close to, and feeling at ease with, others belonging to the same ethnic community.

For some, however, social isolation will persist, whether by choice or force of circumstance. But isolation is not confined to ethnic minorities. It is also something that afflicts the elderly (see Figure 20.9).

The elderly

Not everyone living in an inner-urban area will have followed the common life-cycle pathway leading to the suburbs (see intra-city contrasts later). Many remain there. Perhaps they were born there and took over what was their parents' home. They have been comfortable living in this neighbourly neighbourhood with its nearness to work, shops, social services and entertainment. No need has been felt to uproot. However, gradually all around them begins to change and they find themselves increasingly out of tune with what is going on. Everywhere seems more noisy and hectic. They feel increasingly isolated as other older residents have either died or made retirement moves elsewhere. They have little rapport with the increasing number of new residents moving into the area, particularly if they happen to be immigrants. Concerns about their personal safety and property, and about a rising incidence of crime, add to their discomfort. A decaying urban place all too often can become a seat of political radicalism. In short, the elderly feel increasingly vulnerable and threatened; they feel socially isolated.

> **Key term**
>
> Social isolation: A complete or nearly complete lack of contact with people and society. It differs from loneliness, which is a temporary lack of contact with other people.

It should be stressed that this sort of experience is not confined to inner-urban areas. It also occurs in the suburbs and rural places.

Two other negatives in the urban reality are:

- the pace and stress associated with living in the UK's larger cities
- the social polarisation between the extremes of great wealth and abject poverty.

Inner-city versus suburban places

Possibly the greatest differences in lived experience are those between the central and inner-city on the one hand and the suburbs on the other. Perhaps this is best seen in the experience of working adults as they move through the life cycle.

If you are a young adult, fresh out of university or in secure employment, your ambitions are most likely to be to move out of the family dwelling and into accommodation of your own. Inevitably you have to be content with a small amount of living space (in a flat or a room in a subdivided house) in an area where property prices are low. Your residential destination is likely to be in the inner and older parts of urban

> **Synoptic themes:**
>
> **Attitudes**
>
> Attitudes to inner-city and suburban living will vary depending on a range of factors. These include the nature of the involvement (residence, workplace, leisure, etc.) and personal characteristics (age, stage in the life cycle, affluence, type of work, etc.).

> **Key concept: Life cycle**
>
> Probably better referred to as the family life cycle since the concept is based on the idea that most families or households go through a sequence of changes in their lifetime, which are particularly significant in terms of housing needs and housing moves.

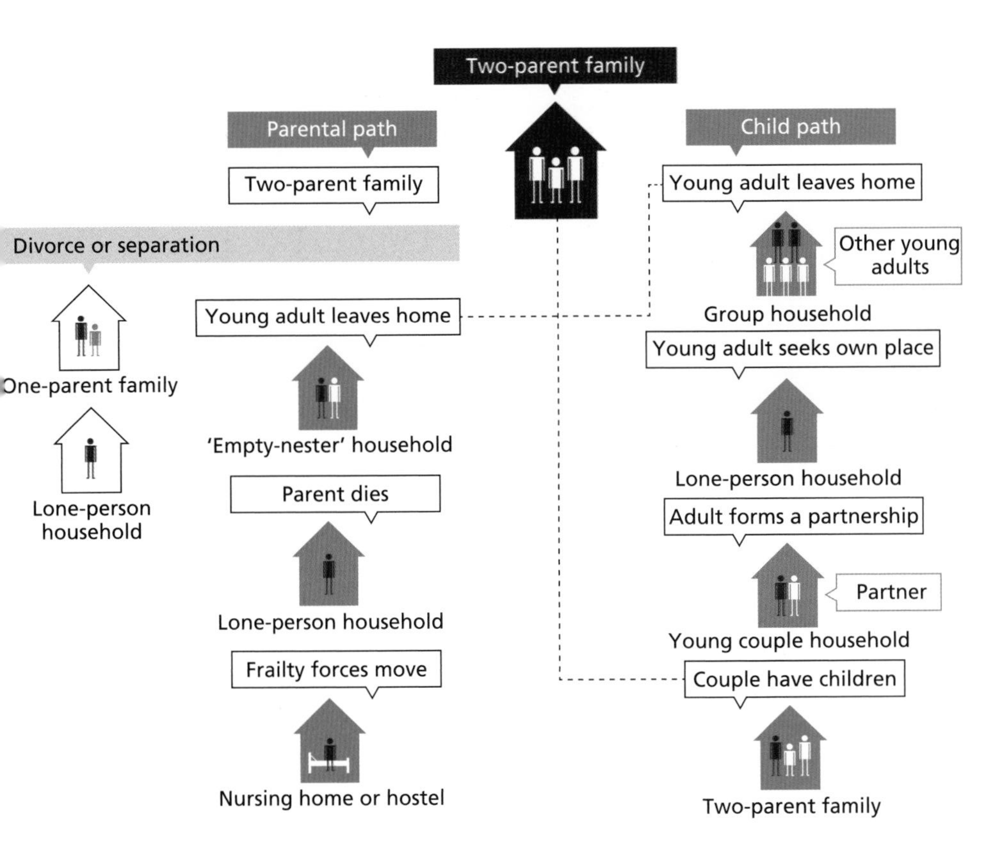

Figure 20.10 Household life cycles

places. This is not considered to be a downside as you will be close to your place of work in the CBD and, after working hours, you will be able to enjoy the exciting nightlife – clubs, wine bars, gigs, theatres, and so on.

This perception that the good life is to be found in the more central parts of urban areas might change when it comes to settling down as a family person. The suburbs are perceived as better fulfilling your needs – for more residential space, good schools for the children and green spaces for them to play in, easy access to good healthcare and basic shops. Not so many years ago, your perception of life in the suburbs was that it would offer a dull lifestyle and you had an image of a characterless environment. Figure 20.10 shows a two-parent household as a starting point of two housing pathways – one for you as a parent and the other for your offspring.

Perhaps you are now well along the career path, your salary is good and the children have left home, so you begin to perceive that the grass might be greener in some commuter village or town. Retirement is also beginning to beckon, so you are not too worried by the costs and time of commuting. Rather, your perception places greater value on such things as a more attractive residential environment, a house with a good-sized garden, proximity to a pleasant golf course and to a nucleus of good-quality shops.

Synoptic themes:

Attitudes

Urban and rural residents may differ in their attitudes to, and perceptions of, places.

So, any debate on the pros and cons of the lived experience in particular places hinges on who you are, your changing perceptions and where you are in the life cycle. It also depends, of course, on what you can afford to pay for housing. House prices will undoubtedly reflect, for example, any widely recognised spatial variations in the quality of the living environment. This applies to urban and rural places alike (Figure 20.11).

Skills focus: Interviewing

See Skill 9 in the Skills focus section (page 350).

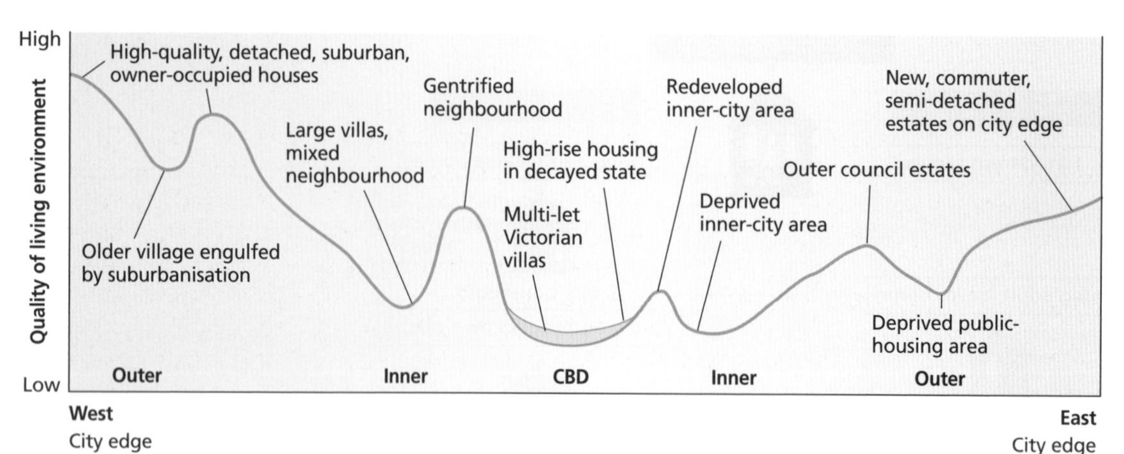

Figure 20.11 Variations in the quality of the living environment across a typical UK city

20.2 Contrasting lived experiences and perceptions of rural places

The same concepts of perception and lived experience come into play in the evaluation of rural places. Again, it all depends on:

- who you are and your perceptions
- your intended use of rural space
- the type of rural area in question. Here it is appropriate to recognised three different types of rural space along the rural–urban continuum (see later).

The rural idyll

In this day and age of counter-urbanisation, it would seem that more and more people in the UK are pursuing their perception of the rural idyll, sometimes referred to as the 'rural myth'. The rural idyll is imagined as involving almost everything that is thought to be negative about urban places. It is therefore a rather nostalgic perception that rural places are problem-free, natural, peaceful, healthy and friendly. It involves a sort of chocolate-box image of green pastures and quaint villages.

Key term

Counter-urbanisation: The movement of people and employment from major cities to smaller settlements and rural places located beyond the city, or to more distant, smaller cities and towns.

Table 20.1 Possible components of the perceived rural idyll

Proximity to nature	Organic farming
Attractive landscape and scenery	Friendly community
Sheltered location	Easy access to services
Peace and quiet	Personal security
No pollution of any kind	Minimal crime

Figure 20.12 Finchingfield (Essex) – a 'chocolate-box' village

Skills focus: Oral accounts and qualitative data

See Skills 2 and 6 in the Skills focus section (pages 346 and 347).

The rural reality of different types of rural living space

It bears repeating the point that in viewing a particular place, it very much depends on who you are and the use you intend to make of that place. Three different types of rural place may be recognised on the basis of their relative accessibility from major towns and cities.

Hardy's 'Wessex'

Perhaps one of the best-known and most widely read purveyors of the rural idyll was the novelist Thomas Hardy (1840–1928). Among his most famous works were those now collectively known as the 'Wessex novels'. While this was a fictitious county, the southern part of it coincided with Hardy's home county of Dorset (Figure 20.13). The best-known novels today are *The Mayor of Casterbridge*, *Tess of the d'Urbervilles*, *Jude the Obscure* and *Far from the Madding Crowd*.

Figure 20.13 Hardy's cottage: quintessential Wessex

In the Wessex novels Hardy gave detailed descriptions of country life throughout the year. These were based on the observations he made as he cycled around Dorset and parts of adjacent counties. Hardy's Wessex has since become recognised as 'the epitome of the vanishing English rural heartland'. This heartland was then, of course, being threatened by urbanisation and the new technologies of the Agricultural Revolution.

Wessex was a 'partly real, partly dream county'. On the one hand, Hardy was perpetuating the rural idyll. On the other hand, his novels constantly remind us of the harshness of rural life, its strange folklore and its often cruel traditions. Hardy was a realist being keenly aware of the warts of rural life and of the progress that was threatening idyllic rural places.

Tourists today are still drawn to Dorset by the Wessex novels. To those unaware of the Hardy connection, the county has much going for it: a fine coastline, rolling chalk downland, desolate heaths, prosperous agriculture and small historic towns.

Synoptic themes:

Attitudes

Attitudes to different types of rural living will vary depending on who you are, your stage in the life cycle and your place preferences.

Accessible rural places are of two types – those that lie within commuting reach of towns and cities, and those that lie beyond that reach but are accessible from urban areas for recreation, leisure and retirement (Figure 20.14). A useful shorthand is to refer to them as having day-tripper access.

Commuter belt

Rural places are the immediate beneficiaries of the processes of suburbanisation and decentralisation. In England, according to the 2011 census, 19.8 million people live in rural areas and, of these, 98 per cent live in accessible rural places.

These are places experiencing quite fast rates of population growth. The growth is partly due to the arrival of large numbers of workers and their families, keen to escape what they perceive to be the downside of urban places. In particular, there are the high costs of housing and other negative aspects of the urban environment. This in-migration means that adults of working age are becoming a conspicuous component of the population and so too their children. These population characteristics are reflected in the population pyramid (Figure 20.15).

With good numbers of young adults in the mix of migrants, population growth is also being fuelled by natural increase. Fertility rates will be fairly buoyant and mortality rates around the national average.

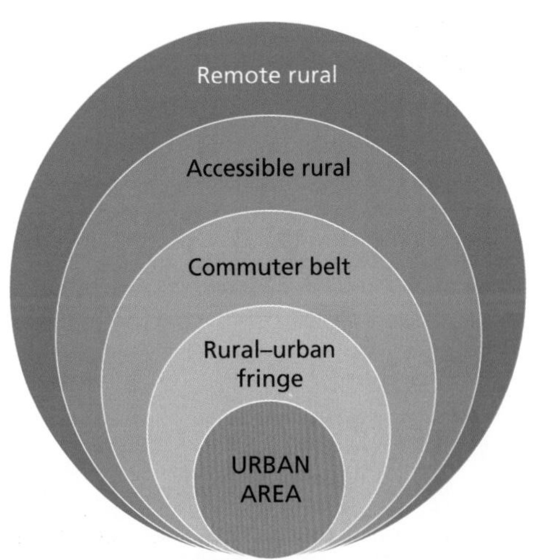

Figure 20.14 Different types of rural space outside the urban area

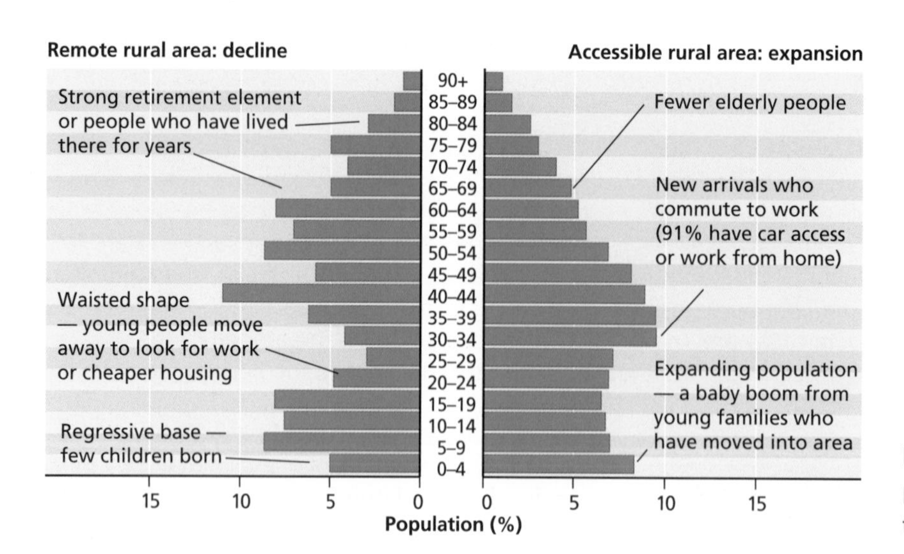

Figure 20.15 Population pyramids of two different types of rural place

Accessible rural

In many of these accessible rural places, perhaps more so in those beyond the reach of easy commuting, there is a strong component of retired people. This component involves both people who have made retirement moves away from urban areas and people who have lived in these places for many years, if not their whole lives. It is interesting to note that, while these accessible rural places are significant migration destinations, the people involved are predominantly 'white British'. It would seem that, as yet, ethnic minorities prefer to remain in urban places. But will this change?

There is another factor encouraging population growth in some of these rural areas further away from urban centres. It is in those honeypot areas that attract urban day-trippers in the context of leisure and recreation (Figure 20.16). It might be the visual appeal of the landscape and of 'chocolate-box' villages, or the availability of recreational opportunities such as walking, boating, fishing and picnicking. Meeting the needs of visitors can generate both jobs and income, so again people in the working age-range may be well represented in the population pyramid.

Figure 20.16 Quay Street, Lymington – a day-tripper honeypot area

It would be wrong to think that these accessible rural places have had things all their own way. Government intervention in the form of planning is very much in evidence. Some rural areas have been earmarked for expansion, such as villages around Cambridge. It is these that are experiencing fast rates of population growth. Others have been given protection by green belt status, as in London's rural–urban fringe. This has had the effect of dampening down population growth and making them into desirable places in which to live. The result is that their populations contain a good many wealthy people in the 50+ age range. Many of those day-tripper honeypot places have also benefited from conservation measures which, in turn, have made them more sought-after locations for more affluent people.

Remote rural

These rural areas have suffered greatly from depopulation. They have been the victims of urbanisation and its associated rural–urban migrations. The push factors behind much of that migration have been the remoteness, the poor quality of life and, in many places, the harsh physical environment. These places have for centuries been gripped by the downward spiral of decline and deprivation (Figure 20.17). Young people in the economically active age range have left in search of secure employment and a better life, leaving behind a residue of elderly people and people with limited means or ambitions.

Figure 20.17 The spiral of decline in remote rural places

However, there are signs in some remote rural places that fortunes may just be undergoing a reversal. Rural areas, most notably in the Highlands and Western Isles of Scotland, are being re-examined. The qualities of their physical environments are being reappraised by the tourist industry. Remoteness is beginning to be given a positive spin as increasing numbers of urban people put a premium on solitude, peace and quiet, as well as on fine scenery. Meeting the needs of 'residential' and 'touring' visitors coming to these places is creating employment and generating income.

The purchasing of second homes in some areas may look to be an encouraging development, but in reality it is proving to be a mixed blessing. Those second homes and holiday lets may bring in temporary residents but they do little to remove the 'ghost-town' effect for much of the year. They do little to stop the closure of local services.

It may be that counter-urbanisation is beginning to reach some remote parts of the UK. People dissatisfied with urban living are moving both their homes and their jobs. The latter is made possible by modern communications technology. For certain types of work, it is no longer necessary to remain in physical contact with colleagues and customers. While there are glimmers of hope in some places, remote rural places are continuing to haemorrhage young people. Their loss is hardly compensated by the return flow of the elderly people who once lived here but who have spent their working lives elsewhere.

Look at the different types of rural area as seen through the eyes of three different age groups in Table 20.2. The table tries to draw together the observations that have been made above (see also Table 19.3 on page 288).

Table 20.2 Contrasting perceptions of three different types of rural space

	Youth	Middle-aged adult	Elderly
		Commuter villages	
Positives	Good school Plenty of friends	Pleasant location in which to live and bring up family	Good access to social services
Negatives	Parents preoccupied with work and getting on	The rising costs of housing	High costs of housing Increasing number of commuters
		Day-tripper villages	
Positives	Casual work in leisure and tourism	Good to visit for leisure and recreation	Plenty of people around during the day
Negatives	Difficult to socialise with friends	Limited employment opportunities	Too busy in honeypot locations
		Remote villages	
Positives	Good outdoor recreation	Good holiday places	Quiet and peaceful
Negatives	Little entertainment Too quiet Not a 'cool' place in which to live	Poor communications Expensive accessibility	Dwindling number of friends

Table 20.2 demonstrates how difficult and dangerous it is to generalise. They fly in the face of the inescapable fact that we are all different and see things in different ways.

Pursuit of the rural idyll in retirement

Today, many retirement moves in the UK are also motivated by pursuit of the rural idyll. Recent research shows that nearly 60 per cent of over 55s in the UK find the idea of retiring to a rural area appealing. Peace and quiet are cited as the most important reason for wanting to move (Table 20.1), but does the reality match the dream? 'Over the Hill?' is a national campaign to encourage people to be proactive about that rural retirement dream. Part of the campaign is to make people aware that rural retirement does have a downside. Sampling rural places and life during holidays hardly provides a sound basis for judging what life there will be like there on a permanent basis.

Services

People in rural areas are often expected to travel much further to access the services and facilities that urban dwellers take for granted (Figure 20.18). 'Rural service deserts' mean that the nearest bank, GP surgery, shop or post office could be several miles away. Country pubs are also less common than they once were. Rural services are more vulnerable to public finance constraints since rural service delivery – even at its most effective – is more expensive than in urban areas.

Transport

Access to services and help is a real issue. Surviving in rural areas is now highly dependent on the car and public transport. For most elderly people, there comes a time when driving a car is no longer safe or indeed legal. Although the free bus may sound an attractive alternative, rural bus services are being withdrawn and schedules reduced.

Housing

For those thinking of moving to the countryside, careful consideration needs to be given to different aspects of the intended new dwelling. Although narrow stairways and latched doors may look rustic and attractive, the potential buyer needs to consider the practicalities of old age. Can the property be 'future proofed'? Could it be adapted to provide disabled access to all rooms? Will the garden and upkeep of the home continue to be manageable? What are the running costs in terms of fuel and keeping warm? Electricity is an expensive way of heating a property; many rural areas have to rely on deliveries of gas cylinders, solid fuel or oil.

Technology

As more and more services are becoming only available online, the ability to access the internet or to communicate via mobile phone is becoming increasingly important. There are many retired people who are frightened by, and have no first-hand experience of, this communications technology. In addition, there are still too many rural areas where mobile phone reception is either patchy or non-existent. Although steps are being taken to extend broadband to the most remote of rural areas, there is still a long way to go.

Isolation and loneliness

Moving to a rural area may mean moving away from friends and family, so retirement moves can often result in feelings of isolation and loneliness. The fact of the matter is that it is not always easy for older people, who are set in their ways, to socialise and make new friends. It can be difficult for them to break into existing rural community networks. But most rural areas do have their clubs, groups and societies.

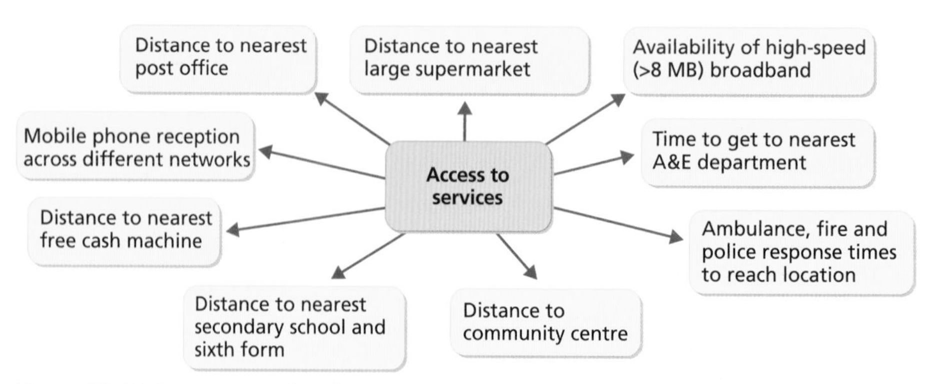

Figure 20.18 Access to services from a rural location

Tensions

A factor that sometimes makes it difficult for retired people to integrate into the rural community is the tension that can exist between the people who have lived in the rural area all their lives and newcomers, be they retirees or second-homers. The resentment is born from the feeling that the newcomers are changing the character of the place. They are seen as 'urban' people who have little idea of what the traditional rural lifestyle entails. The in-comers are regarded as being responsible for the rising costs of housing and the closure of shops, schools and other services.

For anyone who has not considered the issues and discussed and drawn up their own strategy to deal with them, a retirement move can easily become an expensive nightmare. Interestingly, it is the same sort of nightmare that confronts many young people in rural areas, particularly remote ones. The reaction of many is to escape the feeling of entrapment by becoming a rural–urban migrant.

Other spoilers of the rural idyll should be noted, including:

- environmental damage caused by some rural activities – farming, quarrying and forestry
- outbreaks of disease affecting both farm livestock and crops
- the negative impacts of tourism, recreation and leisure (traffic, congestion, etc.)
- the noise and pollution of busy transport lines that run through rural areas linking towns and cities.

20.3 Local place studies (2): evaluating perceptions of living spaces

There are three further lines of investigation that the Specification requires you to pursue with respect to your two chosen places. These are not so much about place characteristics and identity, but rather about:

1 how people perceive those places, and
2 how change might be generating tension and conflict.

Positive and negative images

Not everyone feels at home in the place where they happen to live. Maybe they have been 'forced' to move there or the choice of place was not theirs. In short, people may be divided between those who feel a secure and happy attachment to a place, i.e. have a positive place image, and those who feel alienated and unhappy, i.e. have a negative image. In between, no doubt, there are those whose feelings and perceptions are essentially neutral.

Skills focus: Placecheck

See Skill 12 in the Skills focus section (page 351).

How are the residents of your two places divided between these three crude groupings? Is each of those groupings distinguished by particular types of people? The only sources of relevant information here are interviews and questionnaires. These need to be structured in such a way as to recognise different population groups. Those groups might be the same as used in Chapter 19, namely:

1 secondary school students
2 adults with children living at home, and
3 retired persons.

If there is a significant ethnic component in the population, try to sample three groups of white British and three of other ethnic origin.

Having established which of your population groups an interviewee belongs to, the questions should be along the lines set out in Table 20.3.

Alternatively or additionally, you might follow the Placecheck approach by asking your interviewees to identify specific features which they i) like and ii) dislike about your chosen places.

No matter which approach you adopt, having collected your data it would be interesting to test the strength of

Table 20.3 Some questions about belonging and image

How long have you lived in this place?
Do you feel that you belong to this place?
If 'no', what is it that makes you feel this way?
If 'yes', what is it that make you feel this way?
Have your feelings changed during the time you have lived here?
If they have, have your feelings become more or less positive?
What factors have prompted this change in feeling?
What do you think is the most negative aspect of the place's image today?
What do you think is the most positive aspect of the place's image today?

Skills focus: Testing relationships

See Skill 8 in the Skills focus section (page 348).

relationships in the data between your three groups and what they identify as positive and negative aspects of the place in which they live.

Viewed by the media

Another way of investigating a place is to look at it through different media, such as music, painting, poetry and literature, rather than through residents' reactions to it. Artists of whatever genre are usually very perceptive people. It is highly unlikely that they view a place objectively. However, they can provide insights into the meaning and image of a place. They need to be treated accordingly.

If you happen to find paintings of your places, what sort of image do you think the individual artists are trying to convey – positive or negative? If you have paintings covering a reasonable period of time, are you able to detect a trend in the portrayal of the place? Is the trend positive or negative?

Identifying cultural and demographic issues

Most change is a mixed blessing. There are those who welcome it and those who dislike it. Because of this inescapable fact, change can readily become the generator of tension and the catalyst of conflict. Look at your chosen places. Do you see symptoms of any of the changes outlined in Table 20.4 taking place? If you do, then you are looking at scenarios which most likely involve tensions. In some instances these tensions might be so acute that they generate conflict. Certainly these and other tension scenarios need to be investigated in your local place studies.

Skills focus: Media

See Skill 5 in the Skills focus section (page 347).

Figure 20.19 Portraits of two different places

Table 20.4 Some potential tension scenarios

The influx of people belonging to ethnic minorities – see Glasgow (page 322)
The concentration and segregation of ethnic minorities in particular areas – see Brent Borough (page 286)
The upgrading of residential areas and settlements by the arrival of more wealthy households – see Pepys Estate (page 320)
The building of new houses on greenfield sites to accommodate an increasing population – see Oxford (page 327)
The withdrawal of public services, particularly in rural areas – see Breckland (page 331)

As at the end of Chapter 19, you should be reminded that:

- these two place investigations should be undertaken by you
- you should try to draw out the similarities and differences between your places
- place images are strongly conditioned by the media used to present them
- there are often two sides to place perceptions – the real and the imagined
- you will be expected to feed specific detail from these investigations into your examination answers.

Review questions

1. Examine what you think has most influenced your perceptions of where you live.
2. What do you consider to be the worst aspect of the urban reality? Give your reasons.
3. To what extent do you agree that the benefits of living in the suburbs outweigh the costs?
4. To what extent do you think that the rural idyll is no more? Give your reasons.
5. Examine the reasons for retirement moves away from the city.
6. Suggest possible ways of improving the image of remote rural places.

Further research

Research more about Placecheck: www.placecheck.info

Find out more about your neighbourhood: www.findahood.com

21 Tensions in diverse places

Why are there demographic and cultural tensions in diverse places?

By the end of this chapter you should:

- be aware that culture and society in the UK are becoming more diverse
- understand that the factors underlying segregation vary from place to place and change over time
- understand that the diversity of places and their populations can lead to tensions and conflict.

Key term

Culture: This is the way of life, especially the general customs, values and beliefs, of a particular group of people that are passed on from one generation to next.

Figure 21.1 Another boat overloaded with refugees reaches Italy

Demographic differences between urban and rural places was a theme running through Chapters 19 and 20. In this chapter the spotlight shifts to cultural differences encountered within both types of place.

21.1 An increasingly diverse UK culture and society

Significant internal migration within the UK

This is an important component of population change in local areas within the UK. It has two impacts:

- it changes the total number of people living in an area
- it can alter the structure of an area's population.

Most migration routes involve a two-way traffic. So, the people moving into an area can be quite different from those moving out. For example, the migration of people into London continues to be made up largely of young adults, while the outflow to other parts of the UK involves older adults. This change in composition can happen even if there is little change in population size (in other words, when the inflows and outflows of migrants are almost equal).

Internal migration is said to be a 'zero-sum' phenomenon in that any net migration gain in one area can only occur if there is a net loss of migrants elsewhere. These gains and losses can have immense consequences in terms of labour supply, housing, schools, shops and other services.

The changing distribution of the UK's population was examined in Chapter 19, where it was concluded that migration was a significant factor underlying the changing spatial pattern.

For much of the twentieth century, a major migration within the UK was the so-called 'North–South drift', the general movement of people from the northern parts of the UK to the South East and to London in particular (see Chapter 19, page 278). This really started during the severe economic depression of the 1930s, which particularly hit the northern industrial regions. High levels of unemployment pushed workers and their families towards jobs in the service sector of the South. The drift continued for much of the rest of the twentieth century. It was reinforced by the perception that a better quality of life was to be found in the South.

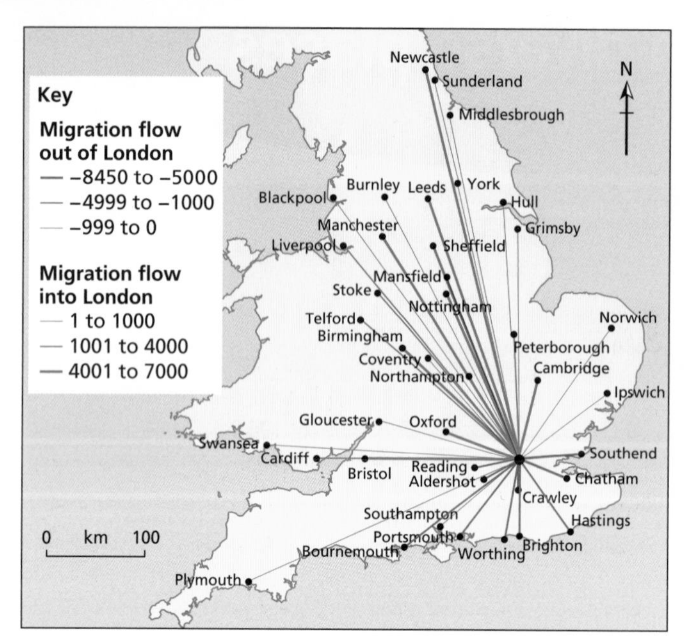

Figure 21.2 Migration into and out of London, 2009–12 (Source: Adapted from data from the Office for National Statistics licensed under the Open Government Licence v.3.0.)

Another noteworthy migration in the twentieth century was suburbanisation, with people moving from inner cities to the suburbs. The decentralisation implied in this process was subsequently extended as increasing numbers of people lengthened their journeys to work by moving out from the suburbs to towns and villages located beyond the fringes of the UK's conurbations and cities.

Key term

Suburbanisation: The outward spread of the built-up area, often at lower densities compared with older parts of a town or city. The decentralisation – of people first and then employment – is encouraged by transport improvements.

Skills focus: Flow lines

Flow lines are a cartographic technique for showing the movement of people, traffic, goods, and so on between places. Lines are shown linking origins and destinations; the thickness of the line gives an indication of the volume of movement (see Figure 21.2).

At the turn of the millennium, the character of this decentralising migration began to change once again with the onset of counter-urbanisation. People began moving out of the UK's conurbations to live and work in smaller urban settlements, as well as in rural areas, both 'accessible' and 'remote'. It is hoped that this process might gradually neutralise and perhaps even begin to reverse the North–South drift. However, Figure 21.2 clearly indicates that London is continuing to attract people from other British cities, and that any outward movement is very much confined to the South East.

Today, these same processes are also changing the cultural pattern of the UK. As will be demonstrated in the next section, that pattern initially reflected the locations that happened to become the main reception areas for a large influx of immigrant ethnic groups in the early post-war period.

Significant migration flows into the UK

There were large flows of migrants into Europe shortly after the end of the Second World War in 1945. A lot of labour was needed to repair the huge amount of bomb damage and to help the economic recovery. In the UK, as in most other parts of Europe, the labour shortage was made worse by the fact that so many people, particularly men, had been killed fighting during the war.

Migration from the Indian sub-continent and the West Indies

The post-war labour shortage was resolved by encouraging migrant workers and their families to come to Europe. The UK's post-war immigrants came mainly from what were or had been colonies in Africa and the Caribbean, and from what had been the Indian Empire (now divided into India, Pakistan and Bangladesh). Immigration was encouraged by an Act of Parliament that gave all Commonwealth (ex-colonial) citizens free entry into the UK. The first ship to bring immigrants from Jamaica docked at Tilbury, Essex, in June 1948. The would-be migrants needed little persuasion as they were from places with high unemployment and a poor quality of life. The prospect of a better life was a strong pull.

Many of the early immigrants found work in public transport, such as on the buses and London Underground. The work was poorly paid. Enterprising immigrants from the Indian sub-continent subsequently acquired small shops and restaurants.

By 1971 there were over 3 million immigrants in the UK who were foreign born. The bulk of these had come from Commonwealth countries. It was at this point that the government decided that the country had enough labour and controls were introduced to reduce the number of migrant arrivals from overseas. Figure 21.3 shows a slight easing back in the growth of the foreign-born population between 1971 and 1981.

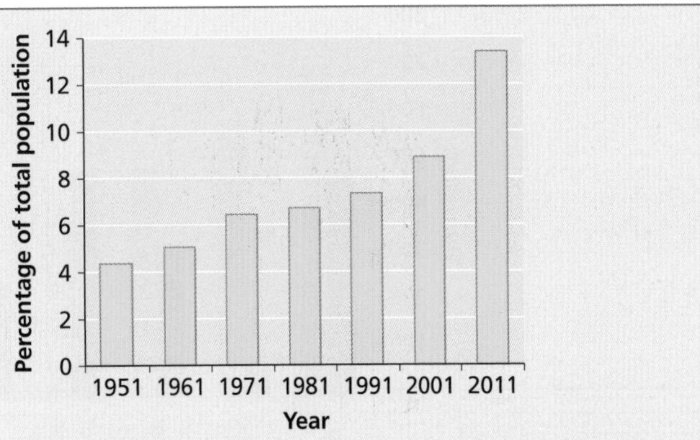

Figure 21.3 UK residents born abroad, 1951–2011 (Source: Adapted from data from the Office for National Statistics licensed under the Open Government Licence v.3.0.)

The Commonwealth immigrants first settled in the major cities of London, Birmingham and Manchester, where there were plenty of job opportunities. But because most of those jobs were poorly paid, the immigrants could only find accommodation in the most deprived inner-city areas. This residential pattern, established in the late 1950s and 1960s, largely persists to the present day. The distribution has developed a sort of inertia.

In the 1990s, however, the UK once again found itself short of labour, so restrictions on immigration were relaxed. This happened to coincide with the collapse of communism in Eastern Europe, which released huge numbers of people 'hungry' for work and a decent wage. In 2004, eight of these former communist countries joined the EU (known as the A8 countries). Their inhabitants were then able to take advantage of one of the benefits of EU membership – the free movement of labour between member states. In 2007 EU membership increased once again with the admission of two more Eastern European countries, Bulgaria and Romania (known as the A2 countries). Figure 21.3 clearly shows a surge in the foreign-born population since 1991.

The global recession that hit the country in 2008 changed the immigration situation overnight. It drastically cut job vacancies and levels of unemployment rose. Clearly, the government needed to do something. The volume of immigration had to be curbed in some way. However, the government was unable to act against the right to move freely between EU countries. There was little it could do to stop the continuing influx of economic migrants. The best it could do was to make it more difficult for migrants to enter the country from outside the EU. Non-EU would-be migrants would now be required to apply for a work visa. But as the UK's boundaries were tightened, so the volume of illegal immigration increased and continues to increase.

Synoptic themes:

Players

National governments are usually the main gatekeepers in migration matters. But the EU's founding principle of a free movement of labour between member countries now means that the UK government is no longer in total control of its borders – an issue that figured in the 2016 referendum.

Today, the UK's economy continues its slow recovery and unemployment is falling, but widespread concern remains about the continuing volume of immigration. This concern was heightened by the 2011 census, which showed that the number of foreign-born people in the UK had increased by 3 million during the previous ten years and that they now accounted for thirteen per cent of the total population. Small wonder that immigration was one of the major political issues during the 2015 general election.

The main suppliers of immigrants are still the Commonwealth countries (Figure 21.4). Of the European sources, Poland is by far the largest, followed by Ireland and Germany. Flows from the Philippines continue to be quite high as large numbers are still recruited as domestic servants. Of course, the volume of immigration is also swollen by refugees fleeing from war-torn countries such as Somalia, Iraq, Syria and Afghanistan.

Perceived job opportunities and the prospect of a decent wage represent the most powerful of the pull factors attracting migrants to the UK.

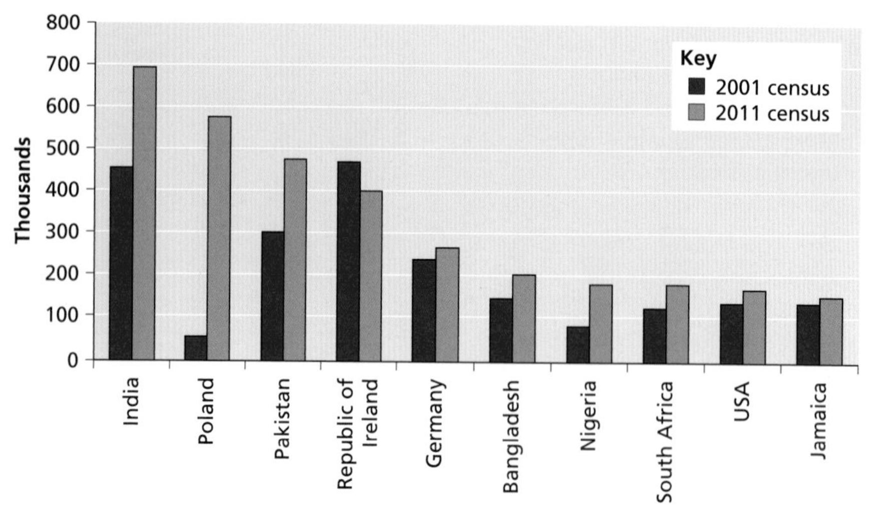

Figure 21.4 Top ten countries of origin of the population of England and Wales born outside the UK, 2001 and 2011 (Source: Adapted from data from the Office for National Statistics licensed under the Open Government Licence v.3.0.)

What sort of reception a migrant receives in UK seems to depend on the skills they have to offer. A doctor from South Africa or nurse from Poland might expect a warm welcome and a reasonable living. However, unskilled migrants and those being trafficked all too readily become the victims of discrimination and exploitation. The jobs they have access to are often unpleasant ones with long, antisocial working hours. Examples include jobs in the hotel and catering trades, and in office and toilet cleaning. They are jobs that British workers tend to shun. Migrants are poorly paid for such work and often at rates below the minimum wage level. As a consequence, these particular migrant workers find they cannot afford even the basic costs of accommodating and feeding themselves and their families. A downward spiral leading to poverty and deprivation is thus set in motion. In short, what started as a journey in pursuit of a dream all too often turns out to be a journey of nightmares.

It is important to remember that migration is a two-way traffic, however. In 2014, 89,000 people emigrated from the UK to other parts of the EU, and 101,000 to non-EU countries. It is reasonable to assume that included in these figures were:

- migrants who came to the UK in pursuit of a better life but who, for various reasons, are returning to their homelands
- economic migrants who came to the UK only on a short-term basis and who have made enough money to finance their dreams back home.

Ethnicity

Ethnicity is a significant aspect of population structure in many societies today. Not only does the ethnic mix within a population vary from place to place, it also changes over time, reflecting the migrational history of a place. In 1950, the UK was almost totally 'white British'. Since then, as has just been noted, there have been waves of immigration coming mainly from Commonwealth countries, but more recently from Eastern Europe.

Table 21.1 Ethnicity in the UK, 2011

Ethnic group	Population (millions)	Percentage of total population
White	55.0	87.1
Asian or Asian British	4.4	6.9
Black or black British	1.9	3.0
Mixed	1.3	2.0
Other ethnic group	0.6	0.9
Gypsy/traveller	0.06	0.1

According to the 2011 census, white people still account for 87 per cent of the population. Of these, 51.8 million or 81.9 per cent of the UK population are white British (born in the UK) and 3.2 million or 5.2 per cent are white but born outside the UK (for example, born in Ireland and other European countries, as well as in Commonwealth countries such as Australia, Canada and New Zealand). The acronym BAME (black, Asian and minority ethnic) is often used when referring to the non-white communities in the UK.

Distribution of immigrants in England and Wales

During the last intercensal period (2001–11), the percentage of the total population of England and Wales that was white British fell from 87 to 82 per cent. But the distribution of the immigrants was not evenly spread across the country (Figure 21.5). It was the ethnically diverse regions, such as London, that showed the greatest increase in ethnicity. Here the white British component decreased by 14.9 per cent. In the West Midlands, the white British component decreased by seven per cent, while in Wales and the North East it decreased by 2.8 per cent.

Two factors largely explain this changing pattern of ethnicity:

- Job opportunities for immigrants – these remain concentrated in London, the South East and West Midlands. The North East, along with Wales and the South West, continue to suffer from relatively high levels of unemployment.
- Immigrants tend to first settle in those places where people from the same ethnic group are already concentrated. It may be that they also have relatives and friends in those places.

In short, increased ethnicity tends to reinforce the existing distribution pattern.

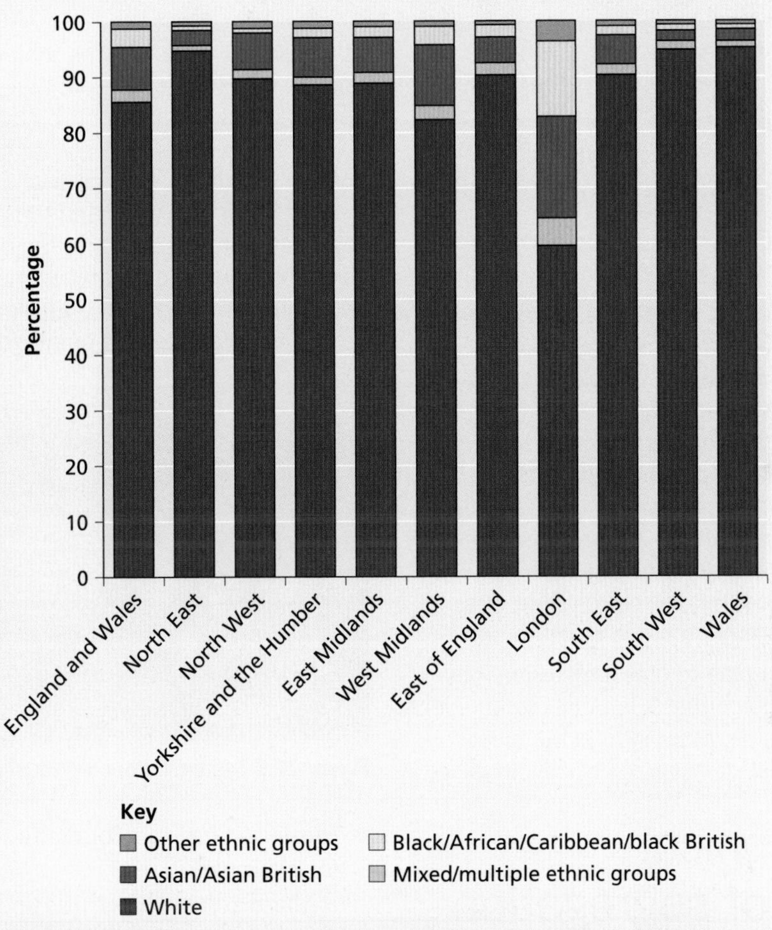

Figure 21.5 Ethnic groups in England and Wales, 2011 (Source: Adapted from data from the Office for National Statistics licensed under the Open Government Licence v.3.0.)

Skills focus: Measuring diversity

See Skill 11 in the Skills focus section (page 350).

There can be no doubt that the arrival of a large number of international migrants has proudly and lastingly impacted on UK society and, particularly, its culture (see page 188).

Immigration and rural areas

The Specification states that 'some international migrants choose to live in rural areas for specific reasons'. This statement is actually misleading on two counts, namely that:

- very few immigrants actually *live* in the rural areas of the UK (but they may work there)
- the migrants involved in rural employment have little *choice*; they are driven to work there by their economic circumstances and the labour market.

There is little evidence of a significant presence of international migrants in the UK's rural areas, although there is evidence that village stores and post offices in rural areas are increasingly being owned and run by people of Asian extraction.

Migrants in Boston, Lincolnshire

The 2011 census revealed that the 'rural' town of Boston in Lincolnshire had the highest percentage of Eastern European immigrants in the UK in its 65,000 population. In fact, one in every ten people living in the town was from 'new' EU countries such as Poland, Romania, Latvia and Lithuania. They have not been drawn here by jobs actually in Boston, but rather in the fields outside the town (Figure 21.6).

They are attracted by the fact that they can work long hours and earn, by Polish standards, large sums of money. The work, often referred to as 'picking, packing and plucking', is physically hard, dirty and, by UK standards, poorly paid. It is for this reason that such work tends to be avoided by local people. Rarely is it heard that these immigrants are 'stealing' English jobs. For sure, these immigrant workers are being exploited by their gangmasters. There is no trade union protection. Local farmers are pleased that the recruitment of a cheap and hardworking labour force is quite so easy.

Most migrants in Boston plan to stay long term. They have come to the conclusion that it is better to take jobs in the rural areas of the UK rather than in the cities, because of the lower cost of living, especially housing. But while the work is rural, their preference is for small town living. Here in Boston, they can begin to feel at home. The high street now boasts a bustling Lithuanian supermarket, a Polish restaurant, a Lithuanian cake shop, a Polish pub and several European-labelled stores. The immigrants and their families are currently segregated in the town's areas of poor housing. But, given that the children are attending local schools and becoming proficient in the English language, there must be a long-term prospect of integration.

Figure 21.6 Eastern European labour at work in the fields

21.2 Geographical expressions of segregation

Figure 21.7 shows that the impacts of migration fall into four categories; in each box some examples of specific UK impacts are given.

The socio-cultural impacts of immigration that come into focus here are:

- those that are to do with ethnicity (a key component of both culture and society)
- those that have a geographical or spatial dimension, as for example:
  - the reasons why ethnic and other minority groups become concentrated in particular areas
  - the degree to which places reflect the ethnicity of their populations.

Segregation

On arrival in the UK, the most immediate need of migrants is shelter (that is, housing) and food. Most often they will go initially to places where friends or

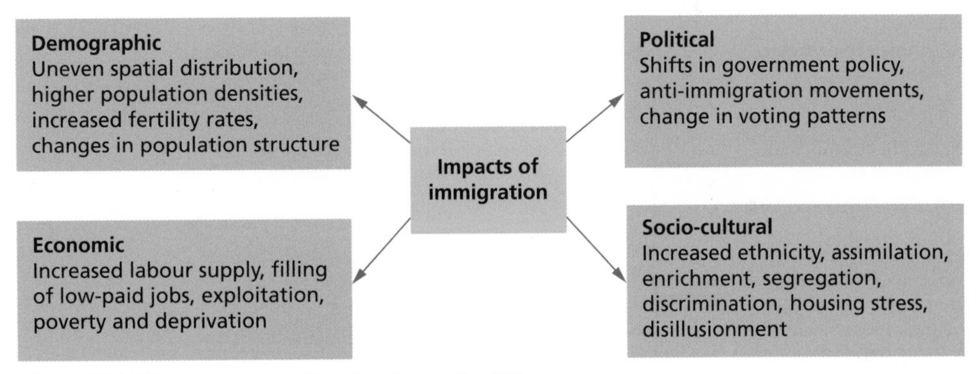

Figure 21.7 Major impacts of immigration in the UK

amily, who have already come from the same country, happen to live. Over time, this process of moving close o earlier migrants from the same source has gathered momentum to the extent that individual groups of migrants have come to be the predominant population n particular places, especially within urban areas. The erm 'ghetto' used to be applied to such concentrations, but use of it is now regarded by some as politically ncorrect; enclave is a possible alternative.

Figure 21.8 shows the collective distribution of ethnic minorities in London. It shows that different ethnic minorities tend to seek out the same areas in which to settle. Here, as in many other cities, it would seem that the magnet for all migrant groups is the availability of relatively cheap housing.

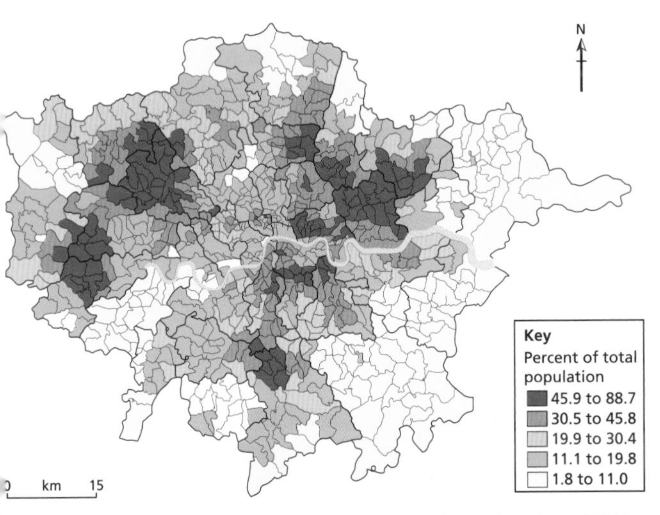

Figure 21.8 The distribution of ethnic minorities in London, 2010

Key term

Enclave: In the present context, an enclave is a group of people surrounded by a group or groups of entirely different people.

There are two contradictory views of the possible reasons for this spatial segregation of minority groups. One argues that it is encouraged by external factors, such as the availability of affordable (cheap) housing. It is more a matter of the host population forcing the segregation. The other view emphasises internal factors. In short, that it is the overriding wish of minority groups to segregate. Segregation and living close together as a group offers protection and security.

Figure 21.9 sets out the two sides of the argument; it is important that you understand both the factors (economic, socio-cultural and political) and how they interact. No matter which side of the argument is taken, most would agree that economic factors are key to understanding the reasons for segregation. The nature of the employment taken up by many immigrants on arrival in the UK is poorly paid. This means that there is little income to be spent on housing. As a consequence, far too many immigrant families find themselves the victims of poverty and deprivation. It is a vicious downward spiral that is extremely difficult to reverse. It is small wonder that some migrants drift into crime (see the next section).

Not all immigrants are poor, however, as shown by the recent arrival of Russians in London and the South East.

While ethnic segregation is driven by cultural affinities, it is also affected by the same factors that underlie the 'normal' spatial segregation of different socio-economic groups. The economic factors include income and type of employment. Improving economic circumstances allow ethnic groups more choice in terms of residential location. That being the case, social and environmental factors then come into play in locational decision making. The choice of residential location will be affected by factors of a more social nature, such as the availability and quality of healthcare and schools, and the incidence of crime.

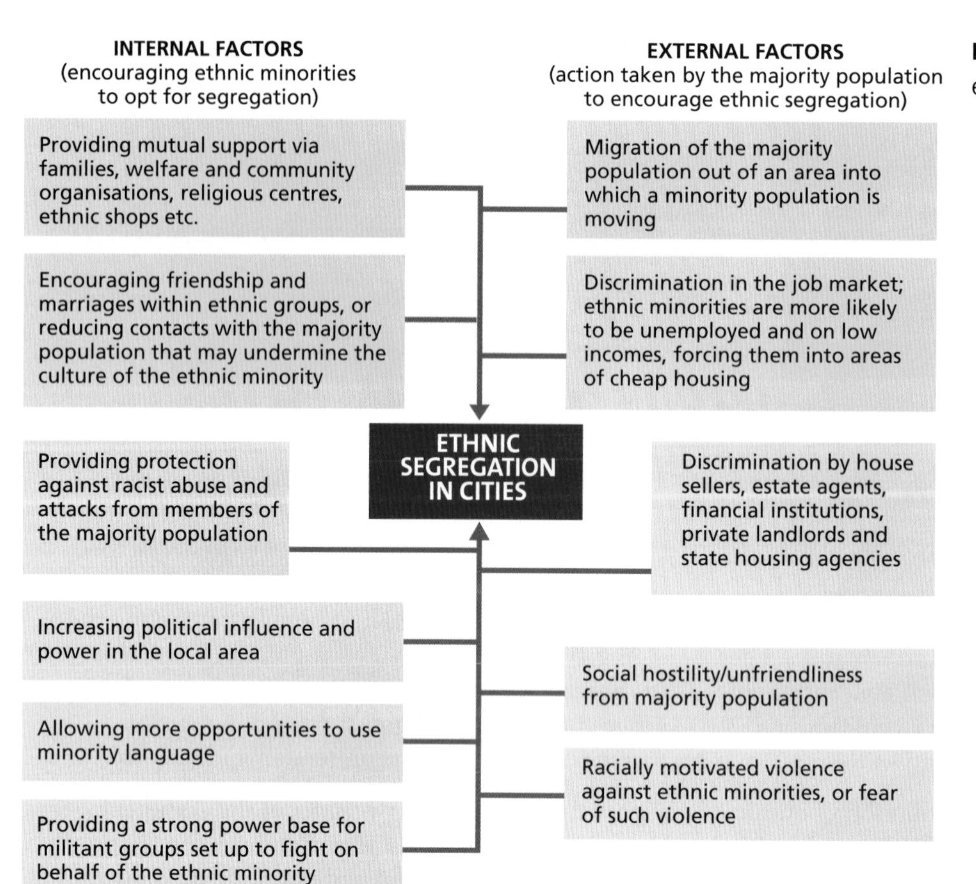

Figure 21.9 Two views of the factors encouraging ethnic segregation

Russian oligarch families in London

Many Russian oligarchs (business magnates) are busily buying swathes of ultra-expensive property in London to protect their wealth against their country's crumbling economy. This has increased recently as a result of economic sanctions taken out by G7 governments in retaliation for Russia's illegal takeover of the Crimea (April 2014) and its continuing military intervention in Ukraine. These sanctions have sent the Russian rouble into free fall; its value fell by over twenty per cent against the US dollar in the early part of 2015.

There are reckoned to be almost a hundred Russian billionaires who made their fortunes in the political and economic chaos that followed the collapse of the Soviet Union in the early 1990s. The properties being bought are in the most expensive areas of London – Belgravia, Mayfair and Kensington – with prices ranging from £6.5 million to £140 million (Figure 21.10). But the oligarchs and their families are not the only Russians buying properties in the UK. There is a growing middle class and, with a flat tax rate in Russia of thirteen per cent, they have plenty of cash. Russian banks are not trusted, so if people can get their money out of the country to a safe haven like the UK, they will. Few rich Russians keep their money in roubles.

Not all of the Russians investing their money in the UK have taken up permanent residence. They still live for much of the year in Russia and still run businesses there. It is the profits made by these businesses that are being transferred the UK. Those opting to live in the UK permanently tend to be the active opponents of the present political regime, who look to London for a sanctuary from the repressive arm of the Russian government.

So, the Russian presence in London is largely one of extremely wealthy people who have little allegiance to the UK other than that it provides them with a safe haven for money.

Figure 21.10 London properties being bought by Russian oligarchs.

Places reflect their ethnicity

Given that ethnic segregation persists, it is hardly surprising that different groups should make their mark on the urban landscapes that they occupy (Table 21.2). Perhaps the most conspicuous are places of worship, from mosques and temples to gospel and pentecostal churches. Restaurants and food stores also reflect the cuisine and food preferences of the local population.

Table 21.2 Some ethnic indicators in the urban landscape

Places of worship	Social clubs
Restaurants – ethnic cuisines	Cultural festivals and ceremonies
Grocery stores – ethnic foods	Cinemas showing ethnic films
Clothes shops – traditional clothing	Non-English signboards and advertising
	Non-English newspapers and magazines

Southall

The extent of ethnic markers in an area can be such as to leave the casual visitor with the impression that they could easily be in a part of Mumbai, Dhaka or Karachi. Figure 21.11 shows four images of Southall, part of the borough of Ealing, which contains the largest Asian community in London. The following extract paints a picture:

'Southall Broadway is the only place where you can see everything Asian, from food, spices, clothes, restaurants, take-aways and jewellers to languages spoken. Looking around you may feel you are in 'Little India', as you won't need to speak English or see many English faces around you (only the tourists). It has more processions, celebrations and festivals than any other town due to the sheer size and mix of its ethnic community. There is even a church, mosque, gurdwaras and a mandir very close to Southall Broadway. The number of radio stations broadcasting from here [in languages other than English] is second to none. The number of newspapers and magazines published from here [in languages other than English] is not small either.'

Source: www.visitsouthall.co.uk

Figure 21.11 Asian landscape markers in Southall, West London

The scale and pace of social change in areas such as Southall has meant that people who would be classified as 'White English/British' now represent a much smaller proportion of the local population than in the past. Among this group, there will be varying perspectives on the social and demographic changes that have occurred. Most younger people will have attended a multi-faith and multi-ethnic school; some will have married people from a different ethnic group and enjoy life in a multicultural society. In contrast, some older residents may feel the scale of cultural change has been too great. This can sometimes lead to a lack of community cohesion and the growth of political movements which aim to restrict further migration into the UK.

Skills focus: Before and after

See Skill 4 in the Skills focus section (page 346).

Changing perceptions and experiences

In the present context, a person's perception of where they live is greatly affected by their experiences of living in that particular place. Those experiences can be both positive and negative. The experiences can also tie a person to a particular place because they perceive it to be safe and they feel secure there. Those positive perceptions are strengthened when the individual feels that:

- they belong to a particular group
- they are living among others belonging to the same group.

There are three factors that particularly encourage immigrants to move out of their original reception or settling locations:

- improving earnings
- feeling more confident and secure in the 'new' society
- a wish to become more assimilated into a host society.

Synoptic themes:

Attitudes

The attitudes and norms of immigrants may change with each successive generation. The changes may reflect increasing degrees of assimilation and feeling part of national society.

Generational shifts

A significant proportion of the Asian, African and Caribbean immigrants who came to the UK in the 1950s and 1960s raised families and became grandparents. But what has happened to their offspring, that is, to the second- and third-generation migrants? Data from the 2011 census indicates that 1.3 million of them now describe themselves as being of 'mixed' ethnicity. That represents two per cent of the total UK population. It indicates a certain amount of intermarriage between ethnic groups; if the intermarriage involved

Jewish immigrants in London

Over a quarter of a million Jewish immigrants have been drawn to the UK over the last 200 years by their wish to find a country where they might live without discrimination and persecution. London, in particular, was a place where they could feel reasonably secure. Here too they could earn a living from their commercial and professional skills, as well as play a part in the cultural life of the country.

Most of the immigrants first settled in the areas of poor housing in the East End. As successive generations established themselves and gained in affluence, so the centre of gravity of the Jewish community moved out to suburban boroughs such as Barnet, Enfield and Harrow to the northwest and Redbridge to the north (Figure 21.12).

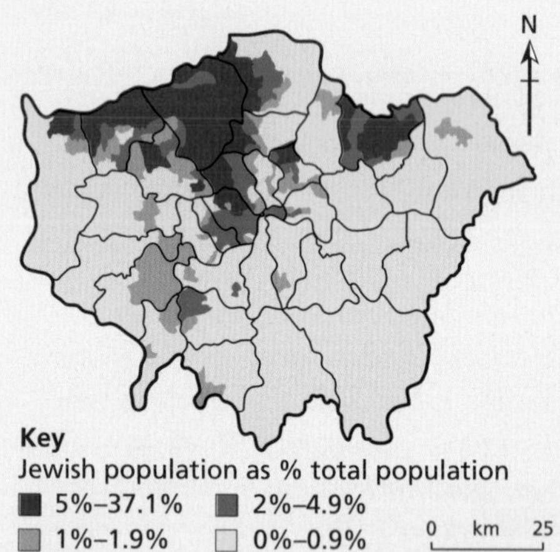

Figure 21.12 Distribution of Jewish population in Greater London, 2011

People who belong to different ethnic groups may have varying experiences of what it is like to live in a particular place. Equally, experiences and perceptions of living spaces change over time as new generations of youngsters from migrant and diaspora communities develop their own hybrid cultural identity while often progressing economically. Changing inter-generational attitudes and norms can be identified among London's Jewish communities over time. Several generations ago, Jewish people arriving in the UK to escape persecution doubtlessly felt relief at having found a place to live where they did not feel threatened. The lived experience of the next generation would most likely have been very different. Many post-war Jewish artists and writers have not only played a part in British society, they have also shaped popular culture, for example Harry Potter star Daniel Radcliffe.

Most recently, some Jewish people's experience of life in London may have become more complicated. There were increased reports of hate crimes against minority ethnic groups in some London neighbourhoods in 2016. The reasons for this are complex and unclear, but there may be heightened tension in some places where a minority of racist people have used the result of the 'Brexit' referendum as an excuse to act deplorably. In contrast to this, the view of many Londoners (whatever community they belong to) is that community cohesion in London has never been stronger.

It is interesting to compare the perceptions and motives of earlier Jewish immigrants with those of Polish immigrants today. The latter have come voluntarily and been drawn by perceived economic opportunities. Their lived experience is one of hard work, poor housing and occasional local hostility. Their motives for coming are treated with suspicion by some people who think they move to the UK to claim benefits rather than with the aim to work and bring in higher wages than they would get in Poland.

Key term

Assimilation: The process by which people of diverse ethnic and cultural backgrounds come to interact and intermix, free of constraints in the life of the larger community or nation.

white British partners then it might be claimed that there has been some measure of assimilation.

The point has been made that first-generation immigrants often suffer in their new homeland, particularly in terms of having to take low-paid jobs and occupying poor housing. In contrast, the offspring of immigrants (the next generation), since they are citizens of the country in which they were born, are seen typically as becoming better off and moving out of poverty. No doubt, access to education has played a part in opening up more employment opportunities.

All is not well, however. Statistics show that black, Asian and other ethnic minorities are twice as likely to be unemployed, half as likely to own their own home and run twice the risk of poor health than their white British counterparts. Statistics also show that the proportion of Muslim children living in overcrowded housing is more than three times the national average. They are twice as likely to live in a house with no central heating. Children from Pakistani and Bangladeshi families suffer twice as much ill-health as their white counterparts. Residential segregation on the basis of ethnicity, particularly in the inner suburbs of the UK's largest cities, remains the rule. Is that by choice or force of circumstance?

An interesting investigation by Gemma Catney analysed what happened to ethnic groups in the local authority areas of England and Wales during the last intercensal period (Figure 21.13). Did they become more or less segregated? The analysis showed that the segregation of all non-white ethnic groups decreased in the majority of local authority areas – an encouraging indicator of integration and assimilation. This was particularly the case with those of Indian and Pakistani extraction. Perversely, white British was the only group to become more segregated.

Another concern is the discovery that many second- and third-generation immigrants only speak or understand a little English. Their poor understanding of the language means that they are unable to take notes or understand basic instructions on training courses. Experts warn that low standards of English in some minority groups widened ethnic divisions and created communities where English is the second language. Schools have been blamed for failing to teach the English language well enough. In addition to this, families have deliberately discouraged their children from learning English in order to protect their culture and ensure that their children can talk to their grandparents.

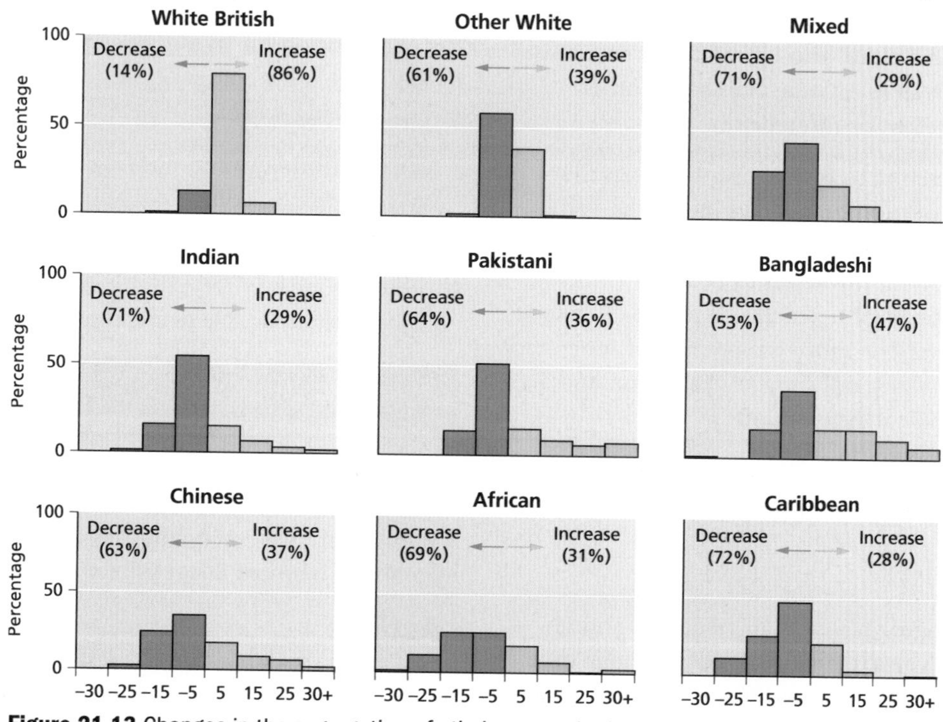

Figure 21.13 Changes in the segregation of ethnic groups in the UK, 2001–11

Table 21.3 Some possible distinguishing characteristics of first-, second- and third-generation immigrants in the UK

First generation	Second generation	Third generation
They feel home is where they were born	Like UK people and want to be British	Successfully integrating into UK society
Identify with their home culture and its traditions and customs	Keen to obtain a more exciting life for themselves	Identify with British culture
Unskilled workers; only came to the UK to earn good money	Happy to take part in Western culture	No connection to the home country of their grandparents
Aim to go back home	Not confident about their identity – neither British nor immigrant	Growing self-confidence
Do not wish to adapt to UK culture	Working-class upbringing	Better chances of employment and promotion at work
Do not want their children to become British	Educated in UK schools and colleges	Enjoy a degree of social mobility
Live in segregated inner-city areas	Less tightly segregated; some upward movement in the housing market	Increasing residential presence in the suburbs
English is second language and only used outside the home	English is first language, but command of it is often very limited	Growing command of the English language

The situation today is a mixed one for those from ethnic minorities. Some are moving up and becoming wealthier. But for others, particularly their children, the future is fairly bleak. They are being drawn into the vicious downward spiral created by the link between poor housing, poor education and future criminality.

Bearing all this in mind, plus poor employment prospects, it is easy to understand how disaffected youths are readily radicalised against UK society. Some are even persuaded to leave the UK and join extremist organisations, such as Al-Qaeda, Daesh and other perpetrators of international terrorism.

Table 21.3 is an attempt to distinguish between the three generations of immigrants coming to the UK originally from Asia, Africa and the Caribbean in the 1950s and 1960s. What this tends to confirm is the slowness of the assimilation process. It may well take at least another generation before the pockets of inner-city concentration show any real signs of dispersion and a change in geographical distribution.

21.3 Change, tension and conflict

'All places are changing places' is an axiom of modern geography. Places, particularly urban ones, are dynamic and constantly evolving. Perhaps the most obvious signs of change are seen in the pattern of land use and in the buildings and spaces that accommodate new activities. Clearly, change is undertaken in the expectation that there will be benefits of various sorts, from business profits to the provision of services, that improve the quality of the lived experience for many, but not all, people.

Competition for space

Underlying much of the change in land uses is the competition for space. In general, it is true to say that the competition for living space is much greater in urban areas. This competition involves two distinct layers of living space:

- competition between housing and other consumers of space (services, commerce and industry)
- competition for housing space within residential areas.

Most urban areas in the UK are expanding. This creates an apparently insatiable demand for more space to

Synoptic themes:

Players

A range of players compete for space, but they differ in terms of their competitiveness in the land market. Some will be more successful or influential than others.

Actions

The different actions of those competing players will have different impacts and outcomes.

ccommodate new housing, new services, new retailing, new industrial and office parks. So every land use is a potential competitor for space. When it comes to bidding on the land market, there is a clear pecking order. Retailing (high street chain stores) and high-order offices (those of TNCs) are usually able to bid highest, with housing and recreation at the bottom end of the bidding league table. So, in a nutshell, if a TNC wants a particular site or piece of land for one of its global network of offices, it will usually come to occupy that site. The only possible intervention in the urban land market is by the local or country authority making use of planning legislation. These powers are particularly useful when space is needed for local or national government purposes.

Four housing sectors compete for space allocated or purchased for residential development. They are:

- owner-occupiers
- property developers acquiring housing to rent to tenants
- housing associations providing affordable housing
- local authorities providing social housing.

In an era of acute housing shortages, this competition immediately creates a situation of winners and losers. The decision as to which of these groups should benefit rests with local and county authorities who, in turn, are guided by government policy and national legislation.

Figure 21.14 High-rise housing in Salford – one solution to the competition for residential space. No doubt who has won!

Key term

Social housing, affordable housing: A key function of social housing is to provide accomodation at affordable rents to people on low incomes. According to Shelter, social housing and affordable housing are one and the same thing. However, some would make the distinction that affordable housing also includes dwellings (usually built by housing associations) for sale at below market prices for first-time buyers.

Further potential tensions are created where consent has been given to build dwellings for sale to owner-occupiers. The type of dwelling (townhouse, apartment block, detached dwellings, etc.) to be built and its price will depend on such considerations as the location and desirability of the site, the price paid by the developers and their market research into what type of property will sell best. In such a situation, there will inevitably be disappointed groups of potential customers, ruled out because they cannot afford the prices or they wanted some other form of dwelling.

Clearly, any new housing development will in some way change the character of a place. To some, the change will be for the better, to others for the worse. So, there are more conflicts, if only of opinion.

Tensions

Tension, perhaps, rather than conflict, is frequently created whenever and wherever there is a change in the use of a particular living space. For example, when residents are displaced by road improvements or independent shops in the high street are pressured to sell up to national retailing chains.

Change within residential living space itself also frequently creates tension, for example, when migration into an area begins to change the basic character of the residential population. Tension is created between two groups: the long-term residents and the new incomers. Much of the tension seems to be rooted in a widespread dislike of change on the part of established residents. They wish to continue in their present comfort zone. There may also be feelings of resentment that they are being dislodged and squeezed out by the newcomers. The more different those two groups are, the greater the tension. It is likely to be intensified where there are ethnic or strong socio-economic differences between the two groups. This sort of tension is just as real and challenging in rural places as it is in urban places.

The next section illustrates some of the tensions just outlined in the context of one urban regeneration scheme implemented in Deptford, London. In fact, the area concerned has undergone two distinct regenerations.

Skills focus: Maps and photographs

See Skill 4 in the Skills focus section (page 346).

Pepys Estate, Deptford

Deptford began to grow into a busy town in the early sixteenth century when a Royal Naval dockyard was established there on the south bank of the River Thames. Within a short time it became the most important dockyard in England (Figure 21.15). The dockyard and associated maritime industries prospered until 1869 when the yard was closed. It was no longer possible to build and launch in the creek the ever-increasing size of vessels required by the Royal Navy. The town, now part of the Greater London conurbation, was badly hit by the economic depression of the 1930s. During the Second World War (1939–45) it was heavily bombed by German aircraft.

The first attempts to revive Deptford were made in the 1950s and 1960s. The strategy was a simple one, namely to bulldoze large areas of slum and bomb-damaged housing and to build large new estates of public housing. It was thought that this would halt the loss of population and improve the image of Deptford.

The Pepys Estate, first occupied in 1966, was dreamt up by the Greater London Council as a super-modern self-contained town. The architects prided themselves on how much of the historic site of Deptford's old naval dockyard they had managed to keep and incorporate into the new estate. The old rum stores became the block looking over the river at Deptford Strand.

Figure 21.16 Pepys Estate in the 1980s

Figure 21.15 The Royal Naval Dockyard at Deptford in the eighteenth century

The estate consisted of three 24-storey tower blocks and ten eight-storey blocks, plus several other four-storey blocks. Altogether the estate provided homes for 1200 households. When completed the estate was seen as a prestigious move forward in the provision of social housing. It quickly became a popular place to move in to. The legacy of this is that many tenants have subsequently remained on the estate, and it is not uncommon to find that people have lived there for 30 years or more. Perhaps the most appealing aspect of the estate is its riverside location and stunning views – of central London upstream, of Canary Wharf on the opposite bank and of Greenwich downstream.

While the modern accommodation was well received at first, residents found it difficult to adjust to vertical living in flats. The majority of homes were built in blocks with long, dark corridors and interlinking bridges. These soon became crime corridors with frequent muggings; the fear of crime greatly increased among residents. The estate was also subject to all manner of vandalism and antisocial behaviour. By the mid-1980s, the planners' dream had turned into a nightmare (Figure 21.16).

Figure 21.17 An upgrade of the old rum store block has created expensive and much sought-after flats

During the 1990s, considerable investment was poured into the estate by the local authority to try to improve it. It did not resolve the fundamental design problems of the development, however, or the poor conditions of residents' homes. In 2000 the Hyde Group entered into a partnership with the local borough to regenerate five blocks of the estate into new, high-quality homes (Figure 21.17). Eventually, ownership of those blocks passed from the borough to the Hyde Group.

Although the Hyde Group is a leading provider of affordable housing, an inevitable outcome of the improvements has been a raising of rents. Rents have also been boosted as more people outside the estate have become aware of its advantages, particularly its location, improved housing and amenities. As rents began to exceed the means of the dwindling number of original residents, properties have become occupied increasingly by more affluent newcomers.

Today the estate is fast becoming a yuppie's paradise. Not only that, but pressure is also mounting to upgrade more of the estate. In the longer term, the present tension between long-established residents and the newcomers will gradually diminish. That is, until such time as the former have disappeared completely, either by death or by removal to cheaper social housing elsewhere. The demographic character of the Pepys Estate will be very different in a generation's time.

As always, there has been a downside to the upgrading of the Pepys Estate. This has been felt by the older and poorer residents:

- originally bombed out of their homes in the Second World War and forced by urban regeneration to adjust to living in tower blocks with their widely recognised social costs: isolation, detachment, crime and anti-social behaviour
- having to cope with the general deterioration of the estate in the 1980s and changing ideas about the residential environment
- now being priced out of their homes by the current round of well-intentioned rebranding and re-imaging.

Key term

Yuppie: Short for 'young urban professional' or 'young upwardly-mobile professional' – a young, university-educated adult who has a well-paid job and who lives and works in a large city.

Skills focus: Oral Accounts

See Skill 2 in the Skills focus section (page 346).

For generations of local residents, urban regeneration has clearly brought disruption, displacement, social exclusion, stress and tension. Sadly, this lived experience has been, and still is, all too common elsewhere in the UK.

Poverty, animosity and social exclusion in Glasgow

Despite a great deal of investment and planning in recent years, Glasgow remains a poverty hotspot in Scotland. It is the most deprived of Scotland's four major cities (Figure 21.18). It contains fifteen per cent of the most deprived areas in Scotland. The poverty and deprivation are shown to be concentrated in particular parts of the city.

From the perspective of those living in these pockets of deprivation, it looks as if they have missed out on the investments that have benefited other parts of the city. It is this sort of perception that fuels the hostility that local residents feel towards various agencies of government.

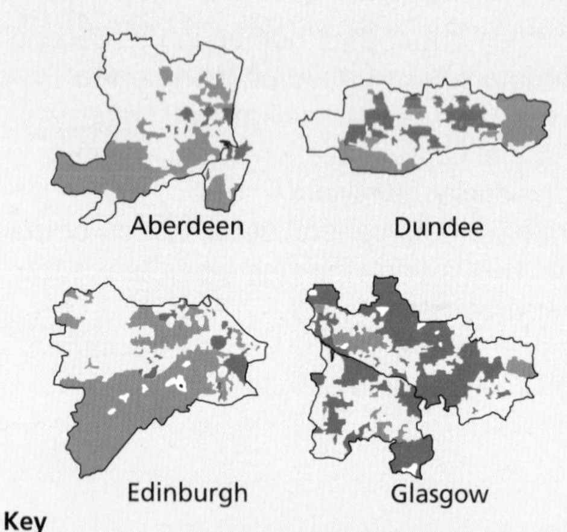

Key

Blue areas are among the 20% least deprived in Scotland

Red areas are among the 20% most deprived in Scotland

Figure 21.18 Deprivation in Scottish cities

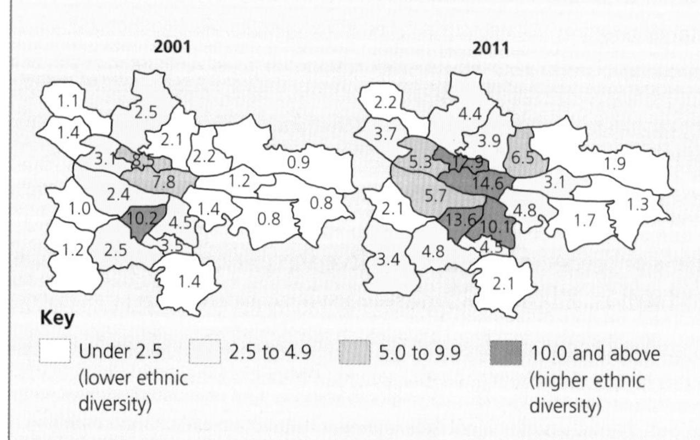

Figure 21.19 Ethnic diversity in Glasgow, by ward, 2001 and 2011

But there are internal tensions within these areas of poverty, particularly in those wards with a high incidence of immigrants. Figure 21.19 shows the distribution of ethnic diversity by means of a reciprocal diversity index. This is derived from calculating how close a ward is to having an equal number of the population in each ethnic group. The higher the index, the greater the ethnic diversity. Clearly, ethnic diversity in Glasgow increased considerably between 2001 and 2011.

Figure 21.20 looks at the changing ethnicity in a different way. It shows that some wards lying close to the city centre contain large numbers of people born outside Scotland, most notably Anderston/City (42 per cent), Hillhead (37 per cent), Pollokshields (30 per cent) and Southside Central (28 per cent). Here many of the white British population feel that the concentration of ethnic minorities is responsible for:

- high rates of unemployment – the perception is that these immigrants are 'stealing' local jobs
- the authorities not bothering too much about investing in the improvement of these hardcore poverty areas
- a dilution of Scottish culture.

The animosity, whether real or imagined, lies at the root of the sense of social exclusion felt by many of the people of Asian and African descent in these wards. They sense that they are being ignored and are not wanted. Such a feeling is reinforced by their poverty and their sense of helplessness in trying to break out of the vicious deprivation cycle.

Ethnic tensions in Glasgow, which are largely anti-Islamic, have increased since the 9/11 attacks on the USA, the 7/7 bombings in London and the Glasgow Airport attack in 2007. The significant number of immigrant arrivals since 2001 has intensified the situation.

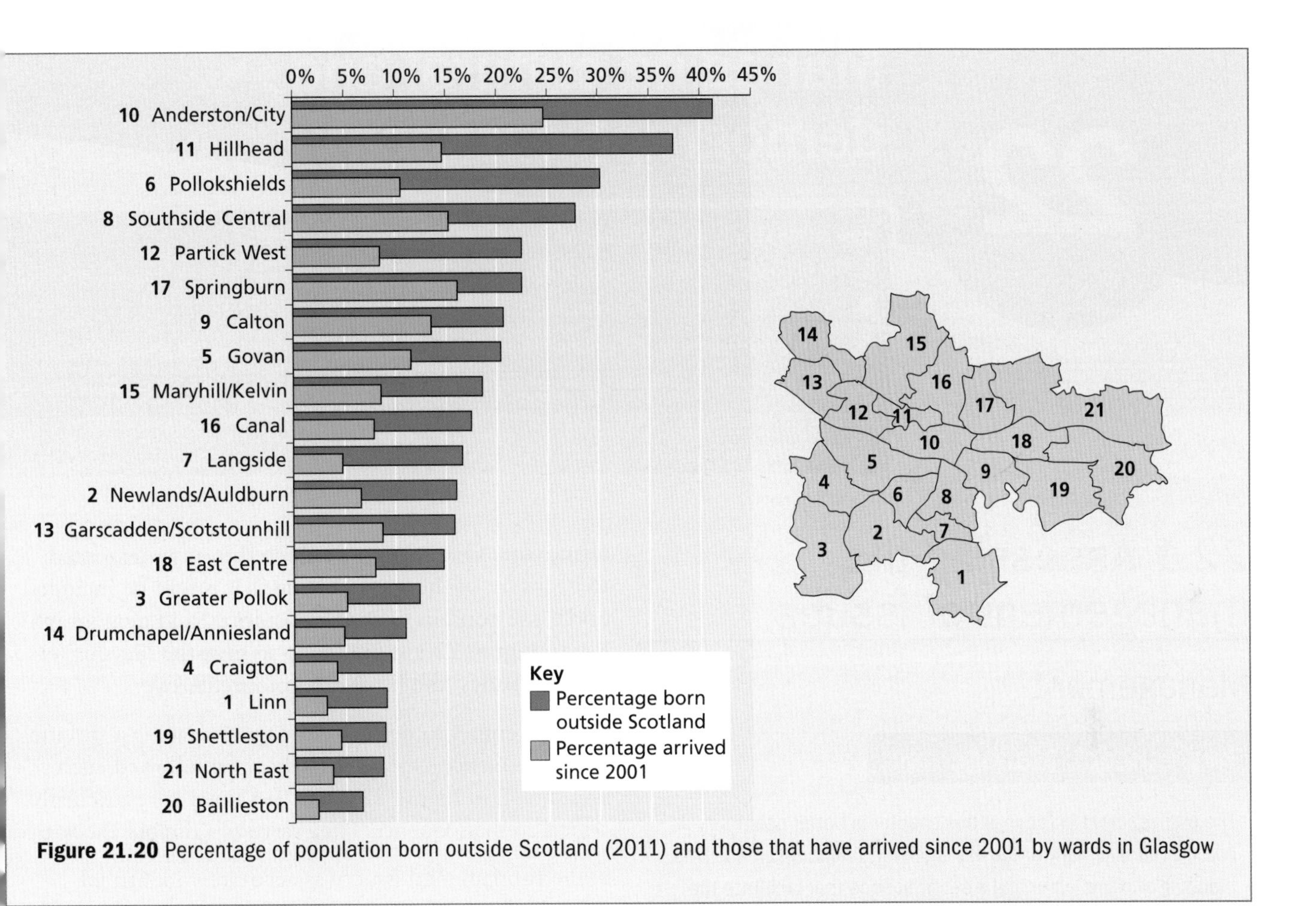

Figure 21.20 Percentage of population born outside Scotland (2011) and those that have arrived since 2001 by wards in Glasgow

Clearly, the challenge in Glasgow and many other British cities is to ensure that spending on improvement of the built environment is prioritised. The worst areas of poverty and deprivation should be given top priority. Only by doing this will people gradually come to accept that social justice has been done.

Synoptic themes:

Players

Planners and developers may make controversial or unpopular decisions. They may also be criticised for failing to act on and remedy particular issues.

Review questions

1. Identify the challenges associated with immigration.
2. Examine the factors encouraging ethnic segregation in the UK's cities.
3. Explain the ways in which ethnic groups make their marks on places.
4. Illustrate the point that ethic groups differ in their perceptions of the UK.
5. Explain what you understand the term 'social exclusion' to mean.

Further research

Find out more about what attracts so much migration to the UK: www.migrationobservatory.ox.ac.uk/topics/migration-to-and-from-uk

Research the reasons why rural places are much less ethnically diverse than urban places: www.openspace.eca.ed.ac.uk/conference/proceedings/PDF/Pendergast.pdf

Read *Getting the measure of rural deprivation in Wales*: www.ocsi.co.uk/news/wp-content/uploads/OCSI-GettingMeasureRuralDeprivationWales.pdf

Use the internet to research the reasons for rural deprivation.

Managing cultural and demographic issues

How successfully are cultural and demographic issues managed?

By the end of this chapter, you should:

- be aware of how the management of change is evaluated
- be aware of some of the cultural and demographic issues confronting the UK and how they are being managed
- understand that different stakeholders will have their own criteria for evaluating the management of change in both urban and rural communities.

22.1 Assessing the management of issues

Management

Key concept: Management

The management in focus in this chapter is better described as *change management* or the *management of change*. It involves planning and implementing a set of actions that facilitate the transition from one situation to another that is hoped for. So, in the context of a geographical issue or problem, management requires setting out and following a series of steps or actions that lead eventually to either a solving or an amelioration of the initial problem. It is important to stress that not all issues can be fully resolved and that, in most cases, the best that can be achieved is some reduction in their seriousness.

It is necessary to start by checking that you understand the concept of management and the meaning of the term issue. Once you have done this, you next need to be aware of the sorts of issue that might be included under the combined heading 'cultural and demographic issues. Figure 22.1 should help to give the 'flavour' of relevant issues; the list is not comprehensive.

When it comes to dealing with and managing a specific issue, a sequence of steps is likely to be followed, such as is shown in Figure 22.2. The last three are particularly critical in the sequence. They serve as a double check that:

- the fundamental issue was accurately identified in the first place and that it has not changed in an unforeseen way
- the strategy and its component actions are proving effective and bringing about change in the required direction.

Monitoring and evaluating how well a particular issue has been or is being managed is very challenging. It is extremely difficult to be completely detached and objective as it all depends on who you are and your particular viewpoint. This vital point will be taken

Ethnicity
- Assimilating ethnic minorities
- Respecting immigrant cultures
- Outlawing discrimination
- Conserving cultural heritage

Population structure
- Anticipating future change
- Encouraging a youthful population
- Coping with an ageing population
- Raising life expectancy

CULTURAL AND DEMOGRAPHIC ISSUES

Migration
- Reducing native versus incomer tensions
- Stemming unwanted outflows
- Controlling immigration
- Improving border security

Quality of life
- Improving access to, and quality of, housing
- Providing healthcare and education
- Reducing poverty and deprivation
- Improving the living environment

Figure 22.1 Possible cultural and demographic issues requiring management

Key term

Issue: This is generally defined as an important topic or problem for debate or discussion. Examples would include immigration, racism and climate change. In this chapter the emphasis is on problems rather than topic issues, and on finding solutions rather than debate or discussion.

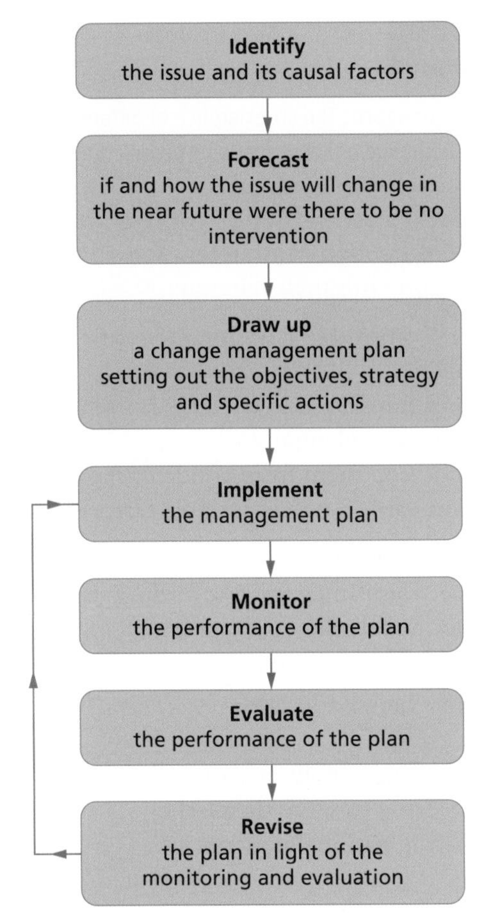

Figure 22.2 Steps in the management of change

up in the remaining sections of this chapter. How the management of an issue is assessed also depends very much on the nature of that issue.

Assessment techniques and measures

Clearly, the monitoring, evaluation and overall assessment of any management plan will require a reliable set of tools – an effective set of discerning measures. In the present context, those measures need to be either demographic or economic.

Table 22.1 Some demographic indicators of change

Total population
Rate of population change
Migration balance
Age structure
Ethnic mix
Family size
Life expectancy

Table 22.2 Some economic indicators of change

Type of employment
Unemployment rate
Household income
% on minimum wage
Dwelling tenure
% on social benefits

Demographic

There is a range of demographic statistical measures that can be used to monitor the progress and achievements of a management of change plan. This is done by simply comparing the data relating to the start situation with that at transitional stages and at the end of the plan. The sort of measures that might be used in these timeline assessments are those that are taken into account by most national censuses (Table 22.1). As well as monitoring what has happened over time, these indicators can be used to compare what has happened in one place with what has happened in another.

Economic

Where the issue is one of poverty, housing or deprivation, there are a good range of performance indicators that can be used to show how situations have changed over a specific period: one, five or ten years (Table 22.2). Again, they can also be used in spatial comparisons.

Social progress

Both sets of measures listed earlier are involved in the broader assessment of what is referred to as social progress. The Index of Multiple Deprivation (IMD) is particularly useful in monitoring inequalities over time, to determine whether they are widening or reducing. The spatial inequalities in focus in this option topic are twofold:

- those that exist between urban and rural places
- those that exist within places (either urban or rural); these inequalities may be either spatial or between social groups.

Key concept: Social progress

The idea that societies can or do improve in terms of their social, political, and economic structures. This may happen as a result of direct human action, through enterprise or activism, or as a natural part of the socio-cultural evolution of society. Social progress may be measured in terms such as levels of poverty and deprivation, educational attainment or socio-economic advancement.

Skills focus: Before and after

See Skill 4 in the Skills focus section (page 346).

Assimilation

When it comes to assessing the management of cultural issues, the task is rather more challenging. This is mainly because these issues, by their very nature, are less tangible and less quantifiable. For example, it is very difficult to determine how a migrant who arrived in the UK, say, five years ago, feels about living in this country now. Do they still have feelings of alienation? To what extent do they feel they have been assimilated?

One route to this sort of data would be via the questionnaire or interview. The other is to set up and consult local community groups. The groups might be made up either exclusively of immigrants or a mix of immigrants and white British people. They have proved to be quite effective in monitoring the changing moods and feelings of minorities.

Information such as that just mentioned would be crucial to what is certainly one of the most pressing of cultural issues in the UK today – the assimilation and integration of diverse ethnic and immigrant groups into UK society. There are few cultural measures *per se.* Rather, we have to rely on some measures that have come under two earlier headings, as well as two new ones (Table 22.3).

An issue arising here concerns the ongoing debate that has been described as assimilation versus multiculturalism. Some are now questioning whether, in fact, the two situations are mutually exclusive.

Table 22.3 Some possible measures for assessing assimilation

Demographic	Mapping changing residential distributions of different ethnic or immigrant groups. Is their segregation becoming less marked? What is the incidence of mixed ethnic marriages?
Economic	Disparities in wage rates and salaries. Are immigrant/ethnic minorities paid the same as white British workers?
Political	Degree to which members of minority groups are engaging in the political process. What percentage of the electorate in those groups is voting at elections? How many adults in those groups are now standing in local and national elections?
Social	The incidence of 'hate crime' and expressions of racism. Is it increasing or decreasing? What trends are seen in the Index of Multiple Deprivation?

Key term

Multiculturalism: The co-existence of different cultural groups. A sharing of living space by people drawn from different cultural backgrounds.

A survey of London Muslims has shown that a strong identification with their religion:

- is quite compatible with their identification as UK nationals
- does not translate into a strong desire to segregate themselves residentially
- does not stop them from condemning attacks on civilians carried out by Islamist terrorists.

Rather than assimilation, it is perhaps more appropriate to use the word 'integrate'. According to a study, integration involves:

- mastering the English language
- getting a better education
- finding a job
- participating in politics
- volunteering to serve the public
- celebrating national holidays.

22.2 Urban stakeholders and their criteria

Introducing stakeholders

Making decisions about the management of change in any place involves not just the actual decision maker but also the people, groups or organisations with an interest or concern in the issue to be managed. These interested or involved parties are collectively referred to as stakeholders. In almost all situations, stakeholders fall into four groups:

- providers
- users or beneficiaries
- governance
- influencers.

To help understand the differences between these groups, let us take the case of a proposal to sell off a school playing field in order to provide affordable housing.

- The *providers* in this case would be the owners of the playing field (most likely the county council), the building contractors who will convert the green space into housing, and the association responsible for running the housing.

- The *users or beneficiaries* fall into the two categories of 'before' and 'after':
  - before – the school's pupils and their parents or guardians; residents immediately adjacent to the playing field who enjoy the amenity of living close to green space
  - after – the occupants of the affordable housing; the local authority from more properties paying council tax.
- *Governance* refers to control (through laws, power or ownership). In this instance the proposal might involve implementing a government policy that encourages county and local authorities to sell off school playing fields. Thus, the county council becomes a double stakeholder as provider and governance.
- *Influencers* can be many and various; in this instance they might include a number of groups opposed to the development, such as teacher–parent organisations and local amenity groups, and those keen to promote the change, such as a political party and homeless families.

What should become clear is that all these stakeholders fall into two groups – those that can directly affect the management or handling of an issue and those that are affected by that management. Each of these two can be further subdivided into:

- those exercising a positive role (i.e., those keen to promote the conversion of the playing field into housing and those who will benefit)
- those whose role is negative (the protestors and those losing out if the scheme goes ahead).

One final general point about stakeholders: each stakeholder has their particular 'vested' interest; their particular perceptions and objectives. Reality is viewed in a biased way; reality is approached from different directions. From this it follows, therefore, that each stakeholder will have their own particular view of what constitutes 'success' and 'failure'. They will have their own criteria for assessing whether a particular issue has been, or is being, managed successfully or not.

Synoptic themes:

Attitudes

Each urban stakeholder will have their own view as to what constitutes 'success' or 'failure'. It all depends on their particular aims and objectives and their assessment of the degree to which those aims and objectives have or have not been realised.

The housing crisis in Oxford – stakeholders and their perceptions

This place context aims to cover, in an integrated way, the three elements of detailed content related to this key idea in the option topic, as set out in the Specification:

- the views of different stakeholders on the living space involved in this issue
- how the stakeholders think the issue should be resolved
- the criteria used by the stakeholders to determine, from their viewpoint, whether or not the outcome will be a satisfactory one.

These three themes are not treated separately in isolation. Rather, they run in an interweaved way through the examination of this particular place issue. In addition, it should be more illuminating to focus on the stakeholders involved in a specific issue and place rather than write about stakeholders in general.

The issue

'The average home in Oxford costs over 11 times the local wage, compared to Stirling in Scotland where homes cost 3.3 times earnings.

Oxford has been named the most expensive city in the UK as rising house prices make city living less affordable, according to a Lloyds Bank report.

… The average house price in Oxford is £340,864 – which is around two-and-a-half times the price of a typical home in Stirling at £132,734.

The report pointed to the large numbers of people living in Oxford who commute to London to work as part of the reason why house prices there are particularly out of step with local wages. A year ago, a house in Oxford cost 9.8 times local earnings.'

Source: www.quantum.international/house-prices-most-and-least-affordable-cities-in-the-uk

The spiralling of housing costs noted in this newspaper extract has arisen because the city of Oxford has nowhere to build new homes. Oxford is an international city; it is successful, vibrant and a national economic asset. It is the focus of a world-class knowledge economy with one of the most important concentrations of high-value businesses in Europe. However, the city's continuing housing crisis through the lack of housing availability, choice and affordability, is significantly undermining its future.

- Oxford needs between 24,000 and 32,000 new homes between now and 2031 to meet the city's growing demand for housing.
- Oxford has overtaken London as the least affordable housing location in the UK. The average cost of buying a house in Oxford is more than eleven times the average salary of an Oxford worker (Table 22.4).
- Recruitment by the city's businesses, universities, hospitals and schools is difficult because of a lack of housing choice and affordability. This adversely affects the city's economy, the quality of services and the lives of those living and working in the city.
- The city's two universities (University of Oxford and Oxford Brookes) are being held back in the global competition for the best research talent; public services, such as health and education, are compromised through the lack of available affordable housing for key staff.
- With over half the city's workforce travelling into Oxford and commuting distances increasing, the pressure on the infrastructure is not sustainable, even with improvements to roads and public transport.

Table 22.4 The ten least and ten most affordable cities in the UK (excluding London), 2014 (Source: Lloyds Bank)

Least affordable	Multiple of local salary needed to buy average home	Most affordable	Multiple of local salary needed to buy average home
Oxford	11.25	Stirling	3.30
Winchester	9.65	Londonderry	3.56
Truro	8.57	Newry	3.90
Bath	8.05	Belfast	4.12
Brighton and Hove	7.94	Bradford	4.15
Chichester	7.71	Lancaster	4.28
Westminster	7.65	Lisburn	4.29
Salisbury	7.40	Salford	4.45
Cambridge	7.32	Glasgow	4.51
Southampton	7.75	Durham	4.60

Search for a solution

The city council is doing its best to find suitable housing sites within the city's administrative boundary. Unfortunately, the city's boundary is tightly drawn around the built-up area. Most agree that there is a shortage of land suitable for housing inside the boundary. So, in 2014 the council set up an independent review of the green belt immediately outside the city's boundaries to see whether there were any areas within it that might be released for new housing (Figure 22.3). Six possible sites were identified (Figure 22.4). They have since been investigated in more detail. Three of those sites have been marked as 'good prospects for development'. The conclusion reached about the other three is that they 'have prospects for some level of development'.

Underlying this housing crisis, there is another issue – affordable housing. Clearly, given the mismatch between house prices and earnings, the case for providing a considerable quantity of such housing cannot be denied. The government has tried for some time to encourage the building of more affordable housing. They have done this by insisting that any planning permission granted for new dwellings should include a prescribed

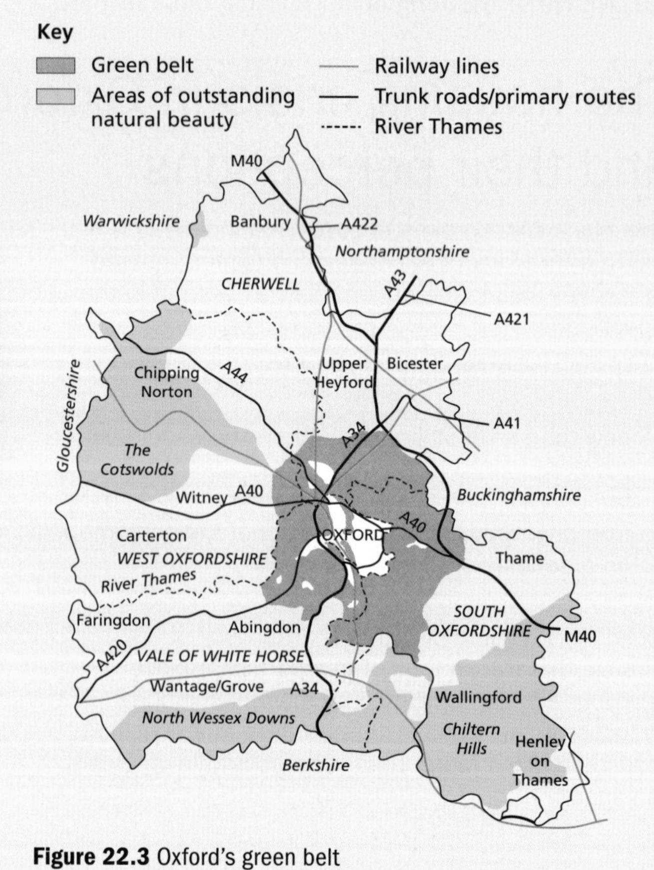

Figure 22.3 Oxford's green belt

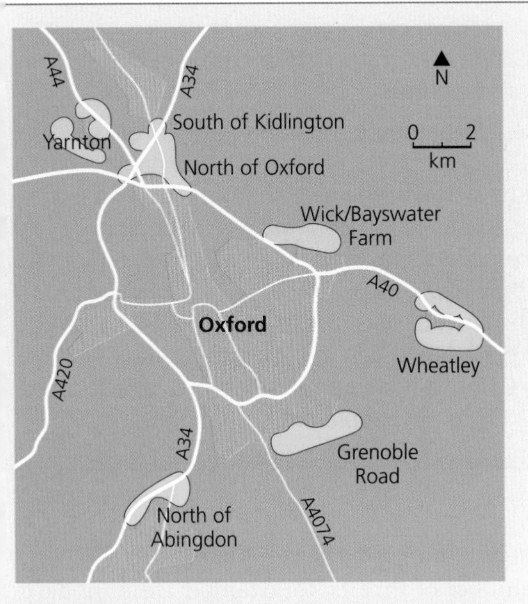

Figure 22.4 The shortlist of six possible sites to help ease Oxford's housing crisis

Skills focus: Newspaper reports – opinions about change

See Skill 3 in the Skills focus section (page 346).

number of affordable dwellings. The larger the number of dwellings in the overall scheme, the larger the required number of affordable units. Such housing is mainly for rent to tenants, but some of it is for sale. Both rents and sale prices are subsidised. The subsidies are mainly paid for by taxpayers and the house-building industry. It has to be said that not everyone supports the idea of affordable housing as a way of housing poorer households.

No doubt, because of the complex legal and planning procedures involved, it will be a long time before anything is agreed and even longer before work starts on the new housing. Even now, however, it is possible to identify some of the main stakeholders involved in the management of Oxford's housing crisis. It is worth trying to identify their particular perceptions and objectives. They fall broadly into two groups.

Stakeholders in Oxford

Residential households of all ethnic and socio-economic groups will perceive that the need is acute, none more so than young adults looking to find their first home. Oxford has quite a significant ethnic population. According to the 2011 census, ethnic diversity is on the increase in Oxford with 22 per cent of residents from a black or Asian minority group. (This can be compared with a figure of 13 per cent for the whole of England.) Another 14 per cent of residents were from white but non-British backgrounds. It is difficult to second guess what the views of these minority groups might be on the housing situation. Their concerns might be the availability of housing adjacent to the areas in which they are concentrated, or the more ambitious might be more concerned about accessibility to better housing in other parts of the city.

It has to be said that not all householders will be keen to have the housing crisis solved. People who have lived in the city for all or most of their working lives and who are shortly to retire will have seen their properties appreciate in price many times over. Downsizing and moving to a more affordable location promises a sizeable boost to their pension situation.

There might also be some residential groups within the city who oppose the continuing growth of the city on the grounds of increased traffic congestion, pollution and pressure on resources.

Other stakeholders include:

- Oxford City Council – it needs to support the city in its search for housing land.
- University of Oxford – it is keen that top-flight researchers are not deterred from coming to study and work in this prestigious institution (Figure 22.5a).
- Social services – the housing situation is making staff recruitment difficult and this, in turn, is threatening service provision.
- City employers – the difficulties with staff recruitment are having an adverse impact on businesses of all sorts. The choice for them is to either pay more to attract staff or move to where labour is cheaper. A major employer is the BMW Mini car plant at Nuffield on the outskirts of the city (Figure 22.5b).

Figure 22.5 Contrasting images of Oxford: a) the city of dreaming spires; b) the home of the BMW Mini

Stakeholders outside Oxford

There are at least four 'external' stakeholders:

- Oxford County Council – needs to help the city find overspill areas.
- People living in areas being shortlisted for new housing – largely against any such moves.
- National and international companies that are anxious to set up close to the universities to benefit from their research, but are keen to avoid the difficulties of recruiting labour.
- The UK government is anxious that the University of Oxford and the city should retain their positive global reputations.
- NGOs are largely concerned about the conservation and protection of heritage, both natural and cultural.

In this case, the perceptions of the majority of stakeholders, particularly within the city, would favour any moves to increase the availability of housing and stem the rise in house prices relative to earnings. Those stakeholders may have slightly different images of the city that they wish to promote, but they are united by the wish to have the housing crisis resolved. The longer the crisis drags on, the greater the risk of damage to Oxford's image and the attractiveness of Oxford as a place for people and investment.

22.3 Rural stakeholders and their criteria

The management of change is a priority in many parts of the UK, no more so than in rural places. Here, there are a number of challenges:

- *Economy* – most rural population depends on agriculture, and rural communities still earn less per head than the urban population. There is a need to diversify the rural economy and create new businesses.
- *Infrastructure* – access to modern infrastructure is vital to ensure a high quality of life, as well as easy access to information and the outside world. Good infrastructure is also the key to economic growth and encouraging in-migration.
- *Affordable housing* – access to a home remains difficult for many due to highly restrictive building policies. It is this that drives young adults to move to urban and suburban areas.
- *Services* – relative to urban places, rural areas are disadvantaged when it comes to education, healthcare, retailing and other services. This can be a powerful push factor behind out-migration.
- *Conservation of the environment* – the countryside faces a number of threats, for example from modern agricultural practices and aspects of tourism.
- *Protection of heritage* – most rural communities have important natural and historic heritage which is important at a national (sometimes international) level. It often provides a resource in the context of tourism.

In this final section, the focus falls on three related aspects of the management of change in rural places:

- the perceptions of local people about their rural living space and the degree to which their views are affected by strategies
- the dimensions of change in a changing rural area
- the management of change as assessed by different stakeholders.

Rather than flitting from one type of rural place to another, the examination of these three different strands, as in Section 22.2, is given greater coherence if the spotlight in focused on just one rural area.

Breckland in East Anglia is the chosen place. It has been selected because it experiences some of the issues and challenges that characterise rural places. Furthermore, Breckland exemplifies all six of the challenges listed at the start of this section.

Synoptic themes:

Futures

Changes in any of these aspects of place will create differing and perhaps unforeseen legacies.

Synoptic themes:

Attitudes

The same comments may be made here about rural stakeholders as were made about urban stakeholders above.

Introducing Breckland

Breckland (often referred to simply as The Brecks) is an area of just over 1000 km^2 located in the heart of Norfolk but abutting the Suffolk border in the south (Figure 22.6). Its distinctiveness as a place lies rooted in its physical environment. Much of the area was once covered by heathland created by the clearance of its original tree cover by prehistoric farmers. Sandy soils overlying a chalk bedrock and a dry climate have subsequently encouraged the persistence of the heathland.

The statistics used in this investigation relate to the Breckland local authority, which is a little more extensive than The Brecks heartland, particularly to the north.

Over the last 100 years, the character of the area has changed considerably, by:

- the planting of Thetford Forest (the largest lowland pine forest in England), created after the end of the First World War in 1918 to provide a strategic reserve of timber; it is now one of the most popular recreational areas in the East of England
- the use of modern farming technology to make the rather infertile land more agriculturally productive
- a tourist industry largely involving day-trip and weekend visitors attracted by the physical landscape, wildlife and cultural heritage
- the use of some areas of heathland as military training grounds
- the expansion of its network of small settlements associated with the growth of its population; the main settlement nodes are the five market towns of Attleborough, Dereham, Swaffham, Thetford and Watton.

The area has a remarkably low population density of less than 100 persons per square kilometre, which compares with an average figure of 340 persons for England and Wales. Having remained fairly stable in the past, during the last intercensal period its population grew by a remarkable 7.5 per cent. Total population is now over 130,000 and is forecast to continue to grow by 9.5 per cent by 2021.

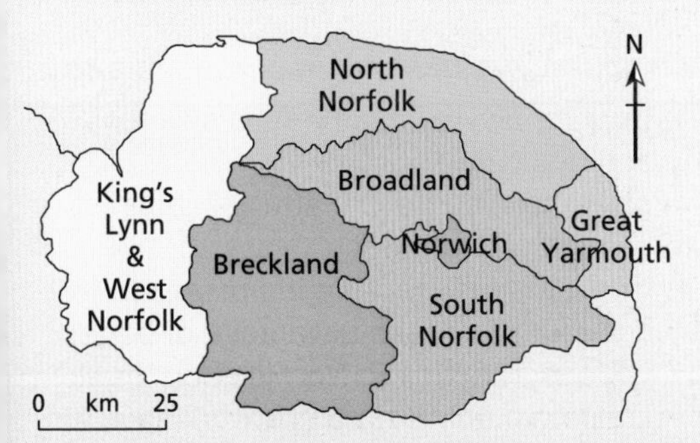

Figure 22.6 The location of Breckland local authority in Norfolk

Figure 22.7 Grimes Graves and the typical Breckland landscape

Today, the appeal of The Brecks as a place in which to live and as a place to visit is based on:

- the low population density and sense of openness
- the rich archaeological heritage, particularly the Neolithic flint mines at Grimes Graves (Figure 22.7)
- the large nature conservation area, such as Special Protection Areas (SPAs) and Sites of Special Scientific Interest (SSSIs), as well as nature reserves run by voluntary organisations such as the Norfolk Wildlife Trust and the RSPB
- the relatively low cost of housing.

Different views of Breckland

From this thumbnail sketch of The Brecks, we can begin to identify some of its main stakeholders. Typical views of both internal and external stakeholders are summarised in Figures 22.8 and 22.9. The sample of external stakeholders is rather institutional and governmental. Their views of Breckland are more focused on their aspirations for the place and what they identify as particular challenges associated with a brighter future.

Dimensions of change in Breckland and their challenges

There is no doubt that Breckland is a changing place – population growth and an economy becoming less dependent on agriculture are just two aspects. But does change necessarily lead to a better future for all?

The popular image of Norfolk is that of a largely agricultural and prosperous county. This popular perception has been called into question, however. A survey, using small area statistics (Output Areas), has revealed small pockets of quite serious multiple

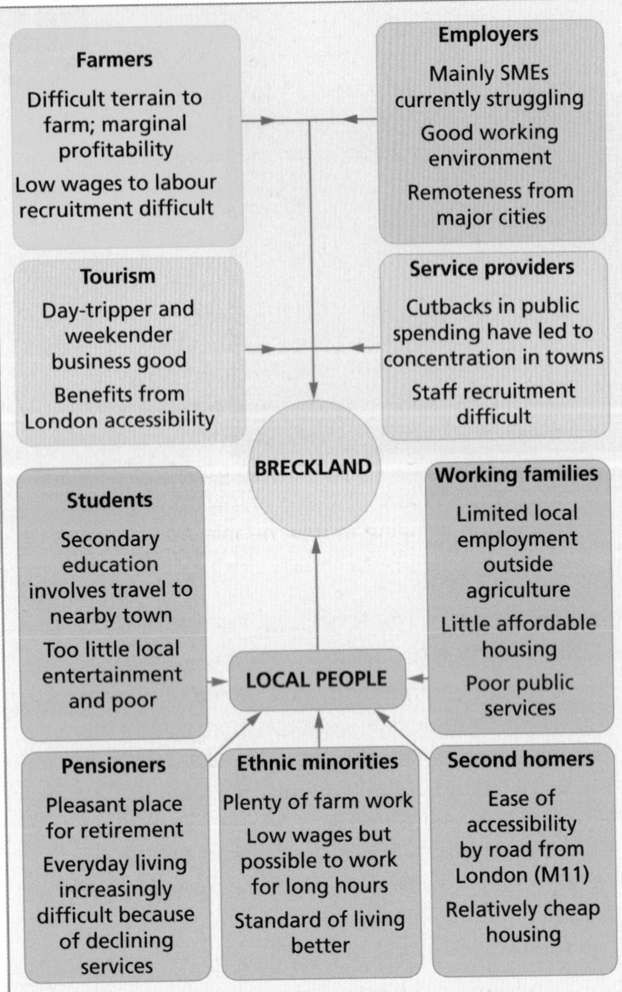

Figure 22.8 Sample views of some internal stakeholders in The Brecks

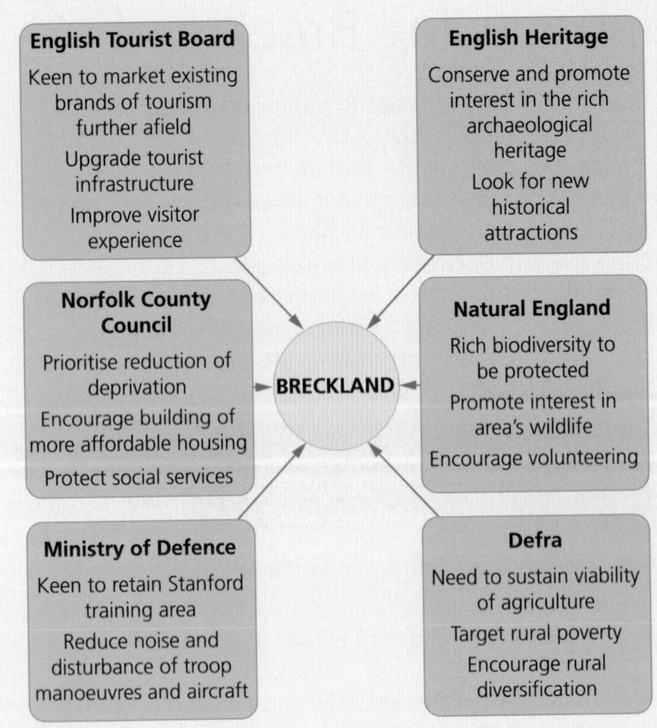

Figure 22.9 Sample views of some external stakeholders in The Brecks

deprivation in rural areas. Because of the aggregated level at which relevant data are usually published, small pockets of deprivation can easily be overlooked.

Not surprisingly, Figure 22.10 shows that deprivation is focused on the main urban settlements of Norfolk – Norwich, Great Yarmouth and King's Lynn. But it is the incidence in Breckland that catches the eye.

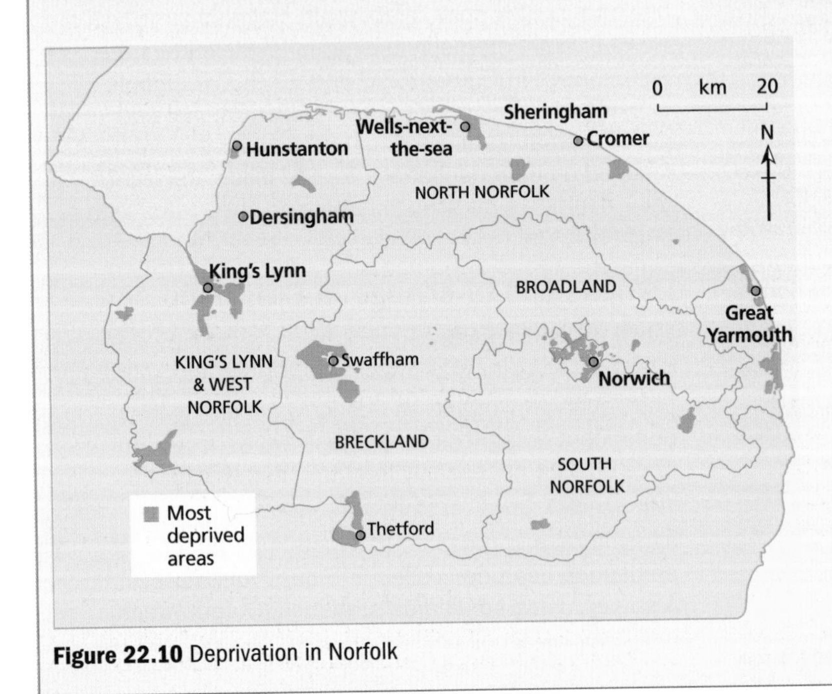

Figure 22.10 Deprivation in Norfolk

The following groups have been identified as suffering most from the poverty and deprivation shown around the small towns of Swaffham and Thetford:

- older people living alone (mainly widows) and older couples often totally dependent on the state pension
- low-paid manual workers and their families; rural employment, as in agriculture and tourism, typically involves low-wage jobs
- people excluded from the labour markets by unemployment, long-term sickness and disability
- self-employed people struggling for commercial survival
- working adults lacking the means to commute and take advantage of better job opportunities in Norwich and Ipswich; London is only a potential workplace for people living in those two cities.

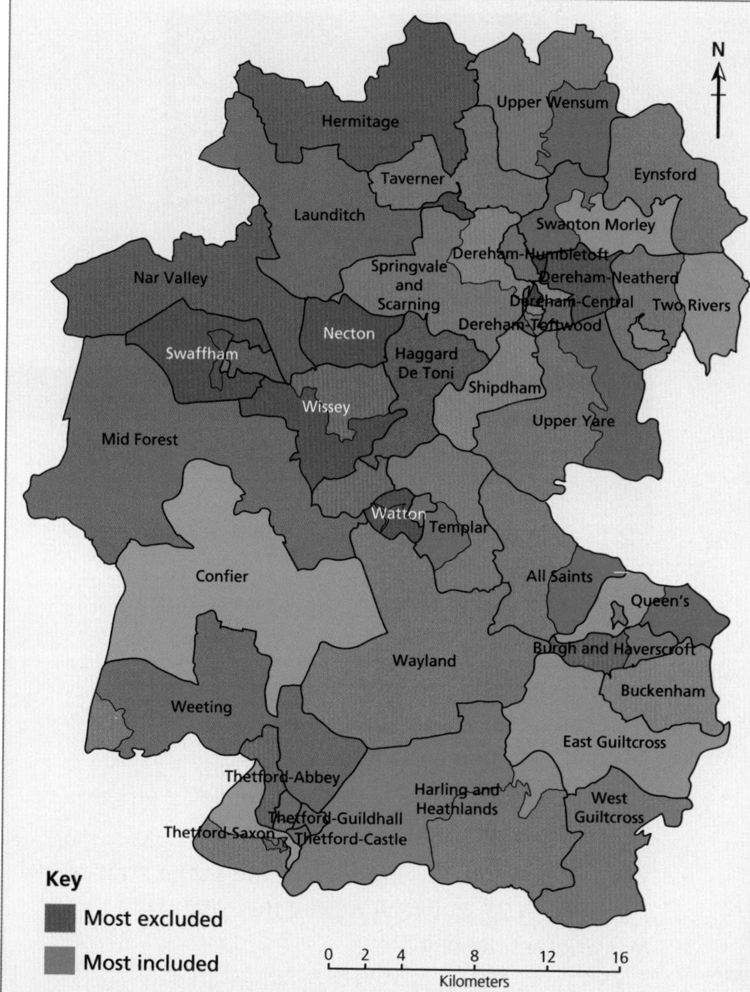

Figure 22.11 Social isolation in Breckland

Figure 22.11 allows us to look at Breckland's deprivation in more spatial detail, but through the indicator of social exclusion. The most striking area here is that involving Swaffham and wards immediately to the east and southeast. Again, there are high levels of exclusion in other towns, such as Thetford, Dereham and Attleborough.

One demographic fact that needs to be stressed is that ethnicity is not a significant characteristic of the rural poor here in Breckland. In 2011 only two per cent of Breckland's population was recorded as BAME. There are certainly plenty of white migrant workers (five per cent of the total population). While they might not live in truly rural areas, they are certainly conspicuous there during the working day, having been transported out of towns such as Swaffham and Thetford by their gangmasters. They come mainly from Portugal and Eastern Europe and are largely involved in the picking, packing and plucking industries of the agricultural sector. Some have settled here with their families, but others are only here on a seasonal basis. However, the fact that these immigrants are willing to work for low wages and occupy poor housing does little or nothing to help relieve the rural deprivation. These workers and their families do, of course, add to the pressure on social services.

Assessment

In conclusion, the question should be asked: what, if anything can be done to rectify this obvious downside of change in Breckland? The short answer is that little can be done during the current global recession to tackle its root causes. However, the upside of change in Breckland is that there is inward internal migration bringing in new people and perhaps new enterprises and skills. Recreation, leisure and tourism are doing well and, in co-operation with some of the external stakeholders, every effort should be made to promote them. They provide a basis for a more prosperous place and a more secure future.

Clearly, there are tangible aspects of change that can be measured. Those measurements will underlie judgements about whether or not change is being managed successfully. To give the flavour of such potential measures, they might relate to trends in:

- the number of deprived households
- the unemployment rate
- new job openings
- visitor numbers
- the extent of protected areas
- the profitability of farms
- the price of housing relative to earnings.

Naturally, the actual choice of discerning measures will vary from stakeholder to stakeholder, depending on their particular engagement with the place.

There are, however, less tangible aspects of Breckland that are much more difficult to assess. A very important one is the perceived quality of the lived experience obtained by local residents, even those in the grip of poverty and deprivation. Another concerns the strength of the attachments made by local people to the place that is Breckland.

Tension and conflict related to demography and culture have been a theme running through this topic. In the example of Breckland there are tensions, mainly related to housing, services, deprivation and social exclusion, but it is interesting to note that there has been little or no conflict in the normal sense of the word. For example, the localised deprivation does not seem to have impacted negatively on tourism and visitor numbers (Figure 22.12). The apparent absence of overt conflict may perhaps be related to the lack of ethnic diversity and to the fact that change in Breckland, unlike that on the Pepys Estate (Deptford, see Chapter 21, page 320), does not involve one group winning at another's expense.

Figure 22.12 Thetford is still a bustling market town, despite the deprivation

A final word on 'place'

It might be almost true to say that places are so diverse as to verge on the unique. Each place is created by a particular blend of internal factors and external linkages. People add to the diversity of place. They not only differ in terms of their demographic and cultural characteristics but perhaps, more importantly, they differ in how they perceive a place, how they engage with it and how they form attachments with the world through it. As individuals, we all perceive our living space in particular ways. Our lived experience of a place is also particular to us as individuals.

Given the diversity that exists within individual places and their populations, it is not surprising that fairly constant change often creates tensions that can readily become compelling issues. Typical issues include rural accessibility, declining services and deprivation, as we have examined in Breckland. At the other end of the rural–urban continuum, issues include those discussed elsewhere in this topic, namely the impact of urban regeneration on long-standing residents (Pepys Estate, Deptford) and greenfield developments to meet acute urban housing shortages (Oxford). The management of such issues can be controversial. The problem is that each and every one of us, whether as individuals or as members of stakeholder groups, has their own perceptions and perspectives. Each of us believes that we are the only ones to see the reality of a situation or place. As a result, often the only way forward in the management of an issue is by consensus and compromise. The worrying question is this: do these two processes necessarily produce the best solution to an issue? What is the 'best solution' anyway?

Review questions

1. Do all issues have to be managed? Give your reasons.
2. Explain what you understand by the term 'social progress'.
3. Explain why the measurement of assimilation is so difficult.
4. Re-read the account of the Pepys Estate in Chapter 21 (page 320). Identify the stakeholders and their perceptions.
5. Who is likely to object most to building on parts of Oxford's green belt?
6. Evaluate the six challenges cited at the start of Section 22.3. Which do you think presents the greatest difficulty?
7. Examine the view that it is impossible to say whether or not a particular issue has been managed successfully.

Further research

Research possible measures of assimilation: www.newstatesman.com/politics/2007/06/integration-ready-values

Find out more about the supply of affordable housing over the last five years: www.gov.uk/government/collections/affordable-housing-supply

Select one of the most affordable cities in Table 22.4 and investigate the specific reasons for it online.

Exam-style questions

AS questions

1 What is the technique most commonly used to show population structure? [1]

2 Study Figure 20.6 (p. 296), which shows house-price-to-earnings ratios between 1985 and 2013.

  a Describe what happened to the national ratio. [2]

  b Suggest one reason why this ratio is an important factor affecting internal migration flows. [3]

3 Explain two reasons why international migrants tend to live in cities. [4]

4 Explain why there are frequently tensions in changing places. [6]

5 Study Figure 22.10 (p. 332), showing the location of the most deprived areas in Norfolk.

  Assess the factors contributing to deprivation in rural places. [12]

A level questions

1 Study Figure 21.8 (p. 313), which shows the distribution of ethnic minorities in London in 2010.

  a Suggest one reason why these ethnic groups have chosen to settle in London. [3]

  b Explain reasons for the concentration of ethnic minorities in particular parts of London. [6]

2 Explain why there are different perceptions of a rural area's attractiveness as a living space. [6]

3 Evaluate the view that management of a rural issue is unlikely to please everyone. [20]

Place Investigations: Topic 4 Shaping Places

Topic 4 of the Specification requires you to study two places. You should have first-hand experience of one and the other should be significantly different. To give you an example, this section attempts to set out a way to investigate a locality and a contrasting one. This checklist and example place investigations are based on the *contrasting* areas of Llanmadoc, Gower Peninsula, near Swansea in Wales, and Harpurhey, Manchester.

The focus of the following section is on Option 4A: Regenerating Places, but this general approach is also useful for Option 4B: Diverse Places.

What do you already know about your local place?

Both are examples of different spirals of decline; their primary function has changed over the last 100 years resulting in loss of community infrastructure.

What is the boundary of your chosen place?

Consider the electoral ward, postcode, LSOA statistics and GIS.

Is there a clear place identity?

Consider brief characteristics of the place, where it is on the rural–urban spectrum and carry out a Placecheck.

Table PI.2 Place boundaries

Llanmadoc	Harpurhey
SA3 1DA The postcode area is smaller than the electoral ward, Gower	M9 4AF The electoral ward is Harpurhey Manchester LSOA of 006B

Table PI.1 Initial perceptions and factual knowledge of two chosen places

Llanmadoc	Harpurhey

A family holiday destination, big caravan parks and holiday homes. Beautiful agricultural landscape, rolling hills and much livestock, with a wild cliff, sandy bays and salt marsh coastline. Scattered small villages, hamlets and isolated farms/houses. Perception of being in remote place, but actually easy access from M4. Seventy per cent of houses are second homes in the Gower Peninsula, and Llanmadoc reflects this. Nearest large town is Swansea, eleven miles away by small winding roads.	Once a separate rural village, industrialised in the nineteenth century as part of Manchester's sprawl. It became famous for its dyeing, engineering, rope making and brewing, plus Manchester Cemetery and Dogs Home. It became densely populated with rows of terraces and some high-rise redevelopment. Developed a reputation for unemployment, derelict land and vandalised buildings. Has a vibrant multi-ethnic community and is only two miles from the main city centre.

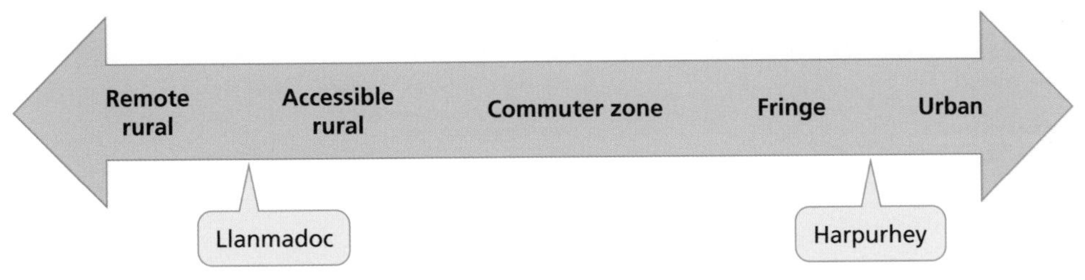

Figure PI.1 The rural–urban continuum and the two chosen places

The results of your Placecheck can be put into a spectrum diagram, as shown in Figure PI.2. The length of the bar shows the strength of like/dislike.

The quiz on your local area from the ONS neighbourhood statistics makes it clear that more in-depth research is needed about the characteristics of your local place. It also helped on boundaries of the area to be studied.

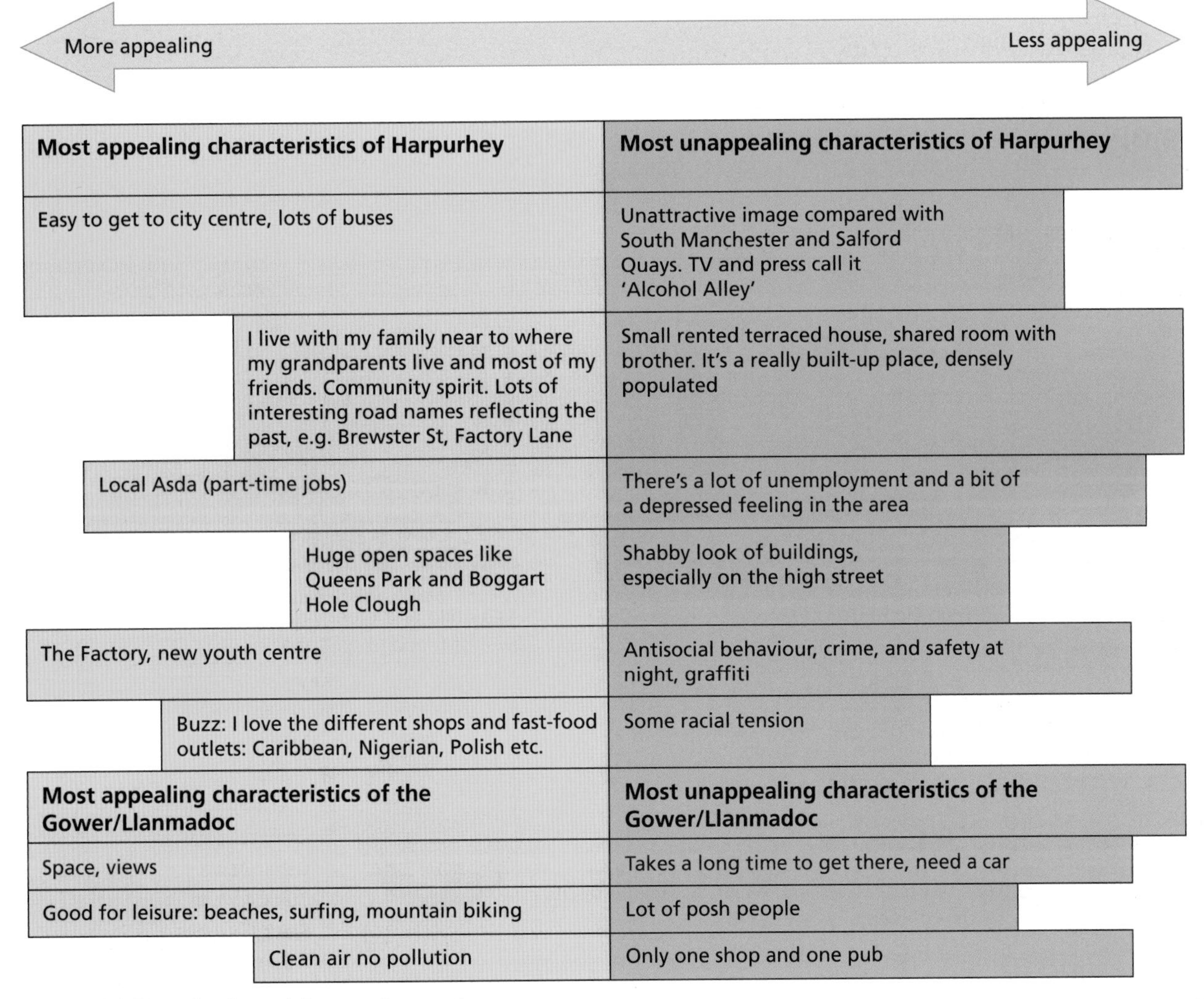

Most appealing characteristics of Harpurhey	Most unappealing characteristics of Harpurhey
Easy to get to city centre, lots of buses	Unattractive image compared with South Manchester and Salford Quays. TV and press call it 'Alcohol Alley'
I live with my family near to where my grandparents live and most of my friends. Community spirit. Lots of interesting road names reflecting the past, e.g. Brewster St, Factory Lane	Small rented terraced house, shared room with brother. It's a really built-up place, densely populated
Local Asda (part-time jobs)	There's a lot of unemployment and a bit of a depressed feeling in the area
Huge open spaces like Queens Park and Boggart Hole Clough	Shabby look of buildings, especially on the high street
The Factory, new youth centre	Antisocial behaviour, crime, and safety at night, graffiti
Buzz: I love the different shops and fast-food outlets: Caribbean, Nigerian, Polish etc.	Some racial tension
Most appealing characteristics of the Gower/Llanmadoc	**Most unappealing characteristics of the Gower/Llanmadoc**
Space, views	Takes a long time to get there, need a car
Good for leisure: beaches, surfing, mountain biking	Lot of posh people
Clean air no pollution	Only one shop and one pub

Figure PI.2 Placechecks and the two chosen places

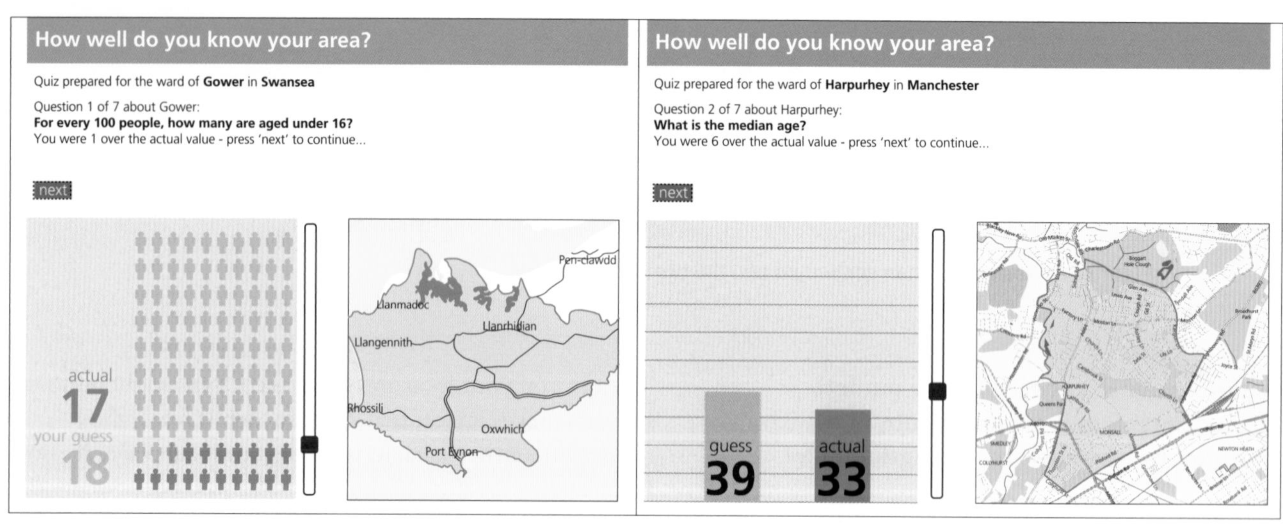

Figure PI.3 Neighbourhood statistics quiz results

Who lives there and in what type of housing?

Table PI.3 Population, education and health overview

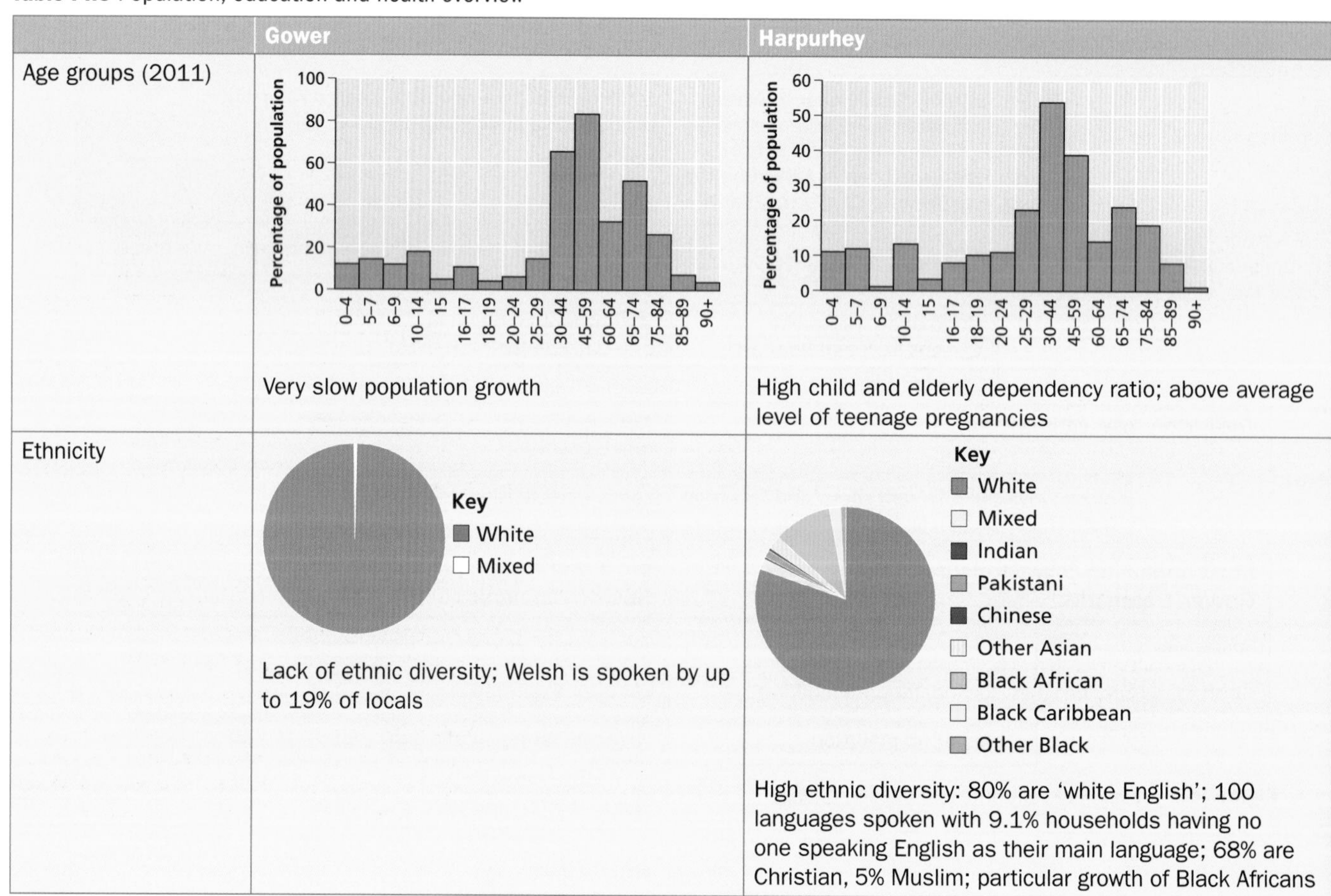

	Gower	Harpurhey
Age groups (2011)	Very slow population growth	High child and elderly dependency ratio; above average level of teenage pregnancies
Ethnicity	Lack of ethnic diversity; Welsh is spoken by up to 19% of locals	High ethnic diversity: 80% are 'white English'; 100 languages spoken with 9.1% households having no one speaking English as their main language; 68% are Christian, 5% Muslim; particular growth of Black Africans

	Gower	Harpurhey
Education		
Degree or equivalent (%)	49	30
Apprenticeship (%)	9	6
2+ A levels (%)	30	13
5+ GCSEs, an A level or 1–2 AS levels (%)	32	22
No GCSEs (%)	53	85
Health		
Very good	194	92*
Good	110	70
Fair	38	50
Bad	19	29
Very bad	4	10
Total population surveyed	365	251

*one of the worst wards for general health and life expectancy in Manchester

Table PI.4 Employment: what do locals do?

ACORN classification	Gower (%)	Harpurhey (%)
AB – higher managerial, administrative, professional	41	4
C1 – supervisory, clerical, and junior managerial/administrative/professional	37	25
C2 – skilled manual workers	12	19
DE – semi-skilled and unskilled manual workers; state benefit/unemployed, lowest grade workers	10	44

Table PI.5 Housing: where do people live?

	Gower	Harpurhey
Average house price (Zoopla ZED Index, 2013)	£346,984	£102,280
Owned outright (%)	83	9
Owned with mortgage (%)	52	22
Rented: from council (%)	5	72
Rented: other social (%)	0	12
Rented: private landlord including letting agents (%)	7	13

Average house prices in Harpurhey are lower than Manchester's average.

What are the place's functions?

Consider the administrative, commercial, financial, retail, culture and residential functions at local, regional, national and international scales.

Table PI.6 Functions overview

	Gower	Harpurhey
Local	Church and community shop serve local people as well as many tourists (mainly during the summer). Some residential function, low density mainly detached houses.	Wide range of local services; Moston Lane known as 'Little Lagos' (Nigerian food shops and restaurants). Sixth form college, youth centre. Large parks serve surrounding postcodes. Important residential function.
Regional/national	Renowned as a holiday destination and conservation area.	Only known for deprivation and antisocial behaviour (because of TV programmes).
International	Marketed as a tourist destination but low numbers from abroad.	Unlikely to be known other than as being part of Manchester (and the Northern Powerhouse).

Table PI.7 Regeneration needs?

Gower	Harpurhey
No area in Gower is in the 10% most deprived areas of Wales. Lack of major crime, a few thefts. Some issues with house prices; 70% second homes, younger locals cannot afford to buy without parental help. Low unemployment, but it does tend to be seasonal because of tourism with long hours and lower rates of pay than many city jobs. Lack of services: one pub, no shop (until 98% of the community clubbed together to create one run by 40 volunteers) and no primary school. No environmental issues.	Harpurhey is 3rd out of 32 Manchester wards in IMD, which is very high, and a national hotspot of high multiple deprivation, especially unemployment and health. It has reduced by one rank (i.e. got marginally better) between 2010 and 2015. Many households are below the poverty line; some evidence of an engrained culture of taking state benefits. High incidence of major crime and antisocial behaviour (ASB). Unemployment 16.2% (national rate 7.5%); only 69% have full-time work; 27.5% of working age claim out-of-work benefits and 9.1% are NEET. Higher than city averages in basic skilled work, services and administration; lower proportions in managerial, professional sectors. Alleyways have been improved and parks enhanced; older retail areas lack investment; new centres have better aesthetic appeal.

Is there a need to regenerate this place?

Consider the population, employment, housing, services, crime, the environment or multiple deprivation – use newspaper reports and other sources.

Is there any statistical evidence showing the need for regeneration in your chosen local place?

To see how Harpurhey compares with the other wards in Manchester on wealth, health and crime, which all affect people's perception and identity with a place, data was collated from the ONS in a table from which scatter graphs could be drawn and Spearman's rank used.

Government and Manchester City Council websites were also used:

- www.gov.uk/government/statistics/english-indices-of-deprivation-2015
- www.manchester.gov.uk/downloads/download/5724/compendium_of_statistics-manchester

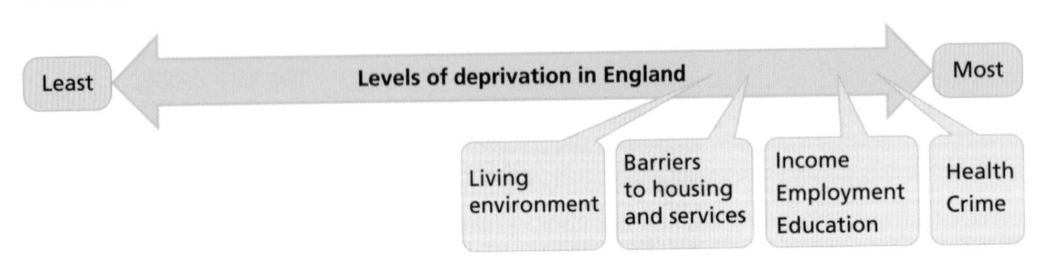

Figure PI.4 Harpurhey deprivation findings

Table PI.8 Manchester wards data collation

North Manchester wards	IMD, 2010	Claiming Jobseekers Allowance, 2015 (%)	Social indicators, 2013		
			Life expectancy (years)	General fertility rate (per 1000)	5 GCSEs at A*–C including English and Maths (%)
1 Ancoats and Clayton	45.64	2.1	74.8	49.4	59.3
2 Bradford	61.30	2.8	75.2	72.2	49.2
3 Charlestown	51.71	2.6	75.1	71.2	37.1
4 Cheetham	48.61	2.2	76.1	86.6	46.2
5 City Centre	17.16	0.3	84.3	10.2	57.1
6 Crumpsall	38.13	3.1	79.9	97.2	48.2
7 Harpurhey	58.04	4.3	75.8	88.0	35.4
8 Higher Blackley	47.30	2.3	78.7	76.0	46.8
9 Miles Platting and Newton Heath	61.12	3.7	73.2	86.7	38.5
10 Moston	38.26	2.2	78.4	73.6	40.0

Main findings

Scatter diagrams were plotted using Excel with best-fit lines to see if any trends and relationships exist between these indicators. An online Spearman's rank coefficient calculator (www.socscistatistics.com/tests/spearman/default2.aspx) helped to show the precise amount of any relationship statistically.

A similar process could also be used for the Gower.

Table PI.9 Main findings using Spearman's rank on Harpurhey

Indicators	Spearman's rank	Statistically significant coefficients r value?
Unemployment and IMD 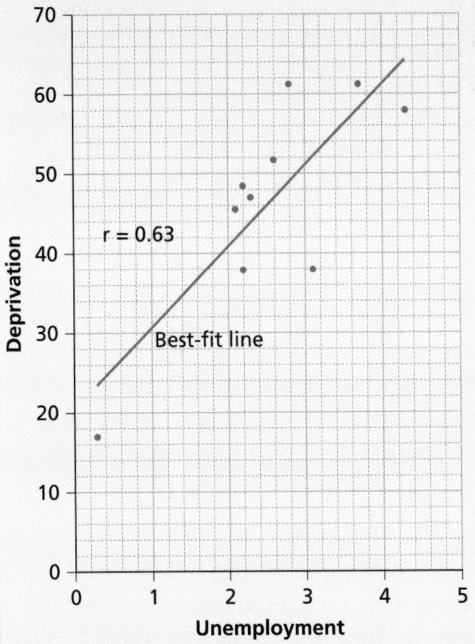	0.63222, a positive relationship. This supports the hypothesis that high unemployment will lead to more deprivation because of less wealth to spend on housing, healthy food, etc.	Yes
IMD and life expectancy 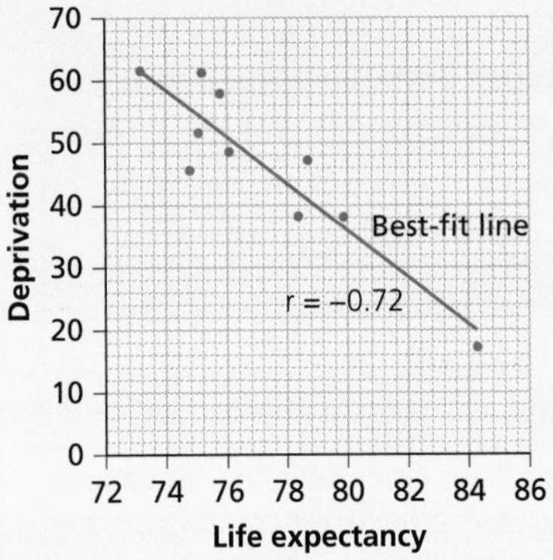	−0.72121, a negative relationship. This supports the hypothesis that high deprivation will lead to a shorter life expectancy because of a poorer environment, health and lifestyle.	Yes
Unemployment and life expectancy 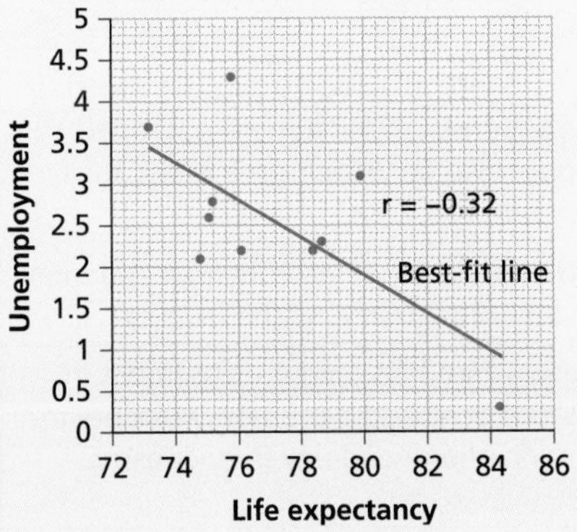	−0.31611, a weak negative relationship. This refutes the hypothesis that life expectancy is related to unemployment, but it may be because of a small sample set.	No

Are there any regeneration schemes in your locality?

Table PI.10 Regeneration schemes

Gower	Harpurhey
The community shop – grants from the Welsh Development Agency and Swansea City Council.	Manchester City Council Local Area Plan: 50 years of intervention. Regeneration schemes include: • 2007 initiatives: £17 million redevelopment of Harpurhey district centre, including a new undercover market • £125 million Manchester–Salford Housing Renewal Pathfinder scheme, one of nine national pilot initiatives with the aim of improving the quality and range of homes • new sixth form college on the brownfield site of Harpurhey Baths • new library • new business park • Factory Youth Zone community centre.

Who are the players or stakeholders, and what is their role?

Table PI.11 Players and stakeholders

Gower	Harpurhey
Local parish council for local planning decisions; Swansea City Council; Welsh Development Agency for aid; National Trust for conservation; local community forum for fundraising and running the shop; businesses such as the Britannia Inn attracting customers; estate agents for second homes, e.g. Rightmove.	Manchester City for regeneration projects; elected city councillors press for aid to Harpurhey and have lobbied the BBC not to make another national TV programme about it; ASDA supermarket is a big employer; local community forum; local businesses, e.g. Wishy Washy laundrette, Moston pound store; Housing developer Redrow.

How effective has regeneration been? What do locals think?

Table PI.12 Local viewpoints

Gower	Harpurhey
Locals very happy with community shop. Some antagonism with second-home owners, but people at the pub are generally very friendly to visitors.	Locals have diverse views. Antagonism with the BBC about its TV shows. Very happy with the Factory Youth Zone and refurbishment of social housing. New affordable homes by Redrow are popular.

Diverse sources of information were also used for Harpurhey:

- A college website described the area and highlighted the fact the famous author Anthony Burgess (*A Clockwork Orange*) lived there. The image it was portraying was that the place had more than local significance because of this.
- Oral history sources: apart from recordings of local families and neighbours there is also an online reminiscence of the area commenting on economic and social changes by Professor Emeritus Cliff Hague (www.befs.org.uk/news/50/49/Harpurhey/d,Blog)
- Online social media, blogs and newspaper articles capturing viewpoints from national and local sources. Table PI.13 shows some of the notes made from articles found in the national and local press on Harpurhey:

Table PI.13 National and local press notes on Harpurhey

National press	Media focus on binge drinking, antisocial behaviour and an alleged benefits culture, especially Moston Lane (branded 'booze alley' with 23 alcohol outlets along 1.5 miles). 200 residents attended a protest meeting against the BBC filming more episodes of *People Like Us,* which gave a negative portrayal of the area. Government point of view: named the most deprived borough in England in 2007.
Local press	Harpurhey has had recent investment in its district centre and the new Manchester Communication Academy (*Manchester Evening News*).

Table PI.14 Summarising influences on local places

Scale of influence	Evidence of influence and the role of players	
	Gower	**Harpurhey**
International	Post-production landscape, less workers in primary sector needed. Rise in tourism.	Manufacturing (dyeing materials, engineering, wireworks and a brewery) closed because of deindustrialisation and global shift due to globalisation. Migration hotspot for over 100 countries/nationalities.
National	Designation as an Area of Outstanding National Beauty and much National Trust land restricts big new developments. Desire for second homes in the village.	UK government: housing, benefits, education and health departments. BBC TV programme *People Like Us* (2013) gave it a negative image, failing to show the positives of cultural diversity. National broadsheet newspapers often sensationalise problems such as binge drinking in the area.
Local	Strong community despite low numbers.	Strong local community groups, e.g. North City Residents Forum. Local authority: Manchester City Council tries to reduce deprivation. Local councillors lobby for more investment and aid. Local police there to keep order. Local estate agents and landlords want to attract people. Local newspaper: the *Manchester Evening News,* may give a positive spin on the area or inflame attitudes with sensationalist headlines (e.g. 'Harpurhey, the worst place in England').

Collating all the previous data, a table of the various influences on the chosen place can be made (Table PI.14).

Conclusion

The ways in which personal identities in a neighbourhood have been affected by demographic and cultural changes can be summarised as shown in Figure PI.5.

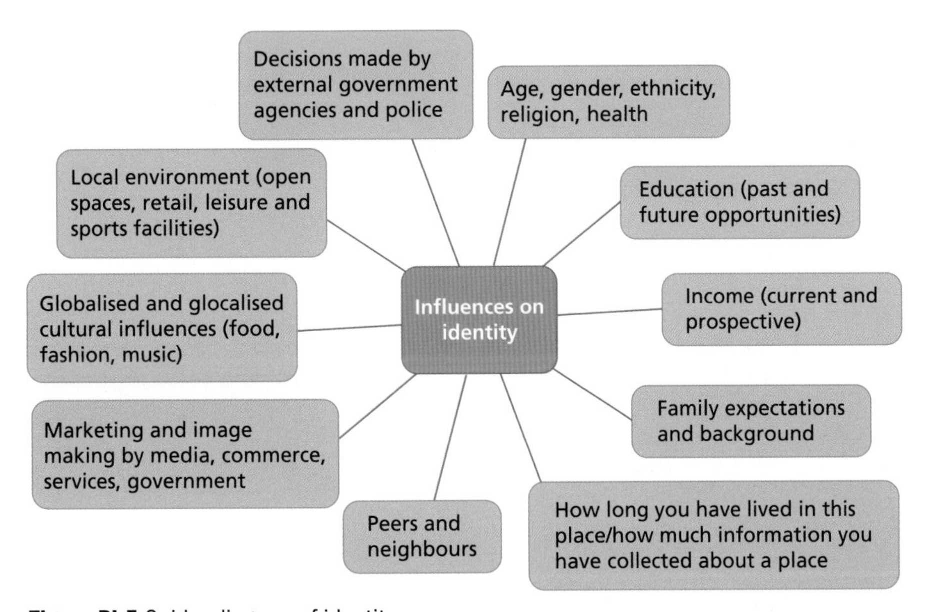

Figure PI.5 Spider diagram of identity

Table PI.15 References

General references	ONS Neighbourhood Statistics: www.ons.gov.uk/ons/index.html
	Quiz using postcodes on places: www.neighbourhood.statistics.gov.uk/HTMLDocs/dvc147/
	This website uses ONS data and graphs some of it: www.streetcheck.co.uk
	This statistics website calculates Spearman's rank data: www.socscistatistics.com/tests/spearman/default2.aspx
	Crime data website: www.police.uk
	Photograph sources: The Francis Frith archive of national photos and your local Facebook page should have many past photos and recorded memories of the locality. These can be compared with current images to see any changes. Tourist hotspots such as the Gower will have many images available online and in publicity leaflets: www.francisfrith.com and www.facebook.com
Gower	http://gov.wales/statistics-and-research/welsh-index-multiple-deprivation/?lang=en
	www.the-gower.com/home/map/interactive.htm
Harpurhey	Manchester City Council website, which has a free GIS mapping facility and updates from the 2011 census, such as the electoral ward registration summary: www.manchester.gov.uk/site/index.php
	Useful population pyramid: www.manchester.gov.uk/download/downloads/id/17872/a26r_harpurhey_2011_02
	www.manchester.gov.uk/downloads/download/4220/corporate_research_and_intelligence_population_publications
	www.manchester.gov.uk/download/downloads/id/22264/chapter_9_community_well-being_and_neighbourhood_satisfaction
	Manchester City Council's ward profile: http://www.manchester.gov.uk/download/downloads/id/17872/a26r_harpurhey_2011_02

Skills Focus: Topic 4 Shaping Places

As you study your chosen Topic 4 option (Regenerating Places or Diverse Places), you will need to acquire and apply geographical skills of both a quantitative (numeric) and a qualitative (more subjective) kind. These skills and associated data are particularly vital to two investigative tasks common to both options, namely researching, describing and analysing:

1. the identity of actual places, and
2. the change taking place within them.

The Specification identifies ten relevant skills in each option, although in the case of regeneration the 'investigation of social media' and the 'interrogation of blogs' are basically one and same. Eight of these skills are stipulated in both options, leaving three that are peculiar to just one of the options. The particular relevance of these skills to specific parts of your option is flagged up at the relevant places in individual chapters, which will point to one or more of the skills descriptions in this section.

Table SF.1 lists all 11 of the skills and distinguishes between quantitative and qualitative techniques. The emboldened skills (1 to 8) are common to both options. Skills 9 and 11 are peculiar to Diverse Places and 10 to Regenerating Places. The table also shows a twelfth technique (Placecheck) which has been added because it is so relevant to an important component of both options, namely the personal investigation of places.

Table SF.1 Topic 4 skills

Qualitative	Quantitative
1 **Social media and blogs**	7 **Use of GIS**
2 **Oral accounts**	8 **Testing relationships**
3 **Newspapers**	10 Index of Multiple Deprivation
4 **Using of 'before' and 'after' evidence**	11 Indices of diversity
5 **Using creative media sources**	
6 **Evaluating qualitative sources**	
9 Interviewing	
12 Placecheck	

It is important to remember the warning in the Specification that these skills are not exclusive to the topics under which they appear. Students will be expected to be able to apply these skills across any suitable topic area through their course of study.

Skills focus 1: Social media and blogs

Social media allows people to create, share and exchange information, ideas, opinions and images in virtual communities. Nationally, the most popular social media websites are Facebook, Google+ and Twitter. Many local places now have online forums and email groups where viewpoints as well as practical lifestyle issues are discussed. In the present context, social media might provide some help with your local place investigations by:

- informing about, and perhaps recommending, useful sources of information
- supplying relevant resources such as photos, video clips, newspaper cuttings and commentary
- informing you of the perceptions and lived experiences of different people
- testing views on the success or otherwise of rebranding and re-imaging projects.

Attachment to place and issues such as regeneration are centred on the perceptions of people (local residents and visitors). Social media allows far more views to be sampled than was ever possible in the past through newspapers, TV and radio programmes. Blogs, online petitions and comments on online news reports may be anecdotal but they can give a more 'rounded' assessment of the sense of loyalty to a place.

Remember, of course, that all material, information and comment from social media comes with a 'health' warning. Can you be sure of the authenticity and soundness of what you are being sent or told? Undoubtedly, social media can be put to best use by inviting others to tell you about their lived experiences in your chosen places. The important point here is to try to structure your collection of comments so that you are sampling the overall population in terms of its component age, ethnic and socio-economic groups. You need to portray that a diverse population inevitably has diverse views about the place in which they happen to live. These may be linked back to their lived experience and attitudes to inequalities.

Skills focus 2: Oral accounts

An oral account is basically a recorded interview or conversation used to capture the points of view and memories of communities or individuals, especially in societies going through significant changes such as de-industrialisation or the segregation of incoming ethnic minorities. Digitally recorded oral accounts obtained from elderly people can also be very helpful in reconstructing a place's past identity.

At the data collection stage, you need to recognise that there are basically two different approaches. Is it better to allow the narrator to talk spontaneously, without any sort of leading, or should their recall be focused on specific aspects and structured more in the manner of an interview?

The British Library (http://sounds.bl.uk/oral-history) and Oral History Society (www.ohs.org.uk) are useful starting points in checking whether or not their archives contain oral accounts relevant to your particular enquiry. Many local museums, records offices and libraries have built up considerable and well-catalogued collections of what is termed 'oral history'. Since all these archives are accessible to the public, you might consider making use of them rather than going out into the field and recording your own.

Like other forms of qualitative data (most notably newspaper reports and media portrayals), oral accounts should be used with the utmost caution. The time lapse between the past and the present can be full of pitfalls, so you need to be assured on a number of issues. For example, can you be fairly certain that the recorded account is a reasonably accurate recall of the past? Remember that the human memory can be very selective and that its quality generally diminishes with age. Some of us tend to see the past through rose-tinted spectacles, while others have a tendency to remember and recall the grim side of reality.

Skills focus 3: Newspaper reports

It is well known that national newspapers display political bias in their coverage of events, both at home and abroad. Compare, for example, the Conservative or right-wing inclinations evident in the *Daily Telegraph* and *Daily Express* with the Labour or left-wing comment provided by the *Guardian* and the *Daily Mirror.* In the middle ground there are newspapers such as the *Independent* and *The Times*, as well as those like the *Daily Mail* and the *Sun*, which seem to oscillate in their political support.

When it comes to local newspapers, however, there is still bias. It is not necessarily political, as such, but rather involves taking sides in the reporting and discussion of local issues. Issues might include whether to use brownfield rather than greenfield sites for the building of affordable housing, or the sustainability of the regeneration of a town centre.

The values, perceptions and possible business interests of the newspaper proprietor or the individual reporter can also be sources of bias.

Skills focus 4: Before and after interpretations

There are two main sources to be employed in the task of showing before and after situations – i.e. how a particular place has changed over a given period: maps (including satellite images) and photographs (including postcards).

Maps and satellite images

The Ordnance Survey has been producing and publishing maps of the UK for a long time. A comparison of the same area on successive editions of the same scale of map can provide plenty of information about many aspects of physical change. Perhaps the most useful maps in place investigations are those at 1:25,000 and 1:10,560. Local libraries, local authority planning departments and county records offices are custodians of old maps.

GOAD maps of retailing areas over time can be particularly useful in the investigation of change in town centres and cities. With regard to satellite images, Google Maps has some images dating back to 2009.

Photographs and postcards

Getty Images holds an online archive of historical photographs. Individuals post images on Flickr and

Facebook. Other websites also hold past images, such as Francis Frith, which has photographs of 7000 towns and villages taken between 1800 and 1970. Many local museums and libraries contain collections of old postcards and photographs of street scenes and buildings taken over 100 years ago. With any luck you will be able to find some for the specific locations you are investigating.

A comparison of today's scene with that of the same sometime in the past will allow you to identify at least five different types of change:

- **Gentrification**: noticeable physical renovation and social upgrading of residential properties and their local services.
- **Filtering down**: a downward spiral involving a steep decline in the physical maintenance of properties; growing presence of squatters and poor housing.
- **Redevelopment**: old buildings completely demolished and replaced by modern ones; physical layout often changed.
- **Change of use**: for example, buildings and spaces are converted from residential to commercial use.
- **Re-imaging**: improvement of the general appearance of an area by cosmetic landscaping, conservation and restoration of old buildings; introduction of new activities.
- **Intensification**: most apparent in suburbs and villages and involving the infilling or development of vacant spaces; the outcome is a raising of population density and the intensity of land use.

Skills focus 5: Evaluating different media

The individual media that you might draw on can be used to convey different place messages. For example:

- **Paintings**: compare Constable's portrayal of the countryside with Hogarth's depiction of urban life.
- **Literature**: compare Jane Austen's descriptions of gentile places with Charles Dickens' graphic descriptions of London and the Medway towns.
- **Photographs**: compare picture postcard images of seaside resorts with today's all-too-graphic images of poverty and slum housing.
- **Newspaper articles and reports**: while these can be very informative and provide interesting analyses of issues, the point has been made in Skills focus 3 that they often suffer from bias related to the author's or editor's perception of reality.
- **Radio and TV**: exactly the same comments apply here to programme producers.
- **Videos and films**: these can certainly be quite powerful in terms of conveying the visual identity of places.

When it comes to evaluating these different media it is vital to remember that they all come from a human mind and hand. They may add 'colour' to your investigations but perhaps they tell you more about the compiler – their perceptions of and attachments to particular places. But also remember, it is not only the artist's perceptions that are involved; there are also yours as the 'receiver' of the message. Nonetheless, they can still throw interesting light on place identity and, particularly, how it has changed over the years.

Skills focus 6: Interpreting qualitative data

Qualitative data is data that is not easily reduced to numbers. Such data tends to relate to concepts, opinions, values and behaviour. Typical data would include interviews, notes based on field observations and copies of documents, as well as audio and video recordings. Types of data also include structured text (articles, reports and newspaper extracts).

It might be helpful to point out some basic differences between qualitative and quantitative data (Table SF.2).

Qualitative data analysis (QDA) is the range of processes and procedures whereby investigation moves from the

Table SF.2 The difference between qualitative and quantitative approaches to data

	Qualitative	Quantitative
Purpose	To describe a situation, to gain insight into a particular place	To measure magnitude; how widespread a practice is
Format	No pre-determined response categories	Pre-determined response categories, standard measures
Data	In-depth explanatory data from a small sample	Wide breadth of data from a statistically representative sample
Analysis	Draws out patterns from insights	Tests hypotheses, uses data to support conclusion
Result	Illustrative explanation and individual responses	Numerical aggregation in summaries, clustered responses
Sampling	Theoretical	Statistical

qualitative data that has been collected into some form of understanding, explanation or interpretation of the places and people being studied.

Getting the understanding, explanation or interpretation correct requires a mix of objectivity, intuition and perceptiveness. A whole series of critical questions need to be confronted, such as:

- Which, if any, of the sources of qualitative data can be allowed to carry more weight?
- Are your interview results more reliable than the opinions reported in local newspapers?
- Do the views represent anti-development NIMBY (not in my back yard) sentiments?
- Does the tourist information brochure give a fair picture of the place?
- What does the local art exhibition tell you about how the artists view their local place?

In short, the interpretation and evaluation of qualitative data can be far more challenging than that of quantitative data. It is extremely difficult to be completely objective at two critical points in an investigation:

- the relative assessment of different sources
- the interpretation of what each data source is showing.

Skills focus 7: Using GIS

Geographic information systems (GIS) are a device for capturing, storing, checking and displaying data by computer. It is geospatial, meaning it shows layers of data on one map, helping to analyse and understand distributions, patterns and relationships. Maps can be used to compare places or the same place over different time periods. Any information with a locational tag can be used, such as latitude and longitude, addresses and postcodes. Socio-economic and environmental data may be captured, for example population, income, education, crime and voting patterns, which may be overlaid on maps with information on satellite images of street layout and terrain.

Businesses use GIS to locate new units such as retail stores. Police and local authorities use GIS to identify hotspots of different types of crime and how these vary over time. Planners use GIS to zone places, target funding and predict future needs.

There are many free data sources to build up a picture of your place's 'health' in terms of employment, health, longevity and overall deprivation. You will need the local postcodes as well as the place name (www.zoopla.co.uk/postcode-finder).

GIS allows us to visualise, question, analyse and interpret data to understand spatial relationships, patterns and trends. GIS is the go-to technology for helping us to make sense of what is happening in geographic space. It is also widely used by almost every economic activity in making better decisions about location.

The GIS you will probably know is Google Earth. Basic layers of data may be built up to create a profile of a chosen place. Use the menu to add layers. Basic health and education data such as the location of hospitals and schools may be plotted.

The Office for National Statistics (ONS) website provides a wide range of demographic data that can be used in your place investigations, especially the graphic deprivation 'swingometer'. Separate statistics are available for Scotland and Wales. Try out the 'How well do you know your area' quiz, which links to 2011 census data: www.neighbourhood.statistics.gov.uk/HTMLDocs/dvc147/index.html.

The investigation of crime is one particular field in which GIS is being used to illuminate such basic questions as:

- Where exactly are crimes committed?
- Does the type of crime vary from place to place?
- Are patterns of crime changing?
- Where do criminals live?

There are now three websites in the UK providing crime statistics for specific areas and that also allow place comparison:

- www.crime-statistics.co.uk – the archive stretches back to 2010; crimes are displayed within a 1 mile radius of the selected postcode.
- www.police.uk – this website allows a comparison of the crime levels in different neighbourhoods.
- http://maps.met.police.uk – a relatively new website that allows users to see what offences (criminal and antisocial) have been reported in local streets.

Skills focus 8: Testing the strength of relationships

A scattergraph is a simple and highly visual method used to investigate the relationships between sets of paired data (variables). Is there a relationship, for example, between life expectancy and the incidence of poverty? Basically, the data of one variable is plotted on two axes. The independent variable is usually plotted on the *y*-axis and the dependent one on the *x*-axis, although there is often a two-way relationship. For example, while low income may be the result of poor qualifications, it can also be the cause of them. Once the graph has been completed, a pattern may become visible that suggests some sort of a relationship exists between the two variables.

A best-fit line (BFL) should then be drawn 'by eye' on the graph. This line should pass through the spread of plotted points, minimising the total distance the points are from the line and with roughly equal deviations on either side of the line (Figure SF.1). Software such as Excel will plot this data using mathematical regression.

A scattergraph is used initially to identify whether or not a relationship exists between two variables. If it does, then the Spearman's rank can be calculated to establish the strength of the relationship and whether it is statistically significant. The calculation of the coefficient is based on the rank differences between the two data sets. Once you have learnt how to calculate this by hand, many websites offer a quick software calculation, for example www.socscistatistics.com/tests/Default.aspx.

The more data you can use, the more confident you can be that your result shows a high positive relationship, a negative relationship or no relationship at all. Lastly you must use a 'significance levels table' to make sure your result is statistically significant, before putting on your geographer's 'hat' to explain it.

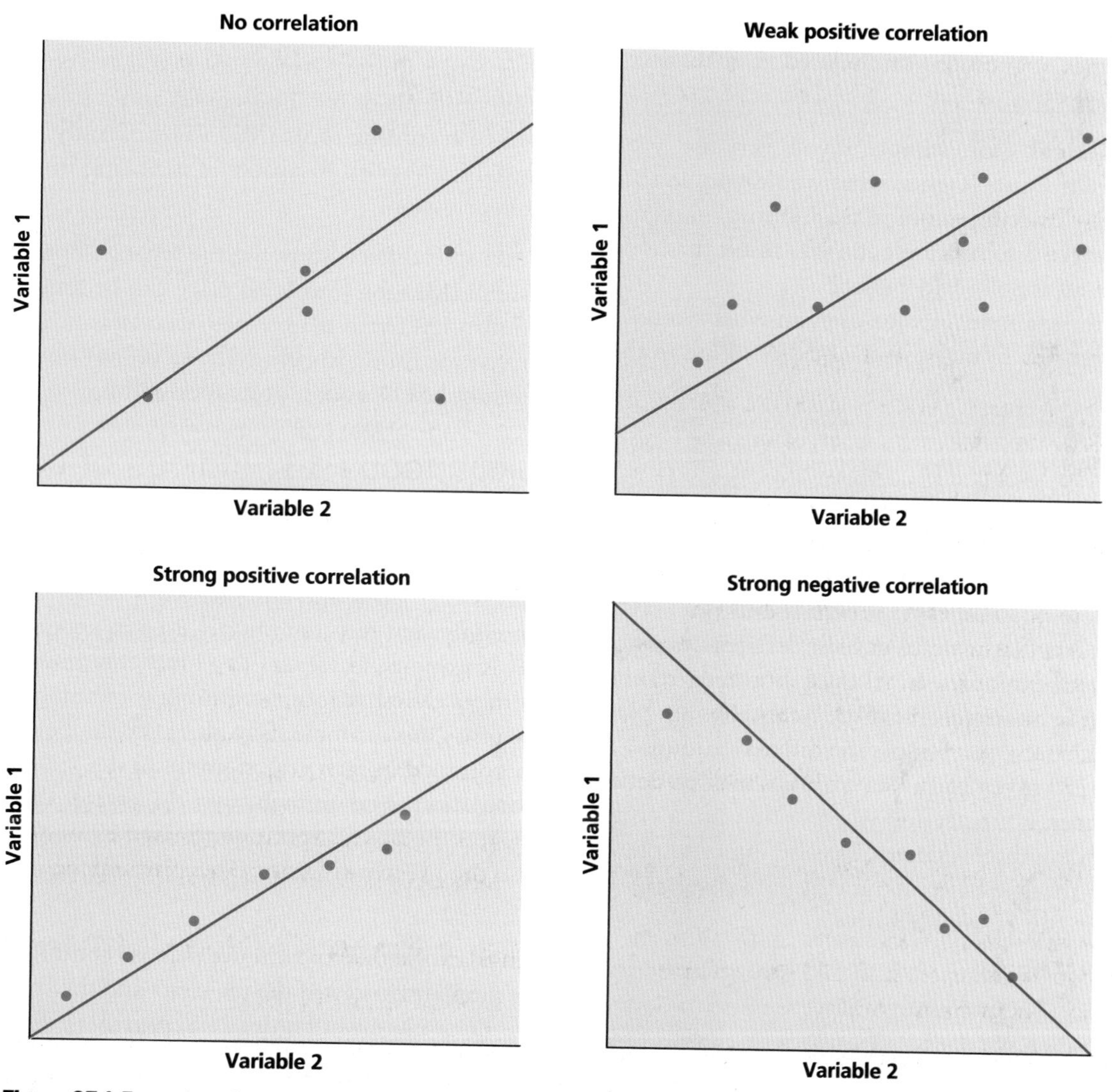

Figure SF.1 Examples of possible conclusions to be drawn from a scattergraph

Skills focus 9: Interviewing

Interviewing people is one way of obtaining information. In general, an interview is more structured than the 'free-flow' oral account. How you conduct an interview depends on:

- What you wish to investigate – if it is a specific issue, then you will need to investigate its roots or causal factors, conflicting views and possible solutions.
- Who you should interview to obtain relevant information or data – in the case of your issue, interviewees might include stakeholders, decision makers or a cross-section of the public. The choice of who to interview is critical.

Most interviews, particularly when dealing with different places, are likely to be of a qualitative kind, designed to extract and identify different viewpoints and perceptions rather than quantitative data. However, different types of questions can be used in qualitative interviews:

- **Hypothetical** – for example, if you were the Minister for Equal Rights, what would you do to stop ethnic discrimination in the housing market?
- **Provocative** – for example, do you think that local government is politically biased?
- **Ideal** – for example, in your opinion what would be the best way to improve the quality of life in this place?
- **Interpretive** – for example, what do you mean when you say this place is a good one in which to live?
- **Leading** – for example, do you think that crime prevention is better than punishing criminals?

None of these question types is perfect and you should be particularly aware of their deficiencies when analysing your responses. In all cases, however, results are likely to be more reliable if the interviews are stratified, whereby you sample the different groups living within a given place. Such groups may be defined by age, gender, ethnicity and whether they are a visitor or resident (length of residence may be important). Remember that questions about cultural values and employment can be very sensitive.

Whatever data is collected, it should give you a better sense of place, because sense of place is more to do with perceptions then statistics.

Skills focus 10: Index of Multiple Deprivation

The Index of Multiple Deprivation (IMD) is a set of measures used by the government to identify relative levels of deprivation in each administrative area. The index, first used in 2007, uses seven different aspects of deprivation (also known as domains). They are:

- income
- employment
- health and disability
- education, skills and training
- barriers to housing and services
- living environment
- crime.

The IMD was revised in 2010 and, in England, is now based on 38 separate indicators organised across the earlier seven domains. The country is divided into over 32,000 small areas or neighbourhoods called Lower-layer Super Output Areas (LSOAs). Each contains about 1500 residents or 650 households and is subsequently ranked according to its IMD value. Clearly, such a fine spatial framework is capable of detecting small pockets of deprivation.

Excel spreadsheets showing the separate domains and overall IMD for your local place can be found for all English LSOAs at http://apps.opendatacommunities.org/showcase/deprivation. If you do not know the relevant LSOA codes, postcodes will do.

Skills focus 11: Indices of cultural and ethnic diversity

Biologists have sophisticated statistical tests for assessing diversity in plant and animal communities. There are no equivalent tests commonly used in geographical investigations, however. The most common and perhaps easiest way of examining something like ethnic diversity, for example, is to calculate the percentage importance of each ethnic group in the make-up of a population. The more groups, the greater the diversity; the less the population is dominated by one group, so also the greater diversity. The same applies to cultural diversity.

The cartographic techniques that are commonly used to show diversity are the pie chart and the compound bar diagrams (as in Figure 21.5, page 311). The location uotient (LQ) is a valuable way of quantifying how oncentrated a particular industry, cluster, occupation r demographic group is in a place as compared to its vider region. It can reveal what makes a particular place ınique' in comparison to the national average. Values reater than 1 show where the incidence is higher than he average for the whole of the area under investigation.

Skills focus 12: Placecheck

Placecheck (www.placecheck.info) is a technique ncreasingly used in place investigations. Unfortunately, t is not mentioned in the Specification. However, the nethod is so relevant to the place focus of both options hat it really cannot be ignored.

Placecheck is a technique developed by Urban Design Skills (www.urbandesignskills.com), a federation of built nvironment professional bodies. It is used to assess the quality of places when drawing up strategies of change. t is widely used, but particularly by planners, to look t a place and to think about how to make it better. It epresents the first step in place improvement. One of ts attractions is that it can be used by anyone, without nuch preparation. It does not need an expensive team of experts.

Placecheck's simple idea is that much of what needs to be known about a place can be seen and understood by looking at it and posing three basic questions during a walkabout:

- What do you like about this place?
- What do you dislike about it?
- What needs changing?

For a more detailed walkabout, UDS has produced a checklist of some 21 questions that add detail to the basic three.

But a walkabout should also take into account the experiences of people who live, work or play there. So these same questions need to be put to a sample of the place's population.

The walkabouts should then be followed by a discussion of the information and perceptions that they turn up, and in turn by some serious thinking focused on the future and appropriate actions. A more detailed breakdown of the 21 questions includes a focus on what makes a place special or unique; how connected, accessible and welcoming it is; how safe and how planet friendly it is. Full details can be found at www.placecheck.info/what-is-a-placecheck.

23 Fieldwork

Understanding the nature of fieldwork, research and the enquiry process

By the end of this chapter you should:

- understand how fieldwork should be carefully planned and executed, following a procedural route to enquiry
- understand the theoretical frameworks that help to separate the different stages in the enquiry process, and how they might link to the fieldwork test
- be able to recognise that geographical fieldwork sometimes produces 'messy', complex and unpredictable outcomes that don't easily fit the expected result.

23.1 The route to enquiry and how fieldwork fits within the AS test

As part of your AS Geography course or first year A level course, you will have to undertake a range of fieldwork opportunities.

The geographical route to enquiry is at the centre of good fieldwork and research. It's essentially a recipe that is followed that allows whole process to be completed from start to finish. It should not be seen as a one-way operation, but moreover a series of linked individual stages that build together and allow reflection from one stage to the next. See Figure 23.1.

Key terms

Fieldwork: Any work carried out in the outdoors. All AS Geography students must undertake two days of fieldwork in contrasting locations.

Enquiry: The process of investigation (through a sequence of stages) to find a probable (or plausible) answer to a question(s) or aim(s) that has been developed.

Each stage is equally important, right from the initial research and context, through to development of the data collection procedure and the overall evaluation. Only at the end can you have an opportunity to reflect on what you have found and, importantly, its wider geographical significance.

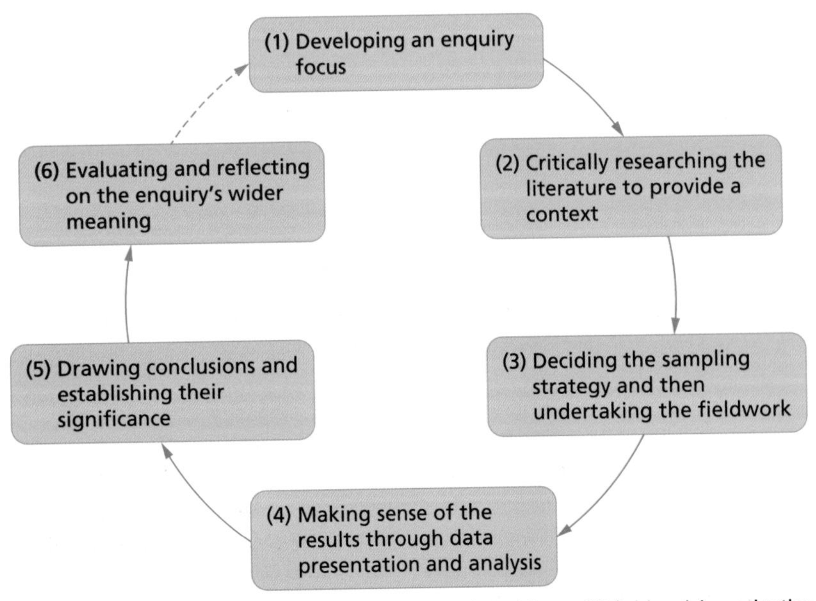

Figure 23.1 The route to enquiry – the planning 'pathway' for an AS fieldwork investigation

Understanding the fieldwork exam

Although you need to undertake a full enquiry process as part of the AS course, the exam for the AS is focused on four optional themes linked to:

1. Glacial landscapes and change (physical).
2. Coastal landscapes and change (physical).
3. Regenerating places (human).
4. Diverse places (human).

So the exam is really a simulation of fieldwork and research. You will be required to answer one question from the physical pair (1 and 2) and one question from the human pair (3 and 4). The fieldwork questions are derived

able 23.1 **Preparing for different types of questions in the exam**

	Unseen and unfamiliar 'ambush' data and information	Familiar data and information
Context	Data and information provided, unseen, in the exam. This is likely to be a new location but the context and approach may be familiar, for example equipment used.	Information and details of places, methods and outcomes, for example, that you will need to remember for the test.
Example lines of questioning	**1** Examine the conclusions that could be drawn from data X. **2** Study Figure X. Design an appropriate sampling framework to establish changes in gradient along the long profile of glacial valley B.	**1** Summarise the most important results and conclusions from your fieldwork investigation. **2** Evaluate your route to enquiry, from the initial research question design through to the development of your conclusions.

rom the topics within the specification content. The questions can be taken from any of the stages of the enquiry process, but probably not across all of them in any one paper. The fieldwork questions will assess:

- your knowledge and understanding of investigating geographical questions and issues
- your interpretation, analysis and evaluation of fieldwork data relating to an unseen fieldwork context
- your ability to construct arguments and draw conclusions in relation to your own fieldwork experience.

There is no need to write up a full piece of coursework for the test as you won't be assessed on it or be handing it in. More information on best practice in readiness for the exam can be found on page 359.

'Familiar' and 'unfamiliar' data and questions

A key part of the AS fieldwork test will be your ability to deal with *other* data presented to you (unfamiliar) as well as evaluating aspects of your own fieldwork and research. Table 23.1 illustrates the differences between these types of questions, and questions that might relate directly to your own personal experience of fieldwork and research.

Stage 1: Developing an enquiry question and focus

A good enquiry depends on having a good focus. A focus might actually be a question, aim, statement or testable hypothesis. See Table 23.2.

A good focus must be directly linked to the overall theme tested within the specification: glacial or coastal landscapes, and regenerating or diverse places. At A level however, you will be undertaking coursework which can have a much wider range of focus. This will be discussed in Book 2.

Figure 23.2 shows an example of a coastal landscape in southern England. Using this as a stimulus, you might begin to go about narrowing the focus into something more like: 'What impacts do different methods of coastal management have on coastal processes and ecosystems at Barton-on-Sea?'

But this is a broad focus – it needs breaking down into questions that are simpler, more workable and linked to the specification and route to enquiry. To carry out this investigation, you could subdivide the main question into a linked sequence of enquiry questions, for example:

Table 23.2 The different approaches to developing a geographical focus

	Aim	Question	Hypothesis
Definition	A statement of what you hope to achieve.	Something you will ask that links closely to the enquiry title (sometimes called a key question).	A testable idea or statement (not a question).
Examples	An investigation to examine the reasons for the changes in retail quality between two different economic zones of town X. An examination into the reasons for population change in rural community Y.	To what extent is there a linear relationship between pedestrian density and shop rateable value in urban area G? How and why does the shape and size of sediments change with distance along coastal stretch F?	There is a scalable relationship between drumlin height and width in lowland glacial landscape D. Spatial patterns of inequality in city R can be directly related to quality of life indices.

Figure 23.2 Coastal landscape at Barton-on-Sea

1. What is the fieldwork evidence and literature research for changing coastal sediments at Barton-on-Sea?
2. What are the changes along an ecosystem transect in the area around Barton-on-Sea?
3. What are the successes of coastal management?
4. What have been the effects of human intervention on coastal processes, communities and ecosystems?

It is important to design the enquiry questions in a logical order, so that each one builds logically into the next. In other words they build into a cohesive set of ideas linked to the overall aim or main question.

Exploring primary data and secondary data

Fieldwork data that you collect yourself (or as part of a group) is called *primary data* – first-hand information that comes from you and people you have worked with. There are many different types of primary data and information, but they are used as part of the evidence base to help reach conclusions and develop better geographical understanding.

Secondary data is information that someone else has collected, for example another person, group or even another organisation. Secondary data is very important in providing background information and a context for the enquiry – it may form part of the literature research for example.

There is an important difference between primary and secondary data, but in reality there is a blurring between the two types. Consider data that might be collected as raw and unprocessed census information, personal opinion on YouTube, or crowd sourced 'big data' from Twitter (see Figure 23.3). These examples don't easily fit in either category and are sometimes called tertiary or hybrid data.

Key term

Tertiary data: Usually refers to sources that collate and summarise other sources (in this respect mostly secondary). Wikipedia is a good example as it provides a summary of a topic with references to the sources that have been used. The internet has blurred the boundary between secondary and primary sources.

Figure 23.3 Primary or secondary? This data is raw: it has not been selected, analysed or pre-processed by another agency or individual

Secondary data can be grouped into three main types: numerical/statistical, graphical/cartographic and written/prose, as shown in Figure 23.4.

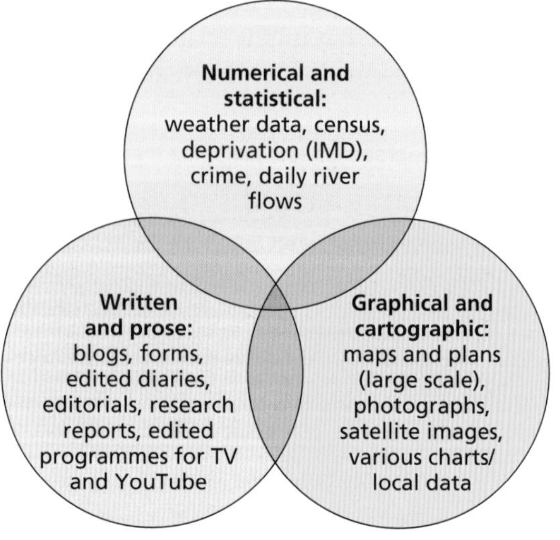

Figure 23.4 The three main types of secondary data

Stage 2: Critical research to understand the context of the enquiry

You are reminded in the specification that you will need to select and undertake informed and critical questioning of data sources and information. So, what does this mean?

'Geographical context' is the geographical background (facts, ideas, theories, models, assumptions, and so on) that may support the aims of the study. The introduction, which may be part of the test, will provide information and brief ideas about the geographical context. This often helps in an understanding of the 'bigger picture' ideas, and often gives a rationale for undertaking the study.

Becoming a geographical researcher

To find out more about the existing ideas that are important to your enquiry (for example, knowledge about any processes or issues, or about the location) you need to become an efficient researcher. Fortunately there is plenty of information available to help you in a variety of formats. Your job is to collect this information carefully and thoughtfully and, most importantly, to be selective. Wikipedia, for example, has a number of what it classifies as 'good articles' (about 1 in 217). It provides a useful framework for what can achieve the good article status. The following box is an adapted summary.

1. **Well written** with clear and concise prose; the spelling and grammar are correct.
2. **Verifiable** with no original research. It contains a list of references (sources of information) and all references (citations) are from reliable sources, including those for direct quotations, statistics, published opinion, counter-intuitive or controversial statements.
3. **Broad in its coverage** in that it addresses the main aspects of the topic and it stays focused without going into unnecessary detail.
4. **Neutral:** it represents viewpoints fairly and without bias, giving due weight to each.
5. **Stable:** it does not change significantly from day to day because of an ongoing content dispute.

You can read more about this at: https://en.wikipedia.org/wiki/Wikipedia:Good_article_criteria

In addition to the ideas in Figure 23.4, you should also consider these overarching ideas:

- The general quality of information and literature you will use to establish a context and how reliable it is. This will include an assessment of the information's provenance: who wrote and researched the article? Are they an interested member of the public or do they work as an academic in a university, or for a large government organisation? Remember that secondary data may contain errors, but they may not be obvious.
- Why does the material exist? For publicity, academic purposes or just general interest?
- How old is the material? If you need up-to-date statistics, go for the most up-to-date reports and opinions available.
- Are there any blogs or forums linked to your particular enquiry? These are widely available and can be used to give a 'profile' of an area, place or issue. This information is often tertiary data and also qualitative.

You will need to carefully document and record researched literature sources (including those from the internet) so that they can be referenced and cited in an exam if required. It would be advisable to practise by using a standard Harvard (name and date) system, for example:

Rushe, D. (2011) Vegas: city of glitter and gloom. *The Guardian*, 18th October, p.29.

Stage 3: Deciding the sampling strategy – frequency and timing of observations

This is one of the most challenging and technical aspects of fieldwork and research, but is often overlooked by

Key term

Qualitative: Refers to descriptive, non-numerical data, for example newspaper articles and interview transcripts, videos, oral histories and recordings.

Key terms

Sample (size): Refers to the number of observations or data points that make up a survey or data set. Very small sample sizes will not reflect the statistical population closely; this makes them unreliable and can lead to incorrect interpretations and explanations. Large samples are much more resource heavy and can take considerable effort and energy to do well.

Quantitative: Numerical data that could include environmental measurements and closed questions (with a rating scale) from a questionnaire. The advantages and disadvantages of these two types of data are shown in Table 23.3.

students. The key concern here is trying to establish a reliable sample.

Sampling (including frequency and pattern of locations to collect data) must be carefully considered. Generally you should start by:

1 defining or delimiting the study area, then
2 designing an appropriate sampling framework (see Figure 23.5; you may choose point, line or area), then
3 determine the sample size and number of sample points – this will determine the frequency of observations, and finally
4 think about the timing of observations (for example, over a few minutes, hourly or monthly). This includes both the time to collect the sample (for example over five minutes) as well as the (time) gap between observations.

Stage 4: Selecting appropriate fieldwork techniques

Fieldwork data collection is the natural follow-on stage from having established a robust sampling framework. This is the 'nuts and bolts' of the actual techniques, rather than where the techniques take place. Of course, this should have already been considered and established. It may be important to be able to *justify* the following in relation to data collection procedures, especially in the context of the AS test format:

- the materials, recording sheets (design) and equipment utilised
- precautions and details of any procedures taken to ensure accuracy, for example repeat sampling/measurements, or handling equipment in a particular way
- the names and types of any alternative equipment/approaches that could be used to collect data with a similar outcome.

You will be using a wide range of field techniques and approaches. It will be important to keep a good record of each technique and how it connected to particular geographical aim. It would also be advisable to make a note of whether it was a qualitative or quantitative technique and the relative advantages and disadvantages of those individual techniques (see Table 23.3).

	Systematic	Stratified	Random
Description	Taking samples at regular known distances, e.g. every fourth step, or at the points of a regular grid over an area. Used when there is an expected change between two locations. Often used along transects.	Selecting a sample to take account of something known about the area or about the people being surveyed, e.g. number of males and females in a town. The adjustment makes the sample fairer and more representative.	Selecting a sample by chance, usually based on published random number tables. This avoids subjectivity and bias in the selection process. Used when the environment or population is expected to be similar everywhere.
Level of difficulty	Straightforward – although interval needs to be determined	More complicated because information about the location is needed to select the sites.	Need random number tables to do it properly.
Visual example	Surveys taken at regular points along a transect (sampling) line. Could also be used for an area.	80% area/ population = 16/20 samples; 20% area/ population = 4/20 samples	Random location within an area

Figure 23.5 Different approaches to sampling: systematic, stratified, random

Table 23.3 Comparing the strengths and weaknesses of qualitative and quantitative data

	Strengths	Weaknesses
Qualitative	People's views and opinions 'flesh out' otherwise dry numerical data Can suggest new research directions, ideas and causal relationships	Can take a long time to collect Analysis can be difficult and conclusions weak Data is subjective and may not be reliable
Quantitative	Precise, numerical data Reliable data, as a result of sampling and collection design Can be analysed statistically Collection can be replicated	Poor collection methods can lead to spurious conclusions Reduces complex opinions and views to numbers Complex analysis produces simplistic black or white conclusions

Skills focus: Technique customisation

There are a number of 'off-the-shelf' tools and tool kits that can be used to measure and record aspects of different environments. In some instances you may be able to use a pre-designed template (for example, an environmental quality recording sheet – there are many examples on the internet and in books ready to be used). Alternatively, you may want to adapt and customise a pre-published one, putting you own stamp on it and making it more fit for purpose. In other instances you may want to devise your own recording or booking sheets from scratch. This would be the case if you had an idea, for example, about an innovative or new approach to recording something in the built or physical environment.

It is also recommended that site descriptions are used; they assist in the recall of particular places and the linked landscape or built environment features.

A fieldwork notebook

The point of a fieldwork notebook is simple. It is to create written documentation of the *geographical* events that were important on a field trip that can be used for future reference, either as part of your analysis and evaluation or simply to preserve a longer-term record of the field trip. It is *not* a recording sheet however. In other words, a fieldwork notebook is not the place to merely write down all of your primary data. It's more sophisticated than that. Remember that you can record field notes orally, for example using a voice-recording app on your phone (Figure 23.6), ready for transcription later.

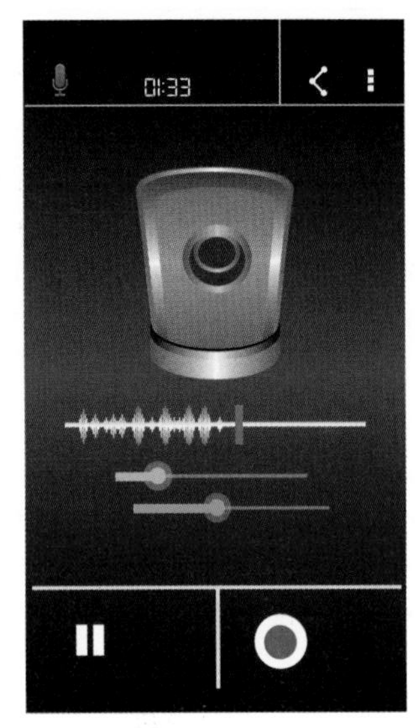

Figure 23.6 An example of a voice-recording app

Stage 5: Making sense of the results through data presentation and analysis

The initial process with any qualitative or numerical data in its raw, unprocessed form is to display and visualise the data. This means using appropriate graphs, diagrams, maps and photographs, for example. ICT is often used at this stage to assist and quicken the procedures of data display. With a quantitative set of data, for instance, there are a number of questions that can be initially asked about any data set:

- What is the range (or spread) of values within the data set?
- Where are most of the values concentrated (is there any clustering)?
- Are there any clear gaps between the concentrations?
- What is the shape of the distribution of values?
- Are there any extreme values (which may include anomalies and/or outliers) and how far separated are these from the normal range of data?

Being able to use the correct language to describe patterns in data (that is, making sense of your findings) is crucial in the context of the fieldwork assessment.

Key terms

Mode: Strictly speaking, this is the number in the data set that appears most frequently. If a data set is organised into groups or classes it can displayed as bars, where the highest bar indicates the modal class or category. It can be a useful indicator to see where most numbers are concentrated, but remember that some data sets can be 'bimodal', that is, have two modes.

Mean: The arithmetic or mathematical average of the values in the sample. It is the most commonly-used measure of central tendency. Its advantage is that all values are taken into account; however, the mean can be influenced by outliers or extreme values. To calculate the mean you need to add all the values together and then divide by the number of values.

Median: Divides the data set into two halves. If you had 41 observations, for instance, the median would rank 21 in the list; if you had an even number of observations, say 40, then you would normally use the mid-point between points 20 and 21.

For example, you should be able to write about modes, means and medians.

Sometimes the visualisation of data, which may form part of the analysis, can be very useful in terms of helping to understand differences and patterns. In Figure 23.7, for instance, two sets of population data have been matched on top of each other.

The following observations and interpretations can be made:

- The light blue population has a larger range and the data is more spread out.
- The dark blue population has a higher minimum value, lower quartile, median, upper quartile and maximum.
- There is some overlap, but all dark blue values are contained in the top 50 per cent of light blue values (since all are above the light blue median).

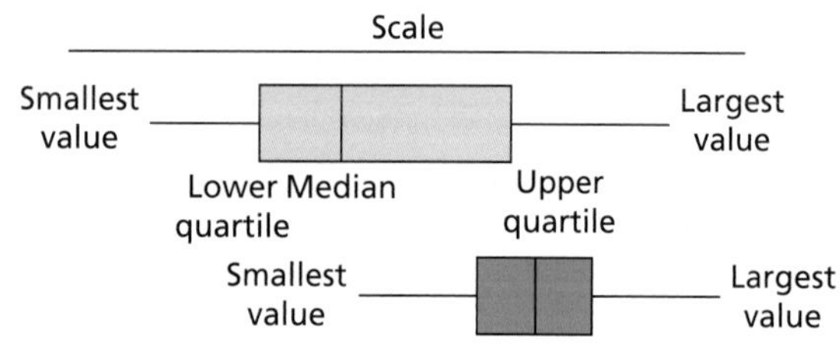

Figure 23.7 Two sets of matched population data

Skills focus: Analysis of qualitative data and information

Softer analysis techniques may be used, such as highlighting of text, coding and annotation of maps and published resources. These are no less sophisticated than statistical analysis, they are just different. Often they can also focus on understanding interaction, that is, how different parts of a system for example link to each other – a way of 'getting behind the numbers'. Analysis of photographs, for example, is a key qualitative analysis research tool yet it is often undervalued by students. When conducted appropriately and with care, analysis of images can reveal important aspects of place, identity and representation that should not be overlooked.

- Dark blue is more symmetrical, with the median being halfway between quartiles, and between minimum and maximum values.

Similarly, you should be able to describe and justify the range of procedures that were used to make sense of any qualitative data.

Stage 6: Drawing conclusions and linking understanding of concepts

Conclusions should do the following:

- Summarise the main geographical outcomes that have been discovered including any evidence (qualitative and quantitative) that backs them up.
- Relate findings back to the aims/focus of the investigation. Has there been an answer to the focus and what does it mean in relation to the literature research?
- Establish geographical links between factors and processes that have been uncovered as part of the fieldwork and research, indicating where there are weak and strong linkages.
- Accept that geography can produce 'messy' outcomes in which the results reveal a complex, unfamiliar and sometimes unpredictable pattern.
- Begin to comment on the geographical significance of your study and whether the results challenge similar studies or are, in fact, in broad agreement.

At all times conclusions must be brief, factual and to the point. This is important when it comes to linking them to a question in the final assessment.

Key terms

Reliability: The consistency or reproducibility of a measurement. A measurement is said to have a high reliability if it produces consistent results under consistent conditions. True reliability cannot be calculated – it can only be estimated based on knowledge and understanding of the topic.

Accuracy: In geography the accuracy of a measurement system/data is the degree of closeness of measurements of a quantity to that quantity's actual (true/real) value. The farther a measurement is from its expected value, the less accurate it is.

Validity: The extent to which a test actually measures what it claims to measure. In other words, the extent to which a concept, conclusion or measurement is well-founded and corresponds accurately to real-world geography. This is perhaps the most important measure of 'trueness'.

Stage 7: Critically reflecting on results and making links to their wider geographical significance

The final part of the route to enquiry is one of reflection. Critical reflection is much more than a list of what went well (or didn't). It's an opportunity to review the whole enquiry process, which may include reference to the fieldwork results as well as the literature research information. Importantly, this stage represents an opportunity to comment on the reliability, accuracy and validity of results; in other words, how much they can be trusted under different circumstances.

Part of Stage 7 may include a brief analysis of how the process could be improved, for example in terms of repeating particular procedures or using different techniques, including those linked to data presentation and analysis.

Try to avoid simplistic and obvious statements in the fieldwork exam (which can sometimes read like excuses), for example:

- 'The work I did on coasts went really very well, although the weather meant that some of my field sketches were difficult to do and the paper got wet.'
- 'I think we worked well as a group, and I got the data; the ruler was good at measuring the length of the pebbles.'

Instead, try to think carefully about any possible sources of error that may have been introduced through the equipment, sampling or operator limitations.

Preparation for the fieldwork assessment

Remember that the AS exam will be asking questions across a range of stages, and both familiar and unfamiliar materials, so there is a requirement to get all your notes 'test ready' (but *not* to write everything up). There are several preparatory steps that can be undertaken. The following is a list of ideas that you can try out as either individuals, in small groups or as a class:

1. Consider doing a virtual fieldtrip, for example using Google Street View to revisit places and locations and remind yourself of the context, processes and possible issues. Make sure you are able to demonstrate a good depth of knowledge about the locations in the test.
2. Keeping a good fieldwork notebook takes practise but should pay dividends. For instance, it can be used as a key revision tool before a fieldwork exam, helping you to remember the places and sites and exactly what you did.
3. Work in groups to create concise notes on the individual stages of the route to enquiry (no more than one side of A4). Work collaboratively so that these can be shared (work should be split between groups).
4. Create a prepared list of the types of questions that could be asked based on familiar aspects of the fieldwork – see Table 23.4.
5. Work together to develop a specialist fieldwork glossary with words and phrases that can be useful for the test.
6. Prepare a 'stats' briefing sheet that justifies as well as describes the procedures used for statistical operations linked to your quantitative data.

Table 23.4 Examples of familiar questions that should be prepared in readiness for the test

Explain **two** reasons why the location chosen was suitable for the investigation.
Assess the usefulness of **two** primary fieldwork methods used to collect data.
Describe how the fieldwork data was collated and analysed using statistics and other methods.
Evaluate the reliability and usefulness of the secondary research sources you used.
Explain why you chose **one** particular method to present your data.
What conclusions can be drawn from the results of the fieldwork and research?

Review questions

1 Explain why there are several stages in the route to enquiry.

2 Why is there no 'magic number' in terms of how many samples should be taken?

3 Outline the advantages and disadvantages of the three different approaches to sampling given in Figure 23.5.

4 Explain how fieldwork and literature research are likely to be tested in the AS exam.

5 Explain how you should go about developing appropriate field research questions.

6 Explain the difference between a 'good' and 'bad' sample in terms of reliability and validity.

7 Describe the range of strategies that can be used to prepare for the AS assessment (including 'familiar' and 'unfamiliar' contexts).

Further research

Look at the online GIS map that shows 2011 local census information for England and suggest how it can be used to help plan a specific geographical question: http://datashine.org.uk

Visit the website of the Royal Geographical Society and the click on 'Our work', 'Schools and education', then 'Fieldwork and local learning', and review the various fieldwork techniques that are on offer: www.rgs.org

Study this global interactive map of air monitoring stations and consider how you could use this type of information to design an enquiry: http://aqicn.org

Research how GIS can be used to help analyse quantitative geographical data and information.

GLOSSARY

Absolute poverty: When a person's income is too low for basic human needs to be met, potentially resulting in hunger and homelessness.

Accretion: This occurs when sediment is added to a landform, such as a river delta, by deposition. It can build up to form new land, allowing a delta to grow out to sea. It tends to be balanced by subsidence, caused by the weight of the newly deposited sediment.

Accuracy: In geography the accuracy of a measurement system/data is the degree of closeness of measurements of a quantity to that quantity's actual (true/real) value. The farther a measurement is from its expected value, the less accurate it is.

Active layer: The top layer of soil in permafrost environments that thaws during summer and freezes during winter.

Albedo: The reflective coefficient of a surface, i.e. the proportion of incident radiation reflected by a surface (very high in the case of snow or ice).

Area Based Initiatives: ABIs aim to improve selected people or places within a specific location and include educational attainment, enhancing crime prevention and reducing unemployment.

Assimilation: The process by which people of diverse ethnic and cultural backgrounds come to interact and intermix, free of constraints, in the life of the larger community or nation.

Backwash: When water runs back down the beach to meet the next incoming wave.

Barrier islands: Offshore sediment bars, usually sand dune covered but, unlike spits, they are not attached to the coast. They are found between 500 m and 30 km offshore and can be tens of kilometres long.

Baseline data: The information used to compare present-day characteristics with, for example, past land-use maps, photographs and statistics.

Beach morphology: The shape of a beach, including its width and slope (the beach profile) and features such as berms, ridges and runnels. It also includes the type of sediment (shingle, sand, mud) found at different locations on the beach.

Benchmark glacier: A designated glacier in which the accumulation and ablation are measured annually using standardised techniques to monitor the impacts of climate change. USGS, for example, studies five such glaciers currently with more chosen for the future to reflect a variety of locational scenarios.

Benefit–cost ratios: The balance between investment and outcomes; a positive ratio is desirable.

Biodiversity: A measure of the variety of organisms present at a particular location.

Blow hole: Forms when a coastal cave turns upwards and breaks through the flat cliff top. Usually this is because of erosion of especially weak strata or the presence of a fault line.

BRICS group: The four large, fast-growing economies of Brazil, Russia, India and China, recently joined at their annual summit meeting by South Africa too.

Brownfield site: Abandoned or derelict urban land previously used by commercial or industrial companies.

Calving: The breaking up of chunks of ice at the glacier snout or ice sheet front to form icebergs as the glacier reaches a lake or the ocean.

Carbon footprint: The amount of carbon dioxide produced by an individual or activity.

Carbon sink: A natural or artificial reservoir that absorbs more carbon than it releases, leading to carbon accumulation.

Catalyst: The method used or event that starts a regeneration scheme, such as the building of a new shopping mall, leisure facility, creation of a country park or holding an event.

Catena: A connected series of related features, such features formed by periglacial processes which change down a slope.

Centripetal migration: Movement of people directed towards the centre of urban areas.

Characteristics: The physical and human aspects that help distinguish one place from another: location, natural features, layout, land use, architecture and cultural traits.

Cliff profile: The height and angle of a cliff face as well as its features, such as wave-cut notches or changes in slope angle.

Coastal accretion: The deposition of sediment at the coast and the seaward growth of the coastline, creating new land. It often involves sediment deposition being stabilised by vegetation.

Commuter villages: Settlements that have a proportion of their population living in them but who commute out daily or weekly, usually to larger settlements either nearby or further afield.

Conflict: In the context of coastal management, conflict means disagreement over how the coast should be protected from threats and which areas should be protected. Conflict often exists between different stakeholders, such as residents versus the local council.

Connections: Any type of physical or online linkages between places. Places may keep some of their characteristics or change them as a result.

Consumer society: A society in which the buying and selling of goods and services is the most important social and economic activity.

Counter-urbanisation: The movement of people and employment from major cities to smaller settlements and rural places located beyond the city, or to more distant, smaller cities and towns.

Crude birth rate: The number of live births per 1000 people per year.

Cryosphere: The parts of the Earth's crust and atmosphere subject to temperatures below 0 °C for at least part of each year.

Cultural change: The modification of a society through innovation, invention, discovery or contact with other societies.

Cultural imperialism: The practice of promoting the culture/language of one nation over another. It is usually the case that the former is a large, economically or militarily powerful nation and the latter is a smaller, less affluent one.

Cultural landscape: The landscape of a place that has been shaped over time in characteristic ways by the combined action of natural and human processes.

Cultural traits: Culture can be broken down into individual component parts, such as the clothing people wear or their language. Each component is called a 'cultural trait'.

Culture: This is the way of life, especially the general customs, values and beliefs, of a particular group of people that are passed on from one generation to the next.

Currents: Flows of seawater in a particular direction driven by winds or differences in water density, salinity or temperature. Some are almost continuous, such as those that form the global thermohaline circulation, and others are more sporadic, such as longshore currents, while some last only for a few hours, such as rip currents.

Deindustrialisation: The decline of regionally important manufacturing industries. The decline can be charted either in terms of workforce numbers or output and production measures.

Demographic: Of or relating to some aspect of a population, for example its size, rate of change, density and composition.

Demographic transition: A model representing changing rates of fertility and mortality over time, their changing balances and their net effect on rates of population growth.

Development: Development is linked to an improving society, enabling people to achieve their aspirations. It includes the provision of social services, acquisition of economic assets, improved productivity and reducing vulnerability to natural disasters. Low levels of development are closely associated with high levels of risk and vulnerability to natural disasters.

Diaspora: The dispersion or spread of a group of people from their original homeland.

Disaster hotspot: A country or area that is extremely disaster prone for a number of reasons, as shown in Figure 3.6.

Disaster: The realisation of a hazard, when it 'causes a significant impact on a vulnerable population' (Degg). The Centre for Research on the Epidemiology of Disasters (CRED) states that a hazard becomes as disaster when: 10 or more people are killed, and/or; 100 or more people are affected.

Dissipation: The term used to describe how the energy of a wave is decreased by friction with beach material during the wave swash up the beach. A wide beach slows waves down and saps their energy so, when they break, most energy has gone.

Dredging: Involves scooping or sucking sediment up from the seabed or a river bed, usually for construction sand or gravel, or to deepen a channel so that large boats can navigate it.

Dynamic equilibrium: The balanced state of a system when inputs and outputs balance over time. If one element of the system changes because of an outside influence, the internal equilibrium of the system is upset and other components of the system change. By a process of feedback, the system adjusts to the change and the equilibrium is regained.

Ecological footprint: A crude measurement of the area of land or water required to provide a person (or society) with the energy, food and resources needed to live, and to also absorb waste.

Economic migrant: A migrant whose primary motivation is to seek employment. Migrants who already had a job may have set off in search of better pay, more regular pay, promotion or a change of career.

Emerging economies: Countries that have begun to experience high rates of economic growth, usually due to rapid factory expansion and industrialisation. There are numerous sub-groups of emerging economies, including Brazil, Russia, India, China and South Africa (the 'BRICS' group). They are sometimes called newly industrialised countries.

Enclave: In the present context, an enclave is a group of people surrounded by a group or groups of entirely different people.

Englacial: Debris transported inside the glacier.

Enquiry: The process of investigation (through a sequence of stages) to find a probable (or plausible) answer to a question(s) or aim(s) that has been developed.

Environmental lapse rate: The rate of atmospheric temperature decreases with altitude at a given time and location.

Environmental refugees: Communities forced to abandon their homes due to natural processes including sudden ones, such as landslides, or gradual ones, such as erosion or rising sea levels.

Epicentre: the location on the Earth's surface that is directly above the earthquake focus, i.e. the point where an earthquake originates.

Equilibrium point: Where losses from ablation are balanced by gains from accumulation in a glacier.

Ethical purchase: A financial exchange where the consumer has considered the social and environmental costs of production for food, goods or services purchased.

Eustatic change: Involves a rise or fall in water level caused by a change in the volume of water. This is a *global* change, affecting all the world's connected seas and oceans.

Extrusion flow: The theory that glacier ice flows faster at depth.

Faults: Major fractures in rocks produced by tectonic forces and involving the displacement of rocks on either side of the fault line.

Fetch: The uninterrupted distance across water over which a wind blows, and therefore the distance waves have to grow in size.

Fieldwork: Any work carried out in the outdoors. All AS Geography students must undertake two days of fieldwork in contrasting locations.

Flagship regeneration projects: Large-scale, prestigious projects, often using bold 'signature architecture'. The hope is to generate a positive spin in a place.

Food miles: The distance food travels from a farm to the consumer. The journey may be short and direct for some local produce, or may take longer, with food often crossing entire continents via a string of depots.

Foreign direct investment: A financial injection made by a TNC into a nation's economy, either to build new facilities (factories or shops) or to acquire, or merge with, an existing firm already based there.

Functions: The roles a place plays for its community and surroundings. Some, usually larger places, offer regional, national or even global functions. Functions may grow, disappear and change over time. There is a hierarchy according to size and number of functions.

Gated communities: Found in urban and rural settlements as either individual buildings or groups of houses. They are landscapes of surveillance, with CCTV and often 24/7 security guards. They are designed to deter access by unknown people and reduce crime.

Glacial surge: This occurs where flow instabilities result in dramatic increases in glacier velocity.

Glacials: Cold, ice-house periods within the Pleistocene.

Glasgow effect: The impacts of poor health linked to deprivation.

Global production network: A chain of connected suppliers of parts and materials that contribute to the manufacturing or assembly of consumer goods. The network serves the needs of a TNC, such as Apple or Tesco.

Governance: 'The sum of the many ways individuals and institutions, public and private, manage their common affairs. It is a continuing process through which conflicting or diverse interests may be accommodated and co-operative action may be taken. It includes formal institutions and regimes empowered to enforce compliance, as well as informal arrangements that people and institutions have either agreed to or perceive to be in their interest.' *(The Commission on Global Governance, 1995)*

Greenhouse conditions: Much warmer interglacial conditions.

Gross domestic product: A measure of the financial value of goods and services produced within a territory (including foreign firms located there). It is often divided by population size to produce a per capita figure for the purpose of making comparisons.

Gross value added: Measures the contribution to the economy of each individual producer, industry or sector. It is used in calculating GDP.

Hazard: 'A perceived natural/geophysical event that has the potential to threaten both life and property' (Whittow). Yet a geophysical hazard event would not be such without, for example, people at or near its location. That is to say, earthquakes would not be hazards if people did not live in buildings that collapse as a result of ground shaking. Many hazards occur at the interface between natural and human systems.

Holocene: The geological epoch that began about 12,000 years ago at the end of the last Pleistocene ice age. Its early stages were marked by large sea level rises of about 35 m and a warming interglacial climate.

Hypocentre is the 'focus' point within the ground where the strain energy of the earthquake stored in the rock is first released. The distance between this and the epicentre on the surface is called focal length.

Ice-house conditions: Very cold glacial conditions.

Inequality: Usually refers to an unfair situation or distribution of assets and resources. It may also be used when people, nations and non-state players (ranging from transnational corporations to international agencies) have different levels of authority, competence and outcomes.

Informal sector: Unofficial forms of employment that are not easily made subject to government regulation or taxation.

Infrastructure: The basic physical systems of a country: economic infrastructure includes highways, energy distribution, water and sewerage facilities, and telecommunication networks; social infrastructure includes public housing, hospitals, schools and universities.

Intensity: A measure of the ground shaking. It is the ground shaking that causes building damage and collapse, and the loss of life from the hazard.

Interdependency: If two places become overreliant on financial and/or political connections with one another, then they have become interdependent. For example if an economic recession adversely affects a host country for migrant workers, then the economy of the source country may shrink too, due to falling remittances.

Interglacials: Warmer periods similar to the present, i.e. greenhouse periods.

Intermodal containers: Large-capacity storage units which can be transported long distances using multiple types of transport, such as shipping and rail, without the freight being taken out of the container.

Internal migrant: Someone who moves from place to place inside the borders of a country. Globally, most internal migrants move from rural to urban areas (rural–urban migrants). In the developed world, however, people also move from urban to rural areas too (a process called counter-urbanisation).

Internal migration: The movement of population within a country, as distinct from the movement of people between countries (international migration).

International migration (immigration and emigration): The movement of people between countries. Immigration refers to the arrival of people from other countries; emigration refers to the departure of people to other countries.

Intervening obstacles: Barriers to a migrant such as a political border or physical feature (deserts, mountains and rivers).

Intra-plate earthquakes: These occur in the middle or interior of tectonic plates and are much rarer than inter-plate (convergent) boundary types.

Isostatic change: A *local* rise or fall in land level.

Issue: This is generally defined as an important topic or problem for debate or discussion. Examples would include immigration, racism and climate change. In this chapter the emphasis is on problems rather than topic issues, and on finding solutions rather than debate or discussion.

Lahar: A Javanese word that describes a mixture of water, mud and rock fragments flowing down the slopes of a volcano.

Least developed countries: The world's very poorest low-income nations, whose populations have little experience of globalisation. A number of these nations are described as 'failed states' by politicians, for example Somalia and South Sudan.

Life expectancy: The average number of years from birth that a person born in a particular year can expect to live. In developed countries, women enjoy greater life expectancy than men by a margin of a few years.

Lithosphere: The surface layer of the Earth is a rigid outer shell composed of the crust and upper mantle. It is on average 100 km deep. The lithosphere is always moving, but very slowly, fuelled by rising heat from the mantle which creates convection currents. The distinction between lithosphere and asthenosphere is one of physical strength rather than a difference in physical composition. The lithosphere is broken into huge sections, which are the tectonic plates.

Littoral cells: All coastlines divide up into distinct littoral cells containing sediment sources, transport paths and sinks. Each littoral cell is isolated from adjacent cells and can be managed as a holistic unit.

Littoral zone: The wider coastal zone including adjacent land areas and shallow parts of the sea just offshore.

Lived experience: The actual experience of living in a particular place or environment. Such experience can have a profound impact on a person's perceptions and values, as well as on their general development and their outlook on the world.

Living space: In a narrow sense, the term living space refers to land given over to housing. In its broader sense, the term embraces all that space given over to the day-to-day needs of a population, from work, shopping and leisure to education, healthcare and entertainment. Housing will certainly be in focus for much of this chapter, but the broader interpretation of living space will prevail.

Location quotient: A mapable ratio which helps show specialisation in any data distribution being studied. A figure equal to or close to 1.00 suggests national and local patterns are similar with no particular specialisation, such as retailing. LQs over 1 show a concentration of that type of employment locally.

Locked fault: A fault that is not slipping because the frictional resistance on the fault is greater than the shear stress across the fault, that is, it is stuck. Such faults may store strain for extended periods that is eventually released in a large magnitude earthquake when the frictional resistance is eventually overcome. The 2004 Indian Ocean tsunami was the result of a mega-thrust locked fault (subducting Indian Plate) with strain building up at around 20 mm per year. It generated huge seismic waves and the devastating tsunami.

Loess: A wind-blown deposit of fine-grained silt or clay in glacial conditions.

Love waves or L waves are surface waves with the vibration occurring in the horizontal plain. They have a high amplitude.

Magnitude: The magnitude of an earthquake is related to the amount of movement, or displacement, in the fault, which is in turn a measure of energy release. The 2004 earthquake in Indonesia was very large (M = 9.3) because a large vertical displacement (15 m) occurred along a very long fault distance, approximately 1500 km. (Earthquake magnitude is measured at the epicentre, the point on the Earth's surface directly above the hypocentre.)

Mass movement: The downslope movement of rock and soil; it is an umbrella term for a wide range of specific movements including landslide, rockfall and rotational slide.

Mean: The arithmetic or mathematical average of the values in the sample. It is the most commonly-used measure of central tendency. Its advantage is that all values are taken into account; however, the mean can be influenced by outliers or extreme values. To calculate the mean you need to add all the values together and then divide by the number of values.

Median: Divides the data set into two halves. If you had 41 observations, for instance, the median would rank 21 in the list; if you had an even number of observations, say 40, then you would normally use the mid-point between points 20 and 21.

Megaproject: A very expensive (over US$1 billion), technically difficult and usually long-term engineering project. Many megaprojects have multiple aims and often large environmental impacts.

Millennium Development Goals: Eight specific objectives for the global community created at the UN Millennium Summit in New York in 2000.

Mode: Strictly speaking, this is the number in the data set that appears most frequently. If a data set is organised into groups or classes it can displayed as bars, where the highest bar indicates the modal class or category. It can be a useful indicator to see where most numbers are concentrated, but remember that some data sets can be 'bimodal', that is, have two modes.

Multiculturalism: The co-existence of different cultural groups. A sharing of living space by people drawn from different cultural backgrounds.

Nationalist: A political movement focused on national independence or the abandonment of policies that are viewed by some people as a threat to national sovereignty or national culture.

Natural change: The outcome of the balance between births (birth rates, fertility) and deaths (death rates, mortality) in a population. Natural increase occurs when births exceed deaths; natural decrease occurs when deaths exceed births.

Natural increase: The difference between a society's crude birth rate and crude death rate. A migrant population, such as that found in developing world megacities, usually has a high rate of natural increase due to the presence of a large proportion of fertile young adults and relatively few older people reaching the end of their lives.

Natural resources: A material source of wealth, such as timber, fresh water, or a mineral deposit, that occurs in a natural state and has economic value. Natural resources may be renewable (sustainably managed forests, wind power and solar energy) or non-renewable (fossil fuels).

Net migration: The overall balance between immigration and emigration.

Névé (or firn): Crystalline or granular snow, especially on the upper part of a glacier, where it has not yet been compressed into ice.

Offshoring: TNCs move parts of their own production process (factories or offices) to other countries to reduce labour or other costs.

Orbital/astronomic forcing: A mechanism that alters the global energy balance and forces the climate to change in response.

Outflanking: Occurs when erosion gets behind coastal defences at the point where they stop, leading to rapid erosion inland and undermining of defences.

Outsourcing: TNCs contract another company to produce the goods and services they need rather than do it themselves. This can result in the growth of complex supply chains.

Paleomagnetism results from the zone of magma 'locking in' or 'striking' the Earth's magnetic polarity when it cools. Scientists can use this tool to determine historic periods of large-scale tectonic activity through the reconstruction of relative plate motions. They create a geo-timeline. The approach was used to provide evidence for Wegener's 1912 theory of plate tectonics.

Paraglacial: Rapidly changing landscapes which were once periglacial or glacial, but are moving towards not-glacial conditions.

Perception: An individual's or group's 'picture' of reality resulting from their assessment of information received.

Phenology: The study of the timing of natural events and phenomena, such as the first day snowdrops appear, in relation to climate.

Place: Geographical spaces shaped by individuals and communities over time.

Plate tectonics: A theory developed more than 60 years ago to explain the large-scale movements of the lithosphere (the upper surface of the Earth). It was based around the evidence from sea floor spreading and ocean topography, marine magnetic anomalies and paleomagnetism and geomagnetic field reversals. A knowledge of Earth's interior and outer structure is essential for understanding plate tectonics (see Figure 1.6).

Pleistocene: A geological period from about 2 million years ago to 11,700 years ago, the early part of the quaternary which included the most recent ice age.

Population density: The number of people per unit area (usually per km^2); i.e. the total population of a given area (country, region or city) divided by its area.

Pore water pressure: The pressure water experiences at a particular point below the water table due to the weight of water above it.

Post-accession migration: The flow of economic migrants after a country has joined the EU.

Postcode lottery: This refers to the uneven distribution of personal health and health services nationally and locally, especially in mental health, early diagnosis of cancer and emergency care for the elderly.

Post-colonial migrants: People who moved to the UK from former colonies of the British Empire during the 1950s, 1960s and 1970s, including economic migrants from the Caribbean (especially Jamaica), India, Pakistan, Bangladesh and Uganda.

Post-glacial isostatic adjustment: Refers to the uplift experienced by land following the removal of the weight of ice sheets. It is sometimes called post-glacial rebound or post-glacial re-adjustment.

Poverty: Poverty is relative to the place and time people live in. The poverty threshold used in the UK is households with an income of less than 60 per cent of the national median, after housing costs are included.

Primary or P waves are vibrations caused by compression, like a shunt through a line of connected train carriages. They spread quickly from the fault at a rate of about 8 km/sec.

Primary productivity: The rate at which energy is converted by photosynthesis; it has a major influence on the level of biodiversity.

Qualitative: Refers to descriptive, non-numerical data, for example newspaper articles and interview transcripts and recordings.

Quality of life: The level of social and economic well-being experienced by individuals or communities measured by various indicators including health, happiness, educational achievement, income and leisure time. It is a wider concept than 'standard of living', which is centred on just income.

Quantitative: Numerical data that could include environmental measurements and closed questions (with a rating scale) from a questionnaire. The advantages and disadvantages of these two types of data are shown in Table 23.3.

Quinary: The highest levels of decision making in an economy – the top business executives and officials in government, science, universities, non-profit organisations, healthcare, culture and the media. It is concentrated in STEM employment (science, technology, engineering and mathematics).

Rebranding: The 'marketing' aspect of regeneration designed to attract businesses, residents and visitors. It often includes re-imaging.

Refugee: People who are forced to flee their homes due to persecution, whether on an individual basis or as part of a mass exodus due to political, religious or other problems.

Regeneration (or place making): Long-term upgrading of existing places or more drastic renewal schemes for urban residential, retail, industrial and commercial areas, as well as rural areas. This sometimes includes conservation to preserve a specific identity. It is connected with rebranding, which centres on place marketing, where places are given a new or enhanced identity to increase their attractiveness and socio-economic viability.

Re-imaging: Making a place more attractive and desirable to invest and live in or visit.

Relative poverty: When a person's income is too low to maintain the average standard of living in a particular society Asset growth for very rich people can lead to more people being in relative poverty.

eliability: The consistency or reproducibility of a easurement. A measurement is said to have a high liability if it produces consistent results under consistent onditions. True reliability cannot be calculated – it can nly be estimated based on knowledge and understanding f the topic.

emittances: Money that migrants send home to their milies via formal or informal channels.

esilience: In the context of hazards and disasters, resilience an be thought of as the ability of a system, community or ciety exposed to hazards to resist, absorb and recover from e effects of a hazard.

eturn period (or recurrence interval): Refers to the equency of a flood of a particular magnitude. A 1:100 ood event will occur, on average, very 100 years (there a one per cent chance of that flood occurring in a given year).

ia: A drowned river valley in an unglaciated area caused by a level rises flooding the river valley, making it much wider an would be expected based on the river flowing into it.

isk: The exposure of people to a hazardous event. More ecifically, it is the probability of a hazard occurring that ads to the loss of lives and/or livelihood.

ural–urban continuum: The unbroken transition from arsely populated or unpopulated, remote rural places to ensely populated, intensively used urban places (town and ty centres).

ample (size): Refers to the number of observations or ata points that make up a survey or data set. Very small mple sizes will not reflect the statistical population closely; is makes them unreliable and can lead to incorrect terpretations and explanations. Large samples are much ore resource heavy and can take considerable effort and nergy to do well.

econdary or S waves move more slowly, however, t around 4 km/sec. They vibrate at right angles to the irection of travel and cannot travel through liquids nlike P waves).

eismic hazards: Generated when rocks within 700 km of e Earth's surface come under such stress that they break and ecome displaced.

ense of place: An overarching impression encompassing e general ways in which people feel about places.

hrinking world: Thanks to technology, distant places start o feel closer and take less time to reach.

ilviculture: The planting of trees for commercial forestry.

ink estate: Housing estates characterised by high levels f economic and social deprivation and crime, especially omestic violence, drugs and gang warfare.

Social housing; affordable housing: A fey function of social housing is to provide accommodation at affordable rents to people on low incomes. According to Shelter, social housing and affordable housing are one and the same thing. However, some would make the distinction that affordable housing also includes dwellings (usually built by housing associations) for sale at below market prices to first-time buyers.

Social isolation: A complete or nearly complete lack of contact with people and society. It differs from loneliness, which is a temporary lack of contact with other people.

Soft power: The global influence a country derives from its culture, its political values and its diplomacy. Much of the USA's soft power has been produced by 'Hollywood, Harvard, Microsoft and Michael Jordan'.

Soil liquefaction: The process by which water-saturated material can temporarily lose normal strength and behave like a liquid under the pressure of strong shaking. Liquefaction occurs in saturated soils (ones in which the pore space between individual particles is completely filled with water). An earthquake can cause the water pressure to increase to the point where the soil particles can move easily, especially in poorly compacted sand and silt.

Solifluction: The gradual movement of soil saturated with melt water down a slope over a permanently frozen soil in tundra regions.

Sovereign wealth funds: Government-owned investment funds and banks, typically associated with China and countries that have large revenues from oil, such as Qatar.

Spatial division of labour: The common practice among TNCs of moving low-skilled work abroad (or 'offshore') to places where labour costs are low. Important skilled management jobs are retained at the TNC's headquarters in its country of origin.

Special Economic Zone: An industrial area, often near a coastline, where favourable conditions are created to attract foreign TNCs. These conditions include low tax rates and exemption from tariffs and export duties.

Stadials and interstadials: Short-term fluctuations within ice-house–greenhouse conditions; stadials are colder periods that lead to ice re-advances.

Subduction zones are broad areas where two plates are moving together, often with the thinner, more dense oceanic plate descending beneath a continental plate. The contact between the plates is sometimes called a thrust or megathrust fault. Where the plates are locked together, frictional stress builds. When that stress exceeds a given threshold, a sudden failure occurs along the fault plane that can result in a 'mega-thrust' earthquake, releasing strain energy and radiating seismic waves. It is common for the leading edge to lock under high friction. The locked fault can hold for hundreds of years, building up enormous stress before releasing. The process of strain, stress and failure is referred to as the elastic-rebound theory.

Subglacial: Debris transported beneath the glacier.

Suburbanisation: The outward spread of the built-up area, often at lower densities compared with the older parts of a town or city. The decentralisation – of people first and then employment and services – is encouraged by transport improvements.

Supraglacial: Debris transported on the surface of the glacier.

Sustainability: The definition of sustainable regeneration varies but in this context it may be thought of as regeneration that creates long-lasting economic, social and environmental benefits for a place.

Sustainable coastal management: Managing the wider coastal zone in terms of people and their economic livelihoods, social and cultural well-being, and safety from coastal hazards, as well as minimising environmental and ecological impacts.

Swash: The flow of water up a beach as a wave breaks.

Tariffs: The taxes that are paid when importing or exporting goods and services between countries.

Tectonic hazard profile: A technique used to try to understand the physical characteristics of different types of hazards, for example earthquakes, tsunamis and volcanoes. Hazard profiles can also be used to analyse and assess the same hazards which take place in contrasting locations or at different times. Hazard profiles are developed for each natural hazard and are based on criteria such as frequency, duration and speed of onset. Figure 2.9, for example, compares the features of earthquakes at two different boundaries.

Tertiary data: Usually refers to sources that collate and summarise other sources (in this respect mostly secondary). Wikipedia is a good example as it provides a summary of a topic with references to the sources that have been used. The internet has blurred the boundary between secondary and primary sources.

Thermohaline circulation: A global system of surface and deep-water ocean currents driven by differences in temperature (thermo) and salinity (haline) between areas of the oceans. Sometimes known as the ocean conveyor.

Topography: The shape and relief of the land.

Trade blocs: Voluntary international organisations that exist for trading purposes, bringing greater economic strength and security to the nations that join.

Transition town: A settlement where individuals and businesses have adopted 'bottom-up' initiatives with the aim of making their community more sustainable and less reliant on global trade.

Transnational corporations: Businesses whose operations are spread across the world, operating in many nations as both makers and sellers of goods and services. Many of the largest are instantly recognisable 'global brands' that bring cultural change to the places where products are consumed.

Trickle-down: The positive impacts on peripheral regions (and poorer people) caused by the creation of wealth in core regions (and among richer people).

Tsunami: The word comes from two Japanese words: *tsu* (port or harbour) and *nami* (wave or sea). Tsunamis are initiated by undersea earthquakes, landslides, slumps and, sometimes, volcanic eruptions. They are characterised by: long wavelengths, typically 150–1000 km; low amplitude (wave height), 0.5–5 m; fast velocities, up to 600 kph in deep water.

Unconsolidated sediment: Material such as sand, gravel, clay and silt that has not been compacted and cemented to become sedimentary rock (it has not undergone the process of lithification) and so is loose and easily eroded.

Urbanisation: An increase in the proportion of people living in urban areas.

Validity: The extent to which a test actually measures what it claims to measure. In other words, the extent to which a concept, conclusion or measurement is well-founded and corresponds accurately to real-world geography. This is perhaps the most important measure of 'trueness'.

Volcanic hazards: Associated with the actual eruption events.

Volcano: A landform that develops around a weakness in the Earth's crust from which molten magma, disrupted from pre-existing volcanic rock, and gases are ejected or extruded.

Water footprint: A measure of the amount of water used in the production and transport to market of food and commodities (also known as the amount of 'virtual water' which is 'embedded' in a product).

Yuppie: Short for 'young urban professional' or 'young upwardly-mobile professional' – a young, university-educated adult who has a well-paid job and who lives and works in a large city.

INDEX